MATERIAL RECYCLING – TRENDS AND PERSPECTIVES

Edited by **Dimitris S. Achilias**

Material Recycling – Trends and Perspectives
Edited by Dimitris S. Achilias

Published by InTech
Janeza Trdine 9, 51000 Rijeka, Croatia

Publishing Process Manager Bojan Rafaj

Technical Editor Teodora Smiljanic

Cover Design InTech Design Team

Additional hard copies can be obtained from orders@intechopen.com

Material Recycling – Trends and Perspectives, Edited by Dimitris S. Achilias
p. cm.
ISBN 978-953-51-0327-1

For EU product safety concerns:
IN TECH d.o.o., Prolaz Marije Krucifikse Kozulić 3, 51000 Rijeka, Croatia,
info@intechopen.com or visit our website at intechopen.com.

Meet the editor

Dimitris S. Achilias occupies the position of Associate Professor in Polymer Chemistry and Technology at the Department of Chemistry, Aristotle University of Thessaloniki, Greece. He obtained his diploma in Chemical Engineering from AUTh in 1985 and his Ph.D. degree in 1991. He has published more than 76 papers in international refereed journals and presented over 100 papers in international and local peer reviewed conferences. He has participated in the writing of 2 educating books in Greek and 4 chapters in international scientific books. He has had up to now over 1400 citations on his work, with an h-index 22 and serves as a reviewer in several journals. His current research interests include chemical recycling, degradation and crystallization kinetics of polymers.

Contents

Preface

If the 20th century could be characterized by the rapid increase in the production and consumption of materials that helped improving the standards of living, then the 21st certainly has many elements to qualify as the century of recycling. Since the duration of life of a number of wastes is very small (roughly 40% have duration of life smaller than one month), there is a vast waste stream that reaches each year to the final recipients creating a serious environmental problem. The presently most common practice of handling such waste streams is to incinerate them with energy recovery or to use them for land-filling. Disposing of the waste to landfill is becoming undesirable due to legislation pressures, rising costs and the poor biodegradability of commonly used materials. Therefore, recycling seems to be the best solution. The major driving force in today's recycling project is not only to re-use the materials but also to produce secondary value-added products, reducing the consumption of natural resources and the amount of energy needed, while lowering CO_2 emissions in the environment.

The word re-cycling comes from the Greek word '*κύκλος*' meaning cycle and is usually used to denote the involvement of materials in a continuous cycle from 'cradle' (resources) to 'grave' (disposal of waste) and back to 'cradle' (re-formation of resources).

The purpose of this book is to present the state-of-the-art for the recycling methods of several materials, including polymers, as well as to propose potential uses of the recycled products. It targets professionals, recycling companies, as well as researchers, academics and graduate students in the fields of waste management and polymer recycling in addition to chemical engineering, mechanical engineering, chemistry and physics.

This book comprises 16 chapters that have been prepared from the contribution of 50 co-authors from almost all around the world. There are contributions from Europe (Belarus, Czech Republic, Greece, Italy, Romania, Slovakia), Asia (Iran, Republic of Korea, Saudi Arabia, Taiwan, Turkey), America (Mexico) and Africa (Tunisia). The book chapters have been organized loosely in three subtopics. The first consisting of 7 chapters deals with a very 'hot' subject both from academic and industrial point of view, that of polymer recycling. The second, including 4 chapters, refers to the recycling of specific large-scale wastes, such as tires, pharmaceutical packaging and hardwood kraft pulp. In the final, counting 5 chapters, possible uses of recycled materials are proposed. A brief description of each chapter follows.

A. Polymer recycling

Chapter 1. Recent advances in the chemical recycling of polymers

This chapter provides a critical review on the methods proposed and/or applied, during mainly the last decade, on the recycling of widely used polymers, including polypropylene (PP), polystyrene (PS), low density polyethylene (LDPE), high density polyethylene (HDPE), poly(vinyl chloride) (PVC), polycarbonate (PC), poly(methyl methacrylate) (PMMA) and nylon. The state-of-the-art of the chemical and thermo-chemical recycling methods of these polymers is illustrated.

Chapter 2. Recent Developments in the Chemical recycling of PET

In this chapter, methods for the chemical recycling of poly(ethylene terephthalate) PET are reviewed. Special emphasis is put on glycolytic depolymerization, one of the oldest, simplest and less capital-investment requiring processes. Supercritical, catalytic and microwave-assisted depolymerization processes are also discussed.

Chapter 3. Overview on Mechanical recycling of post consumed PET bottles by chain extension

This chapter presents an overview on the structural upgrading of post-consumer PET by macromolecular chain extensions at reprocessing (reactive processing). This is a very efficient method for enhancing the properties of mechanically recycled PET.

Chapter 4. Poly(bisphenol A carbonate) recycling: High pressure hydrolysis can be a convenient way.

In this chapter, hydrolysis of PC with sub-critical liquid water under high pressure is investigated as a means of recovering the monomer, bisphenol-A. Both pure PC and DVD wastes were used. A concerted path de-polymerization mechanism is proposed and the process kinetics is characterized and compared with lab-scale experimental data.

Chapter 5. Degradation of plasticized poly(vinyl butyral) during re-processing

This chapter focuses on the possibility and conditions for optimal re-processing of plasticized poly(vinyl butyral) (PVB). The scope is to determine degradation of PVB sheet at different kneading conditions and to estimate the influence of temperature, air oxygen content and mechanical stress on the course of the degradation process.

Chapter 6. Materials and methods for the chemical catalytic cracking of plastic waste

Thermo-chemical methods for the recycling of plastic wastes are discussed in this chapter. In particular the effect of several catalysts on the pyrolysis of polyethylene is examined in detail.

Chapter 7. Pyrolysis of waste polystyrene and high-density polyethylene

In this chapter, the product distribution obtained from pyrolysis of polystyrene and high-density polyethylene, as well as of their mixtures, is investigated. Moreover, the potential use of pyrolysis in the recycling of municipal plastic wastes is illustrated.

B. Recycling of tires, pharmaceutical packaging and hardwood kraft pulp

Chapter 8. Recycling of scrap tires

A comprehensive review of the methods proposed for recycling of scrap tires is presented in this chapter. After a brief history on the production technology of tires, the recycling methods are introduced, including thermo-chemical degradation (pyrolysis), burning for energy recovery, as well as mechanical re-processing and re-using for structural engineering applications.

Chapter 9. Waste Tire Pyrolysis Recycling With Steaming: Heat-Mass Balance & Engineering Solutions for By-Products Quality

This chapter presents detailed material and energy balances as well as engineering solutions for by-product quality concerning the recycling of waste tires using pyrolysis with steaming.

Chapter 10. Study on the feasibility of hazardous waste recycling: the case of pharmaceutical packaging

This chapter is focused on a feasibility study for the management of packaging waste from a pharmaceutical plant, considering waste material characterization, tests on waste processing and set-up of size reduction operations.

Chapter 11. Recycling of the hardwood kraft pulp

This chapter focuses on the recycling process of hardwood kraft pulp fibers. Dimensional, optical and mechanical characteristics of the recycled fibers are presented in detail.

C. Potential uses of recycled wastes

Chapter 12. Reuse of waste shells as a SO2/NOx removal sorbent

In this chapter, the feasibility of using waste oyster shells as a sorbent for removal of both sulfur and nitrogen oxides, at the same time, from exhaust gases, is illustrated. In addition, the calcination and hydration reaction of waste oyster shells were experimented and the preparation method of sorbents was investigated.

Chapter 13. Carbon steel slag as cementitious material for self-consolidating concrete

This chapter focuses on the use of carbon steel slag as a pozzolanic material to partially replace Portland cement in the production of self-consolidating concrete (SCC). Results showed that the design and performance of all the concrete mixtures used were comparable to those of SCC and high performance concrete.

Chapter 14. Possible uses of steelmaking slag in agriculture: an overview

This chapter intends to review the state-of-the-art related to the use of steelmaking slag, mainly coming from the Basic Oxygen Furnace process, in agriculture. Aspects covered include its use as fertilizer, liming agent and amending material for soils.

Chapter 15. From PET Waste to Novel Polyurethanes

This chapter focuses on the synthesis of secondary useful products, such as polyurethanes, from the chemical recycling of PET wastes. Various chemical decomposition methods of PET are described emphasizing in aminolysis in the presence of ethanolamine. The main product was used as chain extender or ring opening agent to obtain new polyurethanes.

Chapter 16. Valorisation of organic wastes by composting process and soil amendment

Last but not least, this chapter assesses the suitability of an organic waste compost to supply some essential plant nutrients such as N, P, K, Fe, Mn, Zn and Cu, also to evaluate and compare the effect of manure and compost on soil chemical properties as well as to elucidate the effect of compost on crop productivity.

I want to express my sincere thanks to all the contributors who provided their expertise and enthusiasm to this project and InTech for making this work possible. I would like also to thank my family for their patience and the time deprived them during the preparation of my chapter and the book editing. Finally, I would like to dedicate this book to the memory of my mother Lilika Achilia who always was a source of inspiration.

Dimitris S. Achilias
Associate Professor, Department of Chemistry,
Aristotle University of Thessaloniki
Greece

Part 1

Recycling of Polymers

1

Recent Advances in the Chemical Recycling of Polymers (PP, PS, LDPE, HDPE, PVC, PC, Nylon, PMMA)

Dimitris S. Achilias et al.*
Laboratory of Organic Chemical Technology, Department of Chemistry,
Aristotle University of Thessaloniki, Thessaloniki
Greece

1. Introduction

During last decades, the great population increase worldwide together with the need of people to adopt improved conditions of living led to a dramatical increase of the consumption of polymers (mainly plastics). The world's annual consumption of plastic materials has increased from around 5 million tones in the 1950s to nearly 100 million tones today. Since the duration of life of plastic wastes is very small (roughly 40% have duration of life smaller than one month), there is a vast waste stream that reaches each year to the final recipients creating a serious environmental problem. The presently most common practice of handling such waste streams is to incinerate them with energy recovery or to use them for land-filling. Disposing of the waste to landfill is becoming undesirable due to legislation pressures (waste to landfill must be reduced by 35% over the period from 1995 to 2020), rising costs and the poor biodegradability of commonly used polymers (Achilias et al., 2009). Therefore, recycling seems to be the best solution.

The recycling of waste polymers can be carried out in many ways. Four main approaches have been proposed presented in Scheme 1 (Karayannidis and Achilias, 2007; Scheirs, 1998):

1. *Primary recycling* refers to the 'in-plant' recycling of the scrap material of controlled history. This process remains the most popular as it ensures simplicity and low cost, dealing however only with the recycling of clean uncontaminated single-type waste.
2. *Mechanical recycling (or secondary recycling)*. In this approach, the polymer is separated from its associated contaminants and it can be readily reprocessed into granules by conventional melt extrusion. Mechanical recycling includes the sorting and separation of the wastes, size reduction and melt filtration. The basic polymer is not altered during the process. The main disadvantage of this type of recycling is the deterioration of product properties in every cycle. This occurs because the molecular weight of the

* Lefteris Andriotis, Ioannis A. Koutsidis, Dimitra A. Louka, Nikolaos P. Nianias, Panoraia Siafaka, Ioannis Tsagkalias and Georgia Tsintzou
Laboratory of Organic Chemical Technology, Department of Chemistry, Aristotle University of Thessaloniki, Thessaloniki, Greece

recycled resin is reduced due to chain scission reactions caused by the presence of water and trace acidic impurities. Strategies for maintaining the polymer average molecular weight during reprocessing include intensive drying, reprocessing with degassing vacuum, the use of chain extender compounds, etc.

3. *Chemical or Feedstock recycling (tertiary recycling)* has been defined as the process leading in total depolymerization of PET to the monomers, or partial depolymerization to oligomers and other chemical substances. The monomers could subsequently re-polymerized to regenerate the original polymer.
4. *Energy recovery (Quaternary recycling)* refers to the recovery of plastic's energy content. Incineration aiming at the recovery of energy is currently the most effective way to reduce the volume of organic materials. Although polymers are actually high-yielding energy sources, this method has been widely accused as ecologically unacceptable owing to the health risk from air born toxic substances e.g. dioxins (in the case of chlorine containing polymers).

Apart from the aforementioned methods, direct reuse of a plastic material (i.e. PET) could be considered as a "zero order" recycling technique (Nikles and Farahat, 2005). In a lot of countries it is a common practice PET-bottles to refilled and reused. However, this should be done with a great care since plastic bottles are more likely than glass to absorb contaminants that could be released back into food when the bottle is refilled. Moreover, refill of a PET-bottle with a high-alcoholic-degree drink may lead to degradation of the macromolecular chains with unexpected results.

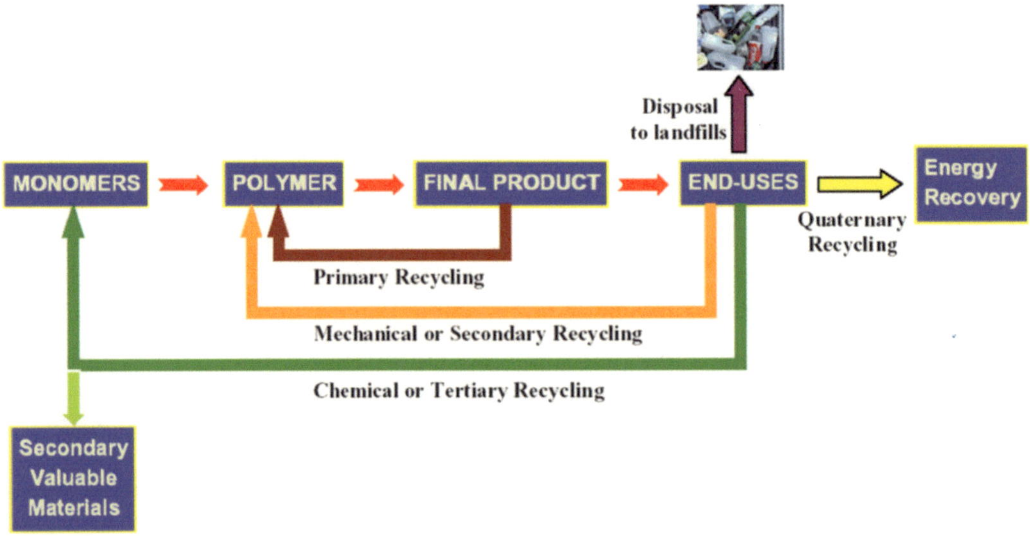

Scheme 1. Polymer Recycling Techniques.

The objective of a plastic management policy, in accordance with the principles of sustainable development (development that meets the needs of present generation without compromising the ability of future generations to meet their needs), should be not only the reuse of polymeric materials but also the production of raw materials (monomers), from which they could be reproduced, or other secondary valuable products, which could be useful as feedstock for a variety of downstream industrial processes or as transportation fuel. In this sense, among the techniques proposed for recycling of waste polymers the most challenging method is chemical or feedstock recycling and various technologies have been successfully demonstrated and continue to be developed (Achilias and Karayannidis, 2004).

The aim of this chapter is to provide a critical review of the methods proposed and/or applied during mainly **the last decade**, on the chemical recycling of polymers. In this way, the state-of-the-art of the chemical recycling methods of several polymers will be presented. Polymers that will be studied include the widely used plastics, based on poly(ethylene terephthalate) (PET), polypropylene (PP), polystyrene (PS), low density polyethylene (LDPE), high density polyethylene (HDPE), poly(vinyl chloride) (PVC), polycarbonate (PC), poly(methyl methacrylate) (PMMA) and nylon.

2. Chemical recycling of poly(ethylene terephthalate) PET

PET is a polyester with functional ester groups that can be cleaved by some reagents, such as water (hydrolysis), alcohols (alcoholysis), acids (acidolysis), glycols (glycolysis), and amines (aminolysis). Thus, chemical recycling processes for PET are divided as follows: (i) Hydrolysis, (ii) Glycolysis, (iii) Methanolysis and (iv) other processes (Scheme 2). According to the reagent used different products are obtained (Karayannidis and Achilias, 2007; Karayannidis et al. 2006; Karayannidis et al., 2005; Karayannidis et al. 2002; Kosmidis et al., 2001). The different process options for chemical recycling of PET waste may be categorized as follows: (i) regeneration of base monomers (methanolysis for dimethyl terephthalate (DMT) and hydrolysis for producing pure Terephthalic acid (TPA) and ethylene glycol (EG)); (ii) conversion into oligomers (glycolysis or solvolysis); (iii) use of glycolyzed waste for value-added products; (iv) conversion into speciality chemicals by aminolysis or ammonolysis; (v) conversion into speciality intermediates for use in plastics and coatings.

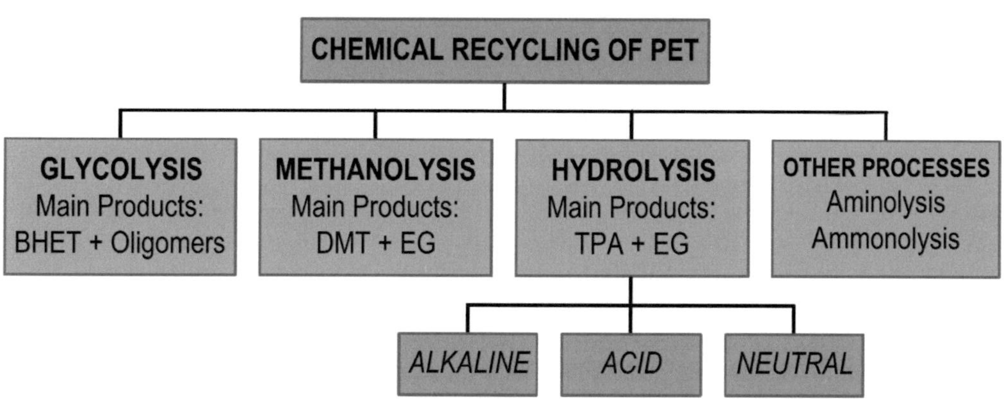

Scheme 2. Chemical recycling techniques of PET.

Recently, recycling of PET using hydrolysis, glycolysis and aminolysis under microwave irradiation has been proposed (Achilias et al., 2010; Achilias et al., 2011; Siddiqui et al., 2010). PET recycling in a microwave reactor has been proved a very beneficial method resulting not only in material recovery but also in substantial energy saving.

This section will not be presented in detail here because it is the subject of another chapter of this book. Interested reader can find extensive details on the techniques used for the chemical recycling of PET in several recent review papers appeared in literature (Scheirs, 1998; Karayannidis and Achilias, 2007).

3. Chemical recycling of polypropylene

3.1 Introduction

Condensation polymers like PET or Nylon, can undergo chemolysis with different reagents to produce mainly the monomers from which they have been produced or other oligomers. In contrast, vinyl polymers, such as polyolefins (PP and PE) cannot be degraded with simple chemicals to their monomers due to the random schission of the C-C bonds. Two main chemical recycling routes are the thermal and catalytic degradation of these polymers. In thermal degradation, the process produces a broad product range and requires high operating temperatures, typically more than 500°C and even up to 900 °C. Thermal cracking of polyethylene and polypropylene is usually carried out either in high temperatures (>700 °C), to produce an olefin mixture (C1–C4) and aromatic compounds (mainly benzene, toluene and xylene) or in low temperature (400–500 °C) (thermolysis) where three fractions are received: a high-calorific value gas, condensable hydrocarbon oil and waxes. In the first case, the objective is to maximize the gas fraction and to receive the olefins, which could be used after separation as monomers for the reproduction of the corresponding polyolefins. Cracking in lower temperatures leaves a waxy product in the reactor that mainly consists of paraffins together with a carbonized char. The gaseous fraction can be used for the supply of the energy required for the pyrolysis after burning. The liquid fraction mainly consists of linear olefins and paraffins with C11–C14 carbon atoms with only traces of aromatic compounds (Aguado and Serrano, 1999). Thermal cracking of polyolefins proceeds through a random scission mechanism in four steps: initiation, depropagation, inter- or intra-molecular hydrogen transfer followed by β-scission and termination. In general, thermal cracking is more difficult in HDPE followed by LDPE and finally by PP.

Due to the low thermal conductivity of polymers together with the endotherm of cracking, thermal pyrolysis consumes large amounts of energy. Thus, catalytic technologies have been proposed to promote cracking at lower temperatures, resulting in reduced energy consumption and higher conversion rates. Furthermore, use of specific catalysts allows the process to be directed towards the formation of a narrower distribution of hydrocarbon products with a higher market value. Heterogeneous catalysis has been investigated extensively using solids with acid properties. Zeolites of the kind employed in the catalytic cracking of hydrocarbon feedstocks (Y, ZSM-5, Beta) as well as other well-known acid solids like silica–alumina, alumina and clays are being the most studied. Mixtures of these catalysts like SAHA/ZSM-5, MCM-41/ZSM-5 have been also used. Cracking with acid catalysts takes place through the formation of carbocations, which requires the presence of strong acidic regions. Acid strength and textural properties are the main parameters dictating the performance of acid solids in the catalytic conversion of polymers. Porosity, surface area characteristics and particle size determine to a large extent the accessibility of bulky polymeric molecules to the internal catalytic acid sites of the solids. Thus, while catalyst HZSM-5 presents bigger reactivity from HMCM-41 in the cracking of HDPE and LDPE, at the decomposition of the large molecules of PP the transformation is almost the same with that of thermal cracking, because cross-section of polymer is very big in order to enter in catalysts' micropores (Achilias et al., 2007).

These facts strongly limit their applicability and especially increase the higher cost of feedstock recycling for waste plastic treatment. Therefore, catalytic degradation provides a means to address these problems. The addition of catalyst is expected to reduce decomposition temperature, to promote decomposition speed, and to modify the products. The catalytic degradation of polymeric materials has been reported for a range of model catalysts centred on the active components in a range of different model catalysts, including amorphous silica–aluminas, zeolites Y, mordinite and ZSM-5 and the family of mesoporous MCM-41 materials. However, these catalysts have been used that even if performing well, they can be unfeasible from the point of view of practical use due to the cost of manufacturing and the high sensitivity of the process to the cost of the catalyst. Another option for the chemical recycling of polymer wastes by using fluidized catalytic cracking (FCC) catalysts is attractive. Therefore, an alternative improvement of processing the recycling via catalytic cracking would operate in mixing the polymer waste with fluid catalytic cracking (FCC) commercial catalysts.

Recently, much attention has been paid to the recycling of waste polymers by thermal or catalytic pyrolysis as a method to recover value added products or energy via the production of high-value petrochemical feedstock or synthetic fuel fractions. The following review is rather selective and not extensive. Detailed reviews on the thermal and catalytic pyrolysis of PP based plastics can be found in an excellent recently published book by Scheirs and Kaminsky, 2006 and in Achilias et al., 2006.

3.2 Pyrolysis

Achilias et al., 2007, studied the technique of pyrolysis of polypropylene in a laboratory fixed bed reactor using as raw materials either model PP or waste products based on these polymer. The conclusions are very interesting. The oil and gaseous fractions recovered presented a mainly aliphatic composition consisting of a series of alkanes and alkenes of different carbon number with a great potential to be recycled back into the petrochemical industry as a feedstock for the production of new plastics or refined fuels. Details are presented in section 5.

Hayashi et al., 1998 studied pyrolysis of polypropylene in the presence of oxygen. The polypropylene was coated on porous α-alumina particles and then pyrolyzed in a flow of helium or a mixture of helium-oxygen at atmospheric pressure. The mass release from PP was dramatically enhanced in the presence of oxygen at temperatures in the range of 200–300ºC. The net mass release rate in the presence of oxygen followed first-order kinetics with respect to the oxygen partial pressure and was controlled by the formation of peroxide on tertiary carbon of PP. The activation energy was 60–70 kJ/mol. The oxidative pyrolysis at 250ºC converted 90% of PP into volatiles which mainly consisted of CS -soluble oils having a number-average chain length of 10.

Dawood et al. 2001 studied the influence of γ-irradiation on the thermal degradation of polypropylene by performing thermogravimetric analysis at three constant heating rates and at a constant temperature. At all the heating rates it can be indicated that the TG curves of the irradiated samples shifted to lower temperatures in comparison with the unirradiated one. The shift clearly increased with increasing irradiation dose, which means that the pyrolysis was enhanced by the irradiation. Since the difference in TG curves between the unirradiated sample and the samples irradiated to 10 and 30 kGy is quite large, small dose of irradiation is judged to be enough to cause a significant enhancement of the pyrolysis activity. The samples irradiated to small doses, 10 and 30 kGy, seem to show a pyrolysis behavior different from the other irradiated samples. At a small heating rate of 3 K/min, the TG curves of 10 and 30 kGy samples are close to the TG curve of 60 kGy sample, whereas the former TG curves are distinctly different from the latter TG curve at 10 K/min. These results may suggest that the mechanism of the increase in pyrolysis activity is different among the irradiated samples. A further examination of the influence of irradiation was performed by pyrolyzing the samples at a constant temperature. Similar to the case of dynamic heating rate, the difference in pyrolysis reactivity between the unirradiated and the 30 kGy irradiated sample is quite large, while the difference between the irradiated samples is small. This supports the suggestion that a small radiation dose is enough to cause a significant enhancement on the pyrolysis activity of PP.

3.3 Co-pyrolysis

Assumpcao et al., 2011, considered co-pyrolysis of PP with Brazilian crude oil by varying the temperature (400◦C to 500◦C) and the amount of PP fed to the reactor. The co-pyrolysis of plastic waste in an inert atmosphere provided around 80% of oil pyrolytic, and of these, half represent the fraction of diesel oil. this technique is a promise for PP waste recycling as it not only minimizes the environmental impact caused by inadequate disposal of this residues, but it also allows the reuse of a non-renewable natural resource (petroleum) through the use of diesel oil fractions obtained in this process. According to the results, the temperature increase has favored the increase of pyrolytic liquid generation and the reduction of the solid formed (Table 1). On the other hand, a huge increase in the PP amount has caused a decrease in total yield (liquid product) (Table 2). In general, it was observed that with temperature increase, there was a small reduction in yield in the diesel distillation range. Moreover, most part of these liquid distillates in a range higher than diesel, corresponding to heavy vacuum gas oil (GOP). This product (GOP) can still be cracked in an FCC generating more profitable products (naphtha and LPG), or can be used as fuel oil. The increase of PP in the reaction favors a yield increase in the diesel distillation range compared to pyrolysis of pure heavy oil, also forming a significant amount of compounds with distillation range lower than diesel.

		Yield of pyrolytic oil present at the liquid boiling point (%)		
Pyrolysis temperature (°C)	Amount of PP (g)	In the Diesel distillation range	Below the Diesel distillation	Above the Diesel distillation
400	0.0	39	n.d.	61
	0.2	40	n.d.	60
	0.4	52	n.d.	48
	0.6	48	1.0	52
	0.8	62	n.d.	38
	1.0	59	n.d.	41
450	0.0	35	n.d.	65
	0.2	30	1.0	69
	0.4	32	n.d.	68
	0.6	37	2.0	61
	0.8	51	1.0	48
	1.0	n.d.	n.d.	n.d.
500	0.0	31	n.d.	69
	0.2	36	1.0	63
	0.4	33	n.d.	67
	0.6	34	6.0	60
	0.8	40	2.0	58
	1.0	n.d.	n.d.	n.d.

Table 1. Yield of pyrolytic oil present at the liquid boiling point during PP co-pyrolysis with Brazilian crude oil (Assumpcao et al., 2011).

Amount of PP (g)	Pyrolysis temperature (°C)	Yield of pyrolytic products (%)		
		Liquid	Solid	Gas
0.0	400	59	30	11
	450	65	26	9
	500	88	10	2
0.2	400	68	28	4
	450	86	11	3
	500	95	4	1
0.4	400	79	13	4
	450	83	10	7
	500	90	3	7
0.6	400	83	14	3
	450	91	5	4
	500	97	2	1
0.8	400	17	81	2
	450	23	74	3
	500	50	49	1
1.0	400	15	84	1
	450	17	82	1
	500	25	73	3

Table 2. Yield of pyrolytic products in mixtures of PP with Brazilian crude oil. (Assumpcao et al., 2011).

Ballice et al., 2002 investigated the temperature-programmed co-pyrolysis of Soma-lignites form Turkey with PP. A series co-pyrolysis operation was performed with lignites and PP using a 1:3, 1:1, 3:1 total carbon ratio of lignites to plastic. A fixed bed reactor was used to pyrolyse small sample of lignites and PP mixture under an inert gas flow (argon). In addition, the performance of the experimental apparatus was investigated by establishing a carbon balance and the degree of recovery of total organic carbon of the samples as aliphatic hydrocarbons and in solid residue was determined. Conversion into volatile hydrocarbons was found higher with increasing PP ratio in lignites–PP system while C16+ hydrocarbons and the amount of coke deposit were lower in the presence of PP. The maximum product release temperature was found to be approximately 440 °C for co-pyrolysis of lignite–PP. Straight- and branched-chain paraffins and olefins from methane to C26, diene and simple aromatic hydrocarbons were determined in co-pyrolysis products. The fraction of *n*-paraffins was higher than that of 1-olefins at a high proportion of lignite in the mixture.

Co-pyrolysis of lignite with PP has been found to give less C16+ *n*-paraffins and 1-olefins than pyrolysis of lignite by increasing PP ratios. Coke deposit in co-processing decreased also by increasing PP ratios. The *n*-paraffins were found to consist of mainly C1–C9, and relatively small amount of C10–C15 and C16+ fractions. The evolution of 1-olefins decreased in co-pyrolysis operation because of the higher hydrogen content in feed by increasing ratios of PP. A slightly synergistic effect were determined in the co-pyrolysis operation and the experimental results indicated that the pyrolysis products of PP are in highly aliphatic character, and during the initial stages of pyrolysis, these pyrolysis products of PP is expected to be a relative poor solvent for the structures of lignite. In addition, relative to liquefaction sources materials such as coals, the dominant components of municipal solid wastes (mainly PE, PS, PET and PP) are hydrogen rich so that co-processing of coal with waste plastics could be a good way to recycle waste plastics into useful products (Table 3).

Hydrocarbon fraction	Lignite	PP	Lignite-PP (1:3)	Lignite-PP (1:1)	Lignite-PP (3:1)
wt.% relative to the 1-olefins					
$C_2 - C_4$	57.8	81.0	58.3	58.2	59.7
C_5-C_9	15.6	19.0	32.6	30.1	26.8
C_{10}-C_{15}	14.7	-	4.2	4.0	3.7
C_{16+}	11.9	-	4.9	8.0	9.8
wt.% relative to all the aliphatic hydrocarbons					
1-Olefins	10.2	10.0	14.4	13.8	13.4

Table 3. 1-Olefins distribution in co-pyrolysis of lignite with PP (Ballice et al., 2002).

Hajekova and Bajus, 2005 investigated the thermal decomposition of polyalkenes as a recycling route for the production of petrochemical feedstock. Polypropylene was thermally decomposed individually in a batch reactor at 450° C, thus forming oil/wax products. Then the product was dissolved in primary heavy naphtha to obtain steam cracking feedstock. The selectivity and kinetics of copyrolysis for 10 mass% solutions of oil/waxes from PP with naphtha in the temperature range from 740 to 820 °C at residence times from 0.09 to 0.54 s were studied.

The decomposition of polyalkene oil/waxes during copyrolysis was confirmed. It was shown that the yields of the desired alkenes propene increased or slightly decreased compared to the yields from naphtha. In addition to the primary reactions, the secondary reactions leading to coke formation have also been studied. Slightly higher formation of coke was obtained at PP wax solution at the beginning of the measurements, on the clean surface of the reactor. After a thin layer of coke covered the walls, the production was the same as that from naphtha. The results confirm the possibility of polyalkenes recycling via the copyrolysis of polyalkene oils and waxes with conventional liquid steam cracking feedstocks on already existing industrial ethylene units.

3.4 Catalyting cracking

A large number of laboratory studies have been conducted for the direct catalytic cracking of different type of plastics. A large variety of catalysts have been used that even if performing well, they can be unrealistic from the point of view of practical use due to the cost of manufacturing and the high sensitivity of the process to the cost of the catalyst. Some of the studies reported in the literature will be reviewed below.

Zhao et al., 1996 have studied the effects of different zeolites as H-Y, Na-Y, L, H-mordenite and Na-mordenite on the catalytic degradation of PP by thermogravimetry under nitrogen flow. It was found that the degradation temperature of PP strongly depended on the type of zeolite used and the amount added. One type of HY zeolite (320HOA) was shown to be a very effective catalyst. Pyrolysis products, which were identified by using a coupled gas-chromatograph-mass-spectrometer, were also affected by the addition of zeolites. Some zeolites did not change the structure of the products but narrowed the product distribution to a smaller molecule region, while the HYzeolite led to hydrocarbons concentrated at those containing 4-9 carbons. Furthermore, some new compounds with cyclic structures were found in the presence of the HY zeolite.

Also, Zhao studied the effect of irradiation on pyrolysis of polypropylene in the presence of zeolites the results revealed that thermal degradation temperature of PP was significantly reduced when PP was irradiated in the presence of a zeolite. The irradiation-induced temperature reduction depended on the zeolite structure and composition, as well as on the morphology of the mixture. Identification of pyrolysis products indicated that, in the absence of zeolite, irradiation resulted only in a change of the product distribution but no formation of new compounds. In the presence of zeolite, however, a series of oxidized products were formed. In addition, the pyrolysis could be performed at a much lower temperature. Irradiation is able to render PP much moresusceptible to thermal degradation when carried out in the presence of zeolite. However, this effect was closely related to the type of zeolites, mixing methods and irradiation conditions.Furthermore, in pyrolysis of properly irradiated PP-zeolite mixtures, new chemicals such as acetone, acetic acid and so on, could possibly be obtained in addition to traditional hydrocarbons.

Ishihara et al., 1993 investigated the catalytic degradation of PP by silica–alumina at temperatures between 180 and 300°C in a semibatch reactor under a nitrogen flow. The production of gas precursors was found essential to decomposition. The most important elementary reaction is the intramolecular rearrangement of chain-end secondary carbonium ions in the liquid fraction to inner tertiary carbon atoms. The catalytic decomposition of polypropylene proceeds as follows: polymer -degraded polymer +oligomer+ liquid+gas. Gas is produced from the chain-ends of the liquid fraction and its components are primarily isobutene and isopentane. The most important elementary reaction in the decomposition is intramolecular rearrangement taking place via a six-membered transition state to inner tertiary carbon atoms (back-biting reactions). The main gas components are produced by the decomposition of the C, fraction formed by the back-biting reaction.

Durmus et al., 2005 studied thermal-catalytic degradation kinetics of polypropylene over BEA, ZSM-5 and MOR zeolites. Degradation rate of the PP over zeolites was studied by thermogravimetric analysis (TGA) employing four different heating rates and apparent activation energies of the processes were determined by the Kissinger equation. The catalytic activity of zeolites decreases as BEA > ZSM-5a (Si/Al = 12.5) > ZSM-5b (Si/Al = 25) > MOR depending on pore size and acidity of the catalysts. On the other hand, initial degradation is relatively faster over MOR and BEA than that over both ZSM-5 catalysts depending on the apparent activation energy. It can be concluded that acidity of the catalyst is the most important parameter in determining the activity for polymer degradation process as well as other structural parameters, such as pore structure and size.

Lin et al., 2005 have investigated the catalytic cracking of PP in a fluidized bed reactor using H-ZSM-5, H-USY, H-mordenite, silica–alumina and MCM-41, with nitrogen as fluidizing gas (Figure 1). PP was pyrolysed over various catalysts using a laboratory fluidised-bed reactor operating isothermally at ambient pressure. The yield of volatile hydrocarbons for zeolite catalysts was higher than that for non-zeolite catalysts. Product distributions with HZSM-5 contained more olefinic materials with about 60 wt% in the range of C3 - C5. However, both HMOR and HUSY produced more paraffin streams with large amounts of isobutene (i-C4) and both catalysts were deactivated during the course of the degradation. SAHA and MCM-41 showed the lowest conversion and generated an olefin-rich product with a rise to the broadest carbon range of C3 - C7.

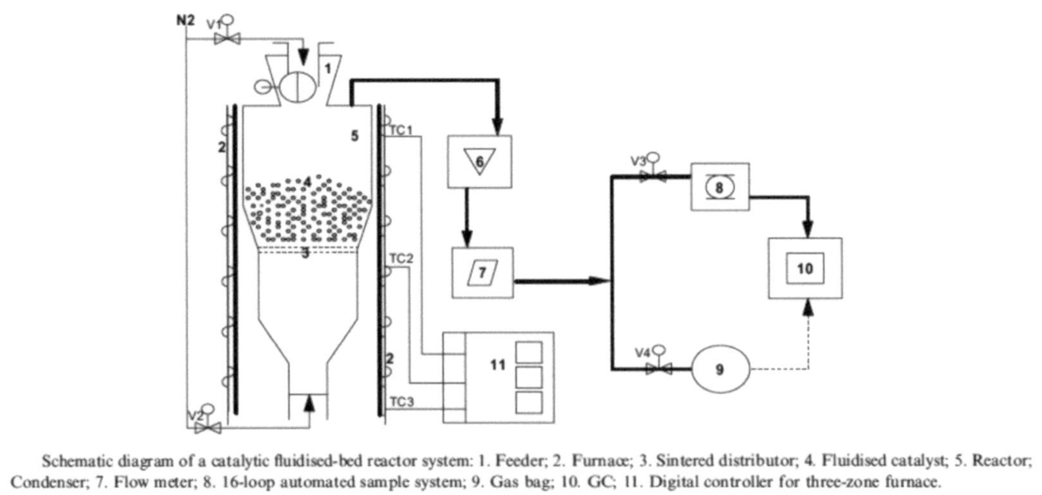

Fig. 1. Schematic diagram of catalytic fluidized-bed reactor system (Lin and Yen 2005).

Greater product selectivity was observed with HZSM-5 and HMOR as catalysts with about 60% of the product in the C3-C5 range and HMOR generating the highest yield of i-C4 for all catalysts studied (Table 4).

Degradation results	Catalyst type				
	HUSY	HZSM-5	HMOR	SAHA	MCM-41
Yield (wt.% feed)					
Gas	89.49	94.77	88.29	86.44	86.19
Liquid	3.75	2.31	4.54	4.73	5.07
Residue	6.76	3.92	7.17	9.83	8.74
Involatile residue	3.32	2.27	4.96	7.49	6.83
Coke	3.44	1.25	3.21	2.34	1.91
Mass balance (%)	93.71	95.32	93.24	89.68	88.46
Distribution of gaseous products (wt % feed)					
Hydrocarbon gases (C_1-C_4)	36.73	67.41	59.86	22.54	25.47
Gasoline (C_5-C_9)	51.83	25.54	27.95	63.65	60.56
BTX	0.93	1.82	0.48	0.25	0.16

Table 4. PP degradation products depending on catalysts type (Lin and Yen, 2005).

The larger pore zeolites (HUSY and HMOR) showed deactivation in contrast to the more restrictive HZSM-5. Observed differences in product yields and product distributions under identical reaction conditions can be attributed to the microstructure of catalysts.Valuable hydrocarbons of olefins and iso-olefins were produced by low temperatures and short contact times.

Lin and Yang, 2007 pyrolysed PP over spent FCC commercial catalyst (FCC-s1) using a laboratory fluidised-bed reactor operating isothermally at ambient pressure. The yield of gaseous and liquid hydrocarbon products at 390°C for spent FCC commercial catalyst (87.8 wt%) gave much higher yield than silicate (only 17.1 wt%). Greater product selectivity was observed with FCC-s1 as a post-use catalyst with about 61 wt% olefins products in the C3 - C7 range.

The use of fluidised-bed reaction system coped with a spent FCC equilibrium catalyst can be a better option from economical point of view since it can gives a good conversion with comparable short reaction time, and even its activity is lower than that of the zeolites (ZSM-5 and HUSY) and silica-aluminas (SAHA), this can be compensated by increasing the catalyst to PP ratio. Product distributions with FCC-s1 catalyst contained more olefinic materials in the range of C3-C7 (about 56 wt% at 390 °C). It was concluded that the use of spent FCC commercial catalyst and under appropriate reaction conditions can have the ability to control both the product yield and product distribution from polymer degradation, potentially leading to a cheaper process with more valuable products (Tables 5 and 6).

Degradation results	Ratio of polymer to catalyst (wt.%)				
	10	20	30	40	60
Yield (wt.% feed)					
Gas	26.7	27.8	26.9	27.6	28.2
Liquid	63.2	61.7	60.9	60.3	59.2
Gasoline (C_5-C_9)	57.4	55.4	54.3	52.9	51.4
Condensate liquid	4.9	5.3	5.4	5.8	6.0
BTX	0.9	1.0	1.2	1.6	1.8
Residue	10.1	10.5	12.2	12.1	12.6
Involatile residue	7.6	8.3	9.9	10.4	10.7
Coke	2.5	2.2	2.3	1.7	1.9
Mass balance (%)	89.5	90.6	91.8	92.5	90.4

Table 5. Products of PP pyrolysis with FCC catalyst (Lin and Yang, 2007).

Degradation results	Fluidizing N_2 rates (mL/min)				
	900	750	600	450	300
Yield (wt.% feed)					
Gas	29.6	28.8	26.9	26.3	26.1
Liquid	60.5	60.4	60.9	60.7	60.1
Gasoline (C_5-C_9)	55.1	54.7	54.3	54.6	53.7
Condensate liquid	4.9	4.9	5.4	4.5	4.3
BTX	0.5	0.8	1.2	1.6	2.1
Residue	9.9	10.8	12.2	13.0	13.8
Involatile residue	7.7	8.5	9.9	10.5	11.2
Coke	2.2	2.3	2.3	2.5	2.6
Mass balance (%)	89.2	89.6	91.8	90.3	94.1

Table 6. Effect of fluidizing N_2 rates on the product yield in PP pyrolysis (Lin and Yang, 2007).

Dawood and Miura, 2002 have studied the effect of exposing PP to γ-irradiation prior to the catalytic pyrolysis over a HY-zeolite using a thermobalance and a semi-batch reactor. A significant increase in the rate of the catalytic pyrolysis was realized when PP was exposed to a small irradiation dose of 10 kGy. The high reactivity of the irradiated PP was conjugated with low yields of residue and coke in addition to enhanced selectivity for light distillate (C7–C10). Examining the effect of pyrolysis temperature revealed that the catalytic pyrolysis preferred high temperature among the investigated temperature range of 325–375 °C. The results presented above clarified that a significant increase in the rate of the catalytic pyrolysis with enhanced selectivity of C7–C10 compounds can be obtained by exposing PP to the ionizing irradiation prior to the catalytic pyrolysis. The results suggested the applicability of the proposed pyrolysis method for enhancing the catalytic conversion of plastic waste into useful hydrocarbons.

Uemichi et al, 1989 investigated the degradation of polypropylene to aromatic hydrocarbons over activated carbon catalysts containing Pt and Fe. The results obtained were compared with those for the degradation of polyethylene. The addition of Pt or Fe to activated carbon resulted in an increase in the yield of aromatics from polypropylene. However, the increase was less than that from polyethylene. Pt metal was more effective than Fe only when the reaction conditions involved a longer contact time. The formation of aromatics was explained by essentially the same mechanism as the case of polyethylene, in which an influence of methyl branching of polyropylene on the aromatization yield and a difference in catalytic activity of the catalysts containing Pt and Fe for a ring expansion reaction were considered.

Park et al., 2008 reported pyrolysis of polypropylene over mesoporous MCM-48 material. Mesoporous MCM-48 materials were employed as catalysts for the degradation of PP. The catalytic activity of Al-MCM-48 was much higher than that of Si-MCM-48. Al-MCM-48 mainly generated C7–C10 hydrocarbons, while Si-MCM-48 exhibited a relatively broader distribution of oil products (C7–C14). Al-MCM-48 showed high catalytic stability for the degradation of PP. In view of these facts, Al-MCM-48 can be considered a promising catalyst for the degradation of other waste plastics (Tables 7 and 8).

	Without catalyst	Catalyst type		
		Si-MCM-48	Al-MCM-48 (Si/Al=60)	Al-MCM-48 (Si/Al=30)
Conversion (%)	3.3	75.3	90.2	95.7
Yield (wt.%)				
Oil	2.1	58.3	72.2	76.5
Gas	1.2	17.0	18.0	19.2

Table 7. Effect of catalyst type on product yield obtained from PP pyrolysis at 380°C for 1 h and PP:catalyst=5:1 (Park et al., 2008).

Catalyst	Product distribution (wt.%)					
	CH_4	C_2H_6	C_2H_4	C_3H_8	C_3H_6	C_{4s}
Without catalyst	0.9	3.6	1.0	2.3	20.4	71.8
Si-MCM-48	0.1	0.1	0.1	0.4	14.0	85.3
Al-MCM-48 (Si/Al=60)	0.1	0.1	0.2	0.4	12.3	86.9
Al-MCM-48 (Si/Al=60)	0.1	0.1	0.2	0.5	13.4	85.7

Table 8. Effect of catalyst type on the product distribution of the gas fraction during PP pyrolysis at 380°C for 1 h and PP:catalyst=5:1 (Park et al., 2008).

Cardona and Corma, 2000 studied the tertiary recycling of polypropylene by catalytic cracking in a semibatch stirred reactor semicontinuous reactor has been presented that allows carrying out efficiently the catalytic cracking of PP. By working with USY zeolites with different unit cell sizes, it has been proven that neither the total amount nor the strength of the acid sites are the most determinant factors for cracking PP. the first cracking event of PP occurs at or close to the external surface. Then the formation of mesopores in the zeolite strongly improves the cracking activity. This has been supported by the results obtained with a Y zeolite synthesized with smaller crystallite sizes (Table 9). Finally it has been shown that amorphous or ordered silica–aluminas are very active catalysts. However, a FCC equilibrium catalyst can be a better option from an economical point of view since it gives a very good selectivity, and even if its activity is lower than that of the silica–aluminas, this can be compensated by increasing the catalyst to PP ratio.

Catalyst	Cumulative selectivity (%)								
	Reaction time = 12 min			Reaction time = 24 min			Reaction time = 72 min		
	Gases	Gasoline	Diesel + gas oil	Gases	Gasoline	Diesel + gas oil	Gases	Gasoline	Diesel + gas oil
Si–Al 13%	5.2	78.6	16.2	4.9	72.2	22.9	6.7	70.3	23.0
Resoc-g	10.5	77.5	12.0	8.1	75.7	16.2	7.4	74.7	17.9
H-USY 500	13.2	81.4	5.4	11.2	81.5	7.3	10.2	82.7	7.1
H-USY 712	6.1	77.8	16.1	5.8	75.8	18.4	6.7	76.2	17.1
H-USY 760	8.9	81.0	10.1	8.6	78.1	13.3	9.5	78.8	11.7

Table 9. Effect of the catalyst type on the product distribution during PP pyrolysis at 380 ^oC (Cardona and Corma, 2000).

Xie et al., 2008 have reported catalytic cracking of polypropylene (PP) over MCM-41 modified by Zr and Mo. The relationship among structure, acidity and catalytic activity of Zr–Mo-MCM-41 was studied. The results showed that Zr–Mo-MCM-41 exhibited high activity for the cracking of PP and good selectivity for producing liquid hydrocarbons of higher carbon numbers. The results were compared with those obtained over HZSM-5, SiO_2–Al_2O_3 and other MCM-41 mesoporous molecular sieves. For the catalytic cracking of PP, Mo enhances the selectivity to high carbon number hydrocarbons and Zr enhances the acidity of catalyst and results in the increasing cracking conversion of PP. Zr–Mo-MCM-41 using $Zr(SO_4)_2$ as Zr source is of the best catalytic activity and selectivity to high carbon number hydrocarbon, which means that Zr–Mo-MCM-41 will probably become good potential catalysts for the cracking of PP (Table 10).

Catalyst	Temperature (^oC)	Conversion (%)	Liquid yield (%)	Gas yield (%)
HZSM-5	400	27.1	50.2	49.8
SiO_2-Al_2O_3	400	25.8	60.3	39.7
Thermal cracking	400	30.4	76.2	23.8
Si-MCM-41	400	39.6	81.4	18.9
Mo-MCM-41	400	57.5	90.0	10.0
Zr-Mo-MCM-41	400	98.6	92.0	8.0
Zr-Mo-MCM-41	380	65.4	81.9	18.1
Zr-Mo-MCM-41	390	84.3	87.7	12.3
Zr-Mo-MCM-41	410	99.6	89.0	11.0

Table 10. Catalytic activities of different catalyst in PP pyrolysis for 30 min at catalyst/ PP = 0.01 (Xie et al., 2008).

Panda et al., 2011 investigated catalytic performances of kaoline and silica alumina in the thermal degradation of polypropylene. Polypropylene was cracked thermally and catalytically in the presence of kaoline and silica alumina in a semi batch reactor in the temperature range 400–550°C in order to obtain suitable liquid fuels. It was observed that up to 450°C thermal cracking temperature, the major product of pyrolysis was liquid oil and the major product at other higher temperatures (475–550°C) are viscous liquid or wax and the highest yield of pyrolysis product is 82.85% by weight at 500°C. Use of kaoline and silica alumina decreased the reaction time and increased the yield of liquid fraction. Again the major pyrolysis product in catalytic pyrolysis at all temperatures was low viscous liquid oil. Silica alumina was found better as compared to kaoline in liquid yield and in reducing the reaction temperature. The maximum oil yield using silica alumina and kaoline catalyst are 91% and 89.5% respectively. On the basis of the obtained results hypothetical continuous process of waste polypropylene plastics processing for engine fuel production can be presented (Table 11).

In conclusion, catalytic pyrolysis of PP reduces environmental impacts, also the time of recycling and results in very useful products, with potential use as fuel replacements.

	Type of catalyst		
	None (thermal cracking)	Kaoline (PP:cat=3:1)	Silica-alumina (PP:cat=3:1)
	T = 500°C	T = 450°C	T = 500 °C
Liquid product (wt.%)	82.85	89.50	91.0
Gaseous product (wt.%)	16.25	9.75	8.0
Solid residue (wt.%)	0.90	0.75	1.0
Density of oil at 15 °C (g/mL)	0.84	0.745	0.770
Viscosity of oil at 30 °C (cSt)	4.31	2.18	2.21

Table 11. Pyrolysis of PP in optimum conditions (Panda and Singh, 2011).

4. Chemical recycling of polystyrene

4.1 Introduction

Polystyrene (PS) is widely used in the manufacture of many products due to its favorable properties such as good strength, light weight, and durability and is the material of choice for packaging various electronics and other fragile items. In general, PS accounts for about 9-10% of the plastic waste in municipal solid waste (MSW). In the past several years, PS has received much public and media attention. Polystyrene has been described by various environmental groups as being nondegradable, nonrecyclable, toxic when burned, landfill-choking, ozone-depleting, wildlife-killing, and even carcinogenic. These misconceptions regarding PS have resulted in boycotts and bans in various localities. Actually, PS comprises less than 0.5% of the solid waste going to landfills.

Polystyrene is used in solid and expanded forms both of which can be recycled. Solid PS components such as coffee cups, trays, etc. can be recycled back into alternative applications such as videocassette cases, office equipments, etc. Expanded PS (EPS) foam waste loses its foam characteristics as part of the recovery process. The recovered material can be re-gassed but the product becomes more expensive than virgin material. Instead it is used in solid form in standard molding applications. Both expanded and solid PS wastes have been successfully recycled in extruded plastic timber-lumber. Recycled PS is used to produce plant pots and desk items such as pen, pencils, etc. As with other types of plastic materials, PS recycling takes place after consideration by the industry of a number of issues including eco-efficiency, availability, corporate social responsibility, product quality=hygiene aspects, and traceability.

More than a thousand tones of PS foam worldwide is being disposed off into environment as MSW. The amount is increasing every year. The booming development of electronic products has sharply increased the quantities of Waste from Electrical and Electronic Equipment (WEEE), amplifying the problem of their disposal. The solution can be found only through a modern Design For Environment (DFE) with a big attention to recycling and disassembly.

4.2 Types of polystyrene accepted for recycling

Expanded polystyrene (EPS) foam packaging, which is the familiar white material, custom molded to cushion, insulate and protect all types of products during transportation, can be recycled. EPS insulation boards used for housing and commercial construction, foodservice products like cups, plates, trays, etc. that are made of PS resin foamed to provide a unique insulating quality and loosefill packaging are accepted for recycling. Non-Foam Polystyrene products also called high impact polystyrene (HIPS), oriented polystyrene (OPS), post consumer products, post industrial products, and styrofoam (A Dow Chemical Company brand trademark for a PS foam thermal insulation product) have also been accepted for recycling (Vilaplana et al., 2006).

4.3 Recycling methods for polystyrene products

Before recycling, the recyclable materials should be rinsed off for the removal of any food or dirt particles, the caps of the plastic bottles and glass jars should be thrown away and the oversized materials like cartons, milk jugs, etc. should be crushed so that they can fit into the bin and into the truck more easily. The volume of EPS is reduced by methods such as solvent volume reduction (dissolved using solvent), heating volume reduction, and pulverizing volume reduction (pulverized).

The processed EPS is used in its reduced state as an ingredient for recycled products or it is burnt to generate heat energy. A large amount of expanded PS is discharged after use at wholesale markets, supermarkets, department stores, restaurants and shops, such as electrical appliances stores, as well as at factories of machinery manufacturers. It is collected through the in-house collection of companies or by resource recycling agents and becomes a recycled resource.

4.3.1 Recycling using the dissolution technique

A rather easy way of recovering polymers from a mixture of different plastics is by using an appropriate solvent to selectively dissolve the polymer and then recovering it by removal of that solvent. In this sense a naturally occurring compound, i.e. limonene (occurring in citrus fruits) has been successfully used to dissolve EPS (Achilias et al., 2009). This solvent has the ability to dissolve EPS in large amounts safely and with negligible degradation of the polymer's performance properties. Conventional melt separation methods cause a large drop in the polymer's molecular weight due to thermal degradation. Consequently dissolved PS can be precipitated through the addition of a non-solvent in the mixture. The solvent is vacuum-evaporated and re-used.

4.3.2 Chemical recycling of PS

One of the attractive chemical recycling processes is the catalytic degradation [Kim et al., 2003] of polystyrene. This process enables to get styrene monomer (S) at relatively low temperature with a high selectivity. Modified Fe-based catalysts were employed for the catalytic degradation of EPS waste, where carbanion may lead to high selectivity of S in the catalytic degradation of PS. The yield of oil (Y_{Oil}) and S (Y_S) were increased in the presence of Fe-based catalysts and with increasing reaction temperature. Y_{Oil} and Y_S were obtained over Fe–K/Al_2O_3 at the relative low reaction temperature (400°C) 92.2 and 65.8 wt. %, respectively. The value of Ea (activation energy) is obtained as 194 kJ/mol for the thermal degradation of EPS. However, the Ea was decreased considerably to 138 kJ/mol in the presence of the catalysts (Fe–K/Al_2O_3).

Bajdur et al., 2002 have synthesized sulfonated derivatives of expanded PS wastes, which may be used as polyelectrolytes. Modification was conducted by means of known methods and products having various contents of sulfogroups in polymer chain were obtained. They have found that the polyelectrolytes have good flocculation properties similar to those of anionic commercial polyelectrolytes. The effect of a base catalyst, MgO, on the decomposition of PS was studied through degradation of both a monodisperse polymer and a PS mimic, 1,3,5-triphenylhexane (TPH), to determine the potential of applying base catalysts as an effective means of polymer recycling [Woo et al., 2000]. The presence of the catalyst increased the decomposition rate of the model compound but decreased the degradation rate of PS as measured by evolution of low molecular weight products. Although the model compound results suggest that the rate of initiation was enhanced in both cases by the addition of catalyst, this effect is overshadowed for the polymer by a decrease in the 'zip length' during depropagation due to termination reactions facilitated by the catalyst. Due to the small size of the model compound, this effect does not impact its observed conversion since premature termination still affords a quantifiable low molecular weight product. A decrease in the selectivity to styrene monomer in the presence of MgO was observed for both PS and TPH. They have discussed the reconciliation of their results with those of Zhang et al., 1995 based on differences in the reactor configuration used.

Degradation of PS into styrene, including monomer and dimer, was studied by Ukei et al., 2000 using solid acids and bases viz. MgO, CaO, BaO, K_2O, SiO_2/Al_2O_3, HZSM5 and active carbonas catalysts. They have found that solid bases were more effective catalysts than solid acids for the degradation of PS into styrene. This was attributed to differences in the degradation mechanisms of PS over solid acids and bases. Among the solid bases employed, BaO was found to be the most effective catalyst and about 90 wt. % of PS was converted into styrene when thermally degraded PS was admitted to BaO powder at 350°C.

Koji et al., 1998 have succeded to obtain PS foam, which can be recycled into styrene by mixing a PS with a basic metal oxide being a catalytic decomposition catalyst and foaming the mixture with an inert blowing agent. When it is wasted, it can be recycled into styrene by decomposing it by heating to 300–450°C in a nonoxidizing atmosphere. The basic oxide is Na_2O, MgO, CaO or the like and among them CaO is desirable. When the basic metal oxide carried by porous inorganic filler is used, it can exhibit improved effectiveness desirably. The blowing agent used is a nitrogen gas, a chlorofluorocarbon, propane or the like.

Lee et al., 2002, have studied several solid acids, such as silica-alumina, HZSM-5, HY, mordenite, and clinoptilolite as catalysts and screened their performances in the catalytic degradation of PS. The clinoptilolite catalysts (HNZ, HSCLZ) showed good catalytic performance for the degradation of PS with selectivity to aromatics more than 99%. Styrene is the major product and ethylbenzene is the second most abundant one in the liquid product. The increase of acidity favored the production of ethylbenzene by promoting the hydrogenation reaction of styrene. Higher selectivity to styrene is observed at higher temperatures. An increase of contact time by reducing nitrogen gas flow rate enhanced the selectivity to ethylbenzene. Thus a designed operation including acidity of catalyst, reaction temperature, and contact time will be necessary to control the product distribution between styrene monomer and ethylbenzene.

Ke et al., 2005 studied the degradation of PS in various supercritical solvents like benzene, toluene, xylene, etc. at 310–370°C and 6.0MPa pressure. It was found that PS has been successfully depolymerized into monomer, dimmer, and other products in a very short reaction time with high conversion. Toluene used as supercritical solvent was more effective than other solvents such as benzene, ethylbenzene, and p-xylene for the recovery of styrene from PS, though the conversions of PS were similar in all the above solvents. The highest yield of styrene (77 wt%) obtained from PS in supercritical toluene at 360°C for 20 min.

Subcritical water is a benign and effective media for polymer degradation. Suyama et al., 2010 have found that on subcritical water treatment in the presence of an aminoalcohol, unsaturated polyesters crosslinked with styrene were decrosslinked, and a linear polystyrene derivative bearing hydroxy-terminated side-chains was recovered. After modification of the hydroxy groups with maleic anhydride, the polystyrene derivative was re-crosslinked with styrene to form a networked structure again. The resulting solid was degradable by subcritical water treatment in the presence of the aminoalcohol to give another polystyrene derivative bearing hydroxy groups. These processes could be repeated successfully, demonstrating the applicability as a novel recycling system of thermosetting resins. The polystyrene derivative was also re-crosslinked again on heating with an alternative copolymer of styrene and maleic anhydride due to the formation of linkage between the hydroxy groups and carboxylic anhydride moieties.

In order to reduce the consumption of energy and get oligostyrene of several thousands of molecular weight, which can be used as a kind of fuel oil, the thermal decomposition of EPS with a-methylstyrene as a chain-transfer agent was studied by Xue et al., 2004, at a temperature about 200°C. Three kinds of organic peroxides were used as radical accelerators. They found that the addition of dicumyl peroxide (DCP) enhanced the thermal decomposition of EPS even at lower temperature, about 140 °C, but the addition of tertbutylcumyl peroxide was less effective than DCP. On the other hand, di-tert-butyl peroxide had almost no effect on the thermal decomposition of EPS.

4.3.3 Mechanism of polystyrene cracking

Radical depolymerization of neat PS samples produces large quantities of monomer (styrene) and chain-end backbiting yields substantial amounts of dimer and trimer. Polymer decomposition proceeds by entirely different processes when a catalyst is present. The formation of the primary PS catalytic cracking volatile products can be explained by initial electrophilic attack on polymer aromatic rings by protons. Protons preferentially attack the *ortho* and *para* ring positions because the aliphatic polymer backbone is an electron-releasing group for the aromatic rings. Most volatile products can be derived from mechanisms beginning with ring protonation. Thermal decomposition of *ortho*-protonated aromatic rings in the polymer chain (1) can lead directly to the liberation of benzene, the primary catalytic cracking product, or may result in chain shortening. Benzene cannot be obtained directly from *para*-protonated aromatic rings in the polymer. However, *para*-protonated rings can react with neighboring polymer chains to yield the same chain scission products that are formed by *ortho*-protonation. The macro cation remaining after benzene evolution (2) may undergo chain shortening β-scission to produce (3) and an unsaturated chain end, rearrange to form an internal double bond and protonate a neighboring aromatic ring (either by intra- or intermolecular proton transfer), cyclize to form an indane structure, or abstract a hydride to produce a saturated chain segment. The substantial quantities of indanes obtained by PS catalytic cracking suggests that cyclization of (2) to form indane structures is a favored process. A consequence of chain unsaturation resulting from (2) might be the formation of conjugated polyene segments that may subsequently cyclize to form naphthalenes. Decomposition of (3), which might be formed from (1) or (2), can result in the formation of styrene or may lead to chain end unsaturation and neighboring ring protonation (Figure 2).

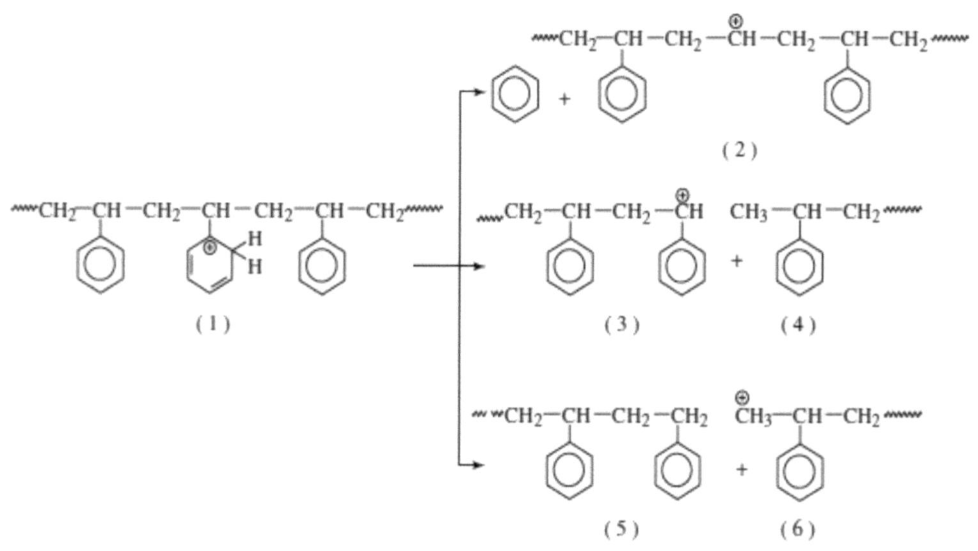

Fig. 2. Mechanism of PS cracking.

Hydride abstraction by (3) would result in a saturated chain end. The lack of significant styrene production from any of the PS-catalyst samples suggests that β-scission of (3) to form styrene is not a dominant decomposition pathway at low temperatures. Chain end unsaturation derived from (3) may result in formation of indenes, which were detected in substantial amounts only when HZSM-5 catalyst was present. The restricted volume of the HZSM-5 channels apparently inhibits hydride abstraction pathways for (3), which results in increased production of indenes and styrene for PS-HZSM-5. Protonation of aromatic rings adjacent to methyl-terminated chain ends (4) can result in the formation of alkyl benzenes, propene, and benzene, depending on how the macro cation decomposes.

4.3.4 Thermo-chemical recycling of PS

Thermochemical recycling techniques such as pyrolysis are usually applied. Thus, PS can be thermally depolymerized at relatively low temperatures in order to obtain the monomer styrene with a high selectivity.

Arandes et al., 2003 have studied the thermal cracking of PS and polystyrene-butadiene (PS-BD) on mesoporous silica which has no measurable acidity. Although the content of PS in domestic plastic wastes is approximately 10wt. %, less attention has been paid to the cracking of dissolved PS than to the cracking of dissolved polyolefins. The kinetic characteristics of PS cracking described by Arandes et al. are different to those of polyolefins and the ideal aim of its valorization is the recovery of the styrene monomer. Bockhorn et al., 1999 and Kruse et al., 2001 made an analysis of the reactions involved in the mechanism of PS cracking and developed a detailed mechanistic model for the polymer degradation. Faravelli et al., 2003 presented a detailed kinetic model for the thermal degradation of PE – PS mixtures.

The technology for thermal cracking that has been more widely studied and that has been tested at larger scale is that based on a fluidized bed, in which the plastics are fed in the solid state and sand is used for helping fluidization [Scheirs and Kaminsky, 2006; Milne et al., 1999; Westerhourt, 1997]. The design of fluidized beds used at laboratory or pilot plant scale has been carried out on the basis, that the kinetics of pyrolysis of plastics is subjected to great uncertainty caused by factors such as heterogeneity of the material, synergy in the cracking of different constituents, and limitations to heat and mass transfer. These factors prevent obtaining kinetics that is reliable for the design of the reactor at temperatures of industrial interest (above 450°C) [Mehta et al., 1995]. Furthermore, this strategy is suitable for its development in a refinery by using the existing equipment and by optimizing the possibilities of incorporating the products either into the market (subsequent to fuel reformulation) or into the production process itself (subsequent to monomer purification). An additional problem in the cracking of PS is the rapid deactivation of the catalyst caused by the coke formed on the acid sites, which is favored by the aromatic nature of styrene and its high C/H ratio.

A conical spouted bed reactor (CSBR) has been used for the kinetic study of PS pyrolysis in the 450–550°C range [Aguado et al., 2003] and the results have been compared with those obtained by thermogravimetry (TGA) and in a microreactor (MR) of very high sample heating rate. The comparison proves the advantages of the gas–solid contact of this new reactor for the kinetic study of pyrolysis of plastics at high temperature, which stem from the high heat transfer rate between gas and solid and from the fact that particle agglomeration is avoided.

A swirling fluidized-bed reactor (0.0508m ID and 1.5m in height) has been developed to recover the styrene monomer [Lee et al., 2003] and valuable chemicals effectively from the PS waste, since it can control the residence time of the feed materials and enhance the uniformity of the temperature distribution. To increase the selectivity and yield of styrene monomer in the product, catalyst such as Fe_2O_3, BaO, or HZSM-5 have been used. It has been found that the reaction time and temperature can be reduced profoundly by adding the solid catalyst. The swirling fluidization mode makes the temperature fluctuations more periodic and persistent, which can increase the uniformity of temperature distribution by reducing the temperature gradient in the reactor. The yields of styrene monomer as well as oil products have increased with increasing the ratio of swirling gas, but exhibited their maximum values with increasing the total volume flow rate of gas.

The thermal degradation of real municipal waste plastics (MWP) obtained from Sapporo, Japan and model mixed plastics was carried out at 430°C in atmospheric pressure by batch operation (Bhaskar et al., 2003). The resources and environmental effects assessed over the life of each of the packaging, includes fossil fuel consumption, greenhouse gas emissions, and photochemical oxidant precursors.

Achilias et al. 2007 investigated catalytic and non-catalytic pyrolysis experiments in a fixed bed reactor using either model polymer or commercial waste products as the feedstock. The liquid fraction produced from all the pyrolysis experiments consisted mainly of the styrene monomer and this was subjected to repolymerization without any further purification in a DSC with AIBN initiator. A basic (BaO) and an acidic commercial FCC catalyst were examined in relation to the yield and composition of gaseous and liquid products. Aromatic compounds identified in the liquid fraction of the thermal and catalytic pyrolysis of model PS and commercial products appear in Table 12.

Results show that this product can be polymerized to produce a polymer similar to the original PS. However, it was found that other aromatic compounds included in this fraction could act as chain transfer agents, lowering the average molecular weight of the polymer produced and contributing to a lower Tg polymer. Therefore, it seems that the polymer can be reproduced but with inferior properties compared to a polymer prepared from neat styrene.

A general model for polymer degradation by concurrent random and chain-end processes was developed by Sterling et al., 2001 using continuous distribution kinetics. Population balance equations based on fundamental, mechanistic free radical reactions were solved analytically by the moment method. The model, applicable to any molecular weight distribution (MWD), reduces to the cases of independent random or chain-end scission. Polystyrene degradation experiments in mineral oil solution at 275–350 °C supported the model and determined reaction rate parameters. The degradation proceeded to moderate extents requiring a MW-dependent random scission rate coefficient. Polystyrene random scission activation energy was 7.0 kcal/mol, which agrees well with other thermolysis investigations, but is lower than that found by pyrolysis due to fundamental differences between the processes. Magnesium oxide, added as a heterogeneous catalyst in solution, was found to have no effect on PS degradation rate.

Thermal and thermo-oxidative degradation of PS in the presence of ammonium sulfate [Zhu et al., 1998] was studied with thermogravimetry and FTIR. TGA results indicated that ammonium sulfate accelerated thermal degradation in nitrogen but delayed thermo-oxidative degradation of PS in air. IR analysis of tetrahydrofuran extracts, from the samples degraded at 340°C and of residues after thermal treatment at 340°C in a furnace, showed that the acceleration of thermal degradation and the suppression of thermo-oxidative degradation were due to sulfonation and oxidation of ammonium sulfate and its decomposition products, and formation of unsaturated structures in the PS chain.

Compounds, chemical formula	Thermal	Catalytic BaO	Catalytic FCC	Plastic container	Plastic glass (EPS)
Styrene (**monomer**) 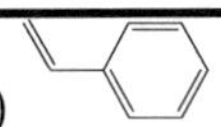	63.9	69.6	45.1	53.3	70.0
Benzene, Toluene 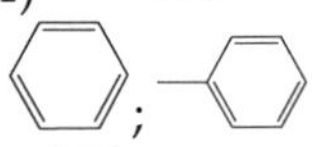	2.0	2.4	5.0	5.6	2.5
Ethylbenzene 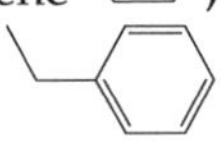	0.5	1.1	-	1.9	1.5
a-Methylstyrene 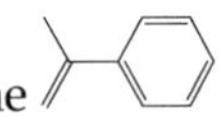	2.1	2.6	6.3	5.9	2.3
Xylene 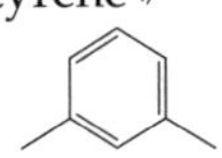	-	-	16.8	-	-
Cumene 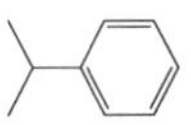	-	-	1.7	0.2	0.3
2,4-Diphenyl-1-butene (**dimer**) 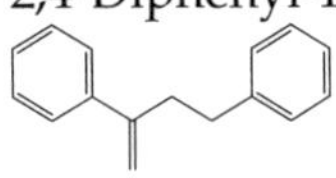	14.0	18.4	1.9	11.9	9.0
Indane, Indene, etc. 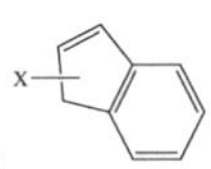	2.0	0.3	5.0	2.5	0.2
1,2-Diphenylethane 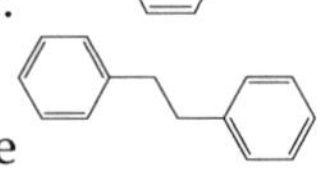	2.2	0.7	0.9	2.1	0.8
1-methyl-1,2Diphenylethan 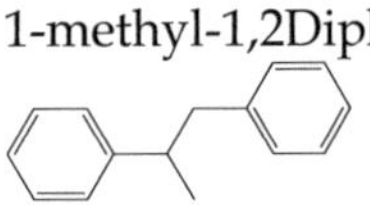	1.1	0.4	0.5	1.4	0.4
1,3-Diphenylpropane 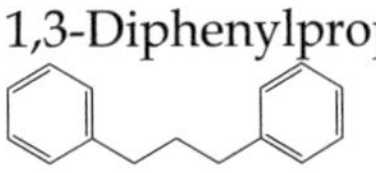	0.6	0.6	0.5	2.8	0.8
1,1′-Diphenylpropene 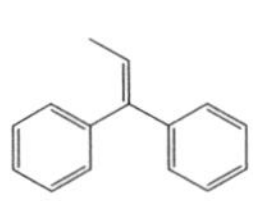	0.7	0.4	1.1	-	0.7
2,4,6-Triphenyl-1-hexene(**trimer**) 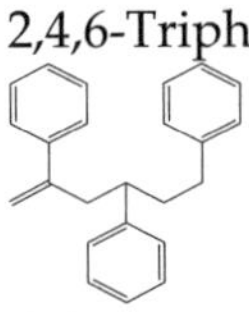	2.2	1.8	0.3	3.5	5.0
Other aromatic compounds	8.7	1.7	14.9	8.9	6.4

Table 12. Aromatic compounds identified in the liquid fraction of the thermal and catalytic pyrolysis of model polystyrene and commercial products based on polystyrene (wt.-% on liquid produced) (Achilias et al., 2007).

Thermal degradation of PS has been investigated in the presence of water [Beltrame et al., 1997] under subcritical conditions (hydrous pyrolysis). The experiments were carried out in closed systems under inert atmosphere, in the temperature range 300–350 °C, at pressures up to 18 MPa, for 1–120 h. The products obtained, separated as gases, volatiles, and heavy compounds. The results showed that the presence of water increases the yields of the volatile products, mainly in the first steps of the pyrolytic process, and leads to higher yields of monomer. This latter observation suggests a lowering of the secondary reactions extent.

The catalytic degradation of waste plastics such as HDPE, LDPE, PP and PS over spent fluid catalytic cracking (FCC) catalyst was also carried out at atmospheric pressure with a stirred semi-batch operation at 400 °C [Lee et al., 2002]. The objective was to investigate the influence of plastic types on the yield, liquid product rate, and liquid product distribution for catalytic degradation. The catalytic degradation of waste PE and PP with polyolefinic structure exhibited the liquid yield of 80–85% and the solid yield of below 1%, whereas that of waste PS with polycyclic structure produced much more liquid, solid products, and much less gas products. Accumulative liquid product weight by catalytic degradation strongly depended on the degradation temperature of the plastics.

In accordance with the option of recycling plastics into fuels by dissolving them in standard feedstocks [Puente et al., 1998] for the process of catalytic cracking of hydrocarbons, FCC, and various acidic catalysts (zeolites ZSM-5, mordenite, Y, and a sulfur-promoted zirconia) were tested in the conversion of PS dissolved into inert benzene at 550 °C in a fluidized- bed batch reactor. Experiments were performed with very short contact times of up to 12 sec. Main products were in the gasoline range, including benzene, toluene, ethylbenzene, styrene, and minor amounts of C9–12 aromatics and light C5– compounds. Coke was always produced in very significant amounts. Even though sulphur promoted zirconia is highly acidic, the low proportion of Brönsted-type acid sites does not allow the occurrence of secondary styrene reactions. It was shown that most favorable product distributions (higher yields of desirable products) are obtained on equilibrium commercial FCC catalysts.

PS can be recycled into styrene monomer in association with some other aromatics, from which styrene can be converted to biodegradable plastic such as polyhydroxyalkanoates (PHA). Recently scientists have achieved 10% yield of PHA from PS [Ward et al., 2006]. The chain length of PHA produced was 10. The yield and composition of oils and gases derived from the pyrolysis and catalytic pyrolysis of PS has been investigated. The pyrolysis and catalytic pyrolysis was carried out in a fixed bed reactor. Two catalysts were used, zeolite ZSM-5 and Y-zeolite and the influence of the temperature of the catalyst, the amount of catalyst loading, and the use of a mixture of the two catalysts was investigated. The main product from the uncatalyzed pyrolysis of PS was oil consisting mostly of styrene and other aromatic hydrocarbons like toluene and ethylbenzene. In the presence of either catalyst an increase in the yield of gas and decrease in the amount of oil production was found, but there was significant formation of carbonaceous coke on the catalyst. Increasing the temperature of the Y-zeolite catalyst and also the amount of catalyst in the catalyst bed resulted in a decrease in the yield of oil and increase in the yield of gas. Oil derived from the catalytic pyrolysis of PS contain aromatic compounds such as single ring compounds like benzene, toluene, styrene, m-xylene, o-xylene, p-xylene, ethylmethylbenzene, propenyl-benzene, methylstyrene; two ring compounds like indene, methylindene, naphthalene, 2-

methylnaphthalene, 1-methylnaphthalene, biphenyl, methylbiphenyl, dimethylnaphthalene, trimethylnaphthalene, tetramethylnaphthalene, ethylbiphenyl; three ring compounds like phenanthrenes and four ring compounds like pyrenes and chrysenes.

4.4 Future prospectus

Some future prospectus of PS recycling include (Maharana et al., 2007).

1. Thermal recycling of PS yields higher percentage of styrene monomer, which can be fermented by bacteria to produce polyhydroxyalkanoates (PHA) – the starting material for the synthesis of biodegradable polymers.
2. The waste PS can be blended with biodegradable polymers to produce biodegradable polymers.
3. Styrene monomer produced by recycling can be grafted onto biodegradable polymers to give biodegradable polymers.

5. Chemical recycling of polyethylene (LDPE and HDPE)

5.1 Introduction

Under the category of chemical recycling of polyethylenes, advanced process (similar to those employed in the petrochemical industry) appear e.g. pyrolysis, gasification, liquid-gas hydrogenation, viscosity breaking, steam or catalytic cracking (Al-Salem *et al,* 2009). Catalytic cracking and reforming facilitate the selective degradation of waste plastics. The use of solid catalysts such as silica alumina, ZSM-5, zeolites, and mesoporous materials for these purposes has been reported. These materials effectively convert polyolefins into liquid fuel, giving lighter fractions as compared to thermal cracking (Al-Salem *et al*, 2009).

In particular, polyethylene has been targeted as a potential feedstock for fuel (gasoline) producing technologies. PE thermally cracks into gases, liquids, waxes, aromatics and char. The relative amounts of gas and liquid fraction are very much dependent on the type of polymer used. Thus, higher decomposition was observed in PP, followed by LDPE and finally HDPE. It seems that less crystalline or more branched polymers are less stable in thermal degradation (Achilias et al., 2007). Many papers have been published recently on this subject and excellent reviews can be found in the book by Scheirs and Kaminsky, 2006 and Achilias et al., 2006.

Polyethylene (as well as other vinyl polymers) degrade via a four step free radical mechanism: radical initiation, de-propagation (as opposed to propagation in the case of polymerization), intermolecular and intramolecular hydrogen transfer followed by β-scission (initial step in the chemistry of thermal cracking of hydrocarbons and the formation of free radicals) and, lastly, radical termination. β-Scission and hydrogen abstraction steps often occur together in a chain propagation sequence. That is, a radical abstracts a hydrogen atom from the reactant to form a molecule and a new radical. A bond β is then broken to the radical centre (β-scission) to regenerate an abstracting radical and to produce a molecule with a double bond (a molecule with a double bond involving the carbon atom that had been the radical centre). Sample size and surface area to volume ratio of the melt have a significant influence on the rate and relative importance of the various mechanisms of polymer degradation. In pyrolysis, which is normally done on micro-scale, only random initiation and intermolecular transfer were reported to be important. Conversely, on milligram scale of polyethylene charges and samples, intermolecular transfer of hydrogen atoms via abstraction by free radicals was considered to be the predominant transfer mechanism to produce volatiles. There is also a growing interest in developing value added products such as synthetic lubricants via PE thermal degradation.

The development of value added recycling technologies is highly desirable as it would increase the economic incentive to recycle polymers. Several methods for chemical recycling are presently in use, such as direct chemical treatment involving gasification, smelting by blast furnace or coke oven, and degradation by liquefaction. The main advantage of chemical recycling is the possibility of treating heterogeneous and contaminated polymers with limited use of pre-treatment. Petrochemical plants are much greater in size (6–10 times) than plastic manufacturing plants. It is essential to utilize petrochemical plants in supplementing their usual feedstock by using plastic solid wastes (PSW) derived feedstock (Al-Salem *et al*, 2009).

5.2 Thermolysis schemes and technologies

5.2.1 Pyrolysis

Thermolysis is the treatment in the presence of heat under controlled temperatures without catalysts. Thermolysis processes can be divided into advanced thermo-chemical or pyrolysis (thermal cracking in an inert atmosphere), gasification (in the sub-stoichiometric presence of air usually leading to CO and CO_2 production) and hydrogenation (hydrocracking).

Thermal degradation processes allow obtaining a number of constituting molecules, combustible gases and/or energy, with the reduction of landfilling as an added advantage. The pyrolysis process is an advanced conversion technology that has the ability to produce a clean, high calorific value gas from a wide variety of waste and biomass streams. The hydrocarbon content of the waste is converted into a gas, which is suitable for utilisation in either gas engines, with associated electricity generation, or in boiler applications without the need for flue gas treatment. This process is capable of treating many different solid hydrocarbon based wastes whilst producing a clean fuel gas with a high calorific value. This gas will typically have a calorific value of 22–30 MJ/m^3 depending on the waste material being processed. Solid char is also produced from the process, which contains both carbon and the mineral content of the original feed material. The char can either be further processed onsite to release the energy content of the carbon, or utilized offsite in other thermal processes (Al-Salem *et al*, 2009).

The main pyrolysis units and technologies on an industrial scale include PYROPLEQ (rotary drum), Akzo (circulating fluidized bed), NRC (melt furnace), ConTherm technology (rotary drum), PKA pyrolysis (rotary drum), PyroMelt (melt furnace), BP (circulating fluidized bed), BASF (furnace) and NKT (circulating fluidized bed). Detailes can be found in Al-Salem *et al*, 2010.

Pyrolysis provides a number of other advantages, such as (i) operational advantages, (ii) environmental advantages and (iii) financial benefits. Operational advantages could be described by the utilisation of residual output of char used as a fuel or as a feedstock for other petrochemical processes. An additional operational benefit is that pyrolysis requires no flue gas clean up as flue gas produced is mostly treated prior to utilisation. Environmentally, pyrolysis provides an alternative solution to landfilling and reduces greenhouse gas (GHGs) and CO_2 emissions. Financially, pyrolysis produces a high calorific value fuel that could be easily marketed and used in gas engines to produce electricity and heat. Several obstacles and disadvantages do exist for pyrolysis, mainly the handling of char produced and treatment of the final fuel produced if specific products are desired. In addition, there is not a sufficient understanding of the underlying reaction pathways, which has prevented a quantitative prediction of the full product distribution (Al-Salem *et al.*, 2009).

5.2.2 Gasification

Air in this process is used as a gasification agent, which demonstrates a number of advantages. The main advantage of using air instead of O_2 alone is to simplify the process and reduce the cost. But a disadvantage is the presence of (inert) N_2 in air which causes a reduction in the calorific value of resulting fuels due to the dilution effect on fuel gases. Hence, steam is introduced in a stoichiometric ratio to reduce the N_2 presence. A significant amount of char is always produced in gasification which needs to be further processed and/or burnt. An ideal gasification process for PSW should produce a high calorific value gas, completely combusted char, produce an easy metal product to separate ash from and should not require any additional installations for air/water pollution abatement (Al-Salem et al, 2009).

Early gasification attempts of plastics, have been reported since the 1970s. The gasification into high calorific value fuel gas obtained from PSW was demonstrated in research stages and results were reported and published in literature for PVC, PP and PET. The need for alternative fuels has lead for the co-gasification of PSW with other types of waste, mainly biomass. Pinto *et al.* (2002, 2003) studied the fluidized bed co-gasification of PE, pine and coal and biomass mixed with PE. Xiao *et al.* (2009) co-gasified five typical kinds of organic components (wood, paper, kitchen garbage, plastic (namely PE), and textile) and three representative types of simulated MSW in a fluidized-bed (400–800 ^{0}C). It was determined that plastic should be gasified at temperatures more than 500 ^{0}C to reach a lower heating value (LHV) of 10,000 kJ/N (Al-Salem *et al*, 2009).

5.2.3 Hydrogenation (hydrocracking)

Hydrogenation by definition means the addition of hydrogen by chemical reaction through unit operation. The main technology applied in PSW recycling via hydrogenation technology is the Veba process. Based upon the coal liquefaction technology, Veba Oel AG converted coal by this process into naphtha and gas oil. Major technologies are summarized in Al-Salem *et al*, 2009.

5.2.4 Other chemical recycling schemes

Degradative extrusion provides an optimum engineering solution especially on a small-industrial scale (10 kg/h). The advantages of degradative extrusion as (i) achieving molecular breakdown of thermoplastics and hence low viscosity polymer melts, (ii) applying a combination of mechanical and chemical recycling scheme prompts the degradation process by introducing steam, gas, oxygen or catalysts, if needed. Another advantageous technology for chemical treatment is catalytic and steam cracking. The concept for both processes is the employment of either steam or a catalyst in a unit operation (Al-Salem *et al,* 2009).

5.3 Polyolefins thermal cracking

Appropriate design and scale (of operation and economy) are of paramount importance when it comes to thermal treatment plants. Thermal degradation behaviour in laboratory scale enables the assessment of a number of important parameters, such as thermal kinetics, activation energy assessment (energy required to degrade materials treated and product formation) and determining reference temperatures of the half life of polymers and maximum degradation point achievable. It is also important to perform pilot scale experiments utilizing a number of rectors and unit operation before commencing with an alteration on a performance scale This will also aid in the determination of the mode of the material processing of the thermal plant (i.e. pulsating, continuous, batch, etc.). Pyrolysis (depolymerization in inert atmospheres) is usually the first process in a thermal plant, and is in need of appropriate end-product design.

A number of studies have been carried out (Achilias et al., 2006) on polyolefins thermal cracking in inert (pyrolysis) and/or partially oxidized atmospheres (e.g. step pyrolysis, gasification). Previous reports focused on kinetic parameters estimations by means of different techniques and experimental conditions. Thermogravimetry is the most commonly used technique for the determination of kinetic parameters, although the experimental conditions utilized are very different, involving broad ranges of temperature, sample amount, heating rates (in the case of dynamic runs), reaction atmospheres and pressures. Almost all of previously published literature shows a power law equation to describe the thermal cracking of polymers and perform isothermal and/or dynamic experiments (Al-Salem *et al.,* 2010).

5.4 Catalytic degradation

Studies concerning the use of different catalysts in the pyrolysis of polyolefins have been conducted by many authors (Achilias et al., 2006). Thus, TG and micro-reactors have been widely used to pyrolyse plastics with zeolite-based acid catalysts (Marcilla et al., 2001; 2004). Catalytic pyrolysis of polyethylene samples has also been carried out in the laboratory scale reactors, such as batch reactors (Seo et al., 2003, Van Grieken et al., 2001), semi-batch reactors (with evacuation of volatile products) (Akpanudoh et al.2005, Cardona and Corma, 2002) and fixed beds (Achilias et al., 2007).

The catalytic degradation of polymeric materials has been reported for a large range of model catalysts, including amorphous silica-alumina, zeolites Y, mordenite and ZSM-5, the family of mesoporous MCM-41 materials (Marcilla *et al.*, 2002; 2003) and a few silico-aluminophosphate molecular sieves (Araujo *et al.*, 2002, Fernandes et al., 2002). Catalytic activity is closely related to the amount of acid sites, pore size and also shape of the catalyst (Park *et al.*, 2008, Serrano *et al.*, 2003). Silicoaluminophosphate (SAPO) molecular sieves represent an important class of adsorbents and catalytic materials generated by the introduction of silicon into its aluminophosphate framework. The medium pore SAPOs are attractive for catalytic applications due to the presence of specific acid sites in its structure which can convert the polymer into useful hydrocarbons (Elordi *et al.*, 2009, Singhal *et al.*, 2010, Park *et al.* 2008). The use of $BaCO_3$ as a catalyst for the thermal and catalytic degradation of waste HDPE was also reported (Rasul Jan *et al.*, 2010).

The catalysts more frequently employed for the cracking of polyolefins are shape-selective zeolites and mesoporous materials, such as HY, HZSM-5, Hβ or MCM-41 (Huang et al., 2009), which undergo inevitable deactivation by coke deposition. Indeed, this deactivation is a major hurdle in the implementation and scale-up of the valorization of plastics by cracking (Marcilla et al. 2007). Microporous zeolites have very high thermal stability and customized acid sites. Thus, the selection of the zeolite should be based on a target selectivity: HZSM-5 zeolite promotes the production of olefins (original monomers), while Hβ and HY zeolites maximize the production of middle distillates (Elordi et al., 2009).

The relevant literature reports well-founded mechanisms for coke formation and protocols for characterizing the coke deposited on zeolites. These studies assay reactions such as the cracking of hydrocarbons (Cerqueira et al., 2005; 2008, Guisnet et al., 2009). Marcilla et al. studied the deactivation of zeolites during the cracking of high-density polyethylene (HDPE), by using mainly a thermobalance as reactor. It should be pointed out that coke formation is strongly affected by the following factors, amongst others: catalyst properties (e.g. shape selectivity, acidity, and concentration of acid sites) (Huang et al, 2009), reactor medium (Aguayo et al., 1997), operating conditions or feedstock properties.

5.5 Reactor types

5.5.1 Pyrolysis in a fluidized bed reactors

Pyrolysis in a fluidized bed reactor and similar devices is the one with most possibilities for large-scale implementation for continuous waste plastic upgrading. The thermal degradation of plastic polymers has been studied first (Predel and Kaminsky, 2000, Berrueco et al. 2002, Mastellone et al., 2002, Mastral et al., 2002), but in situ catalytic pyrolysis has become a relevant research topic (Mastral et al., 2006, Hernández et al., 2007). The conical spouted bed reactor (CSBR) presents interesting conditions for catalytic pyrolysis because of the low bed segregation and lower attrition than the bubbling fluidized bed. The good performance of the CSBR has been proven for the selective production of waxes (Aguado et al., 2002), fuel-like hydrocarbons (Elordi et al., 2007) and monomers (Elordi et al., 2007). This good performance is a consequence of the solid flow pattern, high heat transfer between phases and the smaller defluidization problems when sticky solids are handled. Defluidization is due to the agglomeration of soid particles (sand) coated with melted plastic, constituting a severe problem in fluidized bed reactors. In the CSBR, polyolefins melt as they are fed into the reactor and they uniformly coat the sand and catalyst particles due to their cyclic movement. Vigorous solid flow and the action of the spout avoid the formation of agglomerates. Furthermore, the CSBR has great versatility in terms of gas residence time, which may be reduced to values near 20 ms (Olazar et al., 1999) and, consequently, the yield of polyaromatic compounds is minimized. Besides, the smaller attrition of catalyst particles, due to the absence of a distributor plate, is another advantage over the fluidized bed for its use in catalytic processes. This excellent behaviour of the CSBR has already been recorded in other processes carried out prior to catalytic pyrolysis, such as catalytic polymerization, where a similar problem of fusion of catalyst particles coated with polymer occurs (Olazar et al., 1994). The simple design of a CSBR makes its scaling up straightforward. Furthermore, its throughput by reactor volume unit is higher than that of a bubbling fluidized bed due to the lower amount of sand required for fluidization enhancement.

5.5.2 Batch reactors

Thermal and catalytic degradation of polyethylene was conducted by Seo et al. 2003 at atmospheric pressure in a batch type reactor as is illustrated in Fig. 4. The reactor was a 1.1-liter round shape stainless steel bottle placed in a thermostatic furnace. Experimental procedure is as follows. The reactor system was connected to a nitrogen supply to eliminate air before premixed plastics and catalysts were fed into the reactor. Temperature of the reactor was increased to 450 ^{0}C and held for 30 min until the reaction was completely finished.

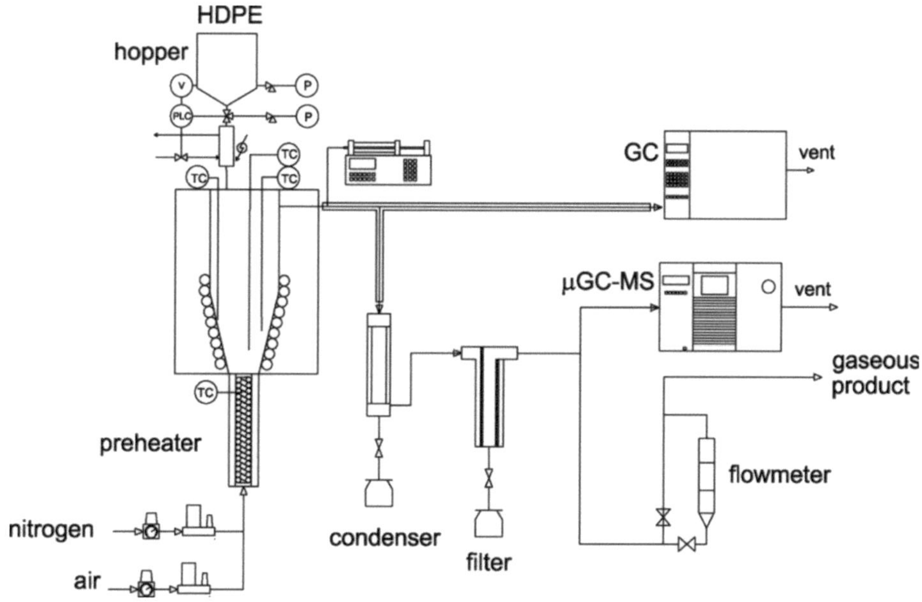

Fig. 3. Schematic representation of the continuous pyrolysis unit (Elordi et al. 2009).

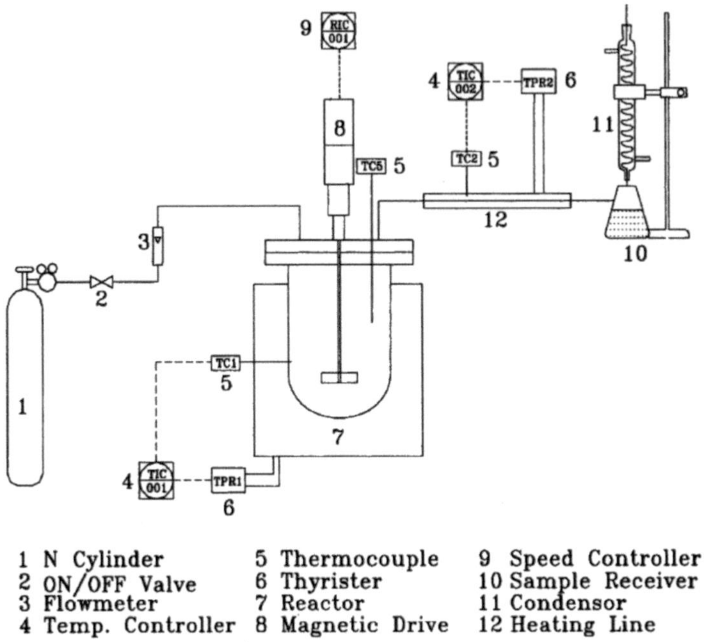

Fig. 4. Schematic diagram of a batch pyrolysis system (Seo et al., 2003).

Another type of laboratory scale batch reactor was used for the catalytic degradation of polyethylene by Van Grieken et al. (2001). The experiments were carried out in a batch reactor provided with a helicoidal stirrer at 120 rpm (Fig. 5). Three temperatures (380, 400 and 420°C) and different reaction times (0–360 min) under nitrogen flow were studied. The effluent from the reactor was connected to a water-cooled trap in order to condense the liquid products, whereas the effluent gas was finally collected in a teflon bag.

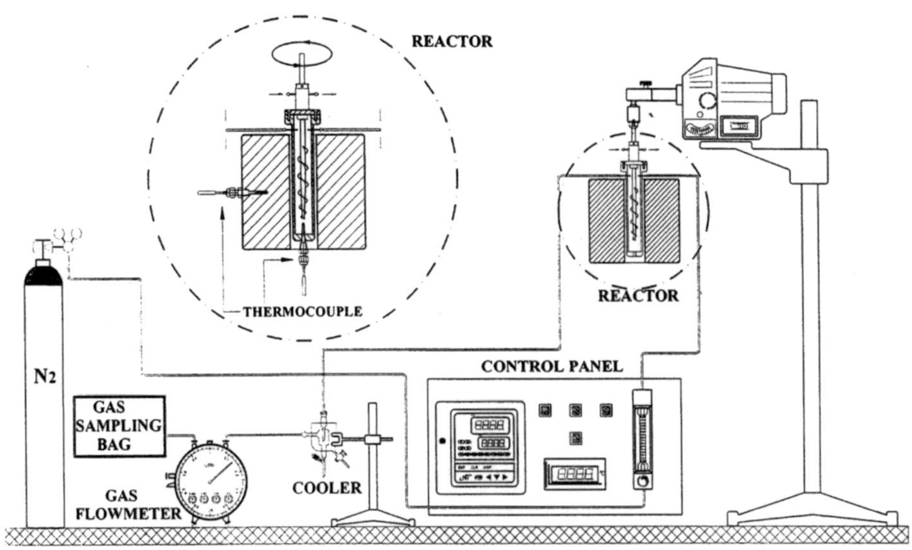

Fig. 5. Scheme of the experimental cracking reaction system (Van Grieken et al., 2001).

5.5.3 Fixed bed reactor

Achilias et al. (2007) used a laboratory-scale fixed bed reactor (Fig. 6) to study the thermal and catalytic degradation of polyethylene. The reactor was filled with the FCC catalyst and the piston was filled with the polymer. The time of the experiment was 17 min and the reaction temperature 450 ^{0}C. Experimental conditions and product yield from the thermal and catalytic pyrolysis of LDPE, HDPE and PP appear in Table 13.

Polymer	Temperature (^{o}C)	Catalyst	Gaseous product (wt.%)	Liquid product (wt.%)	Residue (wt.%)
LDPE	450	-	1.4	22.2	76.4
HDPE	450	-	1.7	21.6	76.7
PP	450	-	4.1	49.3	46.6
LDPE	450	FCC	0.5	46.6	52.9
HDPE	450	FCC	0.5	38.5	61.0
PP	450	FCC	6.2	67.3	26.5

Table 13. Experimental conditions and product yield from the thermal and catalytic pyrolysis of LDPE, HDPE and PP (Achilias et al., 2007).

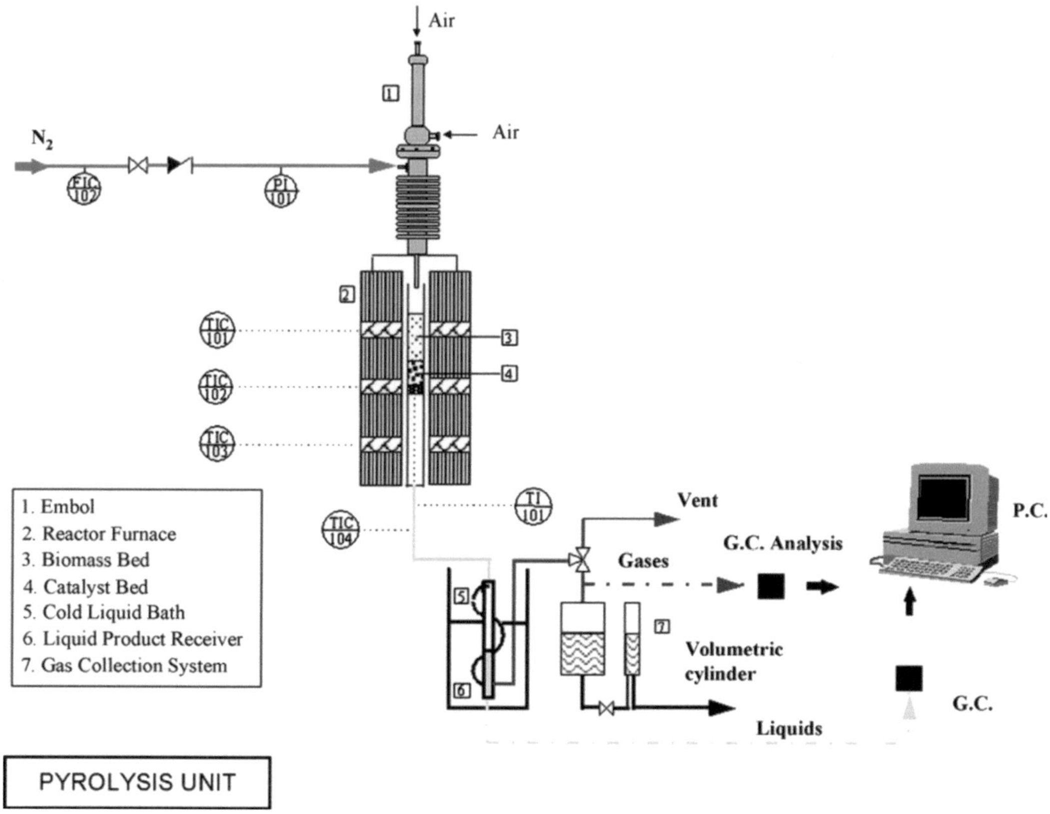

Fig. 6. The fixed bed reactor system (Achilias et al., 2007).

6. Chemical recycling of poly(vinyl chloride)

6.1 Introduction

Poly(vinyl chloride) main applications include food packaging, shoes, flooring ,pipes, clothing (leather-like material), ceiling tiles and multi-layered flooring and windows-windoors frames. It has a lifetime range of 5 years. PVC has the same density as that of PET, a property that made the separation prior to recycling of plastic wastes containing both polymers really difficult. The technique used for the efficient separation is X-ray fluorescence. The chlorine atoms of PVC are detected and the wastes are indicated for separation. Also IR sorting is widely used (Sadat-Shojai and Bakhshandeh, 2011).

In principle, PVC waste can be available in two ways: as a mixed plastic waste (MPW) fraction (with a rather low PVC content), or as a PVC-rich plastics fraction.

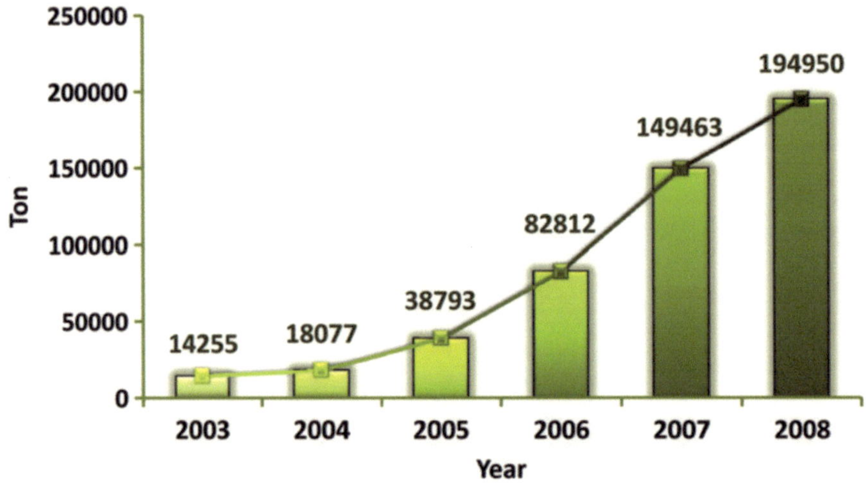

Fig. 7. The tonage of recycled PVC in EUROPE (Sadat-Shojai, Bakhshandeh, 2011).

As it is well-known PVC incineration is connected with problems arising from the high chlorine content of this polymer which yields large amounts of hydrochloric acid (HCl) during thermal decomposition, beside the possibility of formation of persistent and toxic compounds such as toxic dioxines and furans (Garcia et al., 2007; Ulutan, 1998; Ali and Siddiqui, 2005). In addition, when PVC wastes are fired in an incinerator, HCl corrodes the boiler tubes of the incinerator and other equipments. Therefore the steam pressure must be kept relatively low to prevent corrosion of the heat recovery boiler (Yoshioka et al., 1998). One of the best solutions to this problem may be neutralization of HCl with calcium carbonate (lime) and/or sodium hydroxide (caustic soda) to convert the released HCl to the salts. Also special filters can also be used to prevent problems related to atmospheric emissions during the combustion process (Machado et al., 2010).

Beside all above problems, net energy recovered by incineration of PVC-rich waste is not high enough to make it highly economic. As most hydrocarbon polymers, the calorific value from incineration of PVC in an ideal conditions is about 64 MJ/kg, compared to, for example, 17 MJ/kg for paper, or 16 MJ/kg for wood. Moreover, PVC is inherently difficult to combust, so that complete combustion of PVC-rich waste occurs at such high temperatures (>1700 K), that it is economically prohibitive (Xiong, 2010).

Therefore mechanical and/or chemical recycling of PVC plastic wastes seems the logical solution One usual approach for chemical recycling of PVC wastes is currently "thermal cracking" via hydrogenation, pyrolysis or gasification (Ryu et al., 2007; Williams and Williams, 1998; DeMarco et al., 2002; Kaminsky and Kim, 1999; Borgianni et al., 2002).

The main intermediate product of the thermal cracking is a polyene material that continues to degrade by evolution of aromatics and converts to a products which their composition will be strongly determined by processing variable such as type of atmosphere, temperature and residence time. In an inert atmosphere, the degradation products will be hydrochloric acid (HCl), gaseous and liquid hydrocarbons, and char, which among them HCl is a main product and can be reused either in vinyl chloride production, or in other chemical processes (Slapak et al., 1999).

In the case of manufacturing process of vinyl chloride, a gas purification unit must also be added to obtain high purity hydrogen chloride gas.

In a steam atmosphere at high temperatures, the hydrocarbon fraction will be converted into the some other products such as carbon monoxide, carbon dioxide and hydrogen. In a reported process bench-scale bubbling fluidized bed to investigate some processing parameters on the product outcome. The choice of type of bed material is essential for the product outcome, so that the use of catalytic inactive solid quartz as bed material results in the production of large amounts of char and tar, whereas the application of catalytic active material such as porous alumina results in a high conversion of PVC into the syngas. Moreover, according to their results, temperature has a large impact on the composition of the products, so that the carbon to gas conversion improved from about 70% at 1150 K to approximately 100% by increasing the reactor temperature to 1250 K. For chemical recycling of PVC, an increase in efficiency of dehydrochlorination process is usually attributed to the successful recycling (Wu et al., 2009).

It has been also reported that the emission of hydrogen chloride changes significantly with the oxides used indicating the chlorine fixing ability of oxides and also that utilization of poly(ethylene glycol) (PEG) can accelerate dehydrochlorination of PVC, so that at 210 °C for 1 h the dehydrochlorination degree was as high as 74% for PVC/PEG, while for PVC only 50%. Moreover, they demonstrated that for PVC/ PEG the decomposition of PVC shifted to lower temperatures compared with that of pure PVC, suggesting some interactions exist between PEG and PVC that caused the faster dehydrochlorination rate. According to their results, during this process, no waste byproducts such as KCl were produced, and satisfactory recyclability of PEG (10 cycles) can be obtained (Wu et al., 2009).

An alternative method to thermal process of dehydrochlorination is the rather easy process of dehydrochlorination under the influence of alkaline media to recover hydrochloric acid with a possibility that the degradation of PVC by oxygen oxidation in an aqueous alkaline solution to produce various carboxylic acids (Brown, 2002).

Some researchers demonstrated that dehydrochlorination of flexible-PVC occurred first and followed then by oxidation. They reported that the major products were oxalic acid, a mixture of benzenecarboxylic acids, and CO_2. However, the chlorine content could also be recovered in the form of HCl by adjusting the reaction conditions such as alkali concentration (Yoshioka et al., 1998).

Among various methods of thermal cracking, pyrolysis is a more well-known procedure in the chemical recycling of PVC. The process of pyrolysis, which takes places at 500-900°C without any oxygen, is a very suitable recycling method especially in the case of mixed plastic wastes (PVC recycling, 2005). In a typical process, a PVC-rich waste can be pyrolysed to hydrocarbons (oil), soot, hydrochloric acid, chlorinated hydrocarbons, etc., which hydrochloric acid needs to be removed from the pyrolysis gas although this removal process can result in the formation of toxic dioxins in some stages. The main end product of pyrolysis is, however, oil industry (Sadat-Shojai and Bakhshandeh, 2011).

One main problem connected with pyrolysis of PVC and mixed plastics containing PVC materials is corrosion of the process equipment (e.g., pyrolysis reactor and piping) mainly by the formation of the acid gas (HCl). Moreover, many petrochemical specifications limit the amounts of halogens (appeared in the forms of hydrogen chloride and chloro organic compounds) to a very low range in the gas and oil derived from plastic waste. Therefore in the case of mixed plastic wastes (uneconomic to separate to a single polymer) with a low PVC content, the conventional chemical recycling is frequently used only for a waste stream in which the PVC content is less than 30% (for example, the multiple material products) (Duangchang, 2008).

So far, several solutions to such problems have been proposed which some of them have already been put into practical use. For example, milling of PVC with CaO can be an effective way to extract Cl from the waste (Tongamp et al., 2008).

An attempt has also been made to develop a process for recovering metals from alloy-wastes by using a mechanochemical reaction consisting of a co-grinding alloy and PVC waste, followed by washing with water and filtration (Zhang et al., 2007).

Currently, the NKT-Watech pyrolysis process in Europe uses another two-step pyrolysis of PVC wastes in a stirred vessel. Calcium carbonate and filler are used to react with liberated HCl and produce calcium chloride. Then at the increased temperature, the polymer chains break down which produce a solid coke residue. Finally, the residual calcium chloride can be treated to make it suitable for selling (Scheirs, 2010).

Alternative approach is pre-treatment of mixed plastic waste by removing PVC and other halogenated plastics from the feed. Such pre-treatment consists of a dilution of the wastes having excessive chlorine content with less chlorine-containing or chlorine-free polymer mixture. It is also common to dilute the chlorine-containing hydrocarbon feed with chlorine-free petroleum fractions coming from refineries. Another possibility, as a less expensive and more acceptable process, is thermal dehalogenation which takes place either in a liquid or in a fluidized bed pyrolysis.

A new method consisting of copyrolysis of PVC with nitrogen compounds in biowaste to reduce the corrosive effects of the generated HCl was reported. The researchers studied the pyrolysis conditions between PVC and cattle manure via a statistical method and optimized conditions to provide the highest HCl reduction during PVC pyrolysis. They also applied the optimized conditions to a plastic mixture and then determined the quality of the obtained products. They concluded that the lowest heating rate, the highest reaction temperature (450°C), and the PVC:cattle manure ratio of 1:5 are the suitable conditions which provide the highest HCl reduction. However, according to their results the presence of manure decreases the oil yield of pyrolysis by about 17% (Duangchang, 2008).

PVC can also be chemically modified by nucleophilic substitution of chlorine atoms in its structure as it has been described: reactions of rigid PVC with various nucleophiles (Nu) such as iodide, hydroxide, azide, and thiocyanate in ethylene glycol as solvent. Such reactions lead to the substitution of Cl by Nu and finally elimination of HCl, resulting in the dehydrochlorination of the rigid PVC. According to their results, the dehydrochlorination yield increased with an increasing nucleophiles concentration, resulting in a maximum substitution at high nucleophiles amounts. Moreover, when ethylene glycol was replaced by diethylene glycol the dehydrochlorination was found to be accelerated, which may be due to the higher compatibility of diethylene glycol with PVC, making it easier to penetrate the rigid PVC particles (Kameda et al., 2010).

Several different technologies based on depolymerization and repolymerization processes have been developed for chemical recycling of PVC, which unfortunately the most of them are more expensive than the mechanical recycling (LaMantia, 1996).

PVC waste was used in a research wich carried out for the recycling electric arc furnace dust by heat treatment with PVC. The entire process aimed to recover the zinc, lead, and cadmium from the dust and was adjusted so that the residual dust can be injected into the electric arc furnace (Lee et al., 2007).

There are many reports, where thermoanalytical methods especially coupled methods with gas analysis systems, can deliver suitable information for the recycled PVC that needs to be characterized (Matuschek et al., 2000).

6.2 Mixed plastic recycling processes

Recovinyl-Co (UK) deals with post-consumer PVC to reproduce two grades via mechanical recycling. Due to its structure and composition, PVC can easily be mechanically recycled in order to obtain good quality recycling material. Careful and proper sorting is of crucial importance for the optimal recycling of PVC (Recovinyl, 2008).

A pyrolytic process which has proven to be successful for plastic solid waste rich in PVC, is the Akzo process (Netherlands). With a capacity of 30 kg/h, this fast pyrolysis process is based on a circulating fluidised bed system (two reactors) with subsequent combustion. Input to the process is shredded mixed waste including a high percentage of PVC waste. The main outputs consist of HCl, CO, H_2, CH_4 and, depending on the feedstock composition, other hydrocarbons and fly ash (Tukker et al., 1999).

The NRC process is another successful pyrolysis scheme. This process is based on the pyrolysis with subsequent metal extraction technology. The aim is to produce purified calcium chloride instead of HCl. The input to the process is PVC waste (cables, flooring, profiles, etc.). No other plastic solid waste type is fed to the processing, which results in calcium chloride, coke, organic condensate (for use as fuels) and heavy metals for metal recycling, as products (Al-Salem et al., 2009).

The NTK process, depicted below (fig. 8), is a very successful recycling process. The process is based on an initial pre-treatment step that involves separating light plastics (PP, PE, etc.) and other materials, e.g. wood, sand, iron, steel, brass, copper and other metallic pollutants. The PSW waste is then fed to a reactor at a low pressure (2–3 bars) and a moderate temperature (375°C). The process emits neither dioxins, chlorine, metals nor plasticizers. Also, there are no liquid waste streams in the process since all streams are recycled within the system. There is a small volume of carbon-dioxide gas formed by the reaction between lime/limestone and hydrogen chloride. Mixed PVC building waste containing metals, sand, soil, PE, PP, wood and rubber waste have been successfully treated (Al-Salem et al., 2009).

The gasification into high calorific value fuel gas obtained from PVC was also reported by Borgianni et al., 2002.

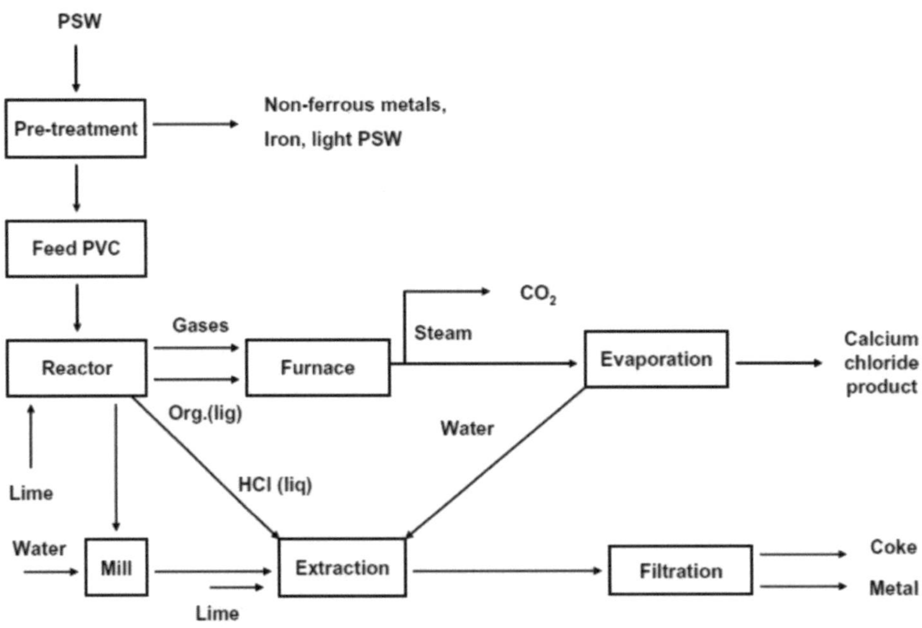

Fig. 8. NTK process diagram (Tukker et al., 1999).

Chemical recycling of PVC has been also attempted. Most of the proposed processes use the rather easy dehydrochlorination of PVC either under the influence of heat of alkaline media. The oxidative degradation of PVC by molecular oxygen in aqueous alkaline solution at temperatures between 150 and 260 °C with oxygen pressures of 1–10 MPa has been reported. The main products are oxalic acid and carbon dioxide, their yield depending on the reaction conditions and the alkali concentration. The maximum yield of oxalic acid was 45%and 42% of the chlorine content could be recovered in the form of HCl (Shin et al., 1998).

Oxidative degradation of rigid-PVC pellets (R-PVC) with oxygen was carried out in 1-25 mol/kg-H_2O (m) NaOH solutions, at 150-260°C and P_{O2} of 1-10 MPa in order to investigate the chemical recycling of PVC materials. The apparent rate of oxidative degradation of R-PVC progressed as a zero order reaction, and the apparent activation energy was 38.5 kJ/mol. The major products were oxalic acid, a mixture of benzene-carboxylic acids, and CO_2. The tin in R-PVC was extracted completely. The possibility of converting PVC materials into raw materials such as carboxylic acids by chemical recycling is reported (Yoshioka et al., 2000).

6.3 Mixed PVC wastes World inititives

Regarding the chemical recycling of mixed plastic wastes with a PVC content of up to several percent, the following initiatives seem to be most realistic for the coming 5 years: **Texaco gasification process** (NL, pilot in the US), Polymer cracking process (consortium project, pilot), **BASF conversion process** (D, pilot but on hold) **Use as reduction agent in blast furnaces** (D, operational), **Veba Combi Cracking process** (D, operational but to be closed by 2000**), Pressurized fixed bed gasification of SVZ** (D, operational). A brief report on these initiatives is presented below (Sadat-Shojai and Bakhshandeh, 2011).

BP Chemicals has led promotion of Polymer Cracking technology for feedstock recycling since its beginnings in the early 1990's. Since the challenge of recycling of plastics is industry wide, support has been provided by a Consortium of European companies to develop the

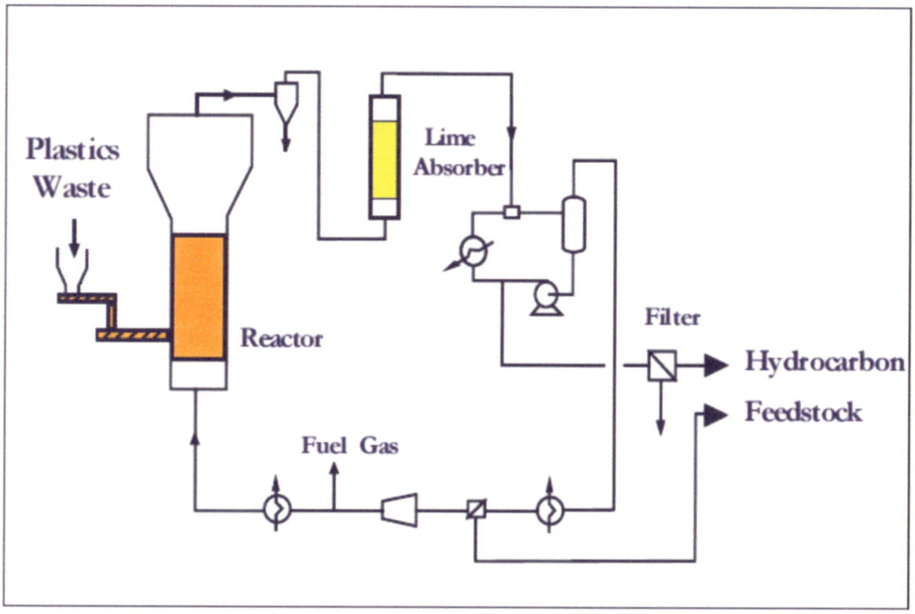

Fig. 9. The BP polymer cracking process (Tukker et al., 1999).

technology - initially including Elf Atochem, DSM, Fina and Enichem. The consortium members at the time of the successful pilot plant trials in 1997 were BP Chemicals, Elf Atochem, EniChem, DSM, CREED and the APME. Some elementary preparation of the waste plastics feed is required, including size reduction and removal of most non-plastics. This prepared feed is fed directly into the heated fluidised bed reactor which forms the heart of the Polymer Cracking process. The reactor operates at approximately 500°C in the absence of air. The plastics crack thermally under these conditions to hydrocarbons which vaporise and leave the bed with the fluidising gas. Solid impurities, including metals from e.g. PVC stabilisers and some coke, are either accumulated in the bed or carried out in the hot gas as fine particles for capture by cyclone. The decomposition of PVC leads to the formation of HCl, which is neutralised by bringing the hot gas into contact with a solid lime absorbent. This results in a $CaCl_2$ fraction that has to be landfilled. The purified gas is cooled, to condense most of the hydrocarbon as valuable distillate feedstock. This is then stored and tested against agreed specifications before transfer to the downstream user plant. The remaining light hydrocarbon gas is compressed, reheated and returned to the reactor as fluidising gas. Part of the stream could be used as fuel gas for heating the cracking reactor, but as it is olefin-rich, recovery options are being considered.

The process shows very good results concerning the removal of elements like chlorine. With an input of 10,000 ppm (or 1%) Cl, the products will contain around 10 ppm Cl. This is somewhat higher than the specifications of 5 ppm typical for refinery use. However, in view of the high dilution likely in any refinery or petrochemical application, BP assumes that this is acceptable.Also, metals like Pb, Cd and Sb can be removed to very low levels in the products. Tests have shown that all the hydrocarbon products can be used for further treatment in refineries (Brophy et al., 1997).

The BASF feedstock recycling process was designed to handle the recycling of mixed plastic waste supplied by the DSD collection system. The process is as follows, before the waste plastics can be fed to the process, a pretreatment is necessary (fig. 10). In this pretreatment the plastics are ground, separated from other materials like metals and agglomerated. The conversion of the pretreated mixed plastic into petrochemical raw materials takes place in a multi-stage melting and eduction process. In the first stage the plastic is melted and dehalogenised to preserve the subsequent plant segments from corrosion. The hydrogen chloride separated out in this process is absorbed and processed in the hydrochloric acid production plant. Hence, the major part of the chlorine present in the input (e.g. from PVC) is converted into saleable HCl. Minor amounts come available as NaCl or $CaCl_2$ effluent (Heyde and Kremer, 1999). Gaseous organic products are compressed and can be used as feedstock in a cracker. In the subsequent stages the liquefied plastic waste is heated to over 400 °C and cracked into components of different chain lengths. About 20-30% of gases and 60-70% of oils are produced and subsequently separated in a distillation column. Naphtha produced by the feedstock process is treated in a steam cracker, and the monomers (e.g. ethylene, propylene) are recovered. These raw materials are used for the production of virgin plastic materials. High boiling oils can be processed into synthesis gas or conversion coke and then be transferred for further use. The residues consist of 5% minerals at most, e.g. pigments or aluminium lids. It seems likely that metals present in PVC-formulations mainly end up in this outlet. The process is carried out under atmospheric pressure in a closed system and, therefore, no other residues or emissions are formed.

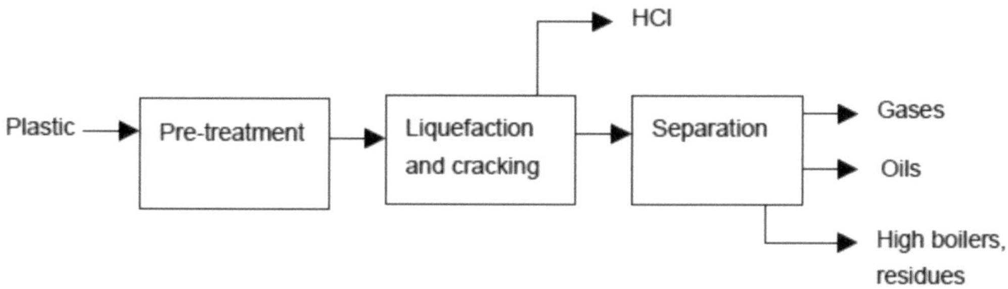

Fig. 10. BASF process [Heyde and Kremer, 1999].

Mixed PVC waste can be used as reduction agent in blast furnaces. For the production of pig iron for steel production, iron ore (Fe_2O_3) has to be reduced to Fe. This process takes place in a blast furnace. Coke, coal and heavy oil are normally used as reducing agents in this process. Iron and steel companies try to lower the consumption of coke, by partly replacing it with coal, gas or fuel oil (30% in weight seems to be the maximum), via coal injection technology. Recently, new developments have started to replace the conventional reducing agents by plastics waste. Though others like British Steel (UK) have done trials as well, the prominent pioneer in this field is Stahlwerke Bremen, Germany. Stahlwerke Bremen is a large German steel manufacturer which operates two blast furnaces to produce over 7000 t/day, or some 3 Million tpa pig iron. Currently, German is the sole country that blast furnace are the only plants using waste this way (DKR/DSD, 1999).

Veba Combi Cracking process. The plant configuration includes a depolymerisation section and the VCC section (fig. 11). Depolymerisation is required to allow further processing in the VCC section. In the depolymerisation section the agglomerated plastic waste is kept between 350-400°C to effect depolymerisation and dechlorination. The overhead product of the depolymerisation is partially condensed. The main part (80 %) of the chlorine introduced with PVC is present as HCl in the light gases. It is washed out in the following gas purification process, yielding technical HCl. The condensate, containing 18 % of the chlorine input, is fed into a hydrotreater. The HCl is eliminated with the formation water. The resulting Cl-free condensate and gas are mixed with the depolymerisate for treatment in the VCC section. The depolymerisate is hydrogenated in the VCC section at 400-450°C under high pressure (about 100 bar) in a liquid phase reactor with no internals. Separation yields a product which after treatment in a fixed-bed hydrotreater is a synthetic crude oil, a valuable product which may be processed in any refinery. From the separation a hydrogenated residue stream also results, which comprises heavy hydrocarbons contaminated with ashes, metals and inert salts. This hydrogenation bitumen is a byproduct which is blended with the coal for coke production (2 wt%). It is most likely that the major part of any metals present in a PVC formulation end up in this residue flow. Light cracking products end up in off-gas (E-gas), which is sent to a treatment section for H_2S and ammonia removal. As indicated above, the main part of the chlorine present in the input (i.e. from PVC) is converted into usable HCl. Some 2% of the chlorine input is bound to $CaCl_2$ in the process by a 4 times leaner than stochiometric amount of CaO.

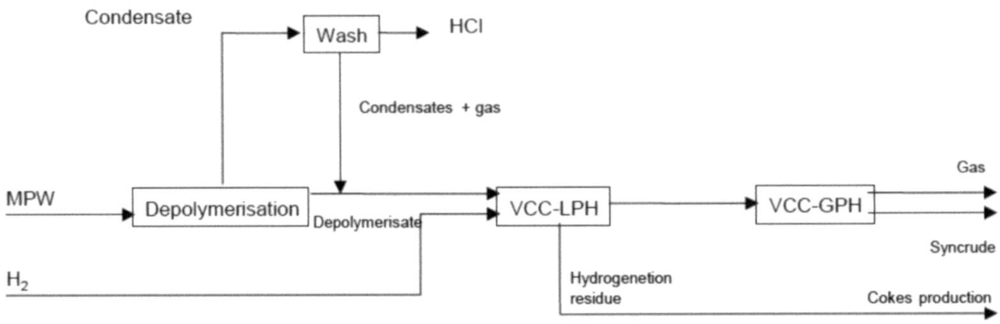

Fig. 11. Veba Combi Cracking process (Sas, 1994).

7. Chemical recycling of polycarbonate

7.1 Introduction

Polycarbonate plastics, $C_{16}H_{14}O_3$ (Fig. 12) are polyesters known for their excellent mechanical properties. Featuring high-impact resistance, UV resistance, and flame retardancy as well as excellent electrical resistance, polycarbonates are used in a wide variety of materials. Polycarbonates do not have their own recycling identification code and therefore fall under the #7 "other" classification. Polycarbonates may be made a variety of ways, the most popular of which from Bisphenol-A (BPA) feedstock. BPA use is highly controversial, and the FDA has recently decided to reopen an inquiry on the safety of BPAs. [Jawad et al., 2009].

This is following an approval in 2008. Nalgene Outdoor Products, the pre-eminent manufacturer of reusable plastic water bottles, is transitioning from polycarbonate bottles to other plastics as well as metal alternatives in the wave of negative consumer perception of BPA.

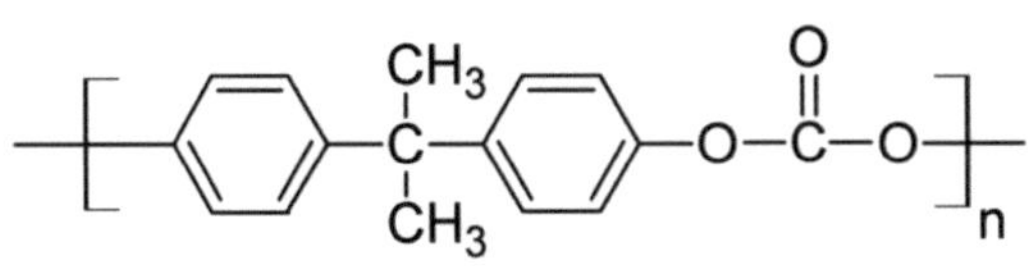

Fig. 12. Structural unit of polycarbonate.

Due to its excellent properties, polycarbonate (PC) is widely used in the manufacture of compact disks, bullet proof windows, food packaging and soft-drink bottles. With the rapid increase in the production and consumption of PC, the chemical recycling of waste PC to obtain valuable products has received greater attention in recent years. Waste PC can be depolymerised through a chemical treatment to produce monomers that can be used to reproduce virgin PC products. Various methods for the chemical recycling of waste PC to recover raw materials have been reported; these methods include thermal pyrolysis [Yoshioka et al., 2005; Chiu et al., 2006; Achilias et al., 2009], alcoholysis [Oku et al., 2000; Hu et al., 1998] and hydrolysis [Grause et al., 2009; Ikeda et al., 2008]. It is difficult to recover pure BPA using thermal pyrolysis, and it can be only obtained using hydrolysis. [Liu et al., 2011].

For the purpose of recycling of polycarbonate (PC), e.g. poly[2,2-bis(4-hydroxyphenyl) propane carbonate], in the form of an essential monomer bisphenol A, there have been reported a number of depolymerization methods.[Hub et al., 1998] Due to the insolubility of PC in water, the aqueous depolymerizations require severe conditions such as long reaction times, high temperatures and pressures. Therefore, instead of using aqueous systems, organic solvent systems such as methylene chloride in combination with ammonia [Fox et al., 1989], a mixed solvent of phenol and methylene chloride in combination with an alkali catalyst [Buysch et al., 1994; Shafer, 1994] have been reported. With organic solvents, however, a tedious product separation process is generally required.

7.2 Recycling techniques

With the rapid increase of production and consumption of PC, the chemical recycling of waste PC has been gaining greater attention in recent years to obtain valuable products. Methanolysis is one of the most important method to recover pure monomer BPA and dimethyl carbonate (DMC). However, due to the insolubility of PC in methanol, the reported methanolysis methods require high temperature and pressure and in presence of a lot amount of concentrated bases or acids. The acid or base catalysts used in traditional methods cannot be reused and result in other disadvantages such as equipment corrosion, tedious workup procedure and environmental problem. Although supercritical method can overcome some of above-mentioned shortcomings, it has its own disadvantages such as severe conditions, so its application is limited. According to a study polycarbonate could be completely decomposed into its monomer, BPA with high pressure (not atmospheric pressure) high temperature steam (300 °C) in five minutes reaction time. It is known that PC can be decomposed into monomer in alkaline alcohol or aqueous solutions. However, the monomer BPA yield has been reported as to be relatively low due to BPA instability in that condition. To develop a high-effective process of PC recycling, a reactive atmosphere must be provided that preserves the stability of BPA and get has high reactivity for PC. To determine the optimum conditions for recycling PC, it is important to know the stability or reactivity of BPA, as well as the decomposition rate of PC.

Alkali-catalysed depolymerization of polycarbonate wastes by alcoholysis in supercritical or near critical conditions has been also studied by other researchers in order to recover the essential monomer BPA and DMC as a valuable by-product (Liu et al., 2009). Some works aimed to develop continuous process and possible scale-up for decomposition of both PC plastic wastes using methanol as solvent/reagent and NaOH as alkali catalyst. Total depolymerization of PC has been achieved working at a temperature range of 75–180 °C and pressures from 2 to 25 MPa.

However, due to the insolubility of PC in water, the aqueous depolymerizations require severe conditions such as long reaction times, high temperatures and pressures. Therefore, instead of using aqueous systems, organic solvent systems such as methylene chloride in combination with ammonia, a mixed solvent of phenol and methylene chloride in combination with an alkali catalyst is also used. An environmentally friendly strategy for methanolysis of polycarbonate to recover bisphenol A and dimethyl carbonate was recently developed in which PC could be methanolyzed in an ionic liquid without any acid or base catalyst under moderate conditions (Liu et al., 2011).

7.2.1 Methanolysis in the presence of ionic liquids

The methanolysis of polycarbonate using ionic liquid [Bmim][Ac] as a catalyst was studied recently by Liu et al., 2011. The effects of temperature, time, methanol dosage and [Bmim][Ac] dosage on the methanolysis reaction were examined. They concluded that methanolysis of PC to obtain its starting monomers, BPA and DMC, could occur in the presence of ionic liquid [Bmim][Ac] under moderate conditions without an acid or base catalyst. The methanolysis conversion of PC was nearly 100% and the yield of BPA was over 95% under the following conditions: m([Bmim][Ac]):m(PC) = 0.75:1, m(methanol):m(PC) = 0.75:1, a reaction temperature of 90 ◦C and a total time of 2.5 h. The ionic liquid [Bmim][Ac] could be reused up to 6 times without an apparent decrease in the conversion of PC and yield of BPA. This strategy could overcome the shortcomings associated with the traditional methods, such as the infeasibility of reusing the catalyst, equipment corrosion, tedious workup procedures and environmental problems. Moreover, the investigation on kinetics indicated that the methanolysis of PC in [Bmim][Ac] was a first-order reaction and the activation energy was 167 kJ/mol. The reaction formula was as follows:

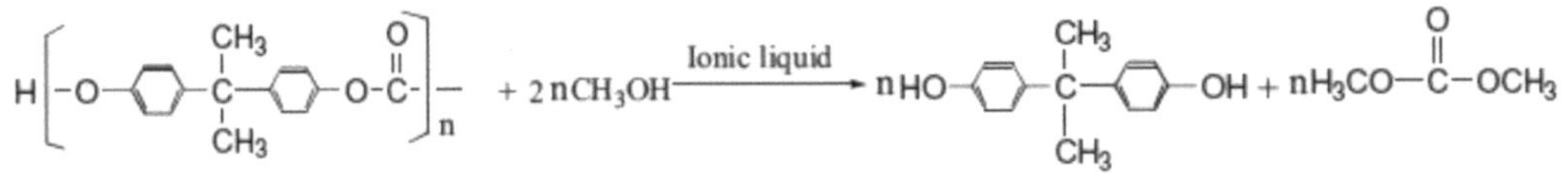

Fig. 13. Methanolysis of PC.

A mechanism for the methanolysis of PC in the presence of ionic liquid [Bmim][Ac] is suggested in Figure 14. After PC was dissolved or swelled in the ionic liquid, it reacted with methanol to form oligomers under ionic liquid catalysis. Then, the resulting oligomers reacted with methanol further to produce the final products, BPA and DMC.

7.2.2 Alkali catalyzed methanolysis

Alkali-catalyzed methanolysis of poly[2,2-bis(4-hydroxyphenyl)propane carbonate] in a mixed solvent of methanol and toluene or dioxane was studied by the team of Oku et al., 2000 at Kyoto Institute of Technology. Treatment of PC pellets in MeOH with a catalytic amount of NaOH at 608C for 330 min yielded only 7% BPA. However, in a mixed solvent of MeOH and toluene, the analogous treatment for 70 min completely depolymerized PC to give free bisphenol A (96%) in a solid form and dimethyl carbonate (DMC) (100%) in solution.

The characteristic feature of the present methanolysis is that PC can be depolymerized to its starting monomer components BPA and DMC by the use of a catalytic amount of alkali-metal hydroxide under mild reaction conditions. The monomers can be obtained almost quantitatively in very pure states and they can be recycled as the monomers of PC and epoxy resins.

The alkali catalysed methanolysis also studied by Liu et al., 2009 but the reaction took place in a reactor with a stirrer and a refluxing condenser. The results did not differ much because of the use of refluxing condenser. The temperature, on the other hand effected the efficiency of methanolysis in both studies with the temperature of 60°C presenting the greater rate of BPA formation.

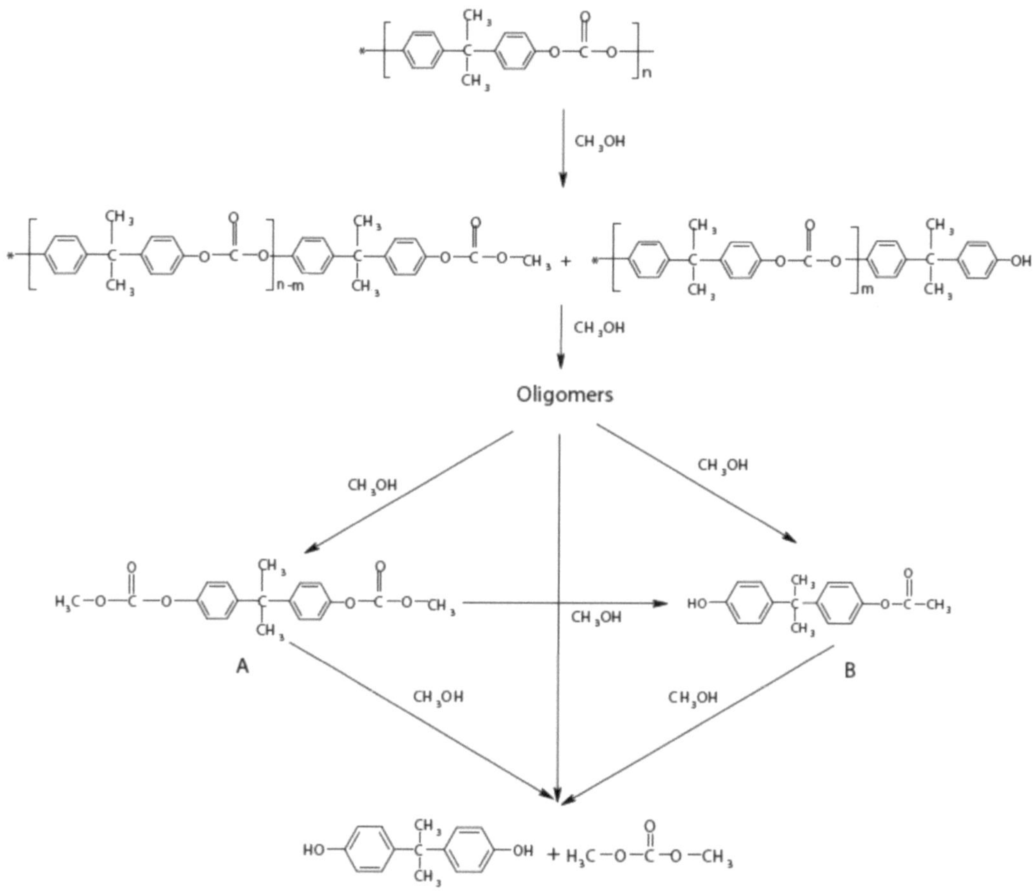

Fig. 14. Methanolysis mechanism for PC in the presence of ionic liquid [Bmim][Ac] [Liu et al., 2011].

7.2.3 Hydrolysis with high temperature steam

Watanabe et al. 2009, found that polycarbonate was rapidly hydrolyzed in high pressure high temperature steam around the saturated pressure of water at 573 K. For 300 s (5 min) reaction time, PC completely decomposed into bisphenol A and the maximum yield of BPA was around 80%. In liquid water phase at 573 K, PC still remained even for 3000 s (50 min). The high yield of bisphenol A in high pressure steam was due to its high stability. The amount of water required for degradation was drastically reduced and thus the high pressure high temperature steam process was energetically and economically preferable.

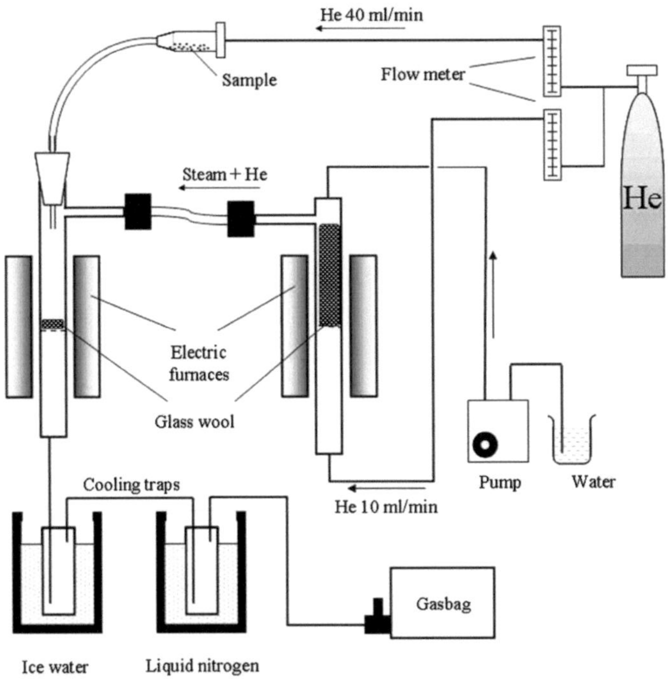

Fig. 15. Apparatus hydrolysis with high temperature steam (Grause et al., 2009).

The reaction mechanism of polycarbonate hydrolysis in high pressure high temperature steam seems to be both surface and bulk erosion.

Another study of polycarbonate hydrolysis has been done by Grause et al., 2009 at Tohoku University. They studied the pyrolytic hydrolysis in the presence of earth-alkali oxides and hydroxides as catalysts. The experiments were carried out in a steam atmosphere in the presence of MgO, CaO, $Mg(OH)_2$ or $Ca(OH)_2$. All of these catalysts accelerated the hydrolysis of PC drastically, with MgO and $Mg(OH)_2$ being more effective than their Ca counterparts. The differences between oxides and hydroxides were negligible indicating the same mechanism for both, oxides and hydroxides. BPA was the main product at 300 ^{o}C, with a yield of 78% obtained in the presence of MgO. At 500 ^{o}C, BPA was mainly degraded to phenol and isopropenyl phenol (IPP). It can be shown that a combined process involving PC hydrolysis at 300 ^{o}C and BPA fission at 500 ^{o}C leads to high yields of phenol and IPP and the drastic decrease of residue. The apparatus of the experimental process is shown in Fig. 15.

7.2.4 Hydrolysis in other solvents but water

Alkali-catalyzed hydrolysis of PC in a solvent in which it can substantially dissolve such as N-methyl-2-pyrrolidone, 1,4-dioxane, tetrahydrofuran or DMF were studied by Liu et al., 2009. The results showed that hydrolysis of PC could take place under moderate conditions. No BPA was detected when hydrolysis of PC was carried out under given conditions in water without co-solvents. However, when the hydrolysis was carried out under the same conditions in presence of such a solvent, the rate of hydrolysis was significantly accelerated.

Also, with the increasing of amount of H_2O, the yield of BPA gradually increased and a maximum yield was obtained when using ratio of PC:H_2O close to 1.5:1. When the amount of water was more than this ratio the BPA yield decreased. Moreover, under the conditions of reaction temperature 100 °C, m(PC):m(H_2O) = 1:0.7, m(PC):m(NaOH) = 10:1, reaction time 8 h and using 1,4-dioxane as solvent, the hydrolysis conversion of PC was almost 100% and the yield of bisphenol A was over 94%.

7.2.5 Noncatalyzed glycolysis of PC in ethylene glycol

Kim et al. 2009 explored the depolymerization of polycarbonate waste by glycolysis using ethylene glycol without catalyst in order to get the monomer bisphenol A. This process can be considered as a green process from the viewpoint of using neither toxic solvents nor alkali catalyst. The maximum yield of BPA of 95.6% was achieved at reaction temperature 220 °C for 85 min with an EG/PC weight.

This reaction mechanism is illustrated in Fig. 16. Ethylene glycol penetrates into the PC polymer particle so that the particles are swollen. The PC is depolymerized in the solid state

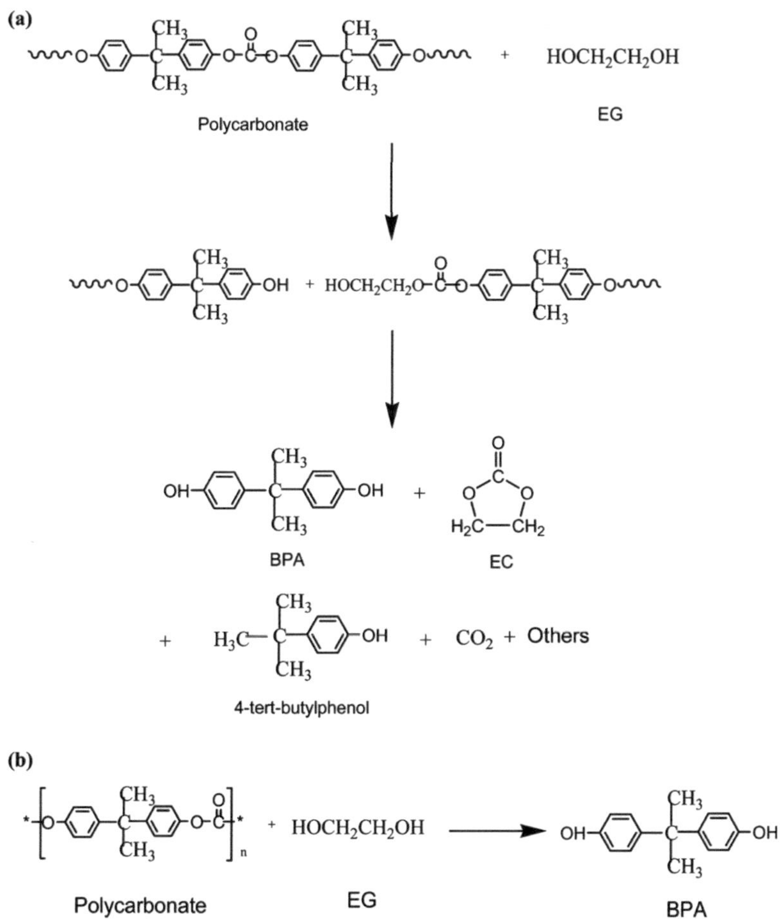

Fig. 16. PC glycolysis reaction pathway (Kim et al., 2009).

by the diffused EG. Random scissions of PC take place to lower the average molecular weight until the resulting oligomer can be dissolved in the bulk EG solution but retains solid state. The solid oligomer dissolves in EG solution, and the size of the PC particle shrinks as the dissolution proceeds, which is a heterogeneous reaction. The dissolved oligomer continues to be depolymerized with EG in the bulk solution to produce its monomer, BPA, which is a homogeneous reaction.

7.3 Pyrolysis of PC based polymers

Achilias et al., 2009, investigated pyrolysis of PC and PC based Waste Electric and Electronic Equipment as a means of chemical recycling of this polymer. A laboratory-scale fixed bed reactor was used and the appropriate pyrolysis temperature was selected after measuring the thermal degradation of model PC by Thermogravimetric analysis. After pyrolysis a large amount of oil was measured, together with a smaller amount of gaseous product, leaving also a solid residue. For both samples (model PC and a compact disc based on PC), the gaseous fraction consisted mainly of CO_2 and CO, whereas in the liquid fraction a large amount of different phenolic compounds, including the monomer bisphenol A, was measured. It seems that recycling of used CDs by pyrolysis is a very promising technique having the potential of producing useful high-value chemicals, which may find applications in the petrochemical industry.

7.4 Problems of PC recycling

Recycled polycarbonate is usually less resilient, have decreased impact resistance when compared with newly manufactured polycarbonate. The addition of fillers and pigments can also decrease the plastic's resilience. This problem can be addressed by the use of chemicals to modify impact resistance in recycled polycarbonate. Up to 15% recycled material can safely be added to the virgin resin without significantly altering properties of the virgin material.

The nature of the compact disc (CD) does not allow it to be easily recycled. The disc is a multi-layer product consisting of PC substrate and three coatings. These coatings, aluminium, lacquer and printing, respectively make up only a small portion of the entire disc. These materials should be separated or recovered in order to recycle the polycarbonate. There are a variety of methods for the removal of paint or plating from engineering plastics, ranging from the chemical to physico-mechanical procedures. Such techniques include chemical stripping or chemical recovery (high-temperature alkaline treatment), melt filtration, mechanical abrasion, hydrolysis, liquid cyclone, compressed vibration, cryogenic grinding, dry crushing and roller pressing.

The disadvantages of PC include high melt viscosity and notch sensitivity. Used PC usually suffered from crazing caused by light, radiation and chemicals present in the service environment, which make the problem of notch sensitivity even worse.

As with other thermoplastics, the level of mechanical and physical properties of polycarbonate depends on the molecular weight. However, production waste, recyclates etc. frequently do not, or no longer, possess the required molecular weights. Direct material recycling of production waste or recyclates is therefore possible only to a very limited extent. When recycling polycarbonate residues, production wastes, remainders, recyclates and similar polycarbonate compositions, it is therefore desirable and essential to increase the molecular weight to a sufficient level for the projected new use. So, for example, low-molecular production scrap from PC production for Compact Discs could be increased to the molecular weight range required for injection molding. Or the average molecular weight of PC recyclate from the de-lamination of Compact Discs should be increased sufficiently to allow the material to be used, as a component in the production of PC/ABS blends.

It was found that, surprisingly, it is possible to condense polycarbonates from waste by simple melting in a vacuum, optionally with bisphenols or suitable oligocarbonates with OH terminal groups, to produce, directly, polycarbonates of higher molecular weights.

8. Chemical recycling of nylon

8.1 Introduction

Nylon is one of the early polymers developed by Wallace Carothers in 1935, at DuPont's research facility. Today, nylon is one of the most commonly used polymers. Nylons, also known as polyamides, can be produced by the reaction of a diamine with a dicarboxylic acid, condensation of the appropriate amino acid, ring opening of a lactam, reaction of a diamine with a diacid chloride, and reaction of a diisocyanate with a dicarboxylic acid. Nylon is a crystalline polymer with high modulus, strength, impact properties, low coefficient of friction, and resistance to abrasion. A variety of commercial nylons are available including nylon 6, nylon 11, nylon 12, nylon 6,6, nylon 6,10, and nylon 6,12. The most widely used nylons are nylon 6,6 and nylon 6. Polyamides are used most often in the form of fibers, primarily nylon 6,6 and nylon 6, although engineering applications are also of importance. Nylon 6,6 is prepared from the polymerization of adipic acid and hexamethylenediamine, while nylon 6 is prepared from caprolactam.

Nylon recycling has increased substantially in the last several years. Most recycling efforts have focused on recovery of carpet. According to the U.S. Department of Energy, about 3.5 billion lb of waste carpet are discarded each year in the United States, with about 30% of them made from nylon 6.

Processing of recyclables is necessary to transform the collected materials into raw materials for the manufacture of new products. In general there are two categories for nylon recycling, chemical and thermal recycling.

- **Chemical recycling.** Involves breaking down the molecular structure of the polymer, using chemical reactions. The products of the reaction then can be purified and used again to produce either the same or a related polymer.
- **Thermal recycling.** Also involves breaking down the chemical structure of the polymer. In this case, instead of relying on chemical reactions, the primary vehicle for reaction is heat. In pyrolysis, for example, the polymer (or mixture of polymers) is subjected to

high heat in the absence of sufficient oxygen for combustion. At these elevated temperatures, the polymeric structure breaks down.

8.2 Depolymerization of nylons

Due to the higher value of nylons in comparison with other polymers used in carpet, nylon carpet has been looked at as a resource for making virgin nylon via depolymerization. Most of polyamides used commercially are nylon 6,6 or nylon 6, and the largest supply of waste for recycling of nylons is obtained from used carpets. The waste carpets are collected, sorted and then subjected to a mechanical shredding process before depolymerization.

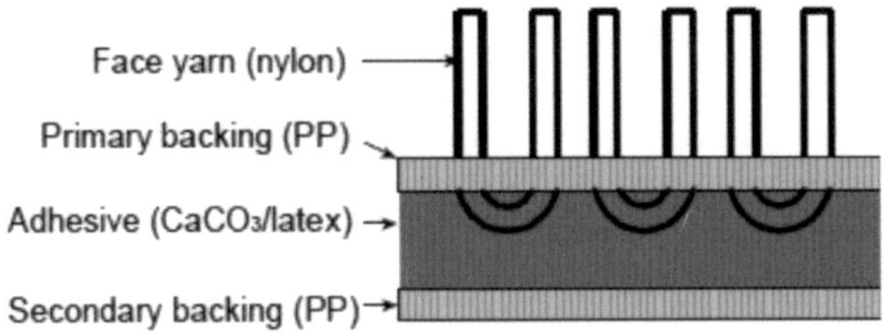

Fig. 17. Typical carpet construction.

Although there are many active recycling operations, they mostly focus their work in the recycling of nylon 6 and nylon 6,6 basically collected from carpet waste. It is obvious the need for further research to develop methods for recycling more commercially used nylons.

8.3 Hydrolysis of nylon 6

A process for depolymerizing nylon 6 scrap using high pressure steam was patented by by AlliedSignal, Inc. (Sifniades et al., 1999]. Ground scrap was dissolved in high-pressure steam at 125-130 psig (963-997 kPa) and 175-180^{0}C for 0.5 hour in a batch process and then continuously hydrolyzed with super-heated steam at 350^{0}C and 100 psig (790 kPa) to form ε-caprolactam at an overall recovery efficiency of 98%. The recovered monomer could be repolymerized without additional purification. Braun et al., 1999 reported the depolymerization of nylon 6 carpet in a small laboratory apparatus with steam at 340^{0}C and 1500 kPa (200 psig) for 3 hours to obtain a 95% yield of crude ε-caprolactam of purity 94.4%. Recently, patents were issued to AlliedSignal for the depolymerization of polyamide-containing carpet (Sifniades et al., 1999).

Acid hydrolysis of nylon 6 wastes [Chaupart, 1998] in the presence of superheated steam has been used to produce aminocaproic acid which under acid conditions is converted to ε-caprolactam, and several patents have been obtained by BASF [Corbin et al., 1999]. Acids used for the depolymerization of nylon 6 include inorganic or organic acids such as nitric acid, formic acid, benzoic acid, and hydrochloric acid [Bajaj and Sharma, 1997]. Orthophosphoric acid and boric acid are typically used as catalysts at temperatures of 250-350^{0}C. In a typical process, superheated steam is passed through the molten nylon 6 waste at 250-300^{0}C in the presence of phosphoric acid. The resulting solution underwent a multistage chemical purification before concentration to 70% liquor, which was fractionally distilled in the presence of base to recover pure ε-caprolactam. Boric acid (1%) may be used to depolymerize nylon 6 at 400⁰C under ambient pressure. A recovery of 93-95% ε-caprolactam was obtained by passing superheated steam through molten nylon 6 at 250-350⁰C. Sodium hydroxide has been used successfully as a catalyst for the base-catalyzed depolymerization of nylon 6.

8.3.1 Catalytic pyrolysis

Catalytic pyrolysis has been studied as a hybrid process for recovering caprolactam from nylon 6 followed by high temperature pyrolysis of the polypropylene into a synthetic natural gas. Czernik *et al.*, 1998 investigated the catalysis of the thermal degradation of nylon 6 with an α-alumina supported KOH catalyst in a fluidized bed reactor. In the temperature range of 330°C to 360°C the yield of caprolactam exceeded 85%. Bockhorn *et al., 2001* use a liquid catalyst composed of a eutectic mixture of 60 mol% NaOH and 40 mol% KOH which melts at 185°C. At 290°C the caprolactam yield exceeded 95%.

8.3.2 Recovery of caprolactam

Approximately 10-12% by weight of oligomers is formed in the synthesis of polycaprolactam (nylon 6). These oligomers are removed by extraction with water or by distillation under vacuum. In the process, two types of liquid wastes are formed: (1) a 4-5% aqueous solution of low-molecular weight compounds, consisting of ca. 75% by weight of caprolactam and ca. 25% by weight of a mixture of cyclic and linear caprolactam oligomers and (2) a caprolactam-oligomer melt, containing up to 98% caprolactam and small amounts of dimer, water, and organic contaminants. The recycle of caprolactam involves two different stages: depolymerization of polymeric waste and purification of the caprolactam and oligomers obtained.

A general recovery of caprolactam from liquid waste generates 20-25% oligomers along with organic and inorganic compounds as impurities. The distillation of caprolactam under reduced pressure produces a residue which consists of inorganic substances such as permanganates, potassium hydrogen sulfate, potassium sulfate, sodium hydrogen phosphate, and sodium phosphate. The larger portion of the residue contains cyclic and linear chain oligomers plus 8-10% of caprolactam. The types and exact amounts of impurities depend on the method used for the purification and distillation of caprolactam.

The cyclic oligomers are only slightly soluble in water and dilute solutions of caprolactam. They tend to separate out from the extracted waste during the process of concentration and chemical purification of the caprolactam. The cyclic oligomers tend to form on the walls of the equipment used in the process equipment. 6-Aminocaproic acid or sodium 6-aminocaproate may also be found in the oligomeric waste especially if sodium hydroxide is used to initiate the caprolactam polymerization.

8.3.3 Applications of depolymerized nylon 6

Chemical recycling of nylon 6 carpet face fibers has been developed into a closed-loop recycling process for waste nylon carpet [Bajaj and Sharma, 1997; Brown, 2001]. The recovered nylon 6 face fibers are sent to a depolymerization reactor and treated with superheated steam in the presence of a catalyst to produce a distillate containing caprolactam. The crude caprolactam is distilled and repolymerized to form nylon 6. The caprolactam obtained is comparable to virgin caprolactam in purity. The repolymerized nylon 6 is converted into yarn and tufted into carpet. The carpets obtained from this process are very similar in physical properties to those obtained from virgin caprolactam.

The "6ix Again" program of the BASF Corp. has been in operation since 1994. Its process involves collection of used nylon 6 carpet, shredding and separation of face fibers, pelletizing face fiber for depolymerization and chemical distillation to obtain a purified caprolactam monomer, and repolymerization of caprolactam into nylon polymer [BASF, 2001].

Evergreen Nylon Recycling LLC, a joint venture between Honeywell International and DSM Chemicals, was in operation from 1999 to 2001. It used a two-stage selective pyrolysis process. The ground nylon scrap is dissolved with high-pressure steam and then continuously hydrolyzed with super-heated steam to form caprolactam. The program has diverted over one hundred thousand tons of post consumer carpet from the landfill to produced virgin-quality caprolactam [Brown, 2001].

8.4 Hydrolysis of nylon 6,6 and nylon 4,6

The depolymerization of nylon 6,6 and nylon 4,6 involves hydrolysis of the amide linkages which are vulnerable to both acid- and base-catalyzed hydrolysis. Polk et al., 1999 reported the depolymerization of nylon 6,6 and nylon 4,6 in aqueous sodium hydroxide solutions containing a phase transfer catalyst. Benzyltrimethylammonium bromide was discovered to be an effective phase-transfer catalyst in 50% sodium hydroxide solution for the conversion of nylon 4,6 to oligomers. The depolymerization efficiency (% weight loss) and the molecular weight of the reclaimed oligomers were dependent on the amount and concentration of the aqueous sodium hydroxide and the reaction time. Nylon 4,6 fibers (M_v = 41,400 g/mole) did not undergo depolymerization on exposure to 100 mL of 25 wt% sodium hydroxide solution at 165^{0}C. Out of 6.0 g of nylon fibers fed for depolymerization, 5.95 g were unaffected. When the concentration of sodium hydroxide was increased to 50 wt%, the depolymerization process resulted in the formation of low molecular weight oligomers. Hence, even in the presence of a phase transfer agent, a critical sodium hydroxide concentration exists between 25 and 50 wt% which is required to initiate depolymerization under the conditions used. Soluble amine salts, were also obtained.

In order to establish the feasibility of alkaline hydrolysis in respect to recycling of nylon 4,6, it was necessary to determine whether the recovered oligomers could be repolymerized to form nylon 4,6. For this purpose, solid state polymerization was performed on nylon 4,6 oligomers formed via alkaline hydrolysis with 50 wt% NaOH at 165^{0}C for 24 hours. Solid state polymerization of the nylon 4,6 oligomers resulted in an increase in intrinsic viscosity from 0.141 to 0.740 dl/g. That corresponds to an increase in viscosity average molecular weight from 1846 g/mole to 16,343 g/mole.

The product of the depolymerization of nylon 6,6 with 50% aqueous sodium hydroxide solution was relatively low molecular weight oligomers. A series of experiments were run in order to examine the applicability and efficiency of benzyltrimethylammonium bromide

[BTEMB] as a phase transfer catalyst in the depolymerization of nylon 6,6. The product of the run with no phase transfer agent showed a 15.9% increase in weight compared to the weight of the original nylon 6,6. The calculated percent increase in weight for a 19-fold decrease in molecular weight (due to the addition of water) would be ca. 1%. Therefore, a large part of the increase must be due to leaching of silicates of the glass container (resin reaction kettle) by the strong alkali (50 wt%) at the temperature of the reaction (130ºC) over 24 hours. The oligomer obtained had a viscosity average molecular weight of 1644 g/mole (the original nylon 6,6 had a molecular weight of 30,944 g/mole). The runs with phase transfer agent produced oligomers with decreases in weight of 40-50%. Although the occurrence of leaching of silicates from the glass container made quantitative assessment difficult, these results suggested that in the absence of phase transfer agent only oligomers are formed; however, soluble low molecular weight products are formed in the presence of phase transfer agent. The oligomers obtained were repolymerized in the solid state by heating at 200ºC in a vacuum. The viscosity-average molecular weight of the solid state polymerized nylon 6,6 obtained was ca. 23,000 g/mole (the molecular weight of the oligomeric mixture was 1434 g/mole).

In order to isolate adipic acid, nylon 6,6 fibers were depolymerized under reflux with a 50% NaOH solution in the presence of catalytic amounts of benzyltrimethylammonium bromide. The oligomers formed in successive steps were depolymerized under similar conditions. However, hexamethylene diamine was not isolated. The overall yield of adipic acid was 59.6%.

8.4.1 Ammonolysis of nylon 6,6

Ammonolysis currently is the preferred route currently in use at the DuPont Company for the depolymerization of nylon 6,6 carpet waste [Kassera, 1998]. McKinney, 1994, has described the reaction of nylon 6,6 and nylon 6,6/nylon 6 mixtures with ammonia at temperatures between 300 and 350ºC and a pressure of about 68 atmospheres in the presence of an ammonium phosphate catalyst to yield a mixture of the following monomeric products: HMDA, adiponitrile, and 5-cyanovaleramide from nylon 6,6 and ε-caprolactam, 6-aminocapronitrile, and 6-aminocaproamide from nylon 6. The equilibrium is shifted toward products by continuous removal of water formed. Most of the monomers may be transformed into HMDA by hydrogenation. Kalfas, 1998 has developed a mechanism for the depolymerization of nylon-6,6 and nylon-6 mixtures by the ammonolysis process. The mechanism includes the amide bond breakage and amide end dehydration (nitrilation) reactions, plus the ring addition and ring opening reactions for cyclic lactams present in nylon 6. On the basis of the proposed mechanism, a kinetic model was developed for the ammonolysis of nylon mixtures.

Bordrero, *et al., 1999* utilized a two step ami/ammonolysis process to depolymerize nylon 6,6. The first step is based on an amminolysis treatment of nylon 6.6 by n-butylamine at a temperature of 300°C and a pressure of 45 atm. Free HMDA and NN'-dibutyladipamide are generated. The second step is ammonolysis of NN'-dibutyladipamide at a temperature of 285°C and a pressure of 50 atm. The end product is adiponitrile (ADN). It is estimated that the yields could be about 48% for ADN and about 100% for HMDA at optimized reaction conditions.

8.4.2 Recovery of nylon 6,6 monomers

Adipic acid and hexamethylene diamine (HMDA) are obtained from nylon 6,6 by the hydrolysis of the polymer in concentrated sulfuric acid (Figure 18). The adipic acid is purified by recrystallization and the HMDA is recovered by distillation after neutralizing the acid. This process is inefficient for treating large amounts of waste because of the required recrystallization of adipic acid after repeated batch hydrolyses of nylon 6,6 waste. In a continuous process, nylon 6,6 waste is hydrolyzed with an aqueous mineral acid of 30-70% concentration and the resulting hydrolysate is fed to a crystallization zone. The adipic acid crystallizes and the crystals are continuously removed from the hydrolysate. Calcium hydroxide is added to neutralize the mother liquor and liberate the HMDA for subsequent distillation.

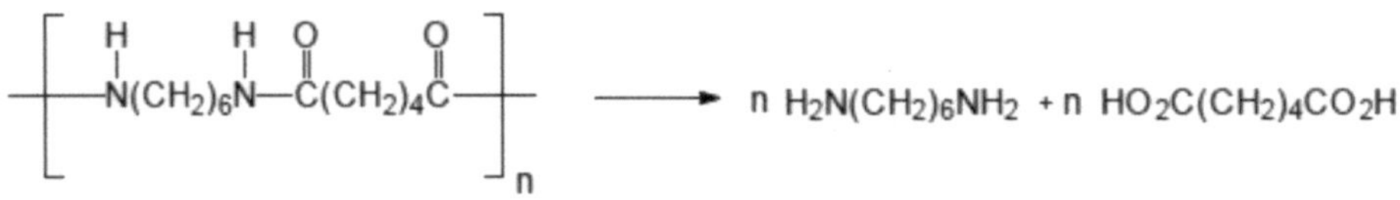

Fig. 18. Depolymerization of nylon 6,6 by hydrolysis.

Continuous recovery requires adipic acid crystals having an average diameter of ca. 40-50 nm. Such crystals are obtained by continuously introducing the hot hydrolysate containing 10-20% adipic acid into an agitated crystallization vessel while maintaining an average temperature of 20-30^{0}C. The slurry obtained from the crystallization vessel is filtered to collect the adipic acid crystals and the filtrate which contains the HMDA acid salt is continuously neutralized with calcium hydroxide. The calcium salt formed is removed by filtration and the HMDA in the filtrate is isolated by distillation.

In the case of nylon 6,6 waste recycled by ammonolysis, nylon is treated with ammonia in the presence of a phosphate catalyst. Reaction occurs at 330^{0}C and 7 MPa. Distillation of the reaction mixture produces ammonia which is recycled and three fractions containing (a) caprolactam, (b) HMDA and aminocapronitrile and (c) adiponitrile. Aminocapronitrile and adiponitrile are hydrogenated to yield pure HMDA and the caprolactam is either converted to aminocapronitrile by further ammonolysis or distilled to produce pure caprolactam. The HMDA produced by this process is extremely pure (>99.8). The main impurities are aminomethylcyclopentylamine and tetrahydroazepine which are expected to be removed more effectively in the larger distillation columns employed in larger plants.

9. Chemical recycling of poly(methyl methacrylate)

Poly(methyl methacrylate) (PMMA) is a major type of thermoplastics used throughout the world in such applications as transparent all-weather sheets, electrical insulation, bathroom units, automotive parts, surface coating and ion exchange resins, etc. The plastics made from PMMA are widely used under the commercial trade names PLEXIGLAS or PERSPEX. In

Western Europe alone approximately 327 000 tones of PMMA are consumed each year with an increasing percentage of approximately 4% per year. In contrast with condensation polymers (e.g. PET), addition polymers, like PMMA, cannot be easily recycled to monomer by simple chemical methods. Instead, thermo-chemical recycling techniques like pyrolysis are usually applied. Thus, various processes for the depolymerization of PMMA have been described in literature. Among them the most prominent ones are the molten metal bath process and the fluidized bed pyrolysis [Kaminsky et al., 2004; Smolders and Baeyens, 2004; Sasse and Emig, 1998]. The first one although widely used in several countries exhibits several serious disadvantages including that the raw condensate MMA may be contaminated by the metal used (usually lead) or other by-products [Sasse and Emig, 1998]. The effect of temperature, addition of filler and amount of feed on the amount and distribution of pyrolytic products was investigated by Kaminsky and co-workers in a fluidized bed reactor [Kaminsky et al., 1991; 2001; 2004, Grause et al., 2006]. Furthermore, PMMA thermal pyrolysis results in a close to 97% recovery of the monomer methyl methacrylate (MMA) at relatively low temperatures (400–500 °C) [Smolders and Baeyens, 2004]. It has been reported that the liquid pyrolysis product was so pure that it could be polymerized again without any further treatment [Kaminsky et al., 1991].

Achilias, 2006 and 2007 investigated the chemical recycling of PMMA using pyrolysis, aiming at the recovery of pure monomer able to be re-polymerized back to polymer. Conventional (thermal) pyrolysis was carried out using either model polymer or a commercial product as feedstock. The experiments were carried out in a laboratory fixed bed reactor at 450 °C, which was found in literature to be the optimum temperature for the maximization of MMA monomer amount [Kaminsky et al., 2001]. The liquid product obtained from both the model and the commercial samples was very high, 99% and 98%, respectively. The monomer recovery was higher by feeding pure PMMA (98.3 wt.%) compared to the commercial sample (94.9 wt.%). In both experiments the gas fraction was very small. Only 0.6 or 1.5 wt.% of gases were produced. Also the residue obtained was very low 0.1 and 0.4 wt.%. Furthermore, the gas composition of both samples was approximately the same with large amounts of CO_2 followed by CO and methane. The former are degradation products of PMMA and MMA due to the existence of oxygen in the macromolecular chain. The liquid fraction mostly consists of the monomer MMA in a large amount (99 and 97 wt.% for the model polymer and the commercial product, respectively) with a small percentage of some other organic compounds mainly esters.

The potential use of the liquid pyrolysis fraction as a raw material for the reproduction of PMMA by polymerization was also investigated (Achilias, 2007).

10. References

Achilias, D.S. Chemical Recycling of Poly(methyl methacrylate), *WSEAS Trans on Environm Develop* 2(2), 85-91 (2006)

Achilias, D.S., Chemical recycling of poly(methyl methacrylate) by pyrolysis. Potential use of the liquid fraction as a raw material for the reproduction of the polymer, *Eur Polym J*, 43 (6), 2564-2575 (2007)

Achilias, D.S.; A. Giannoulis, G.Z. Papageorgiou, Recycling of polymers from plastic packaging materials using the dissolution/ reprecipitation technique, *Polym. Bull.* 63(3), 449-465 (2009).

Achilias, D.S.; Karayannidis, G.P. The chemical recycling of PET in the framework of sustainable development, *Water, Air and Soil Pollution: Focus*, 4, 385-396 (2004).

Achilias, D.S.; E.V. Antonakou, C. Roupakias, P. Megalokonomos, A.A. Lappas, Recycling techniques of polyolefins from plastic wastes, *Global NEST Journal*, 10(1), 114-122 (2008)

Achilias, D.S.; E.V. Antonakou, E. Koutsokosta, A.A. Lappas, Chemical recycling of polymers from waste electric and electronic equipment, *J. Appl. Polym. Sci.*, 114, 212-221 (2009).

Achilias, D.S.; G. P. Tsintzou, A. K. Nikolaidis, D. N. Bikiaris, G. P. Karayannidis, Aminolytic Depolymerization of poly(ethylene terephthalate) Waste in a Microwave Reactor, *Polymer International*, 60, 500–506 (2011).

Achilias, D.S.; H.H. Redhwi, M. N. Siddiqui, A.K. Nikolaidis, D.N. Bikiaris, G.P. Karayannidis, Glycolytic Depolymerization of PET Waste in a Microwave Reactor, *Journal of Applied Polymer Science*, 118 (5), 3066–3073 (2010).

Achilias, D.S., C. Roupakias, P. Megalokonomosa, A.A. Lappas, E.V. Antonakou. Chemical recycling of plastic wastes made from polyethylene (LDPE and HDPE) and polypropylene (PP). J Hazard Mater 149 (2007) 536–542

Achilias, D.S.; Kanellopoulou, I.; Megalokonomos, P.; Antonakou, E.; Lappas, A.; Chemical Recycling of Polystyrene by Pyrolysis: Potential Use of the Liquid Product for the Reproduction of Polymer. Macromol. Mater. Eng. 2007, 292, 923–934.

Achilias, D.S.; P. Megalokonomos, G.P. Karayannidis Current trends in chemical recycling of polyolefins, *J. Environmental Protection and Ecology* 7(2), 407-413 (2006)

Aguado R., M. Olazar, M.J. San José, B. Gaisán and J. Bilbao, *Energy Fuels* 16 (2002), p. 1429.

Aguado, R.; Olazar, M.; Gaisan, B.; Prieto, R.; Bilbao, J. Kinetics of polystyrene pyrolysis in a conical spouted bed reactor. Chem. Engg. J. 2003, 92, 91–99.

Aguado, J.; Serrano, D.P. (Eds.) Feedstock recycling of plastic wastes. The Royal Society of Chemistry, Cambridge UK, 1999.

Aguayo A.T., A.G. Gayubo, J. Ortega, M. Olazar, J. Bilbao, Catal. Today 37 (1997) 239–248.

Akpanudoh N.S., K. Gobin and G. Manos, *J. Mol. Catal. A: Chem.* 235 (2005), p. 67.

Ali MF, Siddiqui MN. Thermal and catalytic decomposition behavior of PVC mixed plastic waste with petroleum residue. J Anal Appl Pyrol 2005;74:282e9.

Al-Salem S.M., P. Lettieri, J. Baeyens. Progress in Energy and Combustion Science 36 (2010) 103–129

Al-Salem S.M., P. Lettieri,. Chemical engineering research and design 88 (2010) 1599–1606

Al-Salem, S.M.; P.Lettieri, J.Baeyens. Recycling and recovery routes of plastic solid waste (PSW): A review. Waste Management 29 (2009) 2625–2643

Arandes, J.M.; Abajo, I.; Lopez-Valerio, D.; Fern_andez, I.; Azkoiti, M.J.; Olazar, M.; Bilbao, J. Transformation of several plastic wastes into fuels by catalytic cracking. Ind. Eng. Chem. Res. 1997, 36, 4523–4529.

Arandes, J.M.; Erena, J.; Azkoiti, M.J.; Olazar, M.; Bilbao, J. Thermal recycling of polystyrene and polystyrene-butadiene dissolved in a light cycle oil. J. Anal. Appl. Pyrol. 2003, 70, 747–760.

Araujo A.S., V.J. Fernandes Jr. and G.J.T. Fernandes, *Thermochim. Acta* 392 (2002), pp. 55–61.

Assumpção, L. C. F. N.; Carbonell, M.M.; Marques, M.R.C. Co-pyrolysis of polypropylene waste with Brazilian heavy oil. *Journal of Environmental Science and Health Part A* (2011) 46, 461–464

Bajaj, P. and Sharma, N. D., "Reuse of Polymer and Fibre Waste" in *Manufactured Fibre Technology* (Ed. Gupta, V. B. and Kothari, V. K.), Chapman & Hall, New York, p. 615, 1997.

Bajdur, W.; Pajaczkowska, J.; Makarucha, B.; Sulkowska, A.; Sulkowski, W.W. Effective polyelectrolytes synthesized from expanded polystyrene wastes. Eur. Polym. J. 2002, 38, 299–304.

Ballice, L.; Reimert, R. Temperature-programmed co-pyrolysis of Turkish lignite with polypropylene. Journal of Analytical and Applied Pyrolysis 65 (2002) 207–219

BASF Corp., "BASF 6ix Again Program", www.nylon6ix.com, 2001.

Beltrame, P.L.; Bergamasco, L.; Carniti, P.; Castelli, A.; Bertini, F.; Audisio, G. Hydrous pyrolysis of polystyrene. J. Anal. Appl. Pyrol. 1997, 40–41, 451–461.

Berrueco C, F.J. Mastral, E. Esperanza and J. Ceamanos, *Energy Fuels* 16 (2002), p. 1148.

Bhaskar, T.; Uddin, M.A.; Murai, K.; Kaneko, J.; Hamano, K.; Kusaba, T.; Muto, A.; Sakata, Y. Comparison of thermal degradation products from real municipal waste plastic and model mixed plastics.J. Anal. Appl. Pyrol. 2003, 70, 579–587.

Bockhorn, H.; Hornung, A.; Hornung, U.; Jakobstroer, P. Modelling of isothermal and dynamic pyrolysis of plastics considering nonhomogeneous temperature distribution and detailed degradation mechanism. J. Anal. Appl. Pyrol. 1999, 49, 53–74.

Bockhorn, H.; S. Donner, M. Gernsbeck, A. Hornung, and U. Hornung, "Pyrolysis of Polyamide 6 under Catalytic Conditions and its Application to Reutilization of Carpets", J. Anal. Appl. Pyrolysis, 58-59, 79-84 (2001).

Bodrero, S.; E. Canivenc, and F. Cansell, "Chemical Recycling of Polyamide 6.6 and Polyamide 6 through a Two Step Ami-/Ammonolysis Process", Presentation at 4th Annual Conference on *Recycling of Fibrous Textile and Carpet Waste*, Dalton, Georgia, May 17-18, 1999.

Borgianni, C., Filippis, P.D., Pochetti, F., Paolucci, M., 2002. Gasification process of wastes containing PVC. Fuel 14 (81), 1827–1833.

Braun D. Recycling of PVC. Prog Polym Sci 2002;27:2171 - 2195.

Braun, M., Levy, A.B., and Sifniades, S., "Recycling Nylon 6 Carpet to Caprolactam", *Polymer- Plastics Technology & Engineering*, Vol 38, No. 3, 1999, 471-484.

Brophy, J.H., S. Hardmann and D.C. Wilson. (1997). Polymer Cracking for Feedstock Recycling of Mixed Plastic Wastes. In: Hoyle, W. and D.R. Karsa, Chemical Aspects of Plastics Recycling. The Royal Society of Chemistry Information Services, Cambridge, UK

Brown, T.; "Infinity Nylon - A Never-Ending Cycle of Renewal", Presentation at 6th Annual Conference on *Recycling of Fibrous Textile and Carpet Waste*, Dalton, Georgia, April 30-May 1, 2001.

Buysch, H. J., Schoen, N. and Kuehling, S., Ger. Offen., DE 4,220,412, 1994.

Cardona S.C. and A. Corma, *Catal. Today* 75 (2002), p. 239.

Cardona, S.C.; Corma, A. Tertiary recycling of polypropylene by catalytic cracking in a semibatch stirred reactor Use of spent equilibrium FCC commercial catalyst. *Applied Catalysis B: Environmental 25 (2000) 151–162*

Castano, P., Elordi, G., Olazar, M., Aguayo, A.T., Pawelec, B., Bilbao J. *Applied Catalysis B: Environmental* 104 (2011) 91–100

Cerqueira H.S., C. Sievers, G. Joly, P. Magnoux, J.A. Lercher, Ind. Eng. Chem. Res. 44 (2005) 2069–2077.

Cerqueira H.S., G. Caeiro, L. Costa, F. Ramoa Ribeiro, J. Mol. Catal. A: Chem. 292 (2008) 1–13.

Chaupart, N., Serpe, G., and Verdu, J., *Polymer*, 39(6-7), 1375 (1998).

Chiu, S.J.; S.H. Chen, C.T. Tsai, Effect of metal chlorides on thermal degradation of (waste) polycarbonate, Waste Manage. 26 (2006) 252–259.

Corbin, T., Handermann; A., Kotek, R., Porter, W., Dellinger, J., Davis, E., "Reclaiming epsiloncaprolactam from nylon 6 carpet",U.S. Patent 5,977,193, November 2, 1999.

Czernik, S.; C. Elam, R. Evans, R. Meglen, L. Moens, and K. Tatsumoto, "Catalytic Pyrolysis of Nylon 6 to Recover Caprolactam", J. Anal. Appl. Pyrolysis, 56, 51-64 (1998).

Dawood, A.; Miura, K. Catalytic pyrolysis of g-irradiated polypropylene (PP) over HY-zeolite for enhancing the reactivity and the product selectivity. *Polym. Degrad. Stabil.* 76 (2002) 45–52

Dawood, A.; Miura, K. Pyrolysis kinetics of g-irradiated polypropylene. *Polym. Degrad. Stabil.* 73 (2001) 347–354

De Marco I, Caballero B, Torres A, Laresgoiti MF, Chomón MJ, Cabrero MA. Recycling polymeric wastes by means of pyrolysis. J Chem Technol Biotechnol 2002;77(7):817e24.

DKR/DSD (1999), Mit Altkunststoffen Ressourcen schonen. Leaflet with information on chemical recycling, DKR, Germany)

Duangchan A, Samart C. Tertiary recycling of PVC-containing plastic waste by copyrolysis with cattle manure. Waste Manage 2008; 28: 2415-21.

Durmus, A.; Koc, S.N.; Pozan, G.S., Kasgoz, A. Thermal-catalytic degradation kinetics of polypropylene over BEA, ZSM-5 and MOR zeolites. *Applied Catalysis B: Environmental* 61 (2005) 316–322

Elordi G, G. López, R. Aguado, M. Olazar and J. Bilbao, *Int. J. Chem. React. Eng.* 5 (2007), A72.

Elordi G, M. Olazar, G. Lopez, M. Amutio, M. Artetxe, R. Aguado, J. Bilbao, J. Anal. Appl. Pyrolysis 85 (2009) 345–351.

Elordi G, M. Olazar, R. Aguado, G. López, M. Arabiourrutia and J. Bilbao, *J. Anal. Appl. Pyrol.* 79 (2007), p. 450.

Faravelli, T.; Bozzano, G.; Colombo, M.; Ranzi, E.; Dente, M. Kinetic modeling of the thermal degradation of polyethylene and polystyrene mixtures. J. Anal. Appl. Pyrol. 2003, 70, 761–777.

Fernandes G.J.T. , V.J. Fernandes Jr. and A.S. Araujo, *Catal. Today* 75 (2002), pp. 233–238.

Fox, D. W. and Peters, E. N., US Patent No. 4,885,407, 1989.

Garcia D, Balart R, Sanchez L, Lopez J. Compatibility of recycled PVC/ABS blends effect of previous degradation. *Polym Eng Sci* 2007;47:789e96.

Grause G, Predel M, Kaminsky W. Monomer recovery from aluminum hydroxide high filled poly(methyl methacrylate) in a fluidized bed reactor. J Anal Appl Pyrol 2006, 75, 236–9.

Grause, G.; K. Sugawara, T. Mizoguchi, T. Yoshioka, Pyrolytic hydrolysis of polycarbonate in the presence of earth-alkali oxides and hydroxides, *Polym. Degrad. Stabil.* 94 (2009) 1119–1124.

Guisnet M., L. Costa, F.R. Ribeiro, J. Mol. Catal. A: Chem. 305 (2009) 69–83.

Hájeková, E., Bajus, M. (2005). Recycling of low-density polyethylene and polypropylene via copyrolysis of polyalkene oil/waxes with naphtha: Product distribution and coke formation. *J. Anal. Appl. Pyrolysis*, Vol. 74, Issue 1-2, 270-281.

Hayashi, J.-I. Pyrolysis of polypropylene in the presence of oxygen. *Fuel Processing Technology* 55 (1998). 265–275.

Hernández M.R., A.N. García and A. Marcilla, *J. Anal. Appl. Pyrol.* 78 (2007), p. 212.

Heyde, M. and Kremer, S. (1999). LCA Packaging Plastics Waste. LCA Documents Vol. 2, No. 5, EcoMed, Landsberg, Germany.

Hu, L.C.; A. Oku, E. Yamada, Alkali-catalyzed methanolysis of polycarbonate. A study on recycling of bisphenol A and dimethyl carbonate, Polymer 39 (1998) 3841–3845.

Huang J., Y. Jiang, V.R.R. Marthala, A. Bressel, J. Frey, M. Hunger, J. Catal. 263 (2009) 277–283.

Hub, L.-C.; Akira Oku, and Etsu Yamada, Alkali-catalyzed methanolysis of polycarbonate. A study on recycling of bisphenol A and dimethyl carbonate, *Polymer* Vol. 39 No. 16, pp. 3841-3845, 1998

Ikeda, A.; K. Katoh, H. Tagaya, Monomer recovery of waste plastics by liquid phase decomposition and polymer synthesis, *J. Mater. Sci.* 43 (2008) 2437–2441.

Ishihara, Y.; Nanbu, H.; Saido, K.; Ikemura, T.; Takesue, T.; Kuroki, T. (1993). Mechanism of gas formation in catalytic decomposition of polypropylene. *Fuel*, 72(8), 1115-1119.

Kalfas, G. A., *Polymer React. Eng.*, 6(1), 41 (1998).

Kameda T, Fukuda Y, Grause G, Yoshioka T. Chemical modification of rigid poly(vinyl chloride) by the substitution with nucleophiles. J Appl Polym Sci 2010;116:36 - 44.

Kaminsky W, Eger C. Pyrolysis of filled PMMA for monomer recovery. *J. Anal. Appl. Pyrol.* 2001, 58–59, 781–87.

Kaminsky W, Frank J. Monomer recovery by pyrolysis of poly(methyl methacrylate). *J. Anal. Appl. Pyrol.* 1991, 19, 311–8.

Kaminsky W, Predel M, Sadiki A. Feedstock recycling of polymers by pyrolysis in a fluidized bed. *Polym Degrad Stabil* 2004, 85, 1045–50.

Kaminsky W, Kim J-S. Pyrolysis of mixed plastics into aromatics. J Anal Appl Pyrol 1999;51(1):127-134.

Kaminsky, W.; Nunez, I.J.; Zorriqueta. Catalytical and thermal pyrolysis of polyolefins. *J. Anal. Appl. Pyrolysis* 79 (2007) 368–374

Karayannidis, G.P. A.K. Nikolaidis, I.D. Sideridou, D.N. Bikiaris, D.S. Achilias, Chemical recycling of PET by glycolysis: Polymerization and characterization of the dimethacrylated glycolysate, *Macromol Mater Eng* 291, 1338-1347 (2006).

Karayannidis, G.P. ; D.S. Achilias Chemical recycling of poly(ethylene terephthalate), *Macromol. Mater. Eng.*, 292(2), 128-146 (2007).

Karayannidis, G.P.; A. Chatziavgoustis and D.S. Achilias, Poly(ethylene terephthalate) recycling and recovery of pure terephthalic acid by alkaline hydrolysis, *Advances in Polymer Technology*, 21(4), 250-259 (2002).

Karayannidis, G.P.; D.S. Achilias, I. Sideridou and D.N. Bikiaris, Alkyd resins derived from glycolized waste poly(ethylene terephthalate), *Eur. Polym. J.*, 41, 201-210 (2005)

Kasserra, H.P. "Recycling of Polyamide 66 and 6", *Science and Technology of Polymers and Advanced Materials*, Ed. P. N. Prasad *et al.*, Plenum Press, New York, p.629-635, 1998.

Ke, H.; Li-hua, T.; Zi-Bin, Z.; Cheng-Fang, Z. Reaction mechanism of styrene monomer recovery from waste polystyrene by supercritical solvents. Polym. Degrad. Stab. 2005, 89, 312–316.

Kim, D., Bo-kyung Kim, Youngmin Cho, Myungwan Han, and Beom-Sik Kim, Kinetics of Polycarbonate Glycolysis in Ethylene Glycol, Ind. Eng. Chem. Res. 2009, 48, 685–691

Kim, D.; Kim, B.-K.; Cho, Y.; Han, M.; Kim, B.S. (2009). Kinetics of Polycarbonate Glycolysis in Ethylene glycol. *Ind. Eng. Chem. Res., 48* (2), 685–691.

Kim, D.; Kim, B.-K.; Cho, Y.; Han, M.; Kim, B.S. (2009). Kinetics of Polycarbonate Methanolysis by a Consecutive Reaction Model. *Ind. Eng. Chem. Res., 48* (14), 6591–6599.

Kim, J.S.; Lee, W.Y.; Lee, S.B.; Kim, S.B.; Choi, M.J. Degradation of polystyrene waste over base promoted Fe catalysts. Catal. Today, 2003, 87, 59–68.

Koji, U.; Yoshio, M.; Naoto, A.; Tamaki, H.; Sanae, H.; Mitsunori, O. Patent number JP10130418, 1998.

Kosmidis, V.; D. S. Achilias and G.P. Karayannidis, Poly(ethylene terephthalate) recycling and recovery of pure terephthalic acid. Kinetics of a phase-transfer catalyzed alkaline hydrolysis, *Macromol. Mater. Eng.,* 286(10), 640-647 (2001).

Kruse, T.M.; Woo, O.S.; Broadbelt, L.J. Detailed mechanistic modeling of polymer degradation: Application to polystyrene. Chem. Eng. Sci. 2001, 56, 971–979.

La Mantia FP. Recycling of PVC and mixed plastic waste. Toronto: ChemTec Publishing; 1996.

Lee, C.G.; Cho, Y.J.; Song, P.S.; Kang, Y.; Kim, J.S.; Choi, M.J. Effects of temperature distribution on the catalytic pyrolysis of polystyrene waste in a swirling fluidized-bed reactor. Catal. Today 2003, 79–80, 453–464.

Lee, G.-S.; Young Jun Song b. Recycling EAF dust by heat treatment with PVC. Minerals Engineering 20 (2007) 739–746

Lee, K.H.; Noh, N.S.; Shin, D.H.; Seo, Y. Comparison of plastic types for catalytic degradation of waste plastics into liquid product with spent FCC catalyst. Polym. Degrad. Stab. 2002, 78, 539–544.

Lee, S.Y.; Yoon, J.H.; Kim, J.R.; Park, D.W. Degradation of polystyrene using clinoptilolite catalysts. J. Anal. Appl. Pyrol. 2002, 64, 71–83.

Lin, Y.-H.; H.-Y. Yen. Fluidised bed pyrolysis of polypropylene over cracking catalysts for producing hydrocarbons. Polymer Degradation and Stability 89 (2005) 101-108.

Lin, Y.-H.; M.-H. Yang. Chemical catalysed recycling of waste polymers: Catalytic conversion of polypropylene into fuels and chemicals over spent FCC catalyst in a fluidised-bed reactor. Polymer Degradation and Stability 92 (2007) 813-821.

Liu, F.-S.; Lei Li, Shitao Yu, Zhiguo Lv, Xiaoping Ge, Methanolysis of polycarbonate catalysed by ionic liquid [Bmim][Ac], J. Hazard. Mater. 189 (2011) 249–254

Liu, F.-S.; Zhuo Li, Shi-Tao Yu, Xiao Cui, Cong-Xia Xie, Xiao-Ping Ge, Methanolysis and hydrolysis of polycarbonate under moderate conditions, J Polym Environ 17:208–211 (2009)

Machado HMAMMS, Rodrigues Filho G, De Assunc RMN, Soares HM, Cangani AP, Cerqueira DA, et al. Chemical recycling of poly(vinyl chloride): application of partially dehydrochlorinated poly(vinyl chloride) for producing a chemically modified polymer. J Appl Polym Sci 2010;115:1474 - 1479.

Maharana, T.; Negi, Y.S.; Mohanty, B. Review Article: Recycling of Polystyrene. Plastics Technology and Engineering, 2007, 46: 7, 729 -736.

Marcilla A, A. Gómez, A. Reyes-Laberta and A. Giner, *Polym. Degrad. Stab.* 80 (2003), pp. 233–240.

Marcilla A, A. Gómez, A.N. García and M.M. Olaya, *J. Anal. Appl. Pyrolysis* 64 (2002), 85–101.
Marcilla A, M. Beltran and J.A. Conesa, *J. Anal. Appl. Pyrol.* 58 (2001), p. 117.
Marcilla A, M.I. Beltran, A. Gomez-Siurana, R. Navarro, F. Valdes, Appl. Catal. A: Gen. 328 (2007) 124–131.A
Marcilla A, M.I. Beltran, F. Hernandez, R. Navarro, Appl. Catal. A: Gen. 278 (2004) 37–43.
Marcilla A, M.I. Beltran, R. Navarro, Appl. Catal. A: Gen. 333 (2007) 57–66.
Marcilla, A.; A. Gomez, J.A. Reyes-Labarta, A. Giner, F. Hernandez. Kinetic study of polypropylene pyrolysis using ZSM-5 and an equilibrium fluid catalytic cracking catalyst. *J. Anal. Appl. Pyrolysis* 68-69 (2003) 46-480
Mastellone M.L., F. Perugini, M. Ponte and U. Arena, *Polym. Degrad. Stab.* 76 (2002), p. 479.
Mastral J, E. Esperanza, P. García and M. Juste, *J. Anal. Appl. Pyrol.* 63 (2002), p. 1.
Mastral J.F., C. Berrueco, M. Gea and J. Ceamanos, *Polym. Degrad. Stab.* 91 (2006), p. 3330.
Matuschek, G. et al., Thermochimica Acta 361 (2000) 77-84.
McKinney, R. J., U.S. Patent 5,302,756 (1994).
Mehta, S.; Biederman, S.; Shivkumar, S. Thermal degradation of foamed polystyrene. J. Mater Sci. 1995, 30, 2944–49.
Milne, B.J.; Behie, L.A.; Berruti, F. Recycling of waste plastics by ultrapyrolysis using an internally circulating fluidized bed reactor. J. Anal. Appl. Pyrol. 1999, 51, 157–166.
Nikles, D.E., M.S. Farahat, New motivation for the depolymerization products derived from PET waste: A review. *Macromol. Mater. Eng.* 2005, 290, 13-30
Oku, A.; S. Tanaka, S. Hata, Chemical conversion of poly(carbonate) to bis(hydroxyethyl) ether of bisphenol A. An approach to the chemical recycling of plastic wastes as monomers, Polymer 41 (2000) 6749–6753.
Olazar M, M.J. San José, G. Zabala and J. Bilbao, *Chem. Eng. Sci.* 49 (1994), p. 4579.
Olazar M., M.J. San José, R. Aguado, B. Gaisán and J. Bilbao, *Ind. Eng. Chem. Res.* 38 (1999), 4120.
Panda, A.K., Singh, R.K. (2011). Catalytic performances of kaoline and silica alumina in the thermal degradation of polypropylene. J. Fuel Chem. Technol., Vol. 39, Issue 3, 198-202.
Park H.J., J.-H. Yim, J.-K. Jeon, J.M. Kim, K.-S. Yoo and Y.-K. Park, *J. Phys. Chem.* 69 (2008), pp. 1125–1128.
Park H.J., Y.K. Park, J.I. Dong, J.K. Jeon, J.H. Yim and K.E. Jeong, *Res. Chem. Interm.* 34 (8–9) (2008), pp. 727–735.
Park, H.-J. et al. *Pyrolysis of polypropylene over mesoporous MCM-48 materia.l* Journal of Physics and Chemistry of Solids 69 (2008) 1125–1128
Pinto, F., Franco, C., Andre, R.N., Miranda, M., Gulyurtlu, I., Cabrita, I.,. Fuel 3 (81), (2002) 291–297.
Pinto, F., Franco, C., Andre, R.N., Tavares, C., Dias, M., Gulyurtlu, I.,. Fuel 15–17 (82), (2003) 1967–1976.
Polk, M. B., LeBoeuf, L. L., Shah, M., Won, C.-Y., Hu, X., and Ding, Y., *Polym.-Plast. Technol. Eng.*, 38(3), 459 (1999).
Predel M. and W. Kaminsky, *Polym. Degrad. Stab.* 70 (2000), p. 373.
Puente, D.L.G.; Sedran, U. Recycling polystyrene into fuels by means of FCC: Performance of various acidic catalysts. Appl. Catal. B Environ. 1998, 19, 305–311.
PVC recycling: A cycle of problems. Available at: http://archive.greenpeace.org/comms/pvctoys/reports/loomingproblems.html (accessed Sep. 2010).

Rasul Jan M., Jasmin Shah, Hussain Gulab. Fuel Processing Technology 91 (2010). 1428–1437

Recovinyl, 2008. Application covered by Recovinyl. Available at: www.recovinyl.com/coveredapplications.

Ryu C, Sharifi VN, Swithenbank J. Waste pyrolysis and generation of storable char. *Int J Energ Res* 2007; 31(2): 177-91.

Sadat-Shojai, M.; G.-R. Bakhshandeh / Polymer Degradation and Stability 96 (2011) 404-415

Sas, H. Verwijdering van huishoudelijk kunststofafval: analyse van milieu-effecten en kosten (Disposal of domestic plastic waste: analysis of environmental effects and costs), CE report, Delft, The Netherlands, 1994)

Sasse F, Emig G. Chemical recycling of polymer materials. *Chem Eng Technol* 1998, 21, 777–89.

Scheirs J. End-of-life environmental issues with PVC in Australia. (2010).

Scheirs, J. *Polymer Recycling, Science, Technology and Applications,* Published by John Wiley & Sons, New York, 1998.

Scheirs, J., Kaminsky, W. (eds.) Feedstock Recycling and Pyrolysis of Waste Plastics. Converting Waste Plastics into Diesel and Other Fuels. Wiley - VCH, 2006

Seo, Y.-H.; Lee, K.-H.; Shin, D.-H. (2003). Investigation of catalytic degradation of high-density polyethylene by hydrocarbon group type analysis. *J. Anal. Appl. Pyrolysis,* 70(2), 383-398.

Serrano D.P., J. Aguado, J.M. Escola, J.M. Rodriguez, L. Morselli and R. Orsi, *J. Anal. Appl. Pyrolysis* 68–69 (2003), pp. 481–494.

Shafer, S. J., US Patent No. 5,336,814, 1994

Shin S-M, Yoshioka T, Okuwaki A. Polym Degrad Stab 1998; 61(2): 349-353.

Siddiqui, M.N.; D. S. Achilias, H.H. Redhwi, D.N. Bikiaris, K.–A. G. Katsogiannis, G. P. Karayannidis, Hydrolytic Depolymerization of PET in a Microwave Reactor, *Macromolecular Materials & Engineering,* 295, 575–584 (2010).

Sifniades, S., Levy, A., Hendrix, J., "Process for depolymerizing nylon-containing waste to form caprolactam",U.S. Patent 5,932,724, August 3, 1999.

Sifniades, S., Levy, A., Hendrix, J., "Process for depolymerizing nylon-containing whole carpet to form caprolactam", U.S. Patent 5,929,234, July 27, 1999.

Singhal, R.; Singhal, C.; Upadhyayula, S. *J. Analyt. Appl. Pyrolysis* 89 (2010) 313–317

Slapak MJP, van Kasteren JMN, Drinkenburg BAAH. Hydrothermal recycling of PVC in a bubbling fluidized bed reactor: the influence of bed material and temperature. Polym Adv Technol 1999;10:596 - 602.

Smolders K, Baeyens J. Thermal degradation of PMMA in fluidized beds. *Waste Manage* 2004, 24, 849–57.

Sterling, W.J.; Walline, K.S.; McCoy, B.J. Experimental study of polystyrene thermolysis to moderate conversion. *Polym. Degrad. Stab.* 2001, 73, 75–82.

Suyama, K.; Masafumi Kubota, Masamitsu Shirai, Hiroyuki Yoshida, Chemical recycling of networked polystyrene derivatives using subcritical water in the presence of an aminoalchocol. *Polym. Degrad. Stab.* 2010, 95, 1588–1592.

Tongamp W, Kano J, Zhang Q, Saito F. Simultaneous treatment of PVC and oyster-shell wastes by mechanochemical means. Waste Manage 2008; 28(3): 484-88.

Tukker, A., de Groot, H., Simons, L., Wiegersma, S., 1999. Chemical recycling of plastic waste: PVC and other resins. European Commission, DG III, Final Report, STB-99-55 Final. Delft, the Netherlands.

Uemichi, Y.; Makino, Y.; Kanazuka, T. Degradation of polypropylene to aromatic hydrocarbons over Pt- and Fe- containing activated carbon. *J. Anal. Appl. Pyrolysis*, 16 (1989) 229-238

Ukei, H.; Hirose, T.; Horikawa, S.; Takai, Y.; Taka, M.; Azuma, N.; Ueno, A. Catalytic degradation of polystyrene into styrene and a design of recyclable polystyrene with dispersed catalysts. Catal. Today, 2000, 62, 67–75.

Ulutan S. A recycling assessment of PVC bottles by means of heat impact evaluation on its reprocessing. J Appl Polym Sci 1998;69:865e9.

Van Grieken R., D.P. Serrano, J. Aguado, R. García and C. Rojo, *J. Anal. Appl. Pyrol.* 58 (2001), 127

Vilaplana, F.; Ribes-Greus, A.; Karlsson, S. Degradation of recycled high-impact polystyrene: Simulation by reprocessing and thermooxidation. Polym. Degrad. Stab. 2006, 91, 2163–2170.

Ward, P.G.; Goff, M.; Donner, M.; Kaminsky, W.; O'Connor, K.E. A two step chemo-biotechnological conversion of polystyrene to a biodegradable thermoplastic. Environ. Sci. Technol. 2006, 40, 2433–2437.

Watanabe, M., Yasuaki Matsuo, Takashi Matsushita, Hiroshi Inomata, Toshiyuki Miyake, Katsuhiko Hironaka, Chemical recycling of polycarbonate in high pressure high temperature steam at 573 K, Polymer Degradation and Stability 94 (2009) 2157–2162

Westerhout, R.W.J.; Waanders, J.; Kuipers, J.A.M.; van Swaaij, W.P.M. Kinetics of the low-temperature pyrolysis of polyethene, polypropene, and polystyrene modeling, experimental determination, and comparison with literature models and data. Ind. Eng. Chem. Res. 1997, 36, 1955–1964.

Williams PT, Williams EA. Recycling plastic waste by pyrolysis. J Inst Energy 1998; 71: 81-93.

Woo, O.S.; Ayala, N.; Broadbelt, L.J. Mechanistic interpretation of base-catalyzed depolymerization of polystyrene. Catal. Today 2000, 55, 161–171.

Wu Y-H, Zhou Q, Zhao T, Deng M-L, Zhang J, Wang Y-Z. Poly(ethylene glycol) enhanced dehydrochlorination of poly(vinyl chloride). J Hazard Mater 2009; 163:1408 - 1411.

Xiao, G.; Ni, M.-J.; Chi, Y.; Jin, B.-S.; Xiao, R.; Zhong, Z.-P.; Huang, Y.-J. (2009). Gasification characteristics of MSW and an ANN prediction model. *Waste Managem.*, Vol. 29, Issue 1, 240-244

Xie, C. et al. Catalytic cracking of polypropylene into liquid hydrocarbons over Zr and Mo modified MCM-41 mesoporous molecular sieve. Catalysis Communications 10 (2008) 79–82

Xiong L. PVC: an enemy of public health and environment. Available at: http:// www.msu.edu/wxiongli/project/PKG875/PKG875.pdf (accessed Sep. 2010).

Xue, F.; Takeda, D.; Kimura, T.; Minabe, M. Effect of organic peroxides on the thermal decomposition of expanded polystyrene with the addition of a -methylstyrene. Polym. Degrad. Stab. 2004, 83, 461–466.

Yoshioka T, Furukawa K, Sato T, Okuwaki A. Chemical recycling of flexible PVC by oxygen oxidation in NaOH solutions at elevated temperatures. J Appl Polym Sci 1998;70: 129-135.

Yoshioka, T.; K. Furukawa, A. Okuwaki. Chemical recycling of rigid-PVC by oxygen oxidation in NaOH solutions at elevated temperatures. Polymer Degradation and Stability 67 (2000) 285-290.

Yoshioka, T.; K. Sugawara, T. Mizoguchi, A. Okuwaki, Chemical recycling of polycarbonate to raw materials by thermal decomposition with calcium hydroxide/steam, Chem. Lett. 34 (2005) 282–283.

Zhang Q, Saeki S, Tanaka Y, Kano J, Saito F. A soft-solution process for recovering rare metals from metal/alloy-wastes by grinding and washing with water. J Hazard Mater 2007;139:438 - 442.

Zhang, Z.; Hirose, T.; Nishio, S.; Morioka, Y.; Azuma, N.; Ueno, A.; Ohkita, H.; Okada, M. Chemical recycling of waste polystyrene into styrene over solid acids and bases. Ind. Eng. Chem. Res. 1995, 34, 4514–4519.

Zhao, W. et al. Effects of zeolites on the pyrolysis polypropylene. Polymer Degradation *and* Stabilily 53 (1996) 129-135

Zhu, X.; Elomaa, M.; Sundholma, F.; Lochmuller, C.H. Infrared and thermogravimetric studies of thermal degradation of polystyrene in the presence of ammonium sulfate. Polym. Degrad. Stab. 1998, 62, 487–494.

2

Recent Developments in the Chemical Recycling of PET

Leian Bartolome[1], Muhammad Imran[2], Bong Gyoo Cho[3], Waheed A. Al-Masry[2] and Do Hyun Kim[1]

[1]Korea Advanced Institute of Science and Technology (KAIST)

[2]King Saud University

[3]Korea Institute of Geoscience and Mineral Resources

[1,3]Republic of Korea

[2]Saudi Arabia

1. Introduction

Poly(ethylene terephthalate), more commonly known as PET in the packaging industry and generally referred to as 'polyester' in the textile industry, is an indispensable material with immense applications owing to its excellent physical and chemical properties. On the other hand, due to its increasing consumption and non-biodegradability, PET waste disposal has created serious environmental and economic concerns. Thus, management of PET waste has become an important social issue. In view of the increasing environmental awareness in the society, recycling remains the most viable option for the treatment of waste PET. Among the various methods of PET recycling (primary or 'in-plant', secondary or mechanical, tertiary or chemical, quaternary involving energy recovery), only chemical recycling conforms to the principles of sustainable development because it leads to the formation of the raw materials from which PET is originally made. Chemical recycling utilizes processes such as hydrolysis, methanolysis, glycloysis, ammonolysis and aminolysis. In a large collection of researches for the chemical recycling of PET, the primary objective is to increase the monomer yield while reducing the reaction time and/or carrying out the reaction under mild conditions. Continuous efforts of researchers have brought great improvements in the chemical recycling processes. This paper reviews methods for the chemical recycling of PET with special emphasis on glycolytic depolymerization with ethylene glycol. It covers the researches, including the works by the authors, on various processes and introduces recent developments to increase monomer yield. Processes including sub- and supercritical, catalytic, and microwave-assisted depolymerization are discussed. This paper also presents the impact of the new technologies such as nanotechnology on the future developments in the chemical recycling of PET.

1.1 PET: Synthesis and properties

PET is a polycrystalline polyester formed from the esterification of terephthalic acid (TPA) with ethylene glycol (EG) or from the transesterification of dimethyl terephthalate (DMT) with EG. Synthesis of PET from either process involves two reaction steps as shown in Fig. 1. The first step (Figs 1a, 1b) is the formation of an intermediate monomer bis(2-hydroxyethyl terephthalate) (BHET) with the release of a small molecule, which is either water or methanol. The second is the polycondensation of BHET to produce PET in melt phase with the release of EG under high vacuum (Scheirs, 1998; Scheirs & Long, 2003).

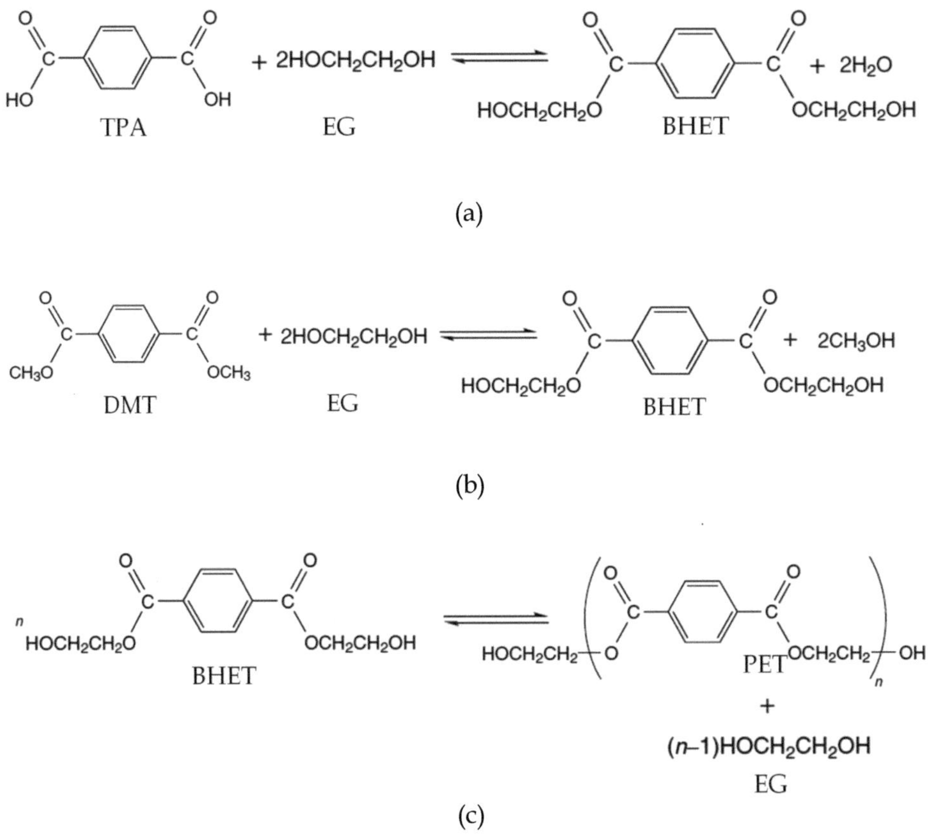

Fig. 1. Reaction scheme for PET synthesis. BHET is first formed from the reaction of either (a) TPA and EG, or (b) DMT and EG, and (c) eventually polymerized to PET.

As a thermoplastic polyester resin, PET exhibits interesting physical and chemical properties. It is an amorphous glass-like material in its purest form. Crystallinity in PET can be enhanced by adding modifying additives or by heat treatment of the polymer melt. PET is classified as a semi-crystalline polymer, and when heated above 72 ^{o}C, it changes from a rigid glass-like state into a rubbery elastic form where the polymer molecular chains can be stretched and aligned in either one direction to form fibers, or in two directions to form films and bottles. If PET is held in the stretched form at temperatures above 72 ^{0}C, it slowly crystallizes and the material starts to become opaque and less flexible. It is then known as crystalline PET. Meanwhile, if the melt is cooled quickly while still in stretched state, the chains are frozen with their original orientation. The resulting material is an extremely tough plastic, typical of a PET bottle (Sinha et al., 2008). Commercial PET melts between 255 and 265 ^0C, while more crystalline PET melts at 265 ^0C. Virgin PET is capable of morphological and structural reorganization, which is attributed to its multiple endothermic transitions. This leads to better crystal structures as temperature increases (Awaja & Pavel, 2005).

1.2 Applications, production and issues

Because of its low cost (Thompson et al., 2009), excellent tensile strength, chemical resistance, clarity, processability, and reasonable thermal stability (Caldicott, 1999), PET has been used in a wide range of applications. The demand and usage of PET worldwide according to application is summarized in Table 1 (Scheirs & Kaminsky, 2006). It is mainly applied in the textile industry, where more than 60% of all the PET produced worldwide is consumed. Enormous amounts are also used for other applications including manufacture of video and audio tapes, X-ray films, thermoformed products (e.g. material handling equipments, house-wares, automobile products, lighting products, sporting goods, etc) and food packaging (Carraher, 2000; ILSI Europe, 2000; Olabisi, 1997). In food packaging, PET has become the choice especially for beverages mainly due to its glass-like transparency coupled with adequate gas barrier properties for retention of carbonation. It provides an excellent barrier against oxygen and carbon dioxide in the carbonated soft drink sector, which has been growing more rapidly than other applications. In addition, it exhibits a high toughness/weight property ratio, which allows lightweight and securely unbreakable containers with large capacity (Welle, 2011).

	1990	**1995**	**2000**	**2005**	**2010**
Fiber	8 900	11 700	18 800	24 200	33 300
PET resin (for bottles)	1 100	3 100	7 100	11 900	18 900
Film	1 000	1 100	1 400	1 400	1 700
Others	700	800	1 100	1900	2 200
Total	**11 700**	**16 700**	**28 400**	**39 400**	**56 100**

Table 1. The global demand and future prediction of PET by application. (Unit in thousand tons).

From its main applications, PET is mainly classified as fiber-grade or bottle-grade. These grades differ mainly in molecular weights, intrinsic viscosity, optical appearance, and production recipes. Fiber-grade PET has a number-average molecular weight (MW_n) of 15,000 to 20,000 g/mol and intrinsic viscosity (IV) of 0.40 to 0.70 dL/g. Bottle-grade PET average molecular weight ranges from 24,000 to 36,000 g/mol and IV from 0.70 to 0.85 dL/g. (Awaja & Pavel, 2005; Gupta & Bashir, 2002).

PET's popularity has risen tremendously since it discovery in the early 1940s. In the year 2000, the global PET production capacity exceeded 33 million metric tons per year (Rieckmann, 2003). The total global consumption has risen from 11.8 million metric tons in 1997 (Paszun & Spychaj, 1997) to 23.6 million in 2005 (Pohler, 2005, as cited in Karayannidis & Achilias, 2007) and 54 million in 2010 (IHS, 2011). It is expected to grow by 4.5% per year from 2010 to 2015. In Europe and America, the rise of PET consumption is mainly maintained by PET bottle production while in Asia, the expansion of PET use is related to the higher production of fibers, due to the shift of fiber production from the industrialized countries to low-wage countries.

Along with the widespread application of PET is the inevitable creation of large amounts of PET waste. PET does not have any side effects on the human body, and does not create a direct hazard to the environment. However, due to its substantial fraction by volume in the waste stream and its high resistance to the atmospheric and biological agents, it is considered as a noxious material (Paszun & Spychaj, 1997). With the increase in the amount of PET wastes, its disposal began to pose serious economical and environmental problems. In view of the increasing environmental awareness in the society, recycling remains the most viable option for the treatment of waste PET. Environmental and economic considerations as well as energy conservation issues pushed the wide-scale recycling of PET (Nir et al., 1993); it was not simply a trend or a new marketing strategy to make a profit (Grasso, 1995, as cited in Shukla & Kulkarni, 2002). The recycling of PET does not only serve as a partial solution to the solid waste problem but also contributes to the conservation of raw petrochemical products and energy. Products made from recycled plastics can result in 50-60% capital saving as compared to making the same product from virgin resin (Sinha et al., 2008). Nevertheless, Welle noted that the main driving force in PET recycling is not cost reduction, but the business sector's embracing of sustainability ethics and the public's concern about the environment (Welle, 2011).

2. PET recycling methods

PET is considered one of the easiest materials to recycle, and is second only to aluminum in terms of the scrap values for recycled materials (Shceirs, 1998). Because of this, PET recycling has been one of the most successful and widespread among polymer recycling (Karayaniddis et al., 2006; Karayaniddis & Achilias, 2007). PET recycling methods can be categorized into four groups namely primary, secondary, tertiary, and quaternary recycling There is also a so called 'zero-order' recycling technique, which involves the direct reuse of a PET waste material (Nikles & Farahat, 2005). There are many other terminologies used for these plastic recycling categories; Hopewell and his colleagues have summarized these different terminologies (Hopewell et al., 2009).

2.1 Primary recycling

Primary recycling, also known as re-extrusion, is the oldest way of recycling PET. It refers to the "in-plant" recycling of the scrap materials that have similar features to the original products. This process ensures simplicity and low cost, but requires uncontaminated scrap, and only deals with single-type waste, making it an unpopular choice for recyclers (Al-Salem, 2009; Al-Salem et al., 2009).

2.2 Secondary recycling

Secondary recycling, also known as mechanical recycling, was commercialized in the 1970s. It involves separation of the polymer from its contaminants and reprocessing it to granules via mechanical means. Mechanical recycling steps include sorting and separation of wastes, removal of contaminants, reduction of size by crushing and grinding, extrusion by heat, and reforming (Aguado & Serrano, 1999). The more complex and contaminated the waste is, the more difficult it is to recycle mechanically. Among the main issues of secondary recycling are the heterogeneity of the solid waste, and the degradation of the product properties each time it is recycled. Since the reactions in polymerization are all reversible in theory, the employment of heat results to photo-oxidation and mechanical stresses, causing deterioration of the product's properties. Another problem is the undesirable gray colour resulting from the wastes that have the same type of resin, but of different color.

2.3 Tertiary recycling

Tertiary recycling, more commonly known as chemical recycling, involves the transformation of the PET polymer chain. Usually by means of solvolytic chain cleavage, this process can either be a total depolymerization back to its monomers or a partial depolymerization to its oligomers and other industrial chemicals. Since PET is a polyester with functional ester groups, it can be cleaved by some reagents such as water, alcohols, acids, glycols, and amines. Also, PET is formed through a reversible polycondensation reaction, so it can be transformed back to its monomer or oligomer units by pushing the reaction to the opposite direction through the addition of a condensation product. These low molecular products can then be purified and reused as raw materials to produce high-quality chemical products (Carta et al., 2003).

Among the recycling methods, chemical recycling is the most established and the only one acceptable according to the principles of 'sustainable development', defined as development that meets the needs of present generation without compromising the ability of future generations to meet their needs (Harris, 2001; World Commission on Environment and Development, 1987), because it leads to the formation of the raw materials (monomers) from which the polymer is originally made. In this way the environment is not surcharged and there is no need for extra resources for the production of PET (Achilias & Karayannidis, 2004).

The reaction mechanism for PET depolymerization consists of three reversible reactions. First, the carbonyl carbon in the polymer chain undergoes rapid protonation where the carbonyl oxygen is converted to a second hydroxyl group. Second, the hydroxyl oxygen of the added hydroxyl-bearing molecule slowly attacks the protonated carboxyl carbon atom. Third, the carbonyl oxygen (which was converted to hydroxyl group in the first step) and a proton are rapidly removed to form water or a simple alcohol and the catalytic proton (Patterson, 2007).

As shown in Fig. 2 (Janssen & van Santen, 1999), there are three main methods in PET chemical recycling depending on the added hydroxyl bearing molecule: glycol for gylcolysis, methanol for methanolysis, and water for hydrolysis. Other methods include aminolysis and ammonolysis. It has been five decades since the start of PET chemical recycling research, when patents were filed by Vereinigte Glanzstoff-Fabriken in the 1950s (Vereinigte Glanzstoff-Fabriken, 1956, 1957). Since then, numerous researches have been done in order to fully understand the chemical pathways of the depolymerization methods, and improve desired products yield from these methods.

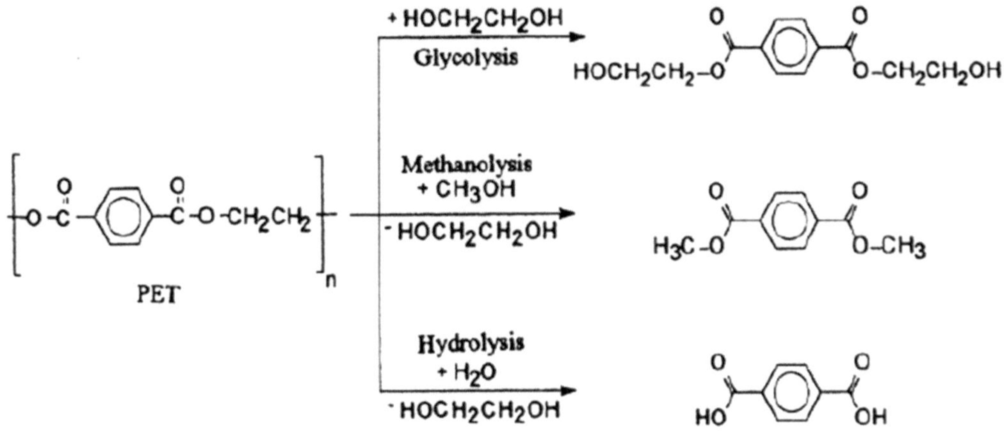

Fig. 2. Different solvolysis methods for PET depolymerization.

2.3.1 Hydrolysis

Hydrolysis involves the depolymerization of PET to terephthalic acid (TPA) and ethylene glycol by the addition of water in acidic, alkaline or neutral environment. The hydrolysis products may be used to produce virgin PET, or may be converted to more expensive chemicals like oxalic acid (Yoshioka et al., 2003). Concentrated sulfuric acid is usually used for acid hydrolysis (Brown & O'Brien, 1976; Pusztaszeri, 1982; Sharma et al., 1985; Yoshioko et al., 1994, 2001), caustic soda for alkaline hydrolysis (Alter, 1986), and water or steam for neutral hydrolysis (Campanelli et al., 1993,1994a; Mandoki, 1986). Hydrolysis is slow compared to methanolysis and glycolysis, because among the three depolymerizing agents (i.e. water, methanol, ethylene glycol), water is the weakest nucleophile. It also uses high temperatures and pressures. Another disadvantage of hydrolysis is the difficulty of recovery of the TPA monomer, which requires numerous steps in order to reach the required purity.

2.3.2 Methanolysis

Methanolysis is the degradation of PET to dimethyl terephthalate (DMT) and EG by methanol. Disadvantages of this method include the high cost associated with the separation and refining of the mixture of the reaction products. Also, if water perturbs the process, it poisons the catalyst and forms various azeotropes. Before, methanolysis and glycolysis were the methods applied on a commercial scale (Paszun, 1997), but today, it is not used for PET production anymore, and the lack of usefulness of recovering DMT rendered the methanolysis of PET to become obsolete (Patterson, 2007).

2.3.3 Glycolysis

As shown in Fig. 3, glycolysis is carried out using ethylene glycol to produce bis(2-hydroxyethyl) terephthalate and other PET glycolyzates, which can be used to manufacture unsaturated resins, polyurethane foams, copolyesters, acrylic coatings and hydrophobic dystuffs. The BHET produced through glycolysis can be added with fresh BHET and the mixture can be used in any of the two PET production (DMT-based or TPA-based) lines.

Diethylene glycol (Karayannidis et al., 2006), triethylene glycol (Öztürk & Güçlü, 2005), propylene glycol (Güclü et al., 1998; Vaidya & Nadkarni, 1987), or dipropylene glycol (Johnson & Teeters, 1991, as cited in Sinha et al., 2008) may also be used as solvent in PET glycolysis.

Besides its flexibilty, glyclolysis is the simplest, oldest, and least capital-intensive process. Because of these reasons, much attention has been devoted to the glycolysis of PET. Numerous works have been published about PET glycolysis, wherein the reaction has been conducted in a wide range of temperature and time. The works involving this process, from 1960, when Challa started to investigate the polycondensation equilibrium of melt glycolysis (Challa, 1960; As cited in Patterson, 2007), up until now when researchers are focused on developing more efficient glycolysis catalysts and investigating on the applications of the glycolysis products, will be discussed in detail in the later part of this work.

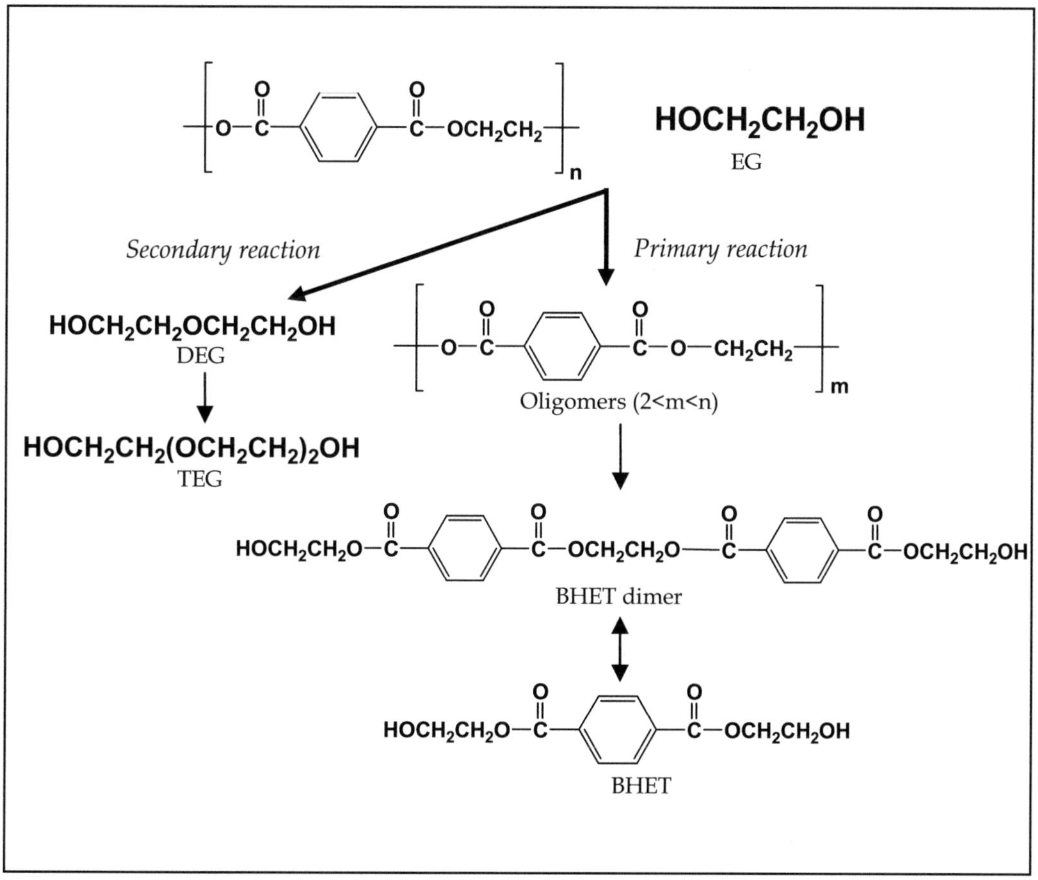

Fig. 3. Reaction scheme for the glycolysis of PET.

2.4 Quaternary recycling

Quaternary recycling represents the recovery of energy content from the plastic waste by incineration. When the collection, sorting and separation of plastics waste are difficult or economically not viable, or the waste is toxic and hazardous to handle, the best waste management option is incineration to recover the chemical energy stored in plastics waste in the form of thermal energy. However, it is thought to be ecologically unacceptable due to potential health risks from the air born toxic substances.

3. Glycolytic depolymerization of PET

Studies on the kinetics of PET glycolysis (Campanelli et al., 1994b; J. Chen & L. Chen, 1999) have shown that glycolysis without a catalyst is very slow and complete depolymerization of PET to BHET cannot be achieved. It also yields an end product that contains significant amount of other oligomers in addition to the BHET monomer. This results in difficulty in recovering the BHET monomer when it is the desired product. Thus, research efforts have been directed towards increasing the rate and BHET monomer yield by developing highly efficient catalysts and other techniques, and optimizing the reaction conditions (e.g. temperature, time, PET/EG ratio, PET/catalyst ratio). Others sought for applications of the glycolysis product without the separation of oligomers (Grzebienek & Wesolowski, 2004). Still others sought for more eco-friendly glycolytic process. Two decades after the beginning of PET glycolysis research, these efforts resulted in the significant increase in BHET monomer yield from just 65% with 8 hours reaction time to at least 90% with a significantly reduced reaction time of around 30 minutes.

3.1 Catalyzed glycolysis

The most studied method of increasing the glycolysis rate is catalysis. PET glycolysis is considered a transesterification reaction. Thus, transesterification catalysts have been applied to increase the reaction rate of PET glyclosis, with metal based catalysts being the most popular. Helwani et al. and Schuchardt et al. have listed all the catalysts that have been used before in other transesterification reactions (Helwani et al., 2009; Schuchardt et al., 1997).

Fig. 4 shows the reaction mechanisms of uncatalyzed glycolysis and that of glycolysis with metal-based catalyst (Shukla & Harad, 2005; Pingale et al., 2010). A free electron pair on the EG oxygen initiates the reaction by attacking the carbonyl carbon of the ester group of the polyester. The hydroxyethyl group of ethylene glycol then forms a bond with the carbonyl carbon of the polyester breaking the long chain into short chain oligomers and finally BHET.

The rate of glycolysis reaction depends on a number of parameters including temperature, pressure, PET/EG ratio, and the type and amount of catalyst. Also, the transformation of dimer to BHET monomer is a reversible process. Prolonging the reaction after the equilibrium of the two is attained will cause the reaction to shift backwards, increasing the amount of dimer at the expense of the BHET monomer. It is thus important to know the optimum conditions of the glycolysis reaction. With metal based catalysts (Fig. 4b), the metal forms a complex with the carbonyl group, facilitating the attack of EG on PET leading to the formation of BHET. A number of glycolytic depolymerization processes have been reported with different catalysts and different reaction conditions. We have listed the catalysts studied to date in Table 2.

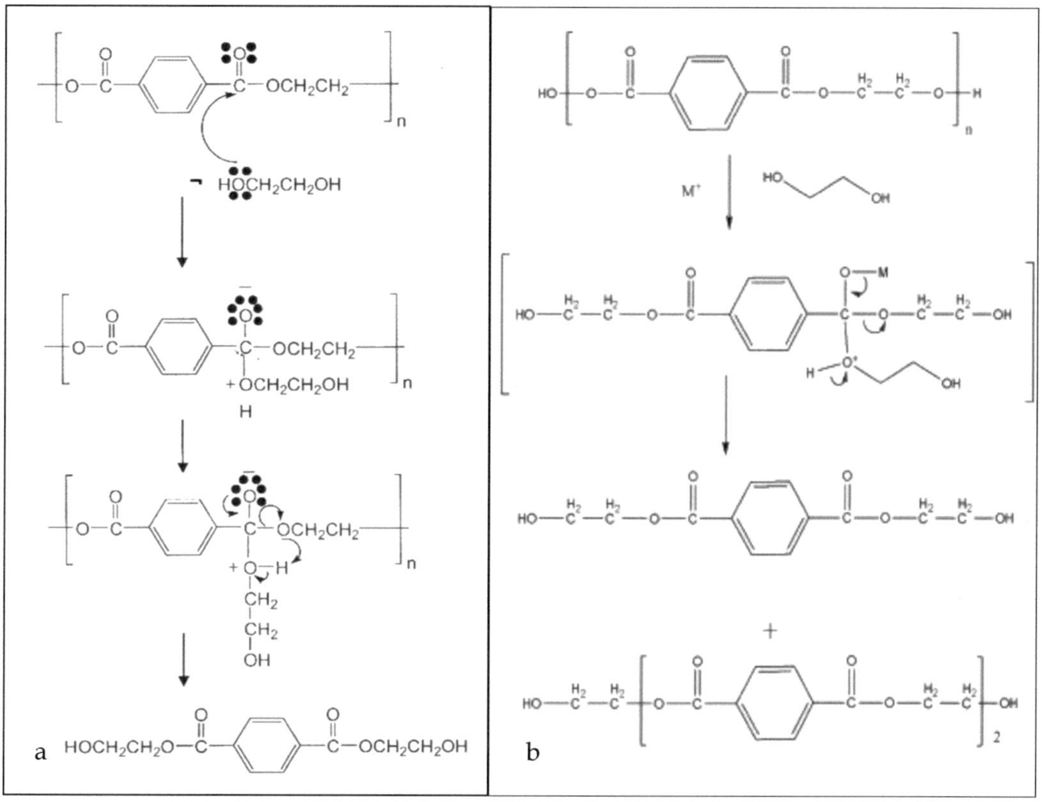

Fig. 4. Reaction mechanism of uncatalyzed (a) and catalyzed (b) PET glycolysis.

3.1.1 Metal salts

The oldest reported catalysts for PET glycolysis are metal acetates. Zinc acetate was first used by Vaidya and Nadkarni for their work dealing with the synthesis of polyester polyols from PET waste (Vaidya & Nadkarni, 1988). In 1989, Baliga and Wong further investigated the use of metal acetates (zinc, manganese, cobalt, and lead) as catalysts. They reported that zinc acetate showed best results in terms of the extent of depolymerization reactions of PET. They also observed that the equilibrium between the BHET monomer and dimer was reached after 8 hours of reaction with the temperature at 190 ^{0}C. This may be considered as the beginning of PET glycolysis catalysts research as several researches followed later.

Ghaemy and Mossaddegh verified the results obtained by Baliga and Wong, and the order of activity of the catalysts (Zn^{+2} > Mn^{+2} > Co^{+2} > Pb^{+2}) (Ghaemy & Mossaddegh, 2005). J. Chen and L. Chen studied the kinetics of PET glycolysis with zinc acetate catalyst at the same temperature, and they found out that the equilibrium between the BHET monomer and the dimer was reached after two hours, as opposed to 8 hours from Baliga and Wong (J. Chen & L. Chen, 1999). Meanwhile, C. Chen studied that of manganese acetate and found out that the best glycolysis condition for the same temperature was the reaction time of 1.5 h with 0.025 mol/kg PET (C. Chen, 2003). Xi et al. investigated the optimum condition of the reaction at 196 ^{0}C. They reported that a 3-hour reaction with EG/PET

Catalyst	BHET Yield, %	Temp, °C	Time, minutes	EG/PET Ratio	PET/Catalyst Weight Ratio	Reference
Zinc acetate	85.6	196	180	5 (w/w)	0.01	*Xi et al., 2005*
Zinc acetate	62.8 (% in the product)	200	150	2.77 (mol/mol)	0.003	*Troev et al., 2003*
Titanium phosphate	97.5 (% in the product)					
Zinc acetate	62.51	190	480	6 (mol/mol)	0.005	*Shukla & Kulkarni, 2002*
Lead acetate	61.65					
Sodium carbonate	61.5					
Sodium bicarbonate	61.94					
Acetic acid	62.42	190	480	6 (mol/mol)	0.005	*Shukla & Harad, 2004*
Lithium hydroxide	63.50					
Sodium sulfate	65.72					
Potassium sulfate	64.42					
β-zeolite	66	196	480	6 (mol/mol)	0.01	*Shukla et al., 2008*
γ-zeolite	65					
Zinc chloride	73.24	197	480	10 (mol/mol)	0.005	*Pingale et al., 2009*
Lithium chloride	59.46					
Didymium chloride	71.01					
Magnesium chloride	55.67					
Ferric chloride	56.28					
Zinc oxide on silica nanoparticle	~85	300	80	11 (mol/mol)	0.01	*Imran et al., 2011*
Magnesium oxide on silica nanoparticle	>90					
Diff. Ionic liquids	No data; 100% conversion	190	120	10 (w/w)	0.05	*Wang et al., 2009*
[bmim]OH	71.2	190	120	10	0.05	*Yue et al., 2011*

Table 2. Catalysts studied for PET glycolysis.

weight ratio of 5, and catalyst/PET weight ratio of 0.01 can deliver 85.6% BHET yield (Xi et al., 2005). Goje and Mishra also studied the optimum conditions of PET glycolytic depolymerization at 197 °C, and they reported 98.66% PET conversion with the reaction time of 90 minutes and PET particle size of 127.5 μm. The optimal PET particle size was the size at which PET weight loss was maximum. They did not measure the BHET yield though, because the reaction pathway they used produced DMT and EG instead of BHET (Goje & Mishra, 2003).

Dayang et al. later used the products from PET glycolysis catalyzed by zinc acete to make themally stable polyester resin via polyesterification with maleic anhydride and crosslinking with styrene (Dayang et al., 2006). The synthesis of unsaturated polyester resin actually dates back to 1964 (Ostrysz et al., 1964, as cited in Paszun & Spychaj, 1997). This unsaturated polyester resin was later reinforced with natural fibers in the study made by Tan et al. to produce a fiber composite with good mechanical properties (Tan et al., 2011).

Although metal salts are effective in increasing the PET glycolysis rate, it should be noted that zinc salts, and presumably other metal salts, have a catalytic effect on glycolysis of PET only below 245 °C, and apparently do not promote any further increase in the reaction rate above that temperature due to mass transfer limitations (Campanelli et al., 1994b). Thus, a need to develop new catalysts that can overcome this limitation. In 2003, Troev et al. introduced titanium (IV) phosphate as a new catalyst. They reported that glycolysis in the presence of the new catalyst was faster compared to that with zinc acetate. Their data showed that at 200°C, 150 minutes reaction time and 0.003 catalyst/PET weight ratio, the glycolyzed products from titanium (IV) phosphate catalyzed reaction consisted of 97.5% BHET, which was significantly higher than that of zinc acetate, which was 62.8 % (Troev et al., 2003).

Since lead and zinc are heavy metals known to have negative effects on the environment, Shukla's group started to develop milder catalysts that are comparatively less harmful to the environment. They started with mild alkalies, sodium carbonate and sodium bicarbonate, and reported that the monomer yields (Refer to Table 2) were comparable with those of the conventional zinc and lead acetate catalysts (Shukla & Kulkarni, 2002). They also reported glacial acetic acid, lithium hydroxide, sodium sulfate, and potassium sulfate to have comparable yields (Table 2) with those of the conventional heavy metal catalysts (Shukla & Harad, 2005). They recently used the recovered BHET monomer to produce useful products such as softeners and hydrophobic dyes for the textile industry (Shuka et al., 2008, 2009). López-Fonseca et al. also used these eco-friendly catalysts in their study of catalyzed glycolysis kinetics (López-Fonseca et al., 2010, 2011). The latest catalysts that Shukla's group developed are inexpensive and readily available metal chlorides, wherein zinc chloride reportedly gave the highest BHET yield equal to 73.24% (Pingale et al., 2010).

3.1.2 High surface area catalysts: Nanocomposites

In 2008, Shukla et al. reported new addition to their set of eco-friendly catalysts in the form of zeolites (Shukla et al., 2008). Zeolites have been used as catalysts in other reactions before, and their catalytic activity can be credited to their large surface area in mesopores and micropores that provide numerous active sites. Their result, however, showed that the BHET yield (Table 2) did not deliver any significant improvement from the other catalysts they previously reported.

Looking back to the number of catalysts previously discussed in this work, it is noticeable that the BHET yield never reached the 90% mark. The restricted amount of BHET yield may be because the reaction was not performed at temperatures above 245°C, since the previously reported catalysts lose their effectiveness at increased temperatures anyway.

With the aim of increasing the BHET monomer yield at reduced reaction time, our group developed catalysts that are highly selective and can work at elevated temperatures - metal oxide catalysts. Metal oxides as glycolysis catalysts could provide a better alternative to conventional catalysts in that they have high mechanical strength, are thermally stable, and are cost effective. Metal oxides were used for other transesterification reactions before (Helwani et al., 2009; Singh & Fernando, 2007), but they had not been applied in PET glycolysis. In order to increase the metal oxide catalysts' efficiency, we tried to increase the surface area of active sites by fabricating them at nanoscale. Besides increasing the surface area of the active sites, it is known that at nanoscale, the intrinsic properties of the catalysts may change, leading to increased effectiveness compared to that of their bulk conterpart (Heiz & Landman, 2007; Niederberger & Pinna, 2009). Fig. 5 (Imran et al., 2011; Wi et al, 2011) shows TEM images of the fabricated 60 nm (a) and 150 nm (b) silica nanoparticle used as supports and the supports with the deposited metal oxide catalysts. The metal oxide catalysts were deposited on the silica nanoparticle supports via a simple ultrasound assisted precipitation method. Good deposition was observed especially for cerium oxide and manganese oxide.

The oxides of zinc, manganese, and cerium deposited on silica nanoparticle support were used as catalysts in a glycolytic reaction performed at 300 °C and 1.1 MPa with EG/PET molar ratio of 11, and PET/catalyst weight ratio of 0.01. The reaction reached equilibrium after 80 minutes, and the highest BHET yield reached more than 90%. Moreover, we found out that the smaller the size of the support is, the better is the distribution of the catalysts on the support. This could be due to the higher chances of contact between the catalyst and the support because of the higher surface-area-to-volume ratio for smaller supports. The better distribution of the catalysts resulted in higher catalytic activity.

3.1.3 Recyclable catalyst: Ionic liquids

It has not been long since ionic liquids were applied as catalyst for PET glycolysis when Wang et al. initiated the study and first reported its use in 2009 (Wang et al., 2009a). The main advantage of ionic liquids over conventional catalysts like metal acetates is that the purification of the glycolysis products is simpler.

They prepared different ionic liquids and performed glycolysis reactions in the presence of these ionic liquids at atmospheric pressure with different temperature and time. 100 % conversion of PET was achieved after 8 hours at a temperature of 180 °C, with the 1-butyl-3-methylimidazolium bromide ([bmim] Br) being the best catalyst in terms of PET conversion and ease and cost of preparation. They concluded that the BHET purity from their method was high. They did not, however quantitatively measure the BHET yield from their experiment. After this, they extended their research by investigating the reusability of the ionic liquid catalysts and kinetics of the PET degradation by ionic liquid alone. They concluded that the catalysts can be used repeatedly, that the degradation reaction is first-order with activation energy equal to 232.79 kJ/mol, and that it can potentially replace the traditional organic solvents used in PET degradation (Wang et al., 2009b). Recently, they successfully applied Fe-containing magnetic ionic liquid as a catalyst for PET glycolysis. They reported that this catalyst has better catalytic activity than the conventional metal salts or the pure ionic liquid with the amount of catalyst affecting the PET conversion and BHET selectivity (Wang et al., 2010). Yue et al. followed this study by using basic ionic liquid, and reported that basic [bmim]OH exhibits higher catalytic activity than [bmim] Br and [bmim] Cl.

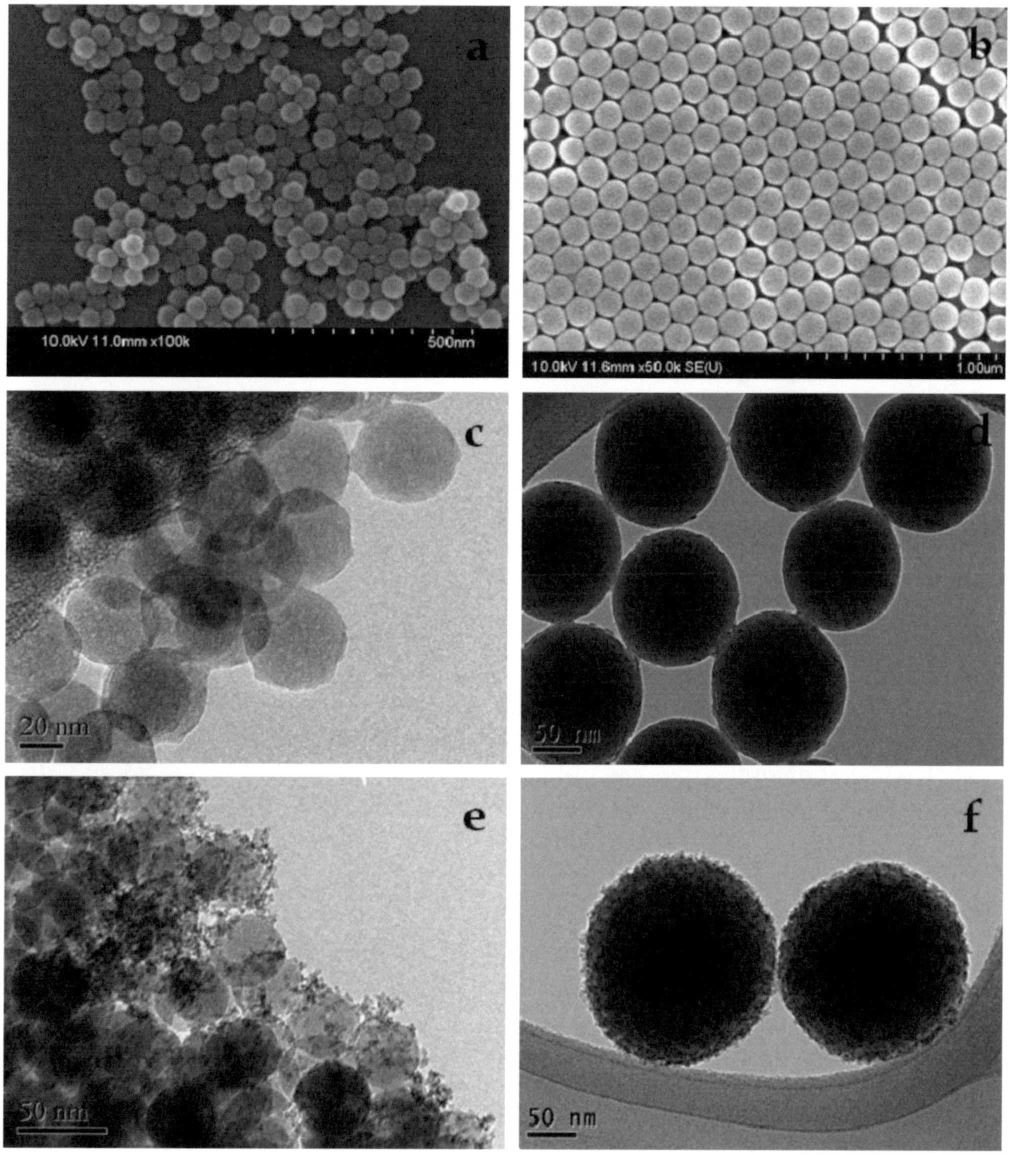

Fig. 5. TEM images of the (a) 60 nm silica support fabricated via water-in-oil microemulsion method, (b) 150 nm silica support from Stober method, (c) ZnO on 60 nm silica support, (d) ZnO on 150 nm silica support, (e) CeO_2 on 60 nm silica support, and (f) Mn_3O_4 on 150 nm silica support.

They attained 100 % PET conversion with 71.2% BHET yield by performing the glycolysis at 190 ^{0}C for 2 hours with EG/PET molar ratio of 10 and catalyst/PET weight ratio of 0.05 (Yue et al., 2011). As can be deduced in this study, the recoverability and reusability of ionic liquid catalyst permits the use of higher amount of catalyst.

3.2 Solvent-assisted glycolysis

In 1997, Güçlü et al. added xylene in the zinc acetate catalyzed PET glycolysis reaction, and obtained 80% BHET yield, which was higher than the yield from that without xylene. The main objective of xylene was initially to provide mixability to the PET-glycol mixture. At temperatures between 170 ^{0}C and 225 ^{0}C, EG dissolves sparingly in xylene while it dissolves readily in PET. Meanwhile, the glycolysis products are soluble in xylene. Therefore, as the reaction progressed, the glycolysis products moved from the PET-glycol phase to the xylene phase, shifting the reaction to the direction of depolymerization (Güçlü et al., 1997). Sole publication is available for this PET glycolysis technique. Further investigations may have been prevented by the reason that organic solvents are harmful to the environment and massive use of these solvents is not a very attractive idea.

3.3 Supercritical Glycolysis

The use of supercritical conditions has been explored earlier in PET hydrolysis (Sato et al., 2006) and methanolysis (Minoru et al., 2005; Yang et al., 2002), but only recently for glycolysis (Imran, et al., 2010). The main advantage of the use of supercritical fluids in a reaction is the elimination of the need of catalysts, which are difficult to separate from the reaction products. It is also environment friendly. Our group investigated the use of EG in its supercritical state (T_c = 446.70 ^{0}C, P_c = 7.7 MPa) (Imran et al., 2010). Supercritical process was carried out at 450 ^{0}C and 15.3 MPa, and the results were compared with those from the subcritical processes carried out at 350 ^{0}C and 2.49 MPa, and 300 ^{0}C and 1.1 MPa. Compared to the subcritical process, the BHET-dimer equilibrium was achieved much earlier for supercritical process: a maximum BHET yield of 93.5 % was reached in mere 30 minutes. Owing to high temperature and pressure, supercritical glycolysis delivered a very high yield of BHET while suppressing the yield of the side products (0.69% DEG yield and almost negligible formation of oligomers, BHET dimer, and TEG). If economically feasible, supercritical glycolysis may be able to replace catalyzed glycolysis.

3.4 Microwave-assisted glycolysis

Beyond eco-friendly catalysts, Pingale and Shukla extended their study to the use of unconventional heating source of microwave radiations. The employment of microwave radiations as heating source drastically decreased the time for the completion of reaction from 8 hours to just 35 minutes. However, it did not increase the BHET monomer yield (Pingale and Shukla, 2008). The use of more efficient catalyst along with microwave irradiation heating may be able to increase the BHET yield while decreasing the reaction time.

4. Conclusion: Challenges and opportunities

From the discovery of PET in 1940s and the start of PET chemical recycling in 1950s that attracted great interest from the research community, PET glycolysis has gone a long way, back when zinc acetate was first used as catalyst to obtain about 60% BHET yield after 8 hours of reaction until when silica nanoparticle-supported metal oxide catalysts were applied to obtain at least 90% yield after 80 minutes. Studies have already dealt with most of the problems dealing with PET glycolysis, including unfeasibility of operation due to long reaction times, low yields, severe conditions, and pollution problems. Researchers have developed catalysts to increase the rate and BHET monomer yield, catalysts that are environmentally friendly, catalysts that can be recovered and reused, a method that does not require catalysts, and many others.

However, PET glycolysis is still far from its peak. Though researchers have found ways to solve each problem separately, there is still no way to solve them all simultaneously. For instance, eco-friendly catalysts deliver lower yields compared to the not-so-eco-friendly ones (e.g. metal oxides). The main challenge that stands now is to deliver an efficient, sustainable, environment friendly, and less energy demanding way to chemically recycle PET. This may be an opportunity for researchers try to develop efficient and highly selective catalysts that can be recovered and reused. There may be many other ways to break the boundaries, and with the rapid advancement of technologies like nanotechnology, solutions may be discovered in the near future. We believe that by exploring the possibilities of technologies that have not yet been applied, great advancements on PET glycolysis can be made. For instance, it has been reported that ultrasound can induce the scission of polymer chains (Kuijpers et al., 2004). Ultrasound assisted depolymerization has been applied to other depolymerization processes before (Sayata & Isayev, 2002; Sayata et al., 2004; Shim et al., 2002), but it has not been explored in PET glycolysis yet. Nanotechnology, which is growing by leaps and bounds may also be exploited to develop more highly efficient glycolytic depolymerization of PET.

5. Acknowledgement

This work was supported by the Resource Recyling R&D Center sponsored by 21C Frontier R&D Program, the Center for Ultramicrochemical Process Systems sponsored by KOSEF, the Basic Science Research Program through a National Research Foundation of Korea (NRF) grant funded by the Ministry of Education, Science and Technology (2010-0025671).

6. References

Achilias, D. & Karayannidis, G. (2004). The chemical recycling of PET in the framework of sustainable development, *Water, Air, & Soil Pollution: Focus*, Vol 4, No. 4-5, (October 2004), pp. 385-396, ISSN 1567-7230

Aguado, J. & Serrano D. (1999). *Feedstock Recycling of Plastic Wastes,* The Royal Society of Chemisty, ISBN 0-85404-531-7, United Kingdom

Al-Salem, S. (2009). Establishing an integrated databank for plastic manufacturers and converters in Kuwait, *Waste Management,* Vol. 29, No. 1, (January 2009), pp. 479-484

Al-Salem, S., Lettieri, J., Baeyens, J. (2009) Recycling and recovery routes of plastic solid waste (PSW): A review, *Waste Management,* Vol. 29, No. 10, (October 2009), pp. 2625-2643

Alter, H. (1986). Disposal and Reuse of Plastics, In: *Encyclopedia of Polymer Science and Engineering,* pp. 103-128, Herman Mark, Wiley Interscience, ISBN 978-0471880981 New York

Awaja, F. & Pavel, D. (2005). Recycling PET. *European Polymer Journal,* Vol. 41, No. 7, (July 2005), pp 1453-1477, ISSN 0014-3057

Brown Jr., G. & O'Brien, R. (1976). Method of recovering terephthalic acid and ethylene glycol from polyester materials. United States Patent 3952053

Caldicott, R. (1999). The Basics of Stretch Blow Molding PET Containers. *Plast. Eng.* Vol. 55, No. 1, (January 1999), pp. 35-40, ISSN 0091-9578

Campanelli, J., Kamal, M., & Cooper, D. (1994b). Kinetics of glycolysis of poly(ethylene terephthalate) melts. *J. Appl. Polym. Sci.,* Vol. 54, No. 11, (December 1994), pp. 1731-1740, ISSN 1097-4628

Campanelli, J., Kamal, M., Cooper, D. (1993). A kinetic study of the hydrolytic degradation of polyethylene terephthalate at high temperatures, *J. Appl. Polym. Sci.,* Vol. 48, No. 3, (April 1993), pp. 443-451, ISSN 1097-4628

Campanelli, J., Kamal, M., Cooper, D. (1994a). Catalyzed Hydrolysis of poly(ethylene terephthalate) melts. *J. Appl. Polym. Sci.,* Vol. 53, No. 8, (August 1994), pp. 985-991, ISSN 1097-4628

Carraher, C. (200). *Polymer Chemistry,* (5th Ed), Marcel Dekker, ISBN 978-0-82470-3622, New York

Carta, D., Cao, G., & D'Angeli, C. (2003). Chemical Recycling of Poly(ethylene terephthalte) (PET) by Hydrolysis and Glycolyis, *Environmental Science And Pollution Research,* Vol. 10, No. 6, pp. 390-394, ISSN 1614-7499

Chen, C. (2003). Study of Glycolysis of Poly(ethylene terephthalate) Recycled from Postconsumer Soft-Drink Bottles. III. Further Investigation. *J. Appl. Polym. Sci.,* Vol. 87, No. 12, (March 2003), pp. 2004-2010, ISSN 1097-4628

Chen, J. & Chen, L. (1999). The Glycolysis of Poly(ethylene terephthalate). *J. Appl. Polym. Sci.,* Vol. 73, No. 1, (April 1999), pp.35-40, ISSN 1097-4628

Dayang, R., Ahmad, I., & Ramli, A. (2006). Chemical Recycling of PET Waste from Softdrink Bottles to Produce a Thermosetting Polyester Resin. *Malaysian Journal of Chemistry,* Vol. 8, No. 1, pp. 22-26

Ghaemy, M. & Mossaddegh, K. (2005). Depolymerization of poly(ethylene terephthalate) fibre waste using ethylene glycol. *Polymer Degradradation and Stability,* Vol. 90, No. 3, (December 2005), pp. 570-576 (2005), ISSN 0141-3910

Goje, A. & Mishra, S. (2003). Chemical Kinetics, Simulation, and Thermodynamics of Glycolytic Depolymerization of Poly(ethylene terephthalate) Waste with Catalyst Optimization for Recycling of Value Added Monomeric Products. *Macromol. Mater. Eng.,* Vol. 288, No. 4 (April 2003) pp. 326-336, ISSN 1439-2054

Grzebienek, K. & Wesolowski, J. (2004). Glycolysis of PET waste and the Use of Glycolysis Products in the Synthesis of Degradable Co-polyesters. *Fibres & Textiles in Eastern Europe,* Vol. 12, No. 2(46), (April/June 2004), pp. 19-22, ISSN 1230-3666

Güçlü, G., Kas¸göz, A., Özbudak, S., Özgümüs, S., & Orbay M. (1998). Glycolysis of poly(ethylene terephthalate) wastes in xylene. *J Appl Polym Sci,* Vol. 69, No. 12, (September 1998), pp. 2311-2319, ISSN 1097-4628

Gupta, V. & Bashir, Z. (2002). PET Fibers, Films, and Bottles, In: *Handbook of Thermoplastic Polyesters,* Stoyko Fakirov, p. 320, Wiley-VCH, ISBN 978352730113, Michigan

Harris, J. (2001), *A Survey of sustainable development: social and economic dimensions,* Island Press, ISBN 978-1-559-63863-0 , Washington, DC

Heiz, U. & Landman U., (2007). *Nanocatalysis*, Springer-Verlag, ISBN 978-1-61583-152-4, New York

Helwani, Z., Othman, M., Aziz, N., Kim, J., & Fernando, W. (2009). Solid Heterogeneous catalysts for transesterification of triglyceride with methanol: a review. *Applied Catalysis A: General,* Vol. 363, No. 1-2, (July 2009), pp. 1-10, ISSN 0926-860X

Hopewell, J., Dvorak, R., & Kosior, E. (2009). Plastics recycling: challenges and Opportunities. *Phil. Trans. R. Soc. B.* Vol. 364, No. 1526 (July 2009) pp. 2115–2126, ISSN 1471-2970

IHS (January 2011). WP Report : Polyethylene Terephthalte (PET), Melt Pahe, In : *World Petrochemicals Report,* August 2011, Available from : http://www.sriconsulting.com/WP/Public/Reports/pet/

ILSI Europe Report Series (2000). *Packaging Materials: 1. Polyethylene Terephthalate (PET) for Food Packaging Applications,* ISLI Press, ISBN 1-57881-092-2, Brussels

Imran, M., Kim, B., Han, M., Cho, B., Kim, D. (2010) Sub- and supercritical glycolysis of polyethylene terephthalate (PET) into the monomer bis(2-hydroxyethyl) terephthalate (BHET). *Polymer Degradradation and Stability,* Vol. 95, No. 9, (September 2010), pp. 1686-1693, ISSN 0141-3910

Imran, M., Lee, K., Imtiaz, Q., Kim, B., Han, M., Cho, B., Kim, D. (2011). Metal-Oxide-Doped Silica Nanopaticles for the Catalytic Glycolysis of Polyethylene Terephthalate. *J. Nanosci. Nanotechnol,* Vol. 11, No. 1, (January 2011), pp. 824-828, ISSN 1533-4899

Janssen, F. & van Santen, R. (1999). *Environmental Catalysis,* Imperial College Press, ISBN 978-1-860-94125-2, London

Karayannidis, G. & Archilias, D. (2007) Chemical Recycling of Poly(ethylene terephthalate). *Macromol. Mater. Eng.,* Vol. 292, No. 2, (February 2007), pp. 128-146, ISSN 1439-2054

Karayannidis, G., Nikolaidis, A., Sideridou, I., Bikiaris, D., & Archilias, D. (2006). Chemical Recycling of PET by Glocolysis: Polymerization and Characterization of the Dimethacrylated Glycolysate. *Macromol. Mater. Eng.,* Vol. 291, No. 11, (November 2006) pp. 1338-1347, ISSN 1439-2054.

Kuijpers, M., Iedema, P., Kemmere, M., & Keurentjes, T. (2004). The mechanism of cavitation-induced polymer scission; experimental and computational verification. *Polymer.* Vol. 45, No. 19, (September 2004), pp. 6461-6467, ISSN 0032-3861

López-Fonseca, R., Duque-Ingunza, I., & de Rivas, B. (2011) Kinetics of catalytic glycolysis of PET wastes with sodium carbonate. *Chemical Engineering Journal,* Vol. 168, No. 1, (March 2011), pp. 312-320, ISSN 1385-8947

López-Fonseca, R., Duque-Ingunza, I., de Rivas, B., Arnaiz, S., & Gutiérrez-Ortiz, J. (2010). Chemical Recycling of post-consumer PET wastes by glycolysis in the presence of metal salts. *Polymer Degradradation and Stability,* Vol. 95, No. 6, (June 2010), pp. 1022-1028, ISSN 0141-3910

Mandoki, J. (1986). Depolymerization of Condensation Polymers. United States Patent 4605762

Minoru, G., Tomoko, I., Mitsuru S., Motonobu, G. & Hirose, T. (2005). Depolymerization Mechanism of Poly(ethylene terephthalate in Supercritical Methanol. *Ind. Eng. Chem. Res.,* Vol. 44, No. 11, (May 2005), pp. 3894-3900, ISSN 1520-5045

Niederberger, M. & Pinna, N. (2009). *Metal Oxide Nanoparticles in Organic Solvents: Synthesis, Formation, Assembly and Application,* Springer-Verlag, ISBN 978-1-84882-670-0, New York

Nikles, E. & Farahat, M. (2005). New motivation for the depolymerization products derived from poly(ethylene terephthalate) (PET) waste, *Macromol. Mater. Eng.*, Vol. 290, No. 1, (January 2005) pp. 13-30, ISSN 1439-2054

Nir, M., Miltz, J., & Ram, J. (1993) Update on plastics and the environment: progress and trends. *Plast. Eng.* Vol. 49, No. 3, (March 1993), pp. 75- 93, ISSN 0091-9578

Olabisi, O. (1997). *Handbook of Thermoplastics,* Marcel Dekker, ISBN 0-8247-9797-3, New York

Öztürk, Y. & Güçlü, G. (2005). Unsaturated Polyester Resins Obtained from Glycolysis Products of Waste PET. *Polymer-Plastics Technology and Engineering,* Vol. 43, No. 5, pp. 1539-1552, ISSN 1525-6111

Paszun, D. & Spychaj, T. (1997). Chemical Recycling of Poly(ethylene terephthalate). *Ind. Eng. Chem. Res.,* Vol. 36, No. 4, (April 1997), ISSN 1520-5045

Patterson, J. (2007) Continuous Depolymerization of Poly(ethylene terephthalate) via Reactive Extrusion, Accessed September 2010, Available from: http://www.lib.ncsu.edu/resolver/1840.16/3783

Pingale, N. & Shukla, S. Microwave assisted ecofriendly recycling of poly(ethylene terepthalate) bottle waste. *European Polymer Journal.* Vol. 44, No. 12, (December 2008), pp. 4151-4156, ISSN 0014-3057

Pingale, N., Palekar, V., & Shukla S. (2010). Glycolysis of Postconsumer Polyethylene Terephthalate Waste. (2010). *J. Appl. Polym. Sci.,* Vol. 115, No. 1, (January 2010), pp. 249-254, ISSN 1097-4628

Pusztaszeri, S. (1982). Method for recovery of terephthalic acid from polyester scrap. United State Patent 4355175

Rieckman, T., Poly(ethylene terephthalate) Polymerization - Mechanism, Catalysis, Kinetics, Mass Transfer and Reactor Design, In : *Modern Polyesters : Chemistry and Technology of Polyesters and Copolyesters.,* J. Scheirs and T. Long, pp 31-106, John Wiley & Sons, Ltd, ISBN 978-0-471-49856-8, New York

Sato, O., Arai, K., & Shirai, M. (2006). Hydrolysis of poly(ethylene terephthalate) and poly(ethylene 2,6-naphthalene dicarboxylate) using water at high temperature: effect of proton on low ethylene glycoly yield. *Catalysis Today,* Vol. 111, No. 3-4, (February 2006), ISSN 0920-5861

Sayata, G. & Isayev, A. (2002). Recycling of Unfilled Polyurethane Rubber Using High-Power Ultrasound. *J Appl Polym Sci,* Vol. 88, No. 4, (April 2003), pp. 980-989, ISSN 1097-4628

Sayata, G., Isayev, A., & Meerwall, E. (2004). Effect of ultrasound on thermoset polyurethane: NMR relaxation and diffusion measurements. *Polymer.* Vol. 45, No. 11, (May 2004), pp. 3709-3720, ISSN 0032-3861

Scheirs, J. & Long T.E. (2003). *Modern polyesters: Chemistry and Technology of Polyesters and Copolyester,* John Wiley & Sons Ltd, ISBN 978-0-471-49856-8, England

Scheirs, J. (1998). *Polymer Recycling: Science, Tecnology and Application,* John Wiley & Sons, ISBN 0-471-97054-9, New York

Scheirs, J., & Kaminsky, W. (2006). *Feedstock recycling and pyrolysis of waste plastics: Converting waste plastics into diesel and other fuels,* John Wiley & Sons Ltd, ISBN 0470021527, West Sussex

Schuchardt, U., Sercheli, R., & Varga, R. (1998). Transesterification of Vegetable Oils: a Review. *J. Braz. Chem. Soc.,* Vol. 9, No. 3, pp. 199-210, (May 1998), ISSN 0103-5053

Sharma, N., Vadiya, A., & Sharma, P. (1985). Recovery of Pure Terephthalic Acid from Polyester Materials. Indian Patent 163385

Shim, E., Sayata, G., & Isayev, A. (2002). Formation of bubbles during ultrasonic treatment of cured poly(dimethyl siloxane). *Polymer.* Vol. 43, No. 20, (September 2002), pp. 5535-5543, ISSN 0032-3861

Shukla, S. & Harad., A. (2005). Glycolysis of Polyethylene Terephthalate Waste Fibers. *J. Appl. Polym. Sci.,* Vol. 97, No. 2, (July 2005), pp. 513-517, ISSN 1097-4628

Shukla, S. & Kulkarni K. (2002). Depolymerization of Poly(ethylene terephthalate) Waste. *J. Appl. Polym. Sci.,* Vol. 85, No. 8, (August 2002), pp. 1765-1770, ISSN 1097-4628

Shukla, S., Harad, A., & Jawale, L. (2008). Recycling of waste PET into useful textile auxiliaries. *Waste Manage,* Vol. 28, No. 1, ISSN 0956-053X

Shukla, S., Harad, A., & Jawale, L. (2009). Chemical recycling of PET waste into hydrophobic textile dyestuffs. *Polymer Degradradation and Stability,* Vol. 94, No. 4, (April 2009), pp. 604-609, ISSN 0141-3910

Shukla, S., Palekar, V., & Pingale, N. (2008). Zeolite Catalyzed Glycolysis of Poly(ethylene terephthalate) Bottle Waste. *J. Appl. Polym. Sci.,* Vol. 110, No. 1, (October 2008), pp. 501-516, ISSN 1097-4628

Singh, A. & Fernando, S. (2007). *Chem. Eng. Technol.,* Vol. 30, No. 12, (December 2007), pp 1716-1720, ISSN 1521-4125

Sinha, V., Patel, M., & Patel, J. (2008). PET waste management by chemical recycling: A review. *J .Polym Environ,* Vol. 18, No.1, (September 2008), pp. 8-25, ISSN 1572-8900

Tan, C., Ahmad, I., & Heng, M. (2011). Characterization of polyester composites from recycled polyethylene terephthalate reinforced with empty fruit bunch fibers. *Materials & Design,* Vol. 32, No. 8-9, (September 2011),pp. 4493-4501 , ISSN 1873-4197

Thompson, R., Swan, S., Moore, C., & vom Saal, F. (2009). Our plastic age. *Phil. Trans. R. Soc. B,* Vol. 364, No. 1526, (July 2009), 1973–1976, ISSN 1471-2970

Troev, K., Grancharov, G., Tsevi, R., & Gitsov, I. (2003). A novel catalyst for the glycolysis of poly(ethylene terephthalate), *J. Appl. Polym. Sci.,* Vol. 90, No. 4, (October 2003), pp. 1148–1152, ISSN 1097-4628

Vaidya, U. & Nadkarni, V. (1987). Unsaturated polyesters from PET waste: Kinetics of polycondensation. *J. Appl. Polym. Sci.,* Vol. 34, No. 1, (July 1987), pp. 235-245, ISSN 1097-4628

Vaidya, U. & Nadkarni, V. (1988). Polyester polyols for polyurethanes from pet waste: Kinetics of polycondensation. *J. Appl. Polym. Sci.,* Vol. 35, No. 3, (February 1988), pp. 775-785, ISSN 1097-4628

Vereinigte Glanzstoff-Fabriken (1956). Conversion of Poly-(ethylene terephthalate) into dimethyl terephthalate. Brit. Patent 755,071

Vereinigte Glanzstoff-Fabriken (1957). Dimethyl terephthalate. Brit. Patent 787, 554

Wang, H., Li, Z., Liu, Y., Zhang, X., & Zhang, S. (2009a). Degradation of poly(ethylene terephthalate) using ionic liquids. *Green Chem.,* Vol. 11, No, 10 (October 2009), pp. 1568-1575, ISSN 1463-9262

Wang, H., Liu, Y., Li, Z., Zhang, X., Zhang, S., & Zhang, Y. (2009b). Glycolysis of poly(ethylene terephthalate) catalyzed by ionic liquids. *European Polymer Journal,* Vol. 45, No. 5, (May 2009), pp. 1535-1544, ISSN 0014-3057

Welle, F. (2011). Twenty years of PET bottle to bottle recycling - An overview. *Resources, Conservation and Recycling,* Vol. 55, No. 11, (September 2011), pp. 865-875, ISSN 0921-3449

Wi, R., Imran, M., Lee, G., Yoon, S., Cho, B., & Kim, D. (2011). Effect of Support Size on the Catalytic Activity of Metal-Oxide-Doped Silica Particles in the Glycolysis of Polyethylene Terephthalate. *J. Nanosci. Nanotechnol,* Vol. 11, No. 7, (July 2011), pp. 6544-6549, ISSN 1533-4899

World Commision on Environmental Development (1987) Our Common Future, Chapter 2: Towards Sustainable Development, In: *Our Common Future, Report of the World Commission on Environment and Development A/42/427,* Accessed August 2011, Available from: http://www.un-documents.net/ocf-02.htm

Xi, G., Lu, M., & Sun, C. (2005). Depolymerization of waste PET into monomer of BHET. *Polymer Degradradation and Stability,* Vol. 87, No. 1, (January 2005), pp. 117-120, ISSN 0141-3910

Yang, Y., Liu, Y., Xiang, H., Xu, Y., & Li, Y. (2002). Study on methanolic depolymerization of PET with supercritical methanol for chemical recycling. *Polymer Degradradation and Stability,* Vol. 75, No. 1, (January 2003), pp. 185-191, ISSN 0141-3910

Yoshioka, T., Ota, M., & Okuwaki, A. (2003). Conversion of a Used Poly(ethylene terephthalate) Bottle into Oxalic Acid and Terephthalic Acid by Oxygen Oxidation in Alkaline Solutions at Elevated Temperatures. *Ind. Eng. Chem. Res.,* Vol. 42, No. 4 (February 2003), pp. 675-679, ISSN 1520-5045

Yoshioko, T., Motoki, T., & Okuwaki, A. (2001). *Ind. Eng. Chem. Res.,* Vol. 40, No. 1, (January 2001), pp. 75-79, ISSN 1520-5045

Yoshioko, T., Sato, T., & Okuwaki, A. (1994). Hydrolysis of PET waste by sulfuric acid acid at 150 ^{0}C for a Chemical Recycling. *J Appl Polym Sci,* Vol. 52, No. 9, (May 1994), pp. 1353-1355, ISSN 1097-4628

Yue, Q., Wang, Z. Zhang, L., Ni, Y., & Jin, Y. (2011). Glycolysis of poly(ethylene terephthalate) using basic ionic liquids catalysts. *Polymer Degradradation and Stability,* Vol. 96, No. 4, (April 2011), pp. 399-403, ISSN 0141-3910

3

Overview on Mechanical Recycling by Chain Extension of POSTC-PET Bottles

Doina Dimonie[1], Radu Socoteanu[2], Simona Pop[1], Irina Fierascu[1], Radu Fierascu[1], Celina Petrea[3], Catalin Zaharia[3] and Marius Petrache[1]

1. Introduction

The post consumer poly(ethylene therephtalate) bottles (POSTC-PET) can be recycled by chemical or / and mechanical processes. The POSTC-PET chemical recycling is wide-spread in Europe and is based on depolycondensation of secondary polymers and usage of the resulted products for the purposes of the fibre and unwoven material industry. The POST - PC mechanical recycling requires a phase transformation (melting) and can be attained without or with polymer up-gradation (Mancini, 1999; Akovali, 1988; Belletti, 1997; Ehrig, 1992; Erema, 2002; Firas, 2005; Sandro & Mari, 1999; Scheirs, 1998; Awaja, 2005).

The well-known worldwide POSTC-PET mechanical and chemical recycling ways are:

1. **Resorption back into the bottles manufacture.** After getting flakes, POSTC-PET is mechanically and /or chemically recycled into bottles for non-food products (soap, cosmetics, and cleaning agents). In 2004, around 50 % of the PET recycled in this way was processed;
2. **Re-use in the fibre and un-woven materials industry** for obtaining insulation membranes for roofs, shoe soles, filters, covers for car inner compartments. This direction is the most popular in Europe (Monika, 2007; Morawiec, 2002);
3. **Processing into thin sheets for thermoforming** refer only to the flakes resulted from POSTC-PET selective collected by colour. It is appreciated that the sheets obtained from such material can undergo a high degree of stretching during thermoforming in order to shape packaging cases such as transport trays for tomatoes, eggs and strawberries etc.;
4. **Up-gradation by melt processing compounding.** Although PET has excellent usage properties, because of certain characteristics such as low glass transition, low crystallization speed (for certain types) and low impact resistance, in order to be

[1]*Research and Development national Institute for Chemistry and Petrochemistry –ICECHIM, Spl.Independentei, sector 6, Bucharest, Romania*
[2]*"IC Murgulescu" Institute, Spl.Independentei, sector 6, Bucharest, Romania*
[3]*"Politehnica" University, Clea Victoriei, Bucharest, Romania*

mechanically recycled into performing products, it is compounded at melt processing. However the results are not spectacular.

5. **Chemical recycling** is related to the recovery of the chemical compounds based on following depolycondensation particular reactions glycolysis, methanolysis, hydrolysis, acidolysis, amynolysis,etc. (Carta, 2001; Karayannidis, 2003; Minoru, 2003;);

In spite of long lasting efforts, because of the low cost and low performance applications of the recycled material, at present, the widely accepted opinion is that the POSTC-PET mechanical recycling without a structural up - gradation is not an efficient procedure.

The chapter presents an overview on the structural up-grading of POSTC-PET by macromolecular chain extension during mechanical recycling (reactive processing), a procedure considered efficient for the enhancement of its properties.

2. Parameters controlling the POSTC – PET mechanical recycling

The main parameters controlling the POSTC-PET mechanical recycling are: the contamination level and the degradation degree.

2.1 POSTC – PET contamination

POSTC-PET contamination can be of the following three types: *macroscopic and microscopic physical contamination and chemical contamination.*

Macroscopic physical contamination of POSTC-PET is easy to clear off as it consists of dust, glass chops, stones, adhesives, product residues, plastics such as PVC and PE, earth impregnation due to improperly storage.

Microscopic physical contamination is more difficult to clear off especially because is partially attached to the bottle because it is about adhesive or other impregnated impurities resulted after abrasion or impact. These impurities break the thread either during granulation in the melt processing or during the spinning in the fibre industry. This leads to decrease the quality and productivity of the recycling.

Chemical contamination is the result of *adsorption* of flavouring, oil, pesticides, household chemicals, and fuel if the bottles were re-filled with such products in a secondary utilization. The proportion of POSTC-PET interaction with these compounds depends on the diffusion behaviour of contaminants and the sorption properties of the polymer. The removing of these contaminants implies *undergoing the reverse processes, namely desorption*. The adsorbed chemical impurities into the polymer *settle on the risk potential of* POSTC-PET mechanically recycled especially if the food packages are targeted. The recycling by desorption can not be considered because of its very low productivity, this process being an extremely slow one. For diminishing as much as possible the impurity content, the POSTC-PET melts are filtered during the mechanical recycling at extrusion, before passing throuth the nozzle, using particular filters (Yang Tang & Menachem, 2008).

POSTC - PET requests a severe control of the contamination level especially if it is recycled into food packaging. Currently, the impurity content limits are established and generally accepted for POSTC-PET recycling as food and non-food packaging (EGPMFC, 1999; Franz,

2004). The following limits of the POSTC-PET impurity residual content have been accepted for recycling as food packaging: 20 ppm or less metal, 10 ppm or less paper and 30 ppm or less polyolefins (Di Lorentzo et al., 2002; Hong JuZhou et al., 2007; Hong Jun Zhou et al., 2007). The framing into these limits depends on the technicality of the applied conditioning solution (sorting - washing etc), and by the legislative effort necessary for: the increasing of the population cooperation, the setting up of the infrastructure to analyze the impurity content down to the parts per million (ppm) level, the inspection on the law observance (Knit, 2002; David, 2001; Novis, 2003;).

2.2 PET degradability

During POSTC-PET conditioning and melt processing, the polymer is degraded by mechanical and thermal agents that act in the presence of water and oxygen. If during the first life the POSTC-PET is exposed to UV radiations rather than to thermo-mechanical and hydrolytic degradation, the photo-oxidation must be considered too (Cioffi et al., 2002; Chen et al., 2002; Raki et al., 2004).

The degradation occurs at the weakest thermodynamic links namely at the ester those between the terephthalic acid and diethylene glycol of POSTC-PET macromolecules (Sandi et al., 2005; Vasiliu et al., 2002). In figs. 1 – 5, the main reactions that characterize the PET thermo-hydrolitic degradation are exposed.

By thermal-oxidative degradation (fig.1, - Awaja & Pavel, 2005), the macromolecular chains break resulting in the formation of volatile products (i.e. acetaldehyde – fig. 5 Alexandru & Bosica, 1966) , 1.8 – 3 % cyclic and linear oligomers (fig.4 - Awaja & Pavel, 2005) and shorter chains with acid carboxylic and vinyl ester end groups. In hydrolytic degradation (fig.2 - (Awaja & Pavel, 2005), the mechanism is similar, with the difference that the end groups of the short macromolecules resulted from degradation are carboxylic acid and hydroxyl ester.

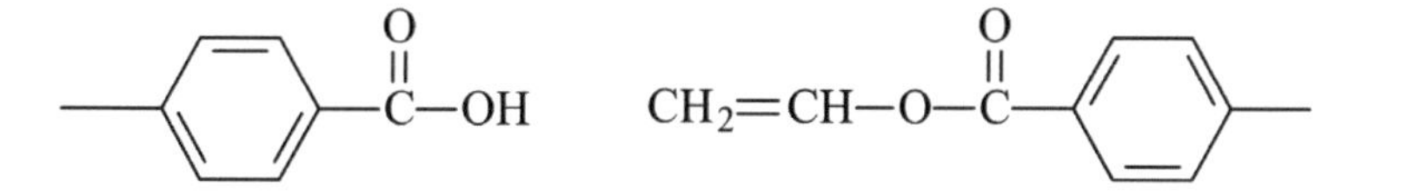

Fig. 1. PET thermal degradation mechanism with the formation of carboxyl acid and vinyl ester end group (Awaja & Pavel, 2005).

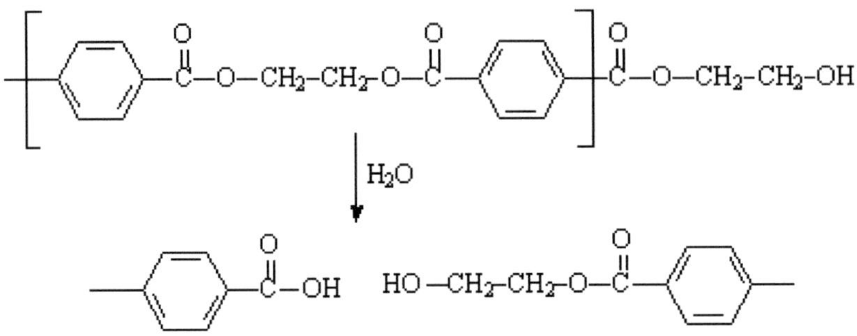

Fig. 2. PET Hydrolysis mechanism resulting in carboxyl acid and hydroxyl ester end group appearance (Awaja & Pavel, 2005).

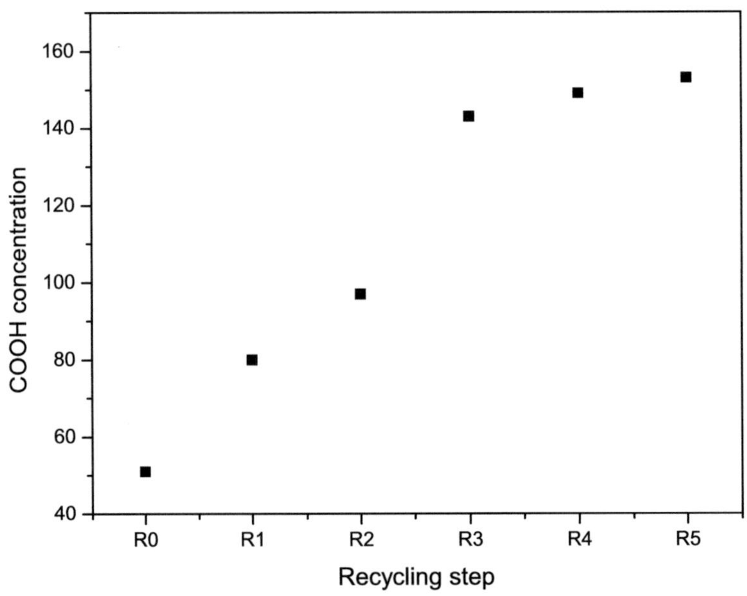

Fig. 3. The dependence of the carboxyl end group by the number of reprocessing of POSTC-PET (Spinace & De Paoli, 2001).

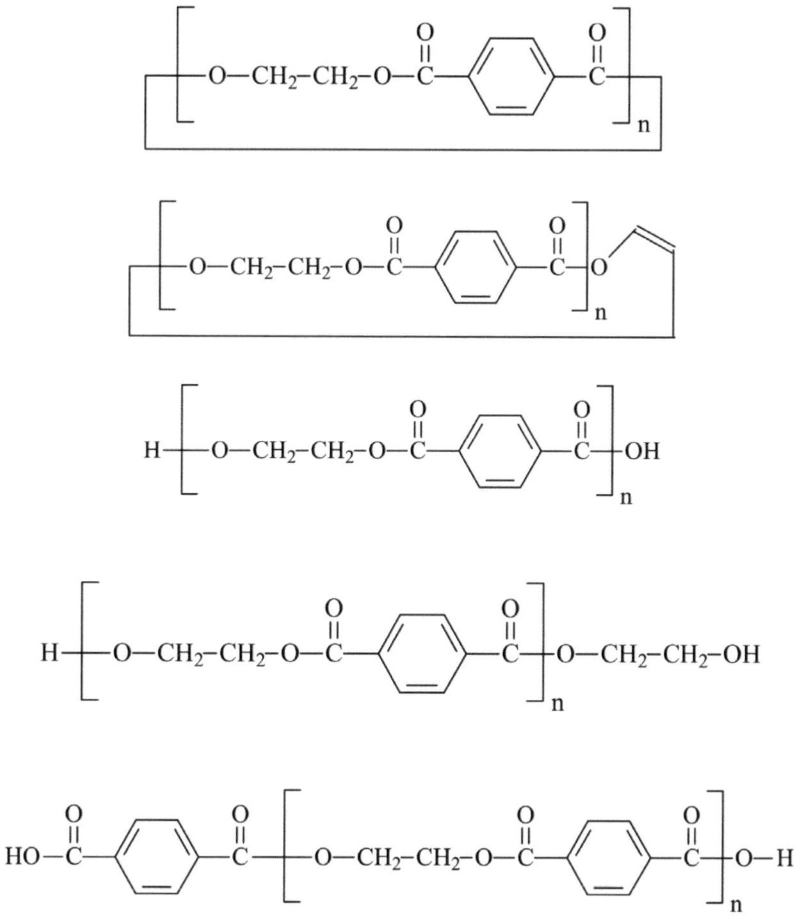

Fig. 4. Cyclic and linear oligomeric compounds resulted from PET degradation (Awaja & Pavel, 2005).

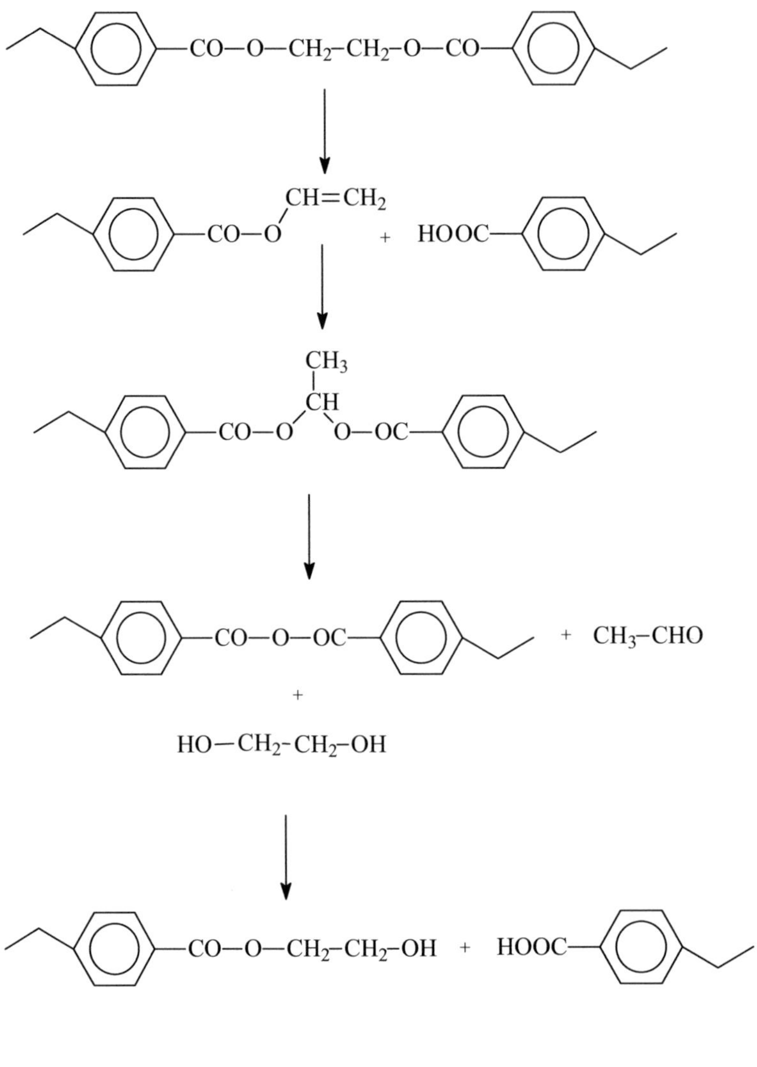

Fig. 5. The acetaldehyde formation during PET thermal degradation (Alexandru & Bosica, 1966).

The increasing of the carboxylic end group with the number of the reprocessing of the POSTC-PET is presented in fig. 3 (Spinace & De Paoli, 2001, Silva Spinace, 2001). The end groups content puts on view the POSTC-PET degradation degree, the carboxyl end groups being correlated with the thermal and hydrolytic degradation and the hydroxylic end groups with hydrolytic ones.

The result of the degradation reactions is a severe drop in the molecular weight which leads to the failing of intrinsic viscosity, melt strength and melt processability and finally, to poor usage properties and a low quality of the products obtained from reprocessed polymers. Because of the severe molecular weight diminishing during POSTC-PET reprocessing, the intrinsic viscosity may decrease from 0.72 dl g-1, the virgin polymer specific value, down to 0,04 - 0,26 dl g-1 (Raki et al., 2004; Zong Zhang et al., 2004;Seo et al., 2006; Cuberes et al., 2000).

Because of the formation of shorter macromolecules as a result of the hydro -thermal degradation, the crystallization capacity of the POSTC-PET increases and its degradability becomes more pronounced. This process known as chemi-crystalization is a complex one because at the beginning it is a chemical one (diminishing the macromolecules length due to degradation) and in the end it is a physical phenomenon (crystallization of the shorter macromolecular chains) (Pralay, 2002;). As a result of an increased crystallinity, the glass transition (Tg), melting temperature (Tm), melting heat and density of the POSTC-PET are greater. Also because of the dependence of the crystallinity on the degradation degree, the colour of POSTC-PET can differ from transparent (un-degraded or poorly degraded), to translucent (small degraded) and opaque (great degraded).

The strong degrading tendency during the melt processing is specific for all polycondensation polymers, not only for PET, and is observed in case of primary polymer melt processing too. The higher the molecular weight of the primary polymer the greater the melt processing degradation.

The structural changes resulted from degradation can be so dramatic that the melt processing of POSTC-PET may become not viable. It is therefore easy to understand why the mechanically recycling of POSTC-PET can consider only applications which do not require high performance properties.

3. The chain extension up-gradation of POSTC-PET

Considering the general opinion according to which POSTC-PEC can be mechanically recycled only into low-property goods, it becomes clear the interest to find new economic solutions for the reprocessing of these materials into products with practical importance. In the last 20 years the researchers have been concerned in the up-gradation of POSTC-PET by increasing the macromolecular weight based on chain extension reactions (Cavalcanti et.al., 2007, Awaja & Pavel, 2005, Villalobos et.al, 2002, Karaianidis, 1993).

The efficiency of these reactions is controlled by many factors. Their presentation begins with emphasis the importance to eliminate humidity by drying before melt processing and to stabilize the POSTC-PET at melt processing.

3.1 Drying /degassing

Before the chain extension, the POSTC-PET is dried to remove the humidity. It was observed that drying before chain extension and degassing and /or operation under vacuum during chain extension are able to decrease the degradation of POSTC-PET during melt processing. The drying of POSTC - PET restrains the hydrolysis during the melt processing but it is not a simple action (Buxton at al., 2002;). The POSTC-PET drying method should be the same as the ones used for primary polymers that means 3 - 7 hours at 140 - 180 °C, in desiccators or standard drying equipments (Xanthos et al., 2000). According to other opinions, the drying at temperatures greater then 160 °C can not be done because at this temperature the polymer hydrolysis becomes active. These opinions argue that the efficient drying temperatures should range between 110 – 140 °C, and the drying time should be greater than 12 hours. The final accepted water level is no more that 50 ppm - 0,01 %. If the POSTC-PET water content is smaller than 100 ppm, then the loss in the intrinsic viscosity during reprocessing shall be less than 0,04 dl g-1 (Denisyuk et al., 2003; Awaja & Pavel, 2005).

3.2 POSTC-PET stabilization

The POST - PET stabilization has the aim to block the polymer's thermo-hydrolitic degradation, to remove the formation of acetaldehyde as a result of degradation and to reduce the influence of the residual PVC. The free radicals resulted from the splitting of the macromolecular chain during degradation and those appeared after the decomposition of hydroperoxides can be captured with phosphorous compounds. Avoiding the degradation, these stabilizers hinder the formation of acetaldehyde (Karayannidis at al., 2003; Swoboda et al., 2008). For the capture of the existing acetaldehyde, compounds such as amino-benzoic acid, diphenylamine, 4,5 dihydroxy benzoic acid are very practical. The PVC traces are inhibited by tin mercaptide, antimony mercaptide and lead phthalate. The only disadvantage in using the stabilizers is the rise in the cost of the POSTC- PET mechanic recycled (Awaja & Pavel, 2005).

3.3 Methods for POSTC-PET up-gradation by chain extension

The macromolecular chain extension is a result of particular post condensation reactions between the degraded polymer and selected chain extenders. Theoretically, these coupling reactions annihilate the effect of degradation as they determine the growth of molecular weight by extension, branching, reticulation. Expertise has shown that it is very difficult to separate these reactions from the degradation that occurs in the same time. The intensity of the degradation is greater or smaller depending on the way the extension reaction is conducted. The reaction of molecular weight increase has to be performed in such a way as to diminish or avoid the degradation. As the high gel content is a disadvantage for the melt processing and for the control of reprocessed POSTC-PET usage properties, the reticulation near degradation should be avoided (Raki et al., 2004;torres et al., 2001; Yilmazer et al., 2000; Inata et al., 1987; Bikiaris & Karayannidis, 1993).

The chain extension reaction is rendered schematically in figure 6 (Villalobos et al., 2006) The most suggestive presentation of the way in which the chain extension reaction can develop depending on the reaction conditions (i.e. concentration of the extender) and how the same material can yield both chain extension and reticulation was accomplished by (Villalobos et al., 2006;) (fig.7). The same figure shows that the chain extension can result in intrinsic viscosity values, proper for various targeted applications.

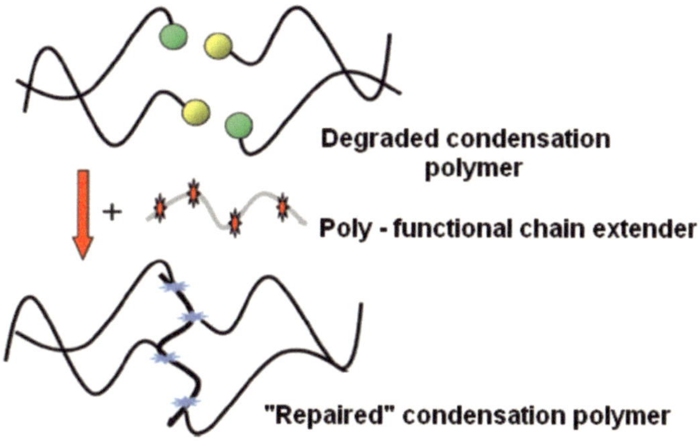

Fig. 6. Schematically presentation of the chain extension reaction (Villalobos et al., 2006).

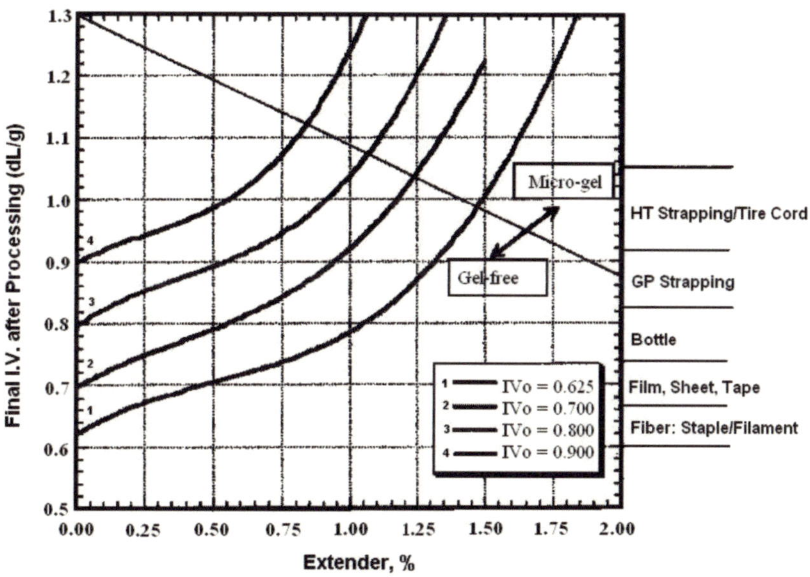

Fig. 7. The dependence of the nature of the reactions in chain extension and the possible applications depending on the accomplished intrinsic viscosity (Villalobos et al., 2006).

Otherwise, in practice, the obtained results demonstrate that the chain extension can be controlled in such a way to ensure the best melt processing properties and the most convenient usage properties for up-graded recycled POSTC-PET. The process can be controlled by monitoring the value of the intrinsic viscosity (ASTM D 4603 -91) and of the carboxyl and hydroxyl end group. An increased carboxyl end group content is associated with a very degraded polymer and the decreasing of the hydroxyl end groups means in progress chain extension reactions (Changli et al., 2006; Bizzaria et al., 2007). The process can be also monitored in terms of melt flow behavior, die swell degree and viscoelastic properties (Yilmazer et al., 2000).

In most cases, chain extension leads to thermal, mechanical and rheological performances equal or even higher than the performances of the primary polymers (Bikiaris at al., 1998). It is appreciated that chain extension is an important way to add value to POSTC-PET and to manufacture products with high technical and economic added value.

The following two alternatives are known for macromolecular chain extension, which are applied to all polycondensation polymers: solid state polymerization (SSP) and reactive processing (RP).

4. Solid State Polymerization (SSP)

Solid state polymerization (SSP) is a coupling reaction between POSTC-PET and extender that takes place in steel reactors, under high vacuum, at temperature above glass transition (Tg) and under melting temperature (Tm), in the catalysts presence (Karayannidis et al., 1991, 2003; Baldi et al., 2006; Flieger et al., 2003; Mano et el., 2004; Cangli et al., 2008; Bikiaris et al., 2003; Gantillon at al,1990, 2004; Rosu et al, 1999; Karayanidis at al., 1991; 1993, 2003). Usually the reaction occurs at temperatures ranging between 200 - 240 °C. These temperatures favour the SSP chain extension in detriment of the degrading ones. In SSP, the temperature control is essential because if the temperature is too low the extension lasts too long, and if the temperature is too high then the POSTC-PET flakes agglomerate and the extension can no longer happen evenly (Lee & Lichtenhan, 1999). In SSP the reaction time is too long (hours) because the reaction speed is controlled by the diffusion of the reaction by-product and the diffusion of the end–groups into the reaction mass (Gantillon et al., 2004; Apoorva, 2002; Yong et al., 2008). A convenient growth of molecular weight is obtained after 8 hours at 230 ^{0}C (Karayannidis et all, 2003). The reactions speed can be increased by the presence of nanomaterials probably because of their nucleation effect (Huimin et al., 2004; Tannenbaum et al., 2002;) The volatiles are constantly removed from the reactor that must operate under vacuum or under an inert gas blanket (Awaja & Pavel, 2005).

To eliminate the negative influence of the residual impurities there exists an alternative solution according to which the POSTC-PET is dissolved first in a selected solvent, then the polymer is recovered by precipitation with methanol and finally the polymer is chain extended according to SSP methods (Karayannidis et al. 2003). SSP can be a proper method to prepare POSTC-PET nanocomposites (Bikiaris et al,2006; Apoorva, 2002).

Although, apparently SSP can be considered a good "bottle to bottle" recycling method, due to the longer reaction time and the high cost of the equipments and of the control devices, the procedure is considered unsuitable for industrial level (Martinez et al., 2008; Cavalcanti et al., 2007; Awaja & Pavel, 2005).

5. Reactive processing

The reactive processing (RP) of POSTC-PET for the extension of the macromolecular chains, takes place in the equipment generally used for primary polymer melt processing, at temperatures ranging between the polymer melting temperature and the degradation those, under particular working conditions to each pair POSTC-PET - chain extender (Akkapeddi, 1988;). The reaction is also used for obtaining those melt properties which make possible the PET melt processing by extrusion –blowing and thermoformation (Lacoste et al., 2005).

Usually for RP the following equipment, that operate as reaction reactors, is used: one or twin screw extruders, Brabender plastographe, capillary rheometers, rheomixers, injection machines s.o, (Ganzeveld, 1993; Torres et al., 2000; Dhavalikar & Xanthos, 2002). In laboratory experiments or at industrial level, the twin screw extruders are preferred because of their effectiveness in achieving a better dispersion of small compounds into the polymeric matrix (Janssen, 1998; Akkapeddi et al., 1988; Rober et al., 2006).

Depending on the chemical structure of the extender and the concrete reaction conditions, the chain extension is more or less accompanied by ramification and reticulation, with gel formation or / and by degradation (Paci & La Mantia, 1998). Nevertheless it is considered that the extender chemical structures can be so conceived and the operating conditions can be found in such a way that, the prevalent reaction during the POSTC-PET reactive processing to be the chain extension those. The "reconstruction" of the macromolecular chains by reactive processing is a simple procedure that takes minutes, and has been performed at industrial level by extrusion and injection (Chem et al., 2002; Cavalcanti et al., 2007; Xanthos et al.,2001).

The obtained experimental results show that based on reactive processing, it is possible to reach an intrinsic viscosity higher than 0,6 dlg-1, the basic quality condition needed for reprocessing POSTC-PET in products for high performing applications. (table 1). It was also reported an intrinsic viscosity higher than 1 dlg-1 (Fumio Asaba, 2002).

Application	Intrinsic Viscosity, dlg-1
Recording tapes	0,60
Fibres	0,65
Bottles for drinks	0,73 - 0,8
Cord for industrial tyres	0,85
Micro and nano foam for multi-layer panels	0,7 - 1,1
Fields with outstanding mechanical properties	> 1

Table 1. Values of PET intrinsic viscosity, specific for different applications.

In most cases, by reactive processing, one can obtain mechanical and rheological performances equal or higher than those of virgin polymers. This is the reason for which the reactive processing of POSTC-PET is seen as an important possibility to add value to the post consumer condensation polymers and to create products with added technical and economical value. The "repaired" POSTC-PET can be used without physically modification in main applications as bottles and foam sheets or after physically modification as compounds, composites and nanocomposites applications that will be detailed in the following.

5.1 Chain extenders / reticulants

The chain extenders ("recycling aids" (Rossi et al., 2002)) are mono, di or polyfunctional (fn) organic liquid or solid compounds, with low molecular weight ($M_n < 3000$) and controlled polydispersity. The typical extender functional group are hydroxyl, carboxyl, anhydride, amine, epoxy, etc (table 2) (Inata, 1985; Inata, 1986;). Oligomeric or multifunctional polymeric extenders are more and more used (Volker et al., 2008).

Extender functional group	Main representatives, reference
Epoxy	Diepoxides (Haralabakopoulos, 1999;), Epoxy/styrene Oligomers which can be used as master - batch (Zammarano et al., 2006) , Epoxy functionalized compounds (Ren at al., 2003; (Dhavalikar, 2002); Polyepoxides (Dhavalikar, 2002;) Di and tri epoxy, glycidil reactive (Dhalikar & Xanthos, 2001; Hambir et al., 2002;Shanti et al.,2002;), Glycidil multifunctional compounds (Bras et al., 2001); Bis(glycidil ester(pyrolellitimides) (Bikiaris et al.,1995;)
Anhydride	Maleic anhydride, Phtalic anhydride (Shivalingappa et al., 2005;), Pyromellitic dianhydride (Kamal et al., 2002; Giusca et al., 2002; Shah et al., 2002; Shanti et al., 2002; Denisyuk et al., 2003; Lacoste et al., 2005;)
Phosphites/phosphates	Triphenyl phosphate(Cavalcanti et al. 2007;), Lactamyl phosphite (Pham Hoai nam et al., 2002; Bikiaris et al., 2006; Aromatic phosphates (Aharoni, 1986)
Oxazoline	2,2 - (1, 4 phenilen) bis 2 oxazoline (Hongyang et al., 2002; Karim et al., 2002; Shyalingappa et al., 2005; Warburton et al., 1990;)
phosphazene	Ciclo-phosphazene , bis-5,6-dihydro-4h-1,3-oxazolines,
lactame	Polyacyllactams (Bureau et al., 2002;
Isocyanate	Isocyanate triglycidil (Knite et al., 2002) 0,9 % Hexamethylene diisocyanate (Teh et al., 2004;)
Alcohol / polyol	Polyol having 3 -6 hydroxyl groups: Glycerol, Trimethylolpropane, Pentaerytritol, Sorbital ; Polyfunctional Alcohol, 0.1 – 2 % pentaerythritol (Denysiuk et al., 2003;)
Hydroxy -acid	Citric acid, Tartric acid, Trihydroxyglutaric acid. (Tang & Menachem, 2007)
Carboxyl / polycarboxylic acids	Trimellitic or Himimellitic acid, Pyromellitic acid (Tang & Menachem, 2007)

Table 2. The possible chain extender for POSTC-PET reactive processing.

Each extender, depending by its own chemical structure, yields typical extention reactions. It seems that di-functional chain extenders like bisepoxy compounds or bis (cyclic carboxylic anhydride or diisocyanate), do not form by-products and lead to strong reticulated POSTC-PET. Polyfunctional extenders having in their molecule at least three functional groups (fn>3) involving a combination of at least one group selected from those presented in table 1 (Arif et al., 2007;) are extremely efficient in case of highly deteriorated macromolecules or when a high level of intrinsic viscosity is targeted. These type of extenders are used in order to avoid combinations among extenders (i.e. pyromellitic dianhydride and pentaerytriol (Forsythe et.al, 2006)). The chain extenders with a higher than 3 functionality (f_n), leads to branched molecules. As a rule, the average functionality of the chain extenders is $f_n \rangle 4$.

The chain extenders can be classified considering the POSTC-PET end functional group with which react to extend the macromolecular chains. It is known that there are chain extenders which react with carboxyl end groups and chain extenders which react with hydroxyl end groups. The chain extenders which react to *carboxyl* end groups yield chain extension reactions in a higher proportion than the branching reactions. The chain extenders which react with POSTC-PET hydroxyl end groups are more efficient in the case of PET with low molecular weight, and the hydroxyl content is higher than the carboxyl one (Cavalcante et al., 2007; Inata et al.1986).

In the following it is presenting a few chain extension mechanisms proper to the most known chain extenders.

Pyromellitic Dianhydride (PMDA) is a tetra functional chain extender (fn = 4), available on the market, thermally stable, which does not lead to secondary products. It is efficient in proportion of 0.2 - 0.3 % and grows the intrinsic viscosity based on the reaction with the POSTC-PET *hydroxide end groups* (fig.8 -(Xantos et al., 2000; Awaja & Turcu, 2005). Depending on the PMDA concentration and the way the reaction is conducted, extremely branched or even reticulated structures can result. PMDA has been used also for primary PET for increasing the melt strength (Inata et al., 1985).

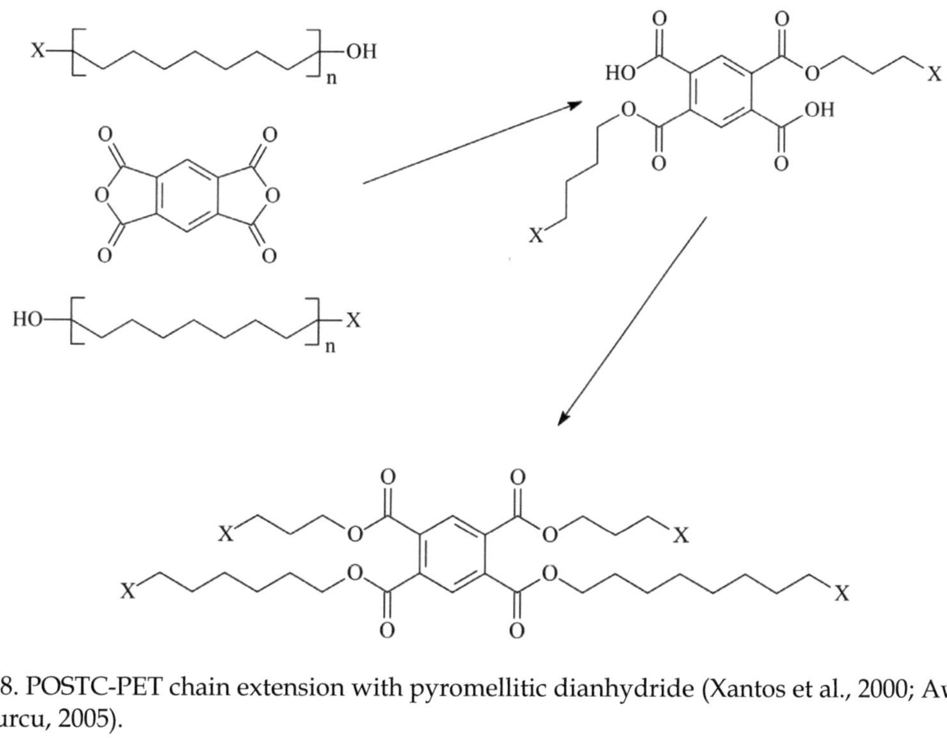

Fig. 8. POSTC-PET chain extension with pyromellitic dianhydride (Xantos et al., 2000; Awaja & Turcu, 2005).

Tri-phenyl phosphit (TPP). The increasing of the intrinsic viscosity is a result of the reaction between the non-participating electrons from phosphorus with the end *carboxyl and hydroxyl* groups of the POSTC-PET (figs 9, 10 Cavalcanti et.al, 2007). The good results are obtained with 1-3%, preferably 1% TPP, at 260 °C. (Cavalcanti et.al, 2005). The main reactions is accompanied by the by - products development. The competition between the chain extension and the formation of by-products is obvious at temperatures ranging from 280 - to 300 ^{0}C.

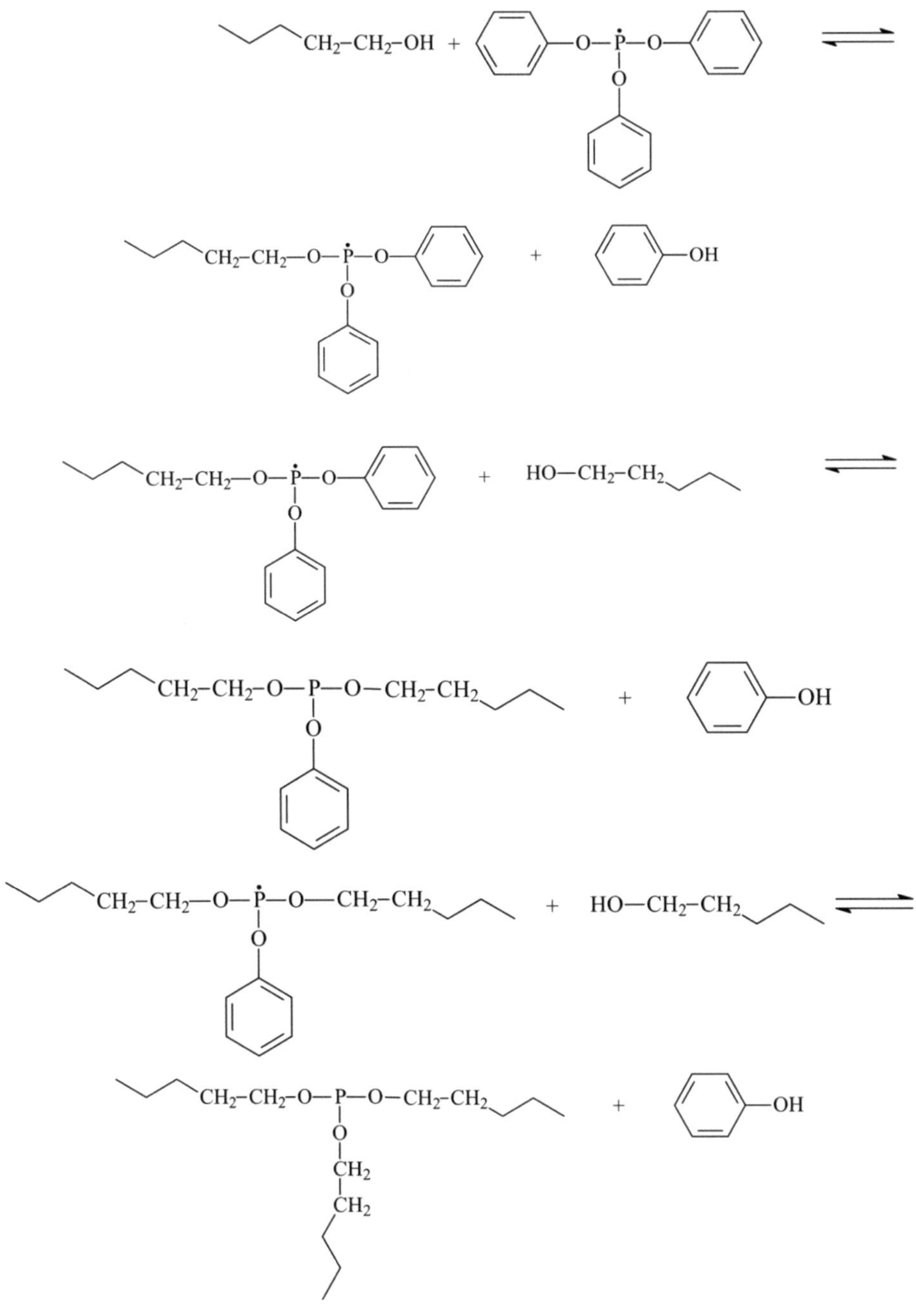

Fig. 9. Chain extension of PET with TPP. Chemical reaction between phosphate and hydroxyl group (Cavalcanti et.al, 2007).

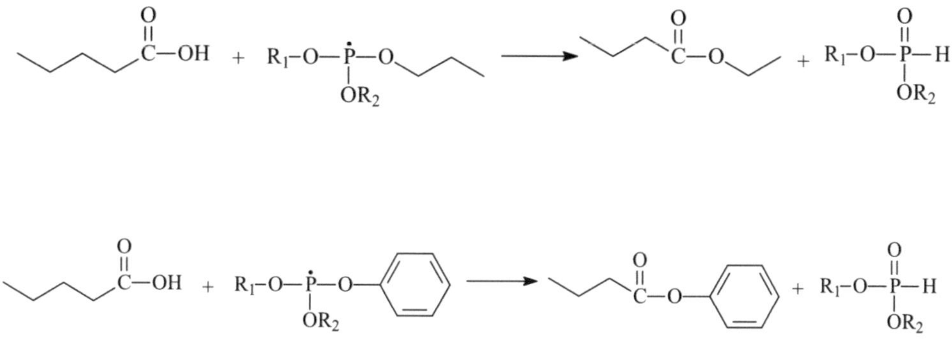

Fig. 10. Extension of PET with TPP. Reaction of phosphite with hydroxyl group (Cavalcanti et.al, 2007).

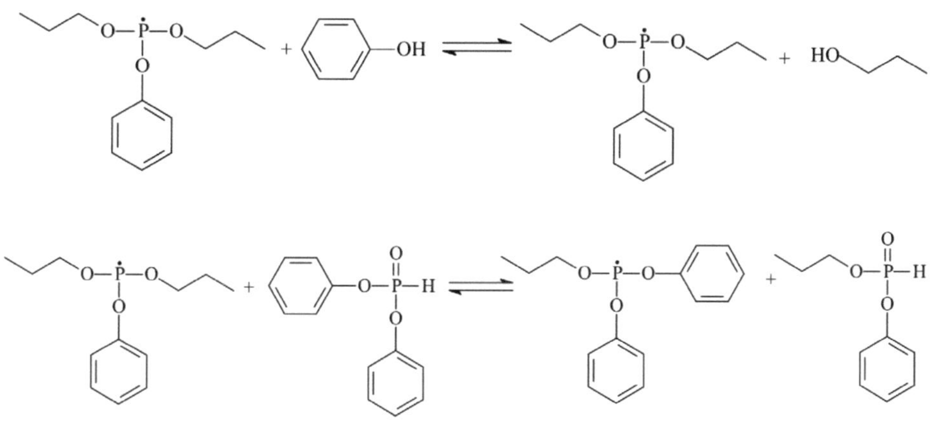

Fig. 11. By-products formation during the chain extension of PET with TPP (Cavalcanti, 2007).

The by-products are dangerous for the reason that, during storage, they act as degrading agent diminishing in this way the stability of "repaired" POSTC-PET. It is demonstrated that if these by-products are extracted with acetone, the degrading during storage is avoided (Cavalcanti et al.,2005).

Epoxy compounds give the esterification of end carboxyl groups (fig.12, (Xanthos et.al, 2000)) and etherification of the end hydroxyl groups (fig.13, (Xanthos et.al, 2000)) from the POSTC-PET macromolecules. In both cases, secondary hydroxyls are formed that can react later with the carboxyl or epoxy groups leading to the formation of branched or reticulate structures (Bikiaris et al., 1995).

Oxazoline compounds such as 2,2′-bis(2-oxazoline) give, with POSTC-PET, the following 3 types of interactions: *blocking reactions* (the molecule of chain extender reacts with the end carboxyl group from a POSTC-PET chain), *coupling reactions* (an extender molecule reacts with 2 polymer chains) and the *absence of any reactions* (Inata, 1987;). *(BO)* yields secondary reactions because the oxzoline ring is sensitive to acids. 2.2 - (1.4 - phenylene) bis(2 - oxazoline) (PBO) is a very reactive compound considering only the carboxyl groups within the macromolecular chains. PBO can be used together with a chain extender which reacts with hydroxyl end group i.e. phtalic anhydride.

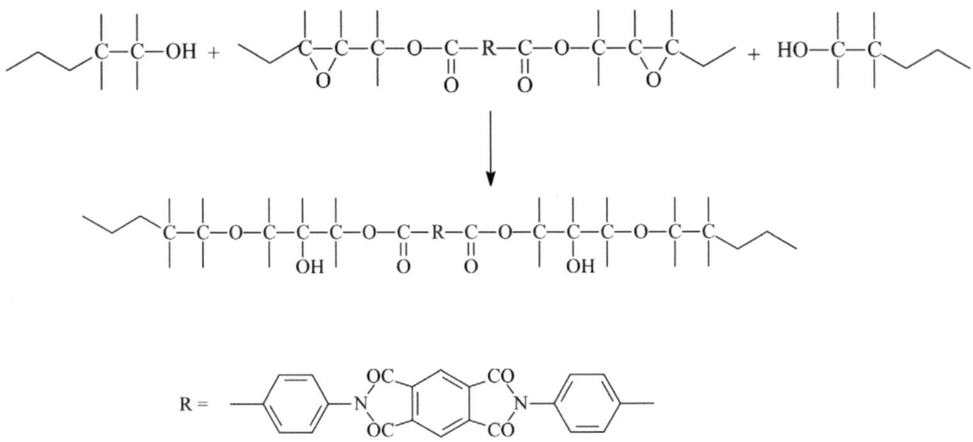

Fig. 12. Initial esterification step in PET chain extension with diepoxide (Xanthos et.al, 2000).

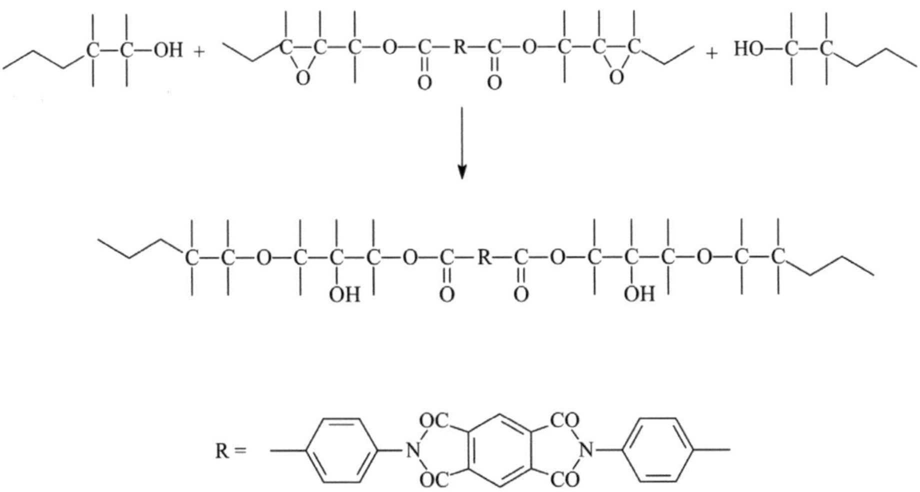

Fig. 13. Initial esterification in the chain extension of PET with diepoxide (Xanthos et.al, 2000).

5.2 Conditions for the chain extension reactions

5.2.1 Reaction parameters - Reaction control

In POSC-PET reactive processing, the chain extension reactions are controlled by the extender concentration, reaction temperature and time and parameters proper to the equipment in which the reactions occur. *The extender concentration is* calculated in relation with the stoichiometry of the extension reaction, considering the *measured* content of hydroxyl and carboxyl end groups (Nair at al., 2002;) In theory, a larger quantity than that resulted from stoichiometry leads to strongly reticulated structures, that means a high gel content. *Reaction time can be* up to 10 min. The measuring of the stationary time in the equipment is important because it controls the development of the chemical reactions (Janssen, 1998). Usually the reaction temperature ranges between 260 ^{0}C and 310 ^{0}C (Bras et al., 2001)

In the case of a Brabender plastometer, the extension is monitored based on the dependence between the motor torque and reaction temperature and time, while in the case of a capillary rheometer, it is recorded the correlation between the nozzle pressure and the swelling extrudate or on the relationship between the melt flow index and the melt strength (Nair et al., 2002;) In modern industrial systems the monitoring of the intrinsic viscosity is automatic. Obviously, the evolution of the chain extension reaction is rounded up with gel measurements and other properties that characterize the "repaired" POSTC-PET in the melt and solid state.

5.2.2 Operation under vacuum or nitrogen blanket

POSTC-PET has always a residual content of humidity. It was underlined that the chain extension reaction is favoured, and the thermal and hydrolytic degradation is diminished if the humidity content and the reaction time are reduced (Haralabakopoulos et al., 1999). If the chain extending reactions take place under vacuum or a nitrogen blanket then the thermal and hydrolytic degradation can be very much diminished or even eliminated. For these reasons the extruders must be equipped with high vacuum degassing areas for volatiles removal. Also the Brabender plastometers must work under a nitrogen blanket. This condition near the procedure price limit the industrial applicability of chain extension on elderly equipments. It is difficult to have industrial devices that work under such conditions (Paci & La Mantia, 1998). Nevertheless the modern POSTC-PET extrusion systems have high vacuum lines for volatiles removal.

5.2.3 The engineering of reactive processing

The POSTC-PET extending chain reactions that take place in an extruder are controlled by the reaction parameters presented in fig.14 (Awaja & Pavel, 2005). For controlling the reactions that occur in such conditions first of all the system has to be stable (Janssen, 1998;). The stability of the twin screw extruders depends on their designing concept (Bulters, 2001; Potente & Flecke, 1997;Shen et al., 2005;).

The fluctuation of the parameters presented in fig.14 is the major cause determining the *thermal, hydrodynamic and chemical instability,* and consequently the fluctuation in the operation of the reactive extruder. All these types of instability were described in detail in (Awaja & Pavel, 2005) where the bi-univocal relations between the parameters presented in fig.14 and the way in which they influence each other were explained.

The concentration of the extender / reticulant and the stationary time within the extruder are two parameters which control the efficiency of the procedure. A longer waiting time in the extruder is the main reasons of the system instability because the longer the waiting time the bigger the thermo - mechanically degradation (Giusca et al., 2002; Hongyang et al., 2002;

Kamal et al., 2002). The system instability can result in various situations. The presence of the branched chains within the polymer structure has a great influence on the crystallization induced by shearing (Hanley, 2007; Rosu et al., 1999; Van Meerveld et al., 2002). The resulted morphology will be heterogeneous if the chemical reactions have fluctuations in their evolution (Rosu et al.1999). The orientation of the macromolecules within a shearing field is directly linked to the increase in viscosity. The orientation degree will be irregular in the case of a random viscosity increase (Soares et al., 2004).

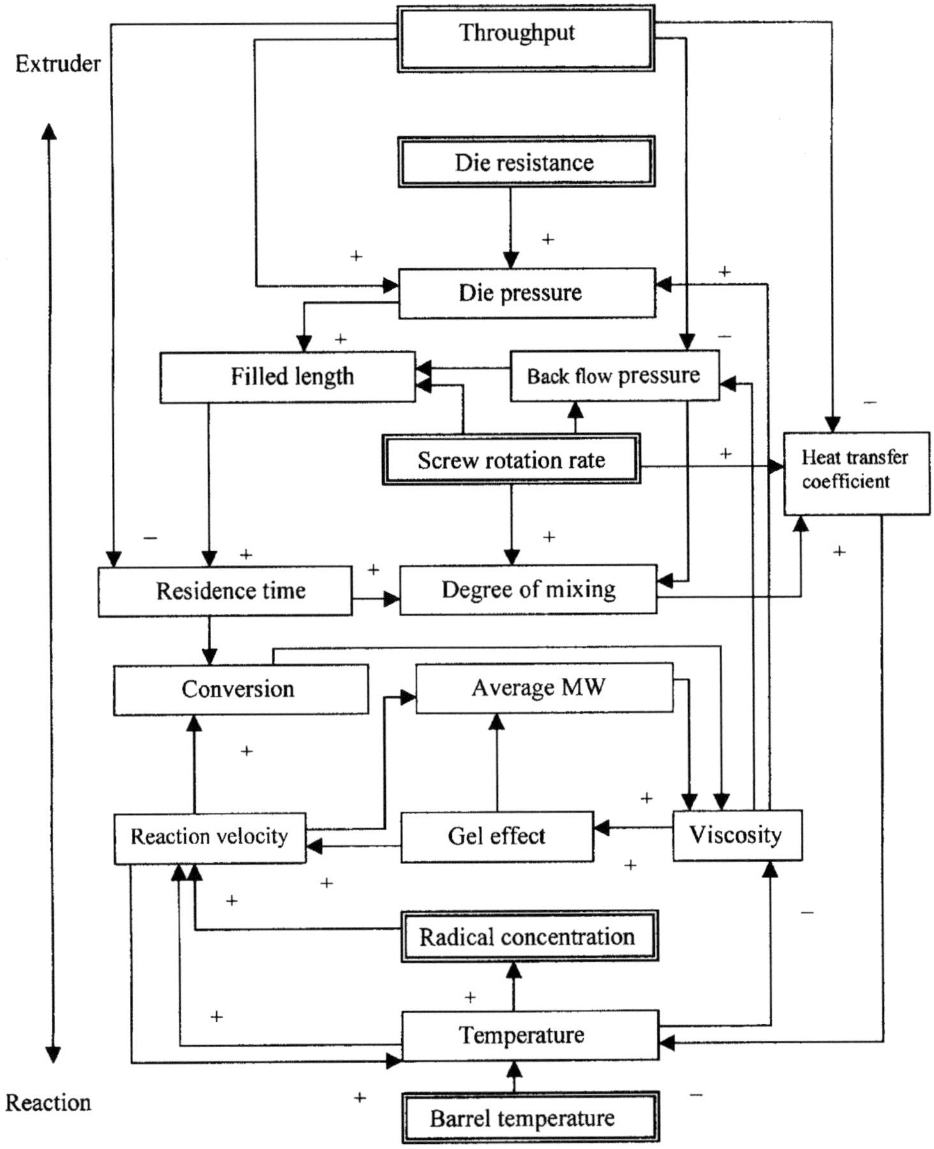

Fig. 14. The factors influencing the stability of an extruder system used for chain extension (+ positive influence; - negative influence) (Awaja & Pavel, 2005).

The counter-pressure and the pressure fluctuation are the most frequent instability described by most of the researchers (Kamal, 2002). Fluctuation can as well occur in the high vacuum degassing system. The pressure in the vacuum system also needs a severe control and a minimal variation (Cavalcanti, 2007).

It is considered that the reaction system specific to the reactive processing is constant if the defining parameters vary within a minimally accepted controllable level. Actually it is considered that the reactive processed is constant when the nozzle pressure, the cylinder temperature and the flow speed are constant.

5.3 SSP and PR comparative economical analyse

A correct approaching of comparative economical analysis for SSP and PR needs details for both procedures and the reaction devices, details about the cost of energy, nitrogen, cooling water, additives and specific labour. In [Vilabados, 2006] it is demonstrated that the chain extension with Joncryl-ADR-4368 (Epoxy/styrene oligomeric extender) by reactive processing in a single screw extruder results in a competitive PR of POSTC-PET. As the reaction uses smaller quantities of energy, water and nitrogen, the reactive processing is more cost-efficient than SSP, which needs catalysts and other special reaction conditions..

5.4 POSTC-PET chain extended applications

The main applications of the "repaired" mechanically recycled POSTC-PET, valuable in practice, are manufacture of: bottles, expanded sheets, multi-layer sheets and foamed panels for constructions and /or compounds composites and nanocomposites for different uses obtained by physical modification.

5.4.1 Bottles

In chap.1 it was underlined that only the colour selected POSTC-PET can be mechanically or thermally recycled into bottles. The chain extension reactions offer o new perspective on this subject. Currently "closing the loop" has become an actual possibility as the bottles and containers can be recycled back as bottles and containers. So, considering the chain extension possibilities it seems that the bottle-to-bottle recycling system is a feasible approach. These bottles can be used for packaging of non-food or food contact products. The re-use of POSTC-PET into food area depends on the potential of the reprocessed material to provide as much safety as the primary polymers do. POSTC-PET can be reprocessed also in multilayer bottles that do not require special safety measures as their inner layer, which comes into contact with the food, is made of primary polymer (Chaiko et al., 2002; kamal et al., 2002; Tannenbaum et al., 2002; Liane et al., 2002; Tjong et al., 2002; Kim et al., 2001; Hu et al., 2002; Lochhead, 2006;)

5.4.2 Sheets and foamed panels

The "repaired" POST-PET can be used for obtaining sheets or multi-layer structures in which at least one layer consists of POSTC-PET (Hong et al., 2007; Yan & Zao, 1988). Sandwich panels (Banosz et al., 1996) and /or high strength uniaxially drawn tapes (Morawiec et al., 2002) can be also attained from "repaired" POST-PET or "repaired" POST-PET foams.

The foaming of the thermoplastic semi-crystalline materials is efficient if at a certain working temperature their melt has high elongation viscosity, elevated strength and enhanced elasticity. The melt of the polymer with low molecular weight and narrow molecular distribution has low viscosity, small strength and reduced elasticity and because of these, the formation and stabilization of the cells cannot be controlled. The increase in molecular weight and polydispersity of POSTC-PET by reactive processing is a way to obtain high property foamed products (Quintans et al.,2004; Forshythe et al.,2006; Fujimoto, 2003; Japon et al., 2004; Kumar et.al, 2001; Place et al., 2003; Warburton et al., 1992).

It was found that the "repaired" recycled RPOSC-PET can be foamed if its apparent viscosity is 0.9 dl g^{-1} (Nair et al., 2002) that was realised by means of extenders with a molecular weight of 50 - 5000 and a functionality of 3 - 6. (Tang & Menachem,2007). In this way, it is possible to produce structures with closed pores which have the right density, pore size, pore distribution, mechanical and thermal properties proper for insulating panels or microcellular foams (Kiatkamjornwong et al., 2001; Xanthos et al., 2004; Chem &Curliss, 2003; Carotenuto et al., 2000). The "repaired" POST-PET can be modified in order to make of cheap composites for expanded panels (Deng at al., 1996).

5.4.3 Compounds, composites and nanocomposites realised by physical modification

In order to improve the melt processability and the utilization properties to POSTC-PET qualify for the desired application, the polymer can be physically modified with: *melt processing agents, agents for improving the mechanical, barrier and optical properties, toughening agents, crystallization and coefficient of friction modifying agent, thermo-oxidative antioxidants and ultraviolet stabilizers* (Smiidt et al., 1999; Salgueiro et al., 2004; Kalpana et al, 2006; debashis et al., 2006; Unnikrishnan & Sabu, 1998;zammarano et al., 2006; Zhang et al., 2001; Zhong et al., 2004;).

Several examples of such modifiers are: *primary* PET (Utraki & Kamal, 2002;), glass fiber (Unnikrishnan & Sabu, 1998; longzhen et al., 2006; Aghlan, 2003; Gersappe, 2002), maleic anhydride grafted styrene - ethylene/butylene - styrene triblock copolymer (SEBS - g-MA) (Javaid, 2006), poly (ε - caprolactone) (Guo, 2002), copolymer having at least one block comprising a vinyl aromatic polymer and at least one block comprising a conjugated diene polymer (Kiatkamjornwong et al., 2002; Shanti, 2002;), polyolefins, recycled polyolefins with proper compatibilization agents (Tortora, 2002; Chen et al., 2002; Chabert et al., 2004; Leszezynsksa et al.,2007; Glasel et al., 1999; Chrissopouloe et al, 2005; Qing-ming et al., 2006; Hadal et al., 2004; Conde et al.,2003; Place et al., 2003; Fujimoti et al., 2003;). Clear blends must be tailored based on branched slow crystallizing PET and faster crystallizing PET (Shriroth et al., 2006; Swoboda et al., 2008; Aghlara,2003). To improve the brittleness, "repaired" PET is modified with an epoxy group containing styrene thermoplastic elastomer and polycaprolactone (Sikdar et al., 2006). To obtain the side material for cooling towers, "repaired" PET is modified with styrenic thermoplastic elastomer (Arif et al., 2007; Zilg et al., 1998;). Nanocomposites ca be achieved with non-modified natural montmorillonite or with ion-exchanged clay modified with quaternary ammonium salt (Pegoretti et al.,2004; Lee & Lichtenhan, 1999; Sharma, 1999; Schmidt et al., 1999; Carotenuto et al., 2000; Aravind, 2007; Bandosz,1996; Bartholome, 2005; Buxton, 2002; Chrissopoulou, 2005; Feng, 2002; Utraki & Kamal, 2002). Nanocomposites can be also obtained with CaCO3 (Di Lorenzo, 2007). Experimental models were conceived for understanding the interaction between the polymer and the ranforsant (Kalpana, 2006; lingaiah, 2005; Tortora et al., 2002). Also, researches are known about the possibility to increase the exfoliation degree of the multilayered silicate, the order degree of the resulted lamellae (Gilmer, 2004; Ren et al., 2003; Rossi, 2002;), the influence of the ranforsant, possible compatibilizer (Giselle, 2005; Hambir, 2002;Pegoretti et al., 2004;Schimidt et al., 1999; The et al., 2004;) The parameters of the reactive processing have a critical role on the obtained results (Hong Jun, 2007).

6. Conclusions

1. The chapter presents an overview on the up-gradation of POSTC-PET by increasing the macromolecular weight based on chain extension reactions, as the most efficient method for adding value to the secondary polymers and for the creation of products with added technical and economical value, for applications within the economy.
2. The post consumer poly(ethylene therephtalate) bottles (PET-PC) can be recycled by chemical or / and mechanical procedures. The PET - PC chemical recycling is based on the depolymerisation of secondary polymers and the use of the depolymerisation products within the fibre and unwoven material industry. The PET - PC mechanical recycling is based on a phase transformation (melting) and can be performed without or with polymer up - grading.
3. The mechanical recycling is controlled by the impurities content and by the reprocessed polymer degradation. The mechanical recycling of PET - PC without up - gradation takes into account the melt processing of the recycled polymer into packages for non - food goods and into thermoformable sheets with a resulted shape adjusted to the transported packed products (eggs, tomatoes, strawberries, apples, so.).
4. In spite of the long efforts performed during the years, because of the low cost and low performance applications of the obtained products, the widely accepted opinion is that the mechanical recycling of PET - PC without up - gradation is not an efficient procedure.
5. The chain extension reaction is favoured, and the thermal and hydrolytic degradation is diminished if the POSTC-PET humidity content is reduced by drying.
6. The POSTC-PET drying is performed as in the case of primary polymers: drying for 3 -12 hours at a temperature of 120 - 180 ^{o}C in desiccators or standard drying equipments. The drying of POSTC - PET restrains the hydrolysis during melt processing.
7. The POSTC-PET stabilization during melt processing is needed to block the polymer thermo-hydrolitical degradation, to remove the formation of acetaldehyde as a result of degradation and to reduce the influence of the residual PVC.
8. The macromolecular chain extension is a result of particular post condensation reactions between the degraded polymer and selected chain extenders. The following two alternatives are known for POSTC-PET macromolecular chain extension, which are applied to all polycondensation polymers: solid state polymerization (SSP) and reactive processing (RP).

9. Although, apparently SSP can be considered a good "bottle to bottle" recycling method, due to the longer reaction times and the high cost of the equipment and of the control devices, the procedure is considered unsuitable for industrial level.
10. Solid state polymerization (SPP) is based on the reaction between the POSTC-PET and the extender that takes place in steel reactors, under high vacuum, at temperature above glass transition (Tg) and under melting temperature (Tm), in the presence of catalysts.
11. The reactive processing (RP) of POSTC-PET, takes place in equipment usually used for primary polymer melt processing, at temperatures ranging between the melting temperature and the degradation those, in working conditions suitable to each pair POSTC-PET - chain extender.
12. The chain extenders ("recycling aids") are mono, di or polyfunctional organic liquid or solid compounds, with low molecular weight ($M_n < 3000$) and controlled polydispersity (PDI › 3). The typical extender functional group are hydroxyl, carboxyl, anhydride, amine, epoxy etc.
13. In POSC-PET reactive processing the chain extension reactions are controlled by the extender concentration, reaction temperature and time and parameters proper to the equipment in which the reactions take place.
14. The extruders used in the reactive processing must be equipped with high vacuum degassing areas for the volatiles removal. These condition limits the industrial applicability of the reaction. The modern POSTC-PET extrusion systems have high vacuum lines for removal of volatiles.
15. It is considered that an extrusion system used for the reactive processing is stable if the defining parameters vary within a minimally accepted controllable interval. Actually it is considered that an extruder system is constant when the nozzle pressure, the cylinder temperature and the flow speed are constant. The pressure in the vacuum system, also needs a severe control and a minimal variation
16. Due to reactive processing, it is possible to reach an intrinsic viscosity higher than 0.6 dlg-1, which is a basic quality condition for the reprocessing of POSTC-PET into products designed for high performance applications. In most cases, by reactive processing to get properties in melt and solid state equal or higher than the ones of the virgin polymers;
17. As the reaction uses smaller quantities of energy, water and nitrogen, the reactive processing is more cost-efficient than SSP, which needs catalyzers and other special reaction conditions.
18. The main applications of the "repaired" mechanically recycled POSTC-PET are: bottles, expanded sheets, multi-layer sheets and foamed panels and compounds, composites and nanocomposites obtained by physical modification used in important applications.

7. References

Aghlara H., (2003), *Capacitors Based on Conducting Polyaniline Films*, Chinese Journal of Physics, Vol. 41, No. 2, April 2003, p. 185-190.

Aharoni S.M., Forbes D.E. & Hammond W.B. , (1986), *High-temperature ractions of hydroxyl an carboxyl PET chain end groups in the presence of aromatic phosphate,* Journal of Polymer Science 24 (1986) 1281 - 1296

Akkapeddi M.K. & Gervasi J., 1988, *Chain extension of PET and nylon in an Extruder, Polymer Preparation,* Division of Polymer Chemistry Meeting 1988

Akovali G., (1998)," *Frontiers in the Science and Technology of Polymer Recycling*", Kluwer Academic Publishers, ISBN 0-7923-5190-8, 1998

Alexandru I & Bosica S, (1966) " Fibre Sintetice - Chimie si Tehnologie, Ed.Tehnica Bucuresti 1966

Aravind D., Zhong-Zhen Yu & Yiu-Wing Mai (2007), *Nanoscratching of nylon 66-based ternary nanocomposites,* Acta Materialia 55 (2007) 635-646

Arif S., Burgess G., Ramani N. & Bruce Harte, (2007), *Evaluation of a biodegradable foam for protective packaging applications,* Packaging Technology and Science 20 (2007) 413-419

Asaba Fumio, (2002), *A new "Bottle to Bottle" mechanical recycling tehnology for PET bottle,* Science Links Japan 54 (2002) 41 - 49

Awaja Firas & Dumitru Pavel,(2005), *Recycling of PET,* Europeean Polymer Journal 41 (2005), 1453 - 1477

Bandosz T.J., Putyera K., JagieŁŁo J. & Schwarz J.A., (1996), *Study of nanocomposites obtained by carbonization of different organic precursors within taeniolite matrices,* Clays and Clay Minerals 44 (1996) 237-243

Bartholome C., Beyou E., Bourgeat-Lami E., Cassagnau P., Chaumont P., Laurent David & Nathalie Zydowicz, (2005), *Viscoelastic properties and morphological characterization of silica/polystyrene nanocomposites synthesized by nitroxide-mediated polymerization,* Polymer 46 (2005) 9965-9973.

Belletti C., G.'*Polyester fiber from 100 % recycled PET bottle*", Chem. Fibers Int 1997, 28 - 30

Bikirias D. & Karayannidis G, *Chain of Polymeresters PET and PBT with N.N - bis(glycidyl ester) pyromellitimides,* Journal Polymers Science 33 (1995) 1705 - 1714

Bikiaris D. & George P. Karayannidis, (1998), *Thermomechanical analysis of chain-extended PET and PTB,* Journal of Applied Polymer Science, 1998

Bikiaris D.N. & Karayannidis G.P., (2003) , *Synthesis and characterisation of branched and pratially crosslinked poly(ethylene terephthalate),* Polymer International 52 (2003) 1230 - 1239

Bikiaris D., Karavelidis V. & Karayannidis G., (2006), *A new approach to preprare poly(ethylene terephthalate)/silica nanocomposites with increased molecular weith and fully adjustable branching or crosslinking by SSP,* Macromolecular Rapid Comunication 2006

Bikiaris D., Papageorgiou G, Pavlidou E., Vouroutzis N., Paraskevas P. & Karayannidis G.P., (2006), *Preparation by Melt Mixing and Characterization of Isotactic Polypropylene/SiO_2 Nanocomposites Containing Untreated and Surface-Trteated Nanoparticles,* Journal of Applied Polymer Science 100 (2006) 2684-2696.

Bizzaria M.T., Andre l.F., m.Giraldi, Cesar m., Jose Velasco, 2007, *Morphology and thermomechanical Properties of recycled Pet- Organoclay nanocomposites*", Journal of Applied polymer Science, vol.14, 1839-1844, 2007

Bulters M. & Elemans PHM, (2001), *The influence of screw design on the stability of a reactive twin screw extrusion process,* ANTEC 2001

Bureau N.M., Denault J., Cole K.C. & Enright G., (2002), *The role of crystallinity and reinforcement in the mechanical behavior of polyamide-6/clay nanocomposites*, Polymer Engineering and Science 42 (2002) 1897-1906

Buxton J. & Balazs A.C., (2002), *Lattice spring model of filled polymers and nanocomposites*, Journal of Chemical Physics 117 (2002) 7649-7658

Carotenuto G., Xuejun X. & Nicolais L., (2000), *Transparent organic-inorganic nanostructured materials: preparation methods*, Polymer News 25 (2000) 6-10

Carta D., (2001)," *Chemical recycling of poly(ethylene terephthalate) (pet) by hydrolysis and glycolysis* ", Environmental Science and Pollution Research Volume 10, Number 6, 390-394, DOI: 10.1065/espr2001.12.104.8

Cavalcanti F.N., Teofilo S., Rabello M. & Silva S.L.M., (2007), *Chain Extension and degradation during Reactive processing of PET in the precence of triphenyl phosphate*, Polymer Engineering and Science, 2007.

Chaiko D.J., (2002), *The colloid chemistry or organoclays*, International SAMPE Symposium and Exhibition (Proceedings) 47 II (2002) 1064-1073

Chabert E., Dendievel R. , Gauthier C. & Cavaillć J.-Y, (2004), *Prediction of the elastic response of polymer based nanocomposites: a mean field approach and a discrete simulation*, Composites Science and Technology 64 (2004) 309-316.

Chen C. & Curliss D., (2003), *Processing and morphological development of montmorillonite epoxy nanocomposites*, Nanotechnology 14 (2003) 643-648

Chrissopoulou K., Alintzi I. , Anastasiadis S.H., Giannelis E.P., Pitsikalis M. , Hadjiechristidis N. & Theophilou N.,(2005), *Controlling the miscibility of polyethylene/layered silicate nanocomposites by altering the polymer/surface interactions*, Polymer 46 (2005) 12440-12451.

Conde A., Durán A. & de Damborenea J.J., (2003), *Polymeric sol-gel coatings as protective layers of alumminium alloys*, Progress in Organic Coatings 46 (2003) 288-296

David J. Hurd, (2001), "*Best Practices and Industry Standards in PET Plastic Recycling*", http://www.napcor.com/pdf/Master.pdf; 2001

Debashis S., Dinesh R. Katti, Kalpana S. Katti & Raúl Bhowmik, (2006), *Insight into molecular interactions between constituents in polymer clay nanocomposites*, Polymer 47 (2006) 5196-5205

Deng Q., Cable K.M., Moore R.B. & Mauritz K.A., (1996), *Small-angle X-ray scattering studies of Nafion® / [silicon oxide] and Nafion® / ORMOSIL nanocomposites*, Journal of Polymer Science, Part B: Polymer Physics 34 (1996) 1917-1923

Dhavalikar R. & Xanthos M., (2002), *Parameters affecting the chain extension and branching of PET in the melt state by polyexposides*, Journal of Applied Polimer Science 87 (2002) 643 - 652

Di Lorenzo M.L., Errico M.E. & Avella M.(2007), *Thermal and morphological characterization of poly(ethyleneterephthalate)/calcium carbonate nanocomposites*, Journal of Materials Science 37 (2002) 2351-2358

Ehrig R.J., (1992), "*Plastics Recycling:Products and Processes*", Hanser Publishers, New York, USA, 1992

Erema (2002), *Plastic recycling systems, the PET pant technology*, 2002

Franz, (2004), "*European Survey on post consumer PET materials to determine contamination level and maximum exposure consumer level from food packages made from recycled PET*" ,Food Additiv Contam, 2004, 21 (3), 265-286

Feng W., Ait-kadi A. & Bernard R., (2002), *Polymerization compounding: Epoxy-montmorillonite nanocomposites*, Polymer Engineering and Science 42 (2002) 1827-1835

Forsythe J.S., Cheah K., Nisbet D.R., Gupta R.k., Lau A. & Donovan A.r., (2006), *Rheological properties of high melt strength poly(ethylene terephthalate) former reactive extrusion*, Journal of Applied Polymer Science 100 (2006) 3646 - 3652

Fujimoto Y. , Ray Sinha S. , Okamoto M., Ogami A., Yamada K. & Ueda K. ,(2003), *Well-controlled biodegradable nanocomposite foams: From microcellular to nanocellular*, Macromolecular Rapid Communications 24 (2003) 457-461

Gantillon B., Spitz R. & McKenna T.F., (2003), *The solid state postcondensation of PET*, Scopus - Macromolecular Materials and Engineering 2003

Ganzeveld K. & Janssen LPBM, (1993), *Twin screw extruders as polymerization rectors for a free radical homo polymerization.* Can J. Chem Eng 38 (1993) 411 - 418

Gersappe D., (2002), *Molecular mechanisms of failure in polymer nanocomposites*, Physical Review Letters 89 (2002) 058301/1-058301/4

Gilmer J. W., Matayabas J.C.Jr., Connell G.W., Owens Jeffrey Todd,Turner Sam Richard & Piner Rodney Layne, (2004), *Process for Preparing an Exfoliated, High I.V. Polymer Nanocomposite with an Oligomer Resin Precursor and an Article Produced Therefrom, 2004*

Giselle Sandi, Riza Kizilel, Kathleen A. Carrado, Rocio Fernández-Saavedra, Norma Castagnola, (2005)*Effect of the silica precursor on the conductivity of hectorite-derived polymer nanocomposites*, Electrochimica Acta 50 (2005) 3891-3896.

Giusca C., Baibarac M., Lefrant S., Chauvet O., Baltog I., Devenyi A. & Manaila R., (2002), *Co-polymer nanocomposites: evidence for interface interaction*, Carbon 40 (2002) 1565-1574.

Gläsel H.J., Hartmann E., Mehnert R., Hirsch D., Böttcher R. & Hormes J., (1999), *Physico-chemical modification effects of nanoparticles in radiation-cured polymeric composites*, Nuclear Instruments and Methods in Physics Research B 151 (1999) 200-206.

Hadal R., Yuan Q., Jog J.P. & Misra R.D.K., (2006),*On stress whitening during surface deformation in clay-coating polymer nanocomposites: A microstructural approach*, Materials Science and Engineering A 418 (2006) 268-281.

Hambir S., Bulakh N. & Jog J.P., (2002), *Polypropylene/Clay nanocomposites: Effect of compatibilizer on the thermal, crystallization and dynamic mechanical behavior*, Polymer Engineering and Science 42 (2002) 1800-1807

Hanley T., Sutton D., Heeley E, Moad G. & Knott R., (2007), *A Small-Angle X-ray Scattering Study of the Effect of Chain Architecture on the Shear-Induced Crystallization of Brached an Linear Poly(ethylene terephthalate)*, Journal of Applied Crystallography, 2007

Haralabakopolulos A., Tsiourvas D. & Paleos C.M., (1999), *Chain extension of poly(ethylene terephthalate) by reactive blending using diepoxides*, Journal of Applied Science 71 (1999) 2121 -2127

Hongyang Y., Jin Zhu, Wilkie C.A. & Morgan A.B., (2002), *Crown ether-modified clays and their polystyrene nanocomposites,* Polymer Engineering and Science 42 (2002) 1808-1814

Hong Jun Zhou, Min Zhi Rong, Ming Qiu Zhang, Wen Hong Ruan & Klaus Friedrich, *Role of reactive compatibilization in preparation of nanosilica/polypropylene composites,* Polymer Engineering and Science, 2007

Huimin Yu, Keqing Han & Muhuo Yu, (2004), *The rate acceleration in solid-state polycondensation of PET by nanomaterials,* Journal of Applied Polymer Science 94 (2004) 971 - 976

Hu Y.S., Rogunova M., Schiraldi D.A., Hiltner A. & Baer E., (2002), *Crystallization Kinetics and Crystalline Morphology of Poly(ethylene naphthalate) and Poly(ethylene terephthalate-co-bibenzoate),* Journal of Apllied Polymer Scince, Vol.86, 98-115 (2002).

Inata H. & Matsumura S., (1985), *Chain extenders for polyesters.Addition type chain extenders reactive with carboxyl end group of polyesters,* Journal of Applied Polymer Science, 30 (1985) 3325 - 3337

Inata H. & Matsumura S., (1986), *Chain extenders for polyesters. Addition type nitrogen-containing chain extenders reactive with hydroxyl end groups for polyesters,* Journal of Applied Polymer Science, 32 (1986) 5193 - 5199

Inata H. & Matsumura S., (1987), *Chain extenders for polyesters. Properties of polyesters chain extended by 2,2 – bis(2-oxazoline),* Journal of Applied Polymer Science, 33 (1987) 3069 – 3079

Janssen LPBM, (1998), *On The Stability of reactive extrusion,* Polymer Engineering Science 1998

Japon S., Leterrier Y. & Jan-Andres E. M., (2004), Recycling of poly(ethylene terephthalate) into closed-cell foams, Polymer Engineering and Science, Volume 40 Issue 8, Pages 1942-1952, 2004

Kalpana S. K, Debashis S., Dinesh R. K., Pijush G. & Devendra V, (2006), *Molecular interactions in intercalated organically modified clay and clay-polycaprolactam nanocomposites: experiments and modeling,* Polymer 47 (2006) 403-414

Kamal R., Borse N.K. & A. Garcia-Rejon, (2002), *The effect of pressure and clay on the crystallization behavior and kinetics of polyamide-6 in nanocomposites,* Polymer Engineering and Science 42 (2002) 1883-1896

Karayannidis G. , Siderdou I., Zamboulis D., Stalidis G. & Bikiaris D., (1991), *Solid-state polycondensation of poly(ethylene terephthalate) films,* Die Angewandte Makromolekulare Chemie 1991

Karaiannidis GP, Kokkalas D.E., Bikiaris D N, 1993 *"Solid State Polycondensation od PRT recycled from postconsumer Soft Drink Bottle".* Journal of Applied Polymer Science, 1993vol.50, 2135 - 2142

Karayannidis G., Sideridou I., Zamboulis D. & Bikiaris D., (2003), *Effect of some phosphorous compunds on the thermo-oxidative stability of poly(ethylene terephthalate),* Die Angewandte Makromolekulare Chemie 208 (2003) 117 - 124

Karayannidis G.P., (2003), Chemical Recycling of PET et by Glycolysis.*Alkyd Resins Derived from the Glycolised PET,* Proceedings of the *8th International Conference on*

Environmental Science and Technology Lemnos Island, Greece, 8 - 10 September 2003 Full paper Vol. B, pp. 401 - 407

Karayannidis G.P., Kokkalas D.E. & Bikiaris D.N, (2003), *Solid-state polycondensation of poly(ethylene terephthalate) recycled from postconsumer soft-drink bottles,* Journal of Applied Polymer Science 56 (2003) 2165 - 2142

Karim A., Amis E., Yurekli E., Krishnamoorti R. & Meredith C., (2002), *Combinatorial methods for polymer materials science: Phase behavior of nanocomposite blend films,* Polymer Engineering and Science 42 (2002) 1836-1840

Kiatkamjornwong, S., Surunchanajirasakul, P. & Tasakorn P., (2001), *Natural rubber-cassava starch foam by compression moulding,* Plastics, Rubber and Composite Processing and Applications 30 (2001) 318-327

Kim M.H. & Gogos C.G. , (2001), *Melting phenomena and mechanism in co-rotating twin-screw extruder,* ANTAC 2001

Kumar V., Waggoner M., Kroeger L., Probert S.M. & Nadella K, (2001),*Microcellular Recycled PET Foams for Food Packaging, 2001*

Lacoste J.F., Bounor-Legare V., Llauro M.F., Monnet C., Cassagnau P. & Michel A., (2005), *Functionalization of poly(ethylene Terephthalate) in the melt state: Chemical and rheological aspects,* Journal of Polymer Science 2005

Lingaiah S., Kunigal N. S., Sadler R. & Sharpe M., (2005), *A method of visualization of dispersion of nanoplatelets in nanocomposites,* Composites Science and Technology 65 (2005) 2276-2280.

Lochhead R.Y., Camille T. Haynes, Stephen R. Jones & Virginia Smith, (2006),*The high throughput investigation of polyphenolic couplers in biodegradable packaging materials,* Applied Surface Science 252 (2006) 2535-2518.

Longzhen Q., Wei C. & Baojun Q, (2006), *Morphology and thermal stabilization mechanism of LLDPE/MMT and LLDPE/LDH nanocomposites,* Polymer 47 (2006) 922-930

Martinez J.G., Benavides R. & Guerrero C., (2008), *Compatibilization of Commingled Plastics with Maleic Anhydride Modified Polyethylenes and Ultraviolet Preirradiation,* Journal of Applied Polymer Science 108 (2008) 2597-2603.

Minoru Genta & Fumitoshi Yano, (2003), *"Development of Chemical Recycling Process for post-Consumer PET Bottles by Methanolysis in supercritical Methanol"* Mitsubishi Heavy industries Ltd., Technical Review Vol.40 Extra No.1 (Jan.2003)

Monika Gneuss, (2007), "Processing PET bottles flakes into nonwovens with fully automatic filtration and on line IV monitoring", http://www.ce-pip.com/docs/Gneuss_bottle_flakes_into_nonwovens.pdf, 19/2007

Morawiec J., Bartczak Z., Pluta M. & Glalesk A. I, (2002), *High-strength uniaxially drawn tapes from scrap recycled poly(ethylene terephthalate),* Journal of Applied Polymer Science 86 (2002) 1426 - 1435.

Novis E., (2003) *"PET recycling increases but tighter regulation needed"*, newsletters - Food, Beverage & Nutrition, 2003.

Paci M. & La Mantia F.P., (1998), *Competition between degradation and chain extension during processing of reclaimed poly(ethylene terephthalate),* Polymer Degradation 61 (1998) 417 - 420

Pralay M., Pham Hoai Nam, Masami O., Tadao K., Naoki H. & Arimitsu Usuki, (2002),*The effect of crystallization on the structure and morphology of polypropylene/clay nanocompositesi*, Polymer Engineering and Science 42 (2002) 1864-1871

Pegoretti A., Kolarik J., Peroni C. & Migliaresi C., (2004), *Recycled poly(ethylene terephthalate)/layered silicate nanocomposites: Morphology and tensile mechanical properties,* Polimer International 2004

Pham Hoai Nam, Pralay Maiti, Masami Okamoto, Tadao Kotaka, Takashi Nakayama, Mitsuko Takada, Masahiro Ohshima, Arimitsu Usuki, Naoki Hasegawa & Hirotaka Okamoto, (2002), *Foam processing and cellular structure of polypropylene/clay nanocomposites* Polymer Engineering and Science 42 (2002) 1910 - 1938

Place I.A., Penner T.L., McBranch D.W. & Whitten D.G., (2003), *Layered nanocomposites of aggregated dyes and inorganic scaffolding,* Journal of Physical Chemistry A 107 (2003) 3169-3177

Potente H. & Flecke, *Analysis and modeling of the residence time distribution in intermeshing co-rotating, twin screw extruders based on finite element simulation,* ANTEC 1997

Raki L., J.J. Beaudoin, L. Mitchell, *Layered double hydroxide-like materials: nanocomposites for use in concrete,* Cement and Concrete Research 34 (2004) 1717-1724.

Ren J., Casanueva B.F., Mitchell C.A. & Krishnamoorti R., (2003), *Disorientation kinetics of aligned polymer layered silicate nanocomposites,* Macromolecules 36 (2003) 4188-4194.

Rossi G.B., Beaucage G., Dang T.D. & Vaia G., (2002), *Bottom-Up Synthesis of Polymer Nanocomposites and Molecular Composites: Ionic Exchange with PMMA Latex,* Nano Letters 2 (2002) 319-323

Rosu R.F., Robert A. Shanks & Sati N. Bhattacharya, (1999),*Synthesis and Characterisation of branched Poly(ethylene terephthalate),* Polymer International 42 (1999) 267 -275

Rosu R.F., Robert A. Shanks & Sati N. Bhattacharya, (1999), *Shear rheology and thermal properties of linear branched poly(ethylene terephthalate) blends,* Polymer International 1999.

Sandro Donnini Mancini & Mari zanin, (1999), *Recyckability of PET from virgin resin,* Materials Research 2 (1999) 33-38

Scheirs J., (1998), *Polymer recycling, science, technology and application.* John Wiley and Sons 1998

Schmidt H.K., Geiter E., Menning M., Krug H., Becker C. & Winkler R.P., (1999), *The sol-gel process for nano-technologies: New nanocomposites with interesting optical and mechanical properties,* Journal of Sol-Gel Science and Technology 13 (1-3) (1999) 397-404

Shanti V. N., Goettler L.A. & Bruce A. Lysek, (2002), *Toughness of nanoscale and multiscale polyamide-6,6 composites,* Polymer Engineering and Science 42 (2002) 1872-1882

Shen L., Wuiwui Chauhari T. & Tianxi L., (2005), *Nanoindentation and morphological studies on injection-molded nylon -6 nanocomposites,* Polymer 46 (2005) 11969-11977.

Soares D.A.W., de Souza P.H.o., Rubinger R.M., de Queiroz A.A.A., Higa O.Z. & de Souza I.R., (2004), *AC Electrical Conductivity of Cr-Doped Polyaniline/Poly(vinyl alcohol) Blends,* Brazilian Journal of Physics, vol. 34, no. 2B, 2004.

Spinace M.A. & De Paoli M.A., (2001), *Characteriztion of Poly(ethylene terephtalate) after multiple processing cycles,* Journal pf Applied Polymer Science 80 (2001) 20-25

Swoboda B., Buonomo S., Leroy E. & Lopez Cuesta J.M., (2008), *Fire retardant poly(ethylene terephthlate)/polycarbonate/triphenylphosphite blends,* Polymer Degradation and Stability 93 (2008) 910 - 917

Tannenbaum R., Reich S., Flenniken C.L. & Goldberg E.P., (2002), *Shape control of iron oxide nanoclusters in polymeric media,* Advanced Materials 14 (2002) 1402-1405

Teh P.L. , Mohd Ishak Z.A., Hshim A.S., Karger-Kocsis J. & Ishiaku U.S., (2004), *Effects of epoxidized natural rubber as a compatibilizer in melt compouned natural rubber-organoclay nanocomposites,* European Polymer Journal 40 (2004) 2513-2521.

Torres N., Robin J.J. & Boutevin B., (2000), *Study of thermal and mechanical properties of virgin and recycled poly(ethylene terephthalate) before and after injection molding,* European Polymer Journal 36 (2000) 2075-2080

Torres N., Robin J.J. & Bountvin B., (2001), *Chemical modification of virgin and recycled poly(ethylene terephthalate) by adding of chain extenders during processing,* Journal of Applied Polymer Science 79 (2001) 1816 - 1824

Tortora M., Gorrasi G., Vittoria V., Galli G., Ritrovati S., Chiellini E., (2002), *Stuctural characterization and transport properties of organically modified montmorillonite/polyurethane nanocomposites,* Polymer 43 (2002) 6147-6157

Unnikrishnan G. & Sabu Thomas, (1998), *Interaction of crosslinked natural rubber with chlorinated hydrocarbons,* Polymer, 1998, Volume 39, Number 17, p. 3933-3938.

Utracki L.A. & Kamal M.R., (2002), *Clay-containing polymeric nanocomposites,* Arabian Journal for Science and Engineering 27 (2002) 43-68

Van Meerveld J., Ottinger H.C. , Hutter M., Flow-induced Crystallization, 2007

Vasiliu E., Wang C.S. & Vaia R.A., (2002), *Preparation of optically transparent films of poly(methyl methacrylate) (PMMA) and montomorillonite,* Materials Research Society Symposium-Proceedings 703 (2002) 243-248

Villalobos M., Awojulu A., Greeley T., Turco G. & Deeter G., (2006), *Oligomeric chain extenders for economic reprocessing and recycling of condensation plastics,* Elvesier 2006

Volker F., Dietrich Scherzer, Villalobos M., (2008), *Multifunctional polymers as chain extenders and compatibilizers for polycondensates and biopolymers, ANTEC (2008) 1682*

Warburton S.C., Donald A.M. & Smith A.C., (1992), *Structure and mechanical properties of brittle starch foams,* Journal of Materials Science 27 (1992) 1469-1474

Xanthos M., Yilmazer U., Dey S.K. & Quintas J., (2004), *Melt viscoelasticity of polyethylene terephthalate resins for low density extrusion foaming,* Polymer Engineering and Science, 2000, vol.40,no.3, 554 - 566

Xanthos M, Young M.W., Karayannidis GP, Bikiaris DN, " *Reactive modifdication of poluethylene terephtalate with polyepoxides",* polym.eng.Sci., 2001, 41(4), 643-55;

Yilmazer U., Xanthos M., Bayram G. & Tan V. , (2000), *Viscoelastic characteristics of chain extended/branched and linera polyethylene terephthalate resins,* Journal of Applied Polymer Science 75 (2000) 1371 - 1377

Yong Tang, Menachem Lewin, (2008*) New aspects of migration and flame retardancy in polymer nanocomposites,* Polymer Degradation and Stability 93 (2008) 1986-1995.

Zammarano M., Séverine B., Jeffrey W.G., Franceschi M., Frederick L. Beyer, Richard H. Harris, Sergio Meriani, *Delamination of organo-modified layered double hydroxides in polyamide 6 by melt processing*, Polymer 47 (2006) 652-662.

Zhang J., Jiang D.D. & Wilkie C.A., (2001), *Thermal and flame properties of polyethylene and polypropylene nanocomposites based on oligomerically-modified clay,* Polymer Degradation and Stability 91 (2006) 298-304.

Zhong Zhang, Jing-Lei Yang & Klaus Friedrich, (2004), *Creep resistant polymeric nanocomposites,* Polymer 45 (2004) 3481-3485.

Zilg C., Reichert R., Dietsche F., Engelhardt T., Muelhaupt R., (1998), *Plastics and rubber nanocomposites based upon layer*

4

Poly(bisphenol A carbonate) Recycling: High Pressure Hydrolysis Can Be a Convenient Way

Giulia Bozzano, Mario Dente and Renato Del Rosso
Politecnico di Milano
Italy

1. Introduction

Polycarbonates are polymers characterized by a carbonate group. Their molecular structure can be of various kinds depending on the unit (G) that is connected with the carbonate. Table 1 reports some of the most used polycarbonates.

group G	*Polymer name*	
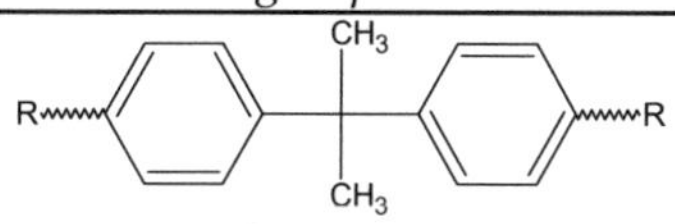	Poly(bisphenol A)carbonate	BAPC
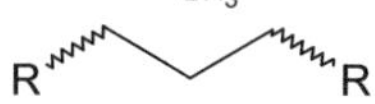	Polytrimethylencarbonate	PTMC
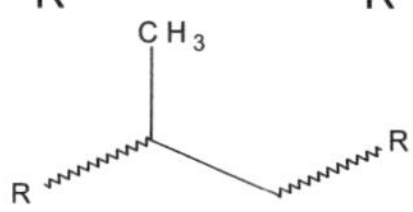	Polypropylencarbonate	PPC
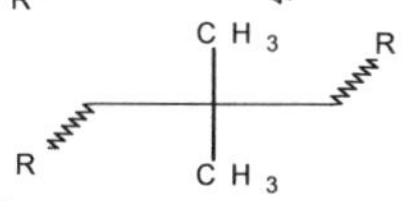	Polyneopentylencarbonate	PNPC
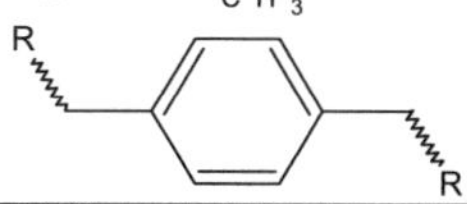	Poly(p-xylilene)carbonate	PPXC

Table 1. most diffused polycarbonates.

Among them, the poly(bisphenol A) carbonate (PC in the following) is the most diffused. It is commonly referred as polycarbonate because of its vast number of industrial applications. It is a lightweight, high-quality plastic. It is well known and appreciated for its transparency, its excellent resistance to impact and its ability to withstand high temperature during the lifespan of the final article. Generally speaking, materials based on polycarbonates are resistant, rigid till 140°C and not fragile under -20°C. They are amorphous and have excellent mechanical properties and dimensional stability. Some restrictions in their use consist in a limited resistance to chemicals and to scrapes, and to color changes after exposition to UV ray. These problems can be solved by means of the proper additives or making use of mechanical mixing with other polymers. The main physical properties are resumed in table 2.

Physical properties		
Traction resistance	70-80	N/mm^2
Impact resistance	60-80	kJ/m^2
Maximum temperature	125	°C
Density	1.2	g/cm^3

Table 2. PC physical properties.

PC also shows a high limiting oxygen index (LOI=27), and can produce a large amount of char on combustion conditions. PC is widely used in mixture with other polymers with the aim of enhancing resistance to external factors. Typical mixtures include PC with poly(butylene-terephthalate) or with ABS. These latter exhibit effective flame retardant properties upon the addition of conventional halogen and/or non halogen flame retardant agents, and this supports their large use in electrical appliances. Its properties make it appropriate for durable goods applications. PC is used in the construction of many everyday products, including CDs and DVDs, dinnerware, computer casings, medical equipments, bicycle helmets, automotive parts, packaging, sports and optical materials. Other applications are for paintings and covertures of buildings.

Two industrially significant syntheses for PC are mainly adopted. They were developed in 1960s . The first one was developed by Bayer in 1962 and consists in a two phase reaction (Schnell et al., 1962). In this process bisphenol A (BPA) is added first to the reactor in methylene chloride (with a monohydroxylic phenol to control molecular weight). Subsequently, phosgene is added to the reactor, along with aqueous sodium hydroxide (HCl scavenger), to produce a biphasic liquid-liquid system. This process allows obtaining high molecular weight polymer with excellent optical clarity and color. Major disadvantage is the use of phosgene and the generation of a large amount of wastewater and methylene chloride to be treated or disposed. The second process, developed by GE in 1964, is a melt transesterification between diphenil carbonate and BPA (Fox, 1964a, Fox 1964b). This results in intermediate molecular weight product with phenol as a condensation byproduct. This route is solvent free and avoids the direct use of phosgene. The high viscosity of the melt limits however the final molecular weight of the polymer.

According to the research "Polycarbonate: 2009 World Market Outlook and Forecast" (http://www.reportlinker.com) the PC market has been shown in not favourable perspectives in recent times. In facts the application of PC in optical media segment shrank as a result of lower demand for CDs and DVDs. Moreover a health concern raised over Bisphenol A (BPA) has negatively influenced the demand as well. PC could disappear from the food and beverage container market in the future. Notwithstanding, the global demand is expected to grow of 6-7% annually, driven by Asia and China in particular (2.4 Mt were produced in 2004). Demand would catch up with production capacity and the market would strengthen. This fact suggests that PC recycling will cover in the future more and more importance. It is therefore necessary to optimize and develop processes for PC wastes treatment. Because it is not a suitable alternative for waste treatment to landfill or incinerate wasted PC products, it is important to find resourceful recycling processes both for environmental protection and for economical benefits purposes. PC recycling can be performed in three main different ways: direct recycle (mechanical recycling or blending with other materials), pyrolysis and chemical treatment.

Mechanically recycled PC is less resilient, because it has decreased impact resistance when compared with newly manufactured polycarbonate. The addition of fillers and pigments can also decrease the plastic's resilience. This problem can be addressed by blending with other materials to modify impact resistance in recycled polycarbonate (Elmaghor et al., 2004).

Thermal degradation or pyrolysis is intrinsically characterized by a low selectivity. This is due to the prevailing radical mechanism of the thermal decomposition that is substantially acting randomly over the backbone of the polymer macromolecules and over the obtained monomer, then producing their further degradation. It is well known that in this kind of radical reactions, all the H active positions can be attacked by metathetical transfer reaction of H and then giving place to different products. Moreover additive reactions can occur on aromatic rings, producing more and more condensed molecules, which are char precursors. Of course, the most active H positions are those of the two methyl groups. It is worthwhile to mention that these kind of attacks are active also on the formed BPA. Moreover the simultaneous crosslinking elementary process, and polycondensation reactions on aromatic rings, give place to the formation of char precursors. These aspects make the simple thermal degradation a not particularly suitable process because of the low selectivity in restoring monomer and the large amount of byproducts.

Several studies of possible depolymerization processes, based on the use of chemicals, have been reported in the literature for decomposing PC to its original monomer, BPA. They are based, for instance, on the use of solvent systems (Pan et al., 2006; Piñero et al., 2005) such as methylene chloride with ammonia, phenol in combination with an alkali catalyst, or via trans-esterification (alcoholysis) in super- or sub- critical conditions. These processes can require a complicate product separation in addition with environmental safety problems related to the use of more or less toxic organic solvents. Also decomposition of PC in sub- and super- critical water has been taken into account by Tagaya et al. (1999).

The chapter firstly resumes and shortly analyzes most of the proposed processes. Then, the results obtained by adopting a PC recycling process based on hydrolysis using subcritical liquid water are reported (Bozzano et al., 2010). In this study both pure PC and CDs wastes are used. The driving concept came from an analogy with the fats hydrolysis producing fatty acids and glycerol in the soap production field or with the hydrolysis of oils (see for instance Khuwijitjaru et al., 2004 and Pinto & Lanças, 2006;). A concerted path de-polymerization mechanism is proposed. A characterization of the process kinetics is presented and compared with lab-scale experimental data. The results show that this process can be a valuable alternative for BPA recovery mainly for its simplicity and absence of toxic agents or non-desired byproducts. It is of interest to mention that similar hydrolysis mechanisms can take place in other fields, like for instance the production of bio-oils from biomasses.

2. Pyrolysis of poly(bisphenol A carbonate)

Pyrolysis is a thermal process taking place typically in the temperature range of 300-1000°C in absence of oxygen. It decomposes organic molecules in gaseous and carbonaceous products. After cooling, the vapors give place to condensed mixtures (the so called tars). Uncondensed products are typically CO, H_2, CH_4 and other hydrocarbons with low molecular weight. The ratio among solid, liquid and gaseous products depends on the pyrolysis conditions in terms of temperatures and residence times. As mentioned before, pyrolysis is one of possible routes for PC recycling (Jang & Wilkie, 2004). The idea is to transform plastic wastes to fuels or valuable products. Several papers related to the study of PC degradation are reported into the literature (Day et al., 1999; McNeil & Rinchon, 1991; McNeil & Basan, 1993; Montaudo & Puglisi, 1992; Montaudo et al., 2002; Oba et al., 2000; Puglisi et al., 1999). Also the pyrolysis in presence of active catalysts has been studied (Ali et al., 2002; Chiu et al., 2006). Figure 1 shows some thermogravimetries of PC from three different research groups. Decomposition occurs in the temperature range of about 300-500°C. The remaining residue is in the range of 20-30 wt%.

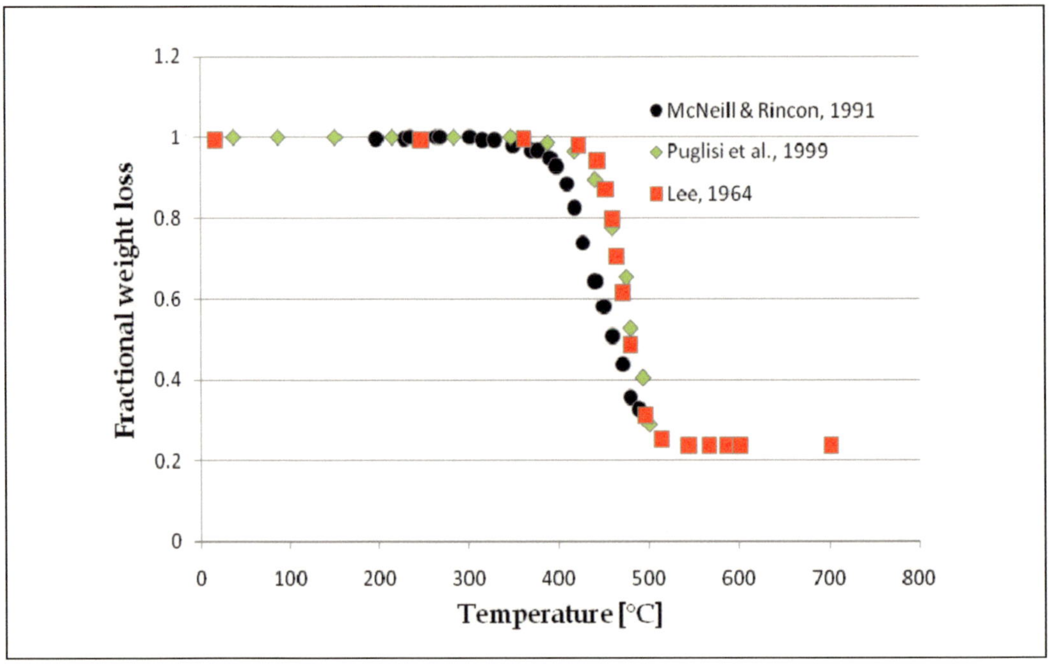

Fig. 1. fractional weight loss during thermal degradation of PC.

At 360°C, gas fraction is composed (wt%) by: 95% CO_2, 3% CO and 2% CH_4. At higher temperature CO and CH_4 increase their amount (Davis & Golden, 1968).

The condensed pyrolysis product has been found to be constituted by resins and a crystalline part. The latter contains BPA which amount depends on the operating conditions (time and temperature). In table 3 phenolic products obtained by different researchers are reported. Non phenolic products are present in minor amounts.

Authors report also the presence of xanthone units, which examples of structures are reported in figure 2. Dibenzofurane and fluorenone are also present in the residue, together with ethers and products coming from crosslinking, substitutive addition and branching reactions. This brief information regarding products obtained from pyrolysis of PC shows that thermal decomposition is not offering a good solution for PC recycling because of the difficulty of separation and low selectivity.

Phenolic products	*Lee(475°C)*	*Davis-Golden (360°C)*
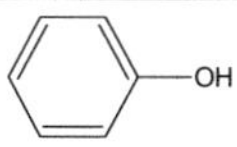	Large amounts	Significative amounts
	Large amounts	Present
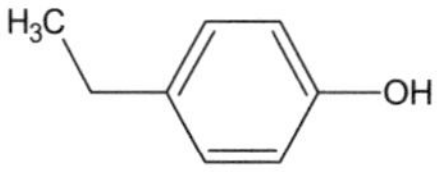	Large amounts	Present
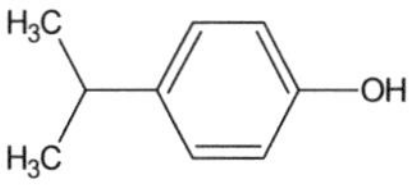	Large amounts	Present
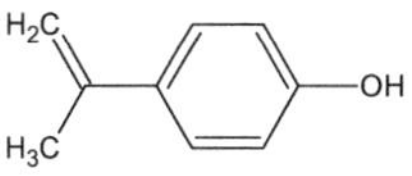	Small amounts	
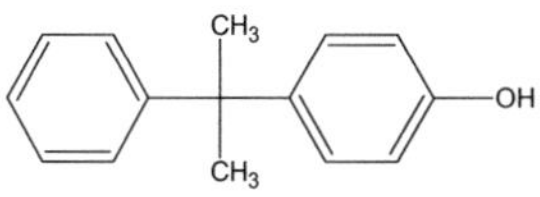		Significative amounts

Table 3. main phenolic products obtained from pyrolysis.

Some researchers (Chiu et al., 2006) have also studied thermal degradation in presence of an active catalyst. The expected advantage is to lower the required temperature, to shorten the degradation time, to increase the extent of the degradation, reduce the portion of solid residue and to narrow the product distribution (Ali et al., 2002). In their paper, Chiu et al. report experiments related to the use of ten different metal chlorides. They found $SnCl_2$ and $ZnCl_3$ to be catalytically active for degradation. They can efficiently perform the expected improvements in thermal treatment of PC and, therefore, some benefits can be obtained. In particular by using these catalysts, liquid degradation products are reduced from 20 to respectively 10 and 8.

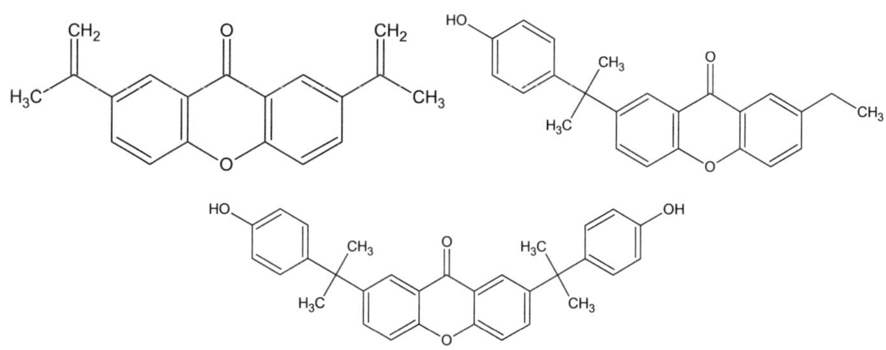

Fig. 2. Examples of xanthone structures.

No chlorinated products were formed. Main obtained liquid products are phenol, p-iso-propylphenol, diphenil carbonate and BPA.

Other researchers (Mitan et al., 2007) proposed to co-pyrolyse at 450°C PC wastes (as for instance DVDs and CDs) with vegetable cooking oil (VCO). They found an increased yield in liquid fraction, at the expense of solid residue, not coming just from the sum of the liquid and solid yields of individual components. This fact was also confirmed by the analysis of degradation rates that results increased by the addition of VCO. Metal contained into DVDs and CDs became 2.5 more concentrated into the solid residue compared to that deriving from thermal treatment of individual reagents.

Grause et al. (2009) studied the pyrolysys of PC with steam in the presence of earth –alkali oxides and hydroxides at temperature between 300°C and 500°C. The experiments were carried out in a semibatch reactor, in steam atmosphere, in the presence of MgO, CaO, $Mg(OH)_2$, $Ca(OH)_2$. All these catalysts accelerated the decomposition, in comparison with the hydrolysis of PC alone, MgO and $Mg(OH)_2$ resulting more effective that CaO and $Ca(OH)_2$. BPA were obtained as the main product at 300°C with a yield of 78% with MgO. Grause et al. (2009) suggested then to further increase the temperature at 500°C in order obtain high yields of phenol and isopropenylphenol and a drastic reduction of the residue.

3. Ester exchange reaction and hydrolysis of poly(bisphenol A carbonate)

Obviously pyrolysis presents some disadvantages such as non-uniform heat flux distribution, low yield of the preferred products and excessive char and gas formation, even if some mitigations towards these aspects have been shown to be possible by using catalysts or by co-pyrolysis. Degradation in liquid solution has been proposed to overcome some of these problems. In this method, the polymer is dissolved in a solvent and degraded thermally in a single phase. Yoshiki et al. (2005) obtained high yields of oil but low selectivity of BPA using tetralin, decalin and ciclohexanol as solvent. Also methylene chloride with ammonia or phenol in combination with an alkali catalyst were applied successfully for PC decomposition (Fox & Peters, 1989; Shafer, 1994). In these cases, however, a complicated product separation is required in addition to environmental problems related to the toxicity of the solvents. Supercritical and subcritical fluids are attractive solvents. They have unique properties, many of their physical and transport properties are intermediate between liquid and gas. The diffusion limited reactions are enhanced because of the increased solubilities and diffusivities. Polymers can therefore be degraded to low molecular weight products by thermolysis in supercritical fluids with high degradation rates. Water and alcohols were mainly used as sub- and supercritical fluids in depolymerization processes of plastics (Goto, 2009). In most of the case the role of the sub- or supercritical fluids is to act both as a solvent reaction medium and as a reactant. In table 4 are reported the critical parameters of water and some simple oxygenated compounds. The critical temperature of methanol is lower than that of water. Properties of water, such as dielectric constant and ion product, change drastically around the critical point (Kruse & Dinjius, 2007). Thus, special effect of water can be expected. When the de-polymerized products are not sufficiently stable, and solvent and supercritical condition are required, alcohols may be better solvents than water, because of their lower critical temperature. During the de-polymerization process, plastic phase is often solid or melt status. In that case, dissolution of solvent into solid or melt phase is essential to enhance the reaction rate, especially at the beginning of de-polymerization. The selection of the appropriate solvent will depend on the evaluation of the efficiency of reaction, separation and purification process.

	Critical temperature [°C]	Critical pressure [MPa]
Water	373.94	22.064
Methanol	239.48	8.097
Ethanol	240.76	6.148
1-propanol	263.62	5.175
Acetone	234.94	4.700
Benzyl alcohol	441.84	4.300

Table 4. critical parameters of water and alcohols.

Pyrolysis of PC in subcritical and supercritical water was studied by Tagaya et al. (1999). Water has the advantage to be in-expensive and non-toxic. In these conditions the decomposition reaction is more selective and the products are reduced to phenol, bisphenol A, p-isopropenylphenol and p-isopropylphenol.

Tagaya et al. (1999) decomposed PC in the temperature range from 230°C to 430°C. No decomposition occurred below 230°C even for a reaction time of 24 h. Production of phenol, BPA, p-isopropenylphenol and p-isopropylphenol begin starting from 240°C. At 250°C a significant decomposition starts after 2 h, while at 300°C, after 2h, 38.1% of the former products were obtained. However estension of reaction time resulted ineffective for increasing the yield. NaCl and CH_3COOH addition was no effective for decomposition reaction. The latter results to be accelerated by the addition of Na_2CO_3. At 300°C no BPA was detected in the product because of its transformation into phenol catalysed by Na_2CO_3. Phenol reached in these conditions, after 10 h, a yield of more than 30% indicating that subcritical water is also an advantageous medium for decomposing polymeric compounds. On the contrary supercritical water, in decomposing polymeric materials, has the disadvantage to cause rapid corrosion of equipments at very severe operating conditions. Chen et al. (2004) observed that high yields of BPA and dimethylcarbonate (DMC) were obtained using methanol. DMC is a valuable byproduct, a nontoxic and environmentally benign carbonylating and methylating agent that can replace hazardous chemicals like for instance phosgene. It is also a raw material useful for producing again polycarbonate resins. Also Dongpil et al. (2009) studied PC methanolysis mechanism in the temperature range 160-220 °C and with reactions time until 180 min. The increase of molar ratio between methanol and PC has been found to improve both DMC and BPA yields but, of course, it demands larger capital and energy costs.

Hu et al. (1998) studied alkali catalysed methanolysis using a mixed solvent of methanol and toluene. They found that by using more than a stoichiometric amount of methanol, DMC was produced in quantitative yields, highlighting the importance of adopting also an optimized ratio between PC and methanol.

Piñero et al. (2005) used methanol-water mixtures and developed (Piñero et al., 2006) a shrinking particle model to describe the reactive dissolution of the BPA-PC particles in semicontinuous de-polymerization of polycarbonate. Jie et al. (2006) studied decomposition in ethanol that has a critical point practically at same temperature and a lower pressure than that of methanol, allowing lower operating temperatures and pressures. Comparing the use of methanol and ethanol as solvents, it is reported that PC completely decomposes in supercritical methanol at 290°C and 9.96 MPa, in supercritical water above 374°C and pressure higher than 22.1 MPa, while in ethanol this is possible above 243 °C and 6.38 MPa producing BPA and diethylcarbonate (DEC). The mechanism consists in the random reaction along the polymer chain of the ester linkage with the solvent , that produces two smaller polymer chains, which can still react by ester exchange reaction until the polymer is completely converted to BPA and another product depending on the solvent.

In all the reported examples of alcoholysis yields of 90% of BPA (in terms of weight of obtained BPA divided the weight of initial PC) were obtained. The activation energy deduced from experimental data are resumed in table 5 (paragraph 6).

4. Hydrolysis near the critical point

In order to extend the results obtained by Tagaya et al. (1999), several experiments have been performed by adopting liquid water in the more convenient subcritical conditions. The results obtained in decomposing PC and producing BPA, will be reported in the following.

4.1 Experimental

4.1.1 Materials

The reagent grade Poly(bisphenol A carbonate) (average MW 64000) [CAS 25037-45-0], Bisphenol A (>99%) [CAS 80-05-7], Phenol (>99%) [CAS108-95-2] and other chemical used were all purchased from Sigma-Aldrich. Water was twice distilled. Commercial Recordable Compact Discs crashed for de-polymerization tests were Verbatim Datalife CD-R.

4.1.2 Apparatus and methods

De-polymerization tests were carried out in a 316 stainless steel tubular batch reactor (internal diameter 7.8 mm, length 150 mm) having an internal volume of 7.1 mL. The tests were performed by first weighting the empty reactor and then charging it with 1 g of BAPC and 1 g of water. The reactor, exactly weighted after charge, was put into a laboratory fan assisted furnace (Heraeus M110 electronic), preheated at the temperature set point, over a support disposed along the symmetry axis of the oven. A thermocouple fixed on the external wall of the reactor measured temperature level. Both horizontal and vertical disposition were studied. At the end of the experiment the reactor was recovered from the oven, rapidly cooled and newly weighted at ambient temperature. Not more than ±0.2 mg of weight difference from initial and final total weight was accepted as a probe of no spill and good capping for the test, otherwise the test was repeated. The reactor was unhead and the evolved gas collected and GC analyzed. The degassed reactor was finally weighted and the difference was assumed as CO_2 produced. The yield was evaluated from this latter information. A Mettler Toledo analytical precision balance (model B154-S) was employed.

By cooling the reactor, the resulting content (at room temperature) is constituted only by condensed phases. The liquid and solid internal material was discharged, identified and quantified by FTIR and GC.

4.1.3 Some analytical aspects

The solid sample obtained from the tests was characterized by Shimadzu FTIR: IRAffinity-1 in the 500 - 4000 cm^{-1} range (KBr disc). Gas chromatographic analyses of solid products were carried out (using methanol as solvent) on a GC-FID HP 6890plus with SLB-5ms capillary SUPELCO column (30 m length, 1 μl injection volume, split ratio 1:10, Helium carrier 1.4 mL/min, constant flow). The temperature was held at 65°C for the first 2 minutes, then increased at 255°C at a heating rate of 10°C/min and kept at this temperature for 15 minutes. The main products were identified and quantified by comparing the retention time with standard compounds.

Figure 3 shows a typical gas chromatographic analysis of the condensed phase discharged at the end of a test at high conversion. As it can be observed BPA is the main product. Gas chromatographic analyses of evolved gas were performed with a column 0.53 mm ID, Molecular Sieve 5A as stationary phase, GC-T as detector, isothermal at T=25°C. Permanent gases and CO_2 were detected.

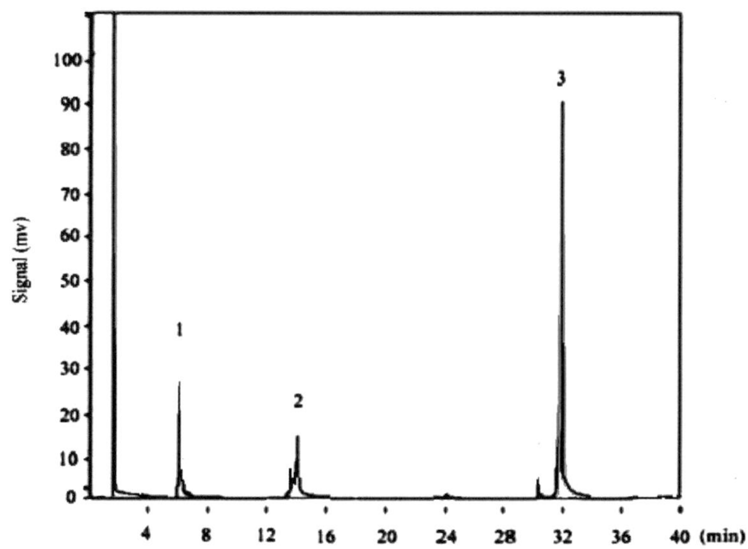

Fig. 3. Typical gas chromatographic analysis of solid product obtained by the tests.

Figure 4 shows FT transmittance plot showing the FTIR analyses of commercial poly(bisphenol A carbonate): (A) starting material obtained from a CD crash, (B) low conversion depolymerization solid discharged material, (C) medium conversion material, (D) solid material discharged at complete depolymerization. The FTIR spectra of reagent grade BPA, p-isopropylphenol and phenol are also reported. The figure 4 D shows the same shape as pure BPA.

4.2 Results and discussion

The experimental tests on hydrolysis have been performed at temperatures ranging from 240 to 290 °C (corresponding to a pressure range of 3.5-8 MPa). The latter temperature has been selected as a maximum, because it is known that BPA, obtained by hydrolysis of PC, starting from 300°C decomposes giving place to phenol (Tagaya et al., 1999).

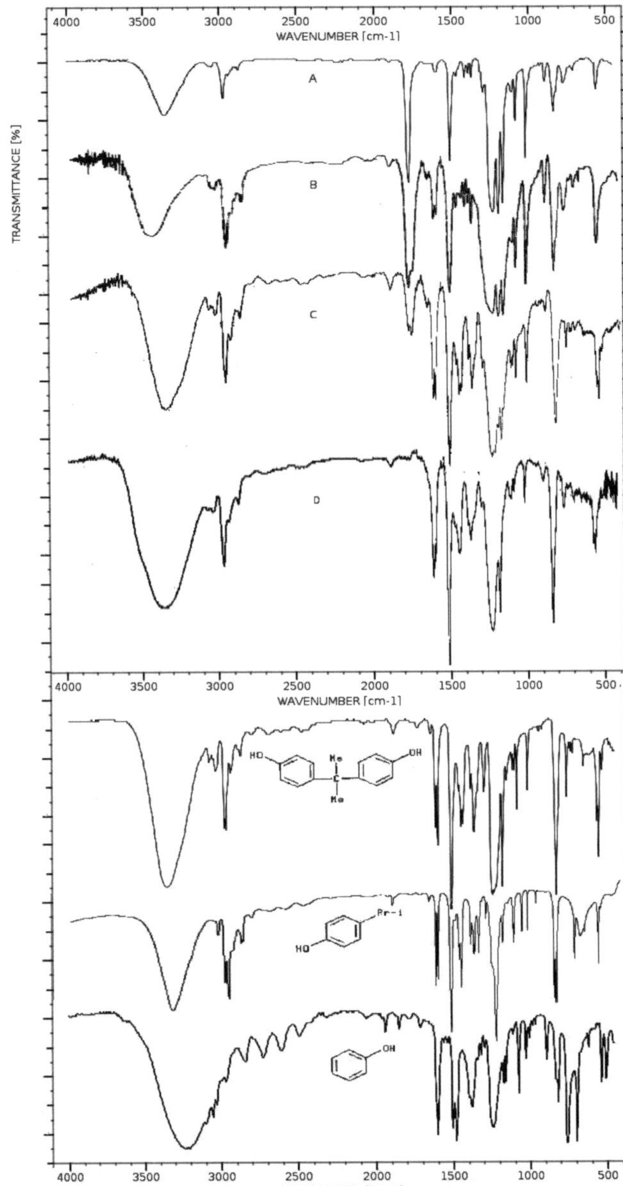

Fig. 4. FT transmittance plot from starting material until 100% conversion compared with BPA, p-isopropylphenol and phenol FTIR spectra.

This was also confirmed by some preliminary tests that have been performed. Different residence times have been adopted. Because of the competition existing among the concerted path hydrolysis reactions and the radical path (non selective) reactions, the selected temperature range has to be considered optimal. In facts the apparent activation energy of the pyrolysis is quite higher (40-80 kJ/mol more). The hydrolysis yields have been evaluated by means of both the amount of CO_2 released and the weight loss of the sample (of course the last one is comprehensive of all the gaseous matter produced). In the preliminary tests it has been verified that the horizontal position of the reactor inside the furnace is largely preferable. In fact the reaction takes place starting from the contact surface between the molten and swollen polymer and the water, producing fractures and pores that favor the further penetration of water (as also reported by Dongpil et al., 2009 in the case of methanolysis). It is important to point out that the horizontal configuration allows enhancing the surface contact between the molten immiscible polymer and the water and, therefore, the apparent global reaction rate.

Figure 5 reports the conversion versus time at different temperatures. One test has been performed for a time as long as 4 h. There is an apparent induction time for the process that can be essentially attributed to the time required for the heating of reagents inside the reactor. In order to understand whether some autocatalytic effect can be related to the products, it has been decided to perform also some tests by adding small amounts of BPA: practically no significant effect has been observed. The result obtained by Kitahara et al. (2009) in subcritical conditions are also reported in the figure.

The obtained data have been used for deducing the reaction rate and the apparent activation energy of the reaction. Figure 6 reports the experimental reaction rate constant as a function of 1/T. This diagram incidentally can be considered preliminary to deduce in a simple way the activation energy of the reaction that results to be about 80 kJ/mol (in agreement with other experiences).

The gases produced by the hydrolysis reaction was evaluated not only in terms of their total amount but, periodically, they were also analyzed by GC in terms of their composition. First of all the analyses have clearly shown that N_2/O_2 ratio, in the effluent gases, were practically the same of the original air inclusion. This result simply means that no oxidative degradation was taking place during hydrolysis. Indirectly, the last considerations are also showing that practically no extra components are entering during the periodical filling and emptying of the reactor. In the experiments with the highest de-polymerization degree the products obtained have been analyzed after mixing with methanol. The results show always a high purity in BPA in comparison with byproducts like phenol, p- propyl and propylidene phenols. This confirms that the polymer decomposition through breakage of the bond C aromatic-C isopropylic (i.e. pyrolysis) is negligible in practice and that this parasitic radical mechanism doesn't occur significantly in the selected operating conditions.

The FTIR analysis on the reaction residue after evaporation of the water shows a spectrum that is coherent with a progressive hydrolysis of the carbonate bond, without appreciable evidences of de-alkylation followed by formation of terminal propylidene and phenol. The IR spectra at higher conversion are practically coincident with those of standard BPA.

On the basis of these evidences, it can be affirmed that the de-polymerization reaction proceeds in the condensed phase and, under the preferred conditions, it regenerates the monomer. It is important to point out that after the tests at the highest conversion level (i.e. BPA yield >90% wt), the melting point of the product resulted over 145-150 °: this is another excellent indication of the substantial purity of the obtained raw monomer.

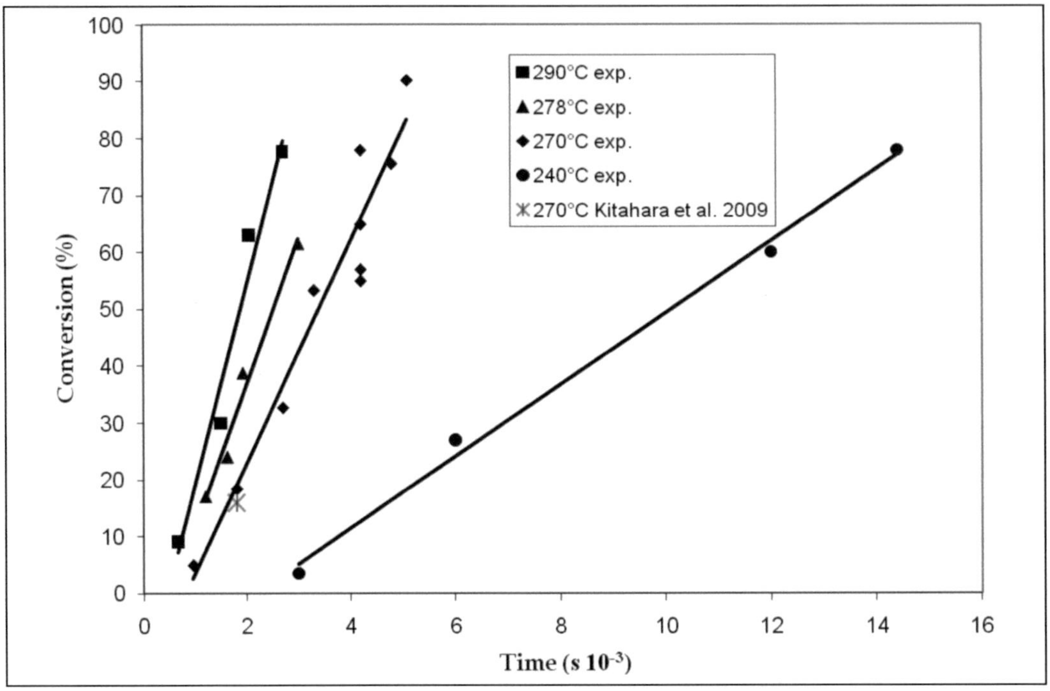

Fig. 5. Yield of BPA versus time at different temperatures.

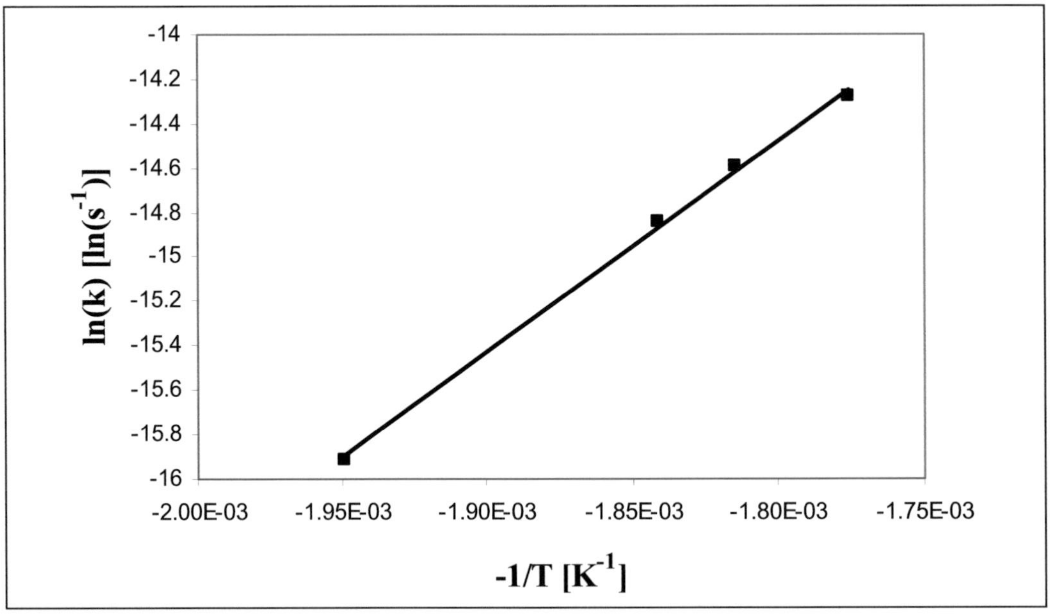

Fig. 6. Experimental kinetic constant (s^{-1}) as a function of -1/T [K^{-1}].

These results seem to be in partial contradiction with some of those obtained by Tagaya et al. (1999) indicating low BPA yields (around 35%) after 2 h reaction at 300°C . In that paper no indication was given on modalities of contact between molten polymer and water. A hypothesis on the difference of the obtained results can be related to a non adequate surface contact in their experiments. On the contrary other data supports our results (Kitahara et al., 2009, see figure 5). In facts, after hydrolysis of PC with water for 30 minutes at 270°C, they obtained a BPA yield of 16% that is substantially in line with our results.

Regarding the experiments performed with pure PC or with crashed CDs, essentially no difference was observed.

5. Hydrolysis kinetic mechanism

It is interesting to observe that PC hydrolysis is mainly dominated by a six center concerted path reaction mechanism. By analogy with other reactions following the same mechanism, involving hydrolysis of esters, the kinetic constant suggested for every elementary de-polymerization act is:

$$k = 10^9 \exp\left(-\frac{84000}{RT}\right) \quad \text{l/mol/s} \tag{1}$$

This figure is coherent also with other activation energies as represented in Table 5.

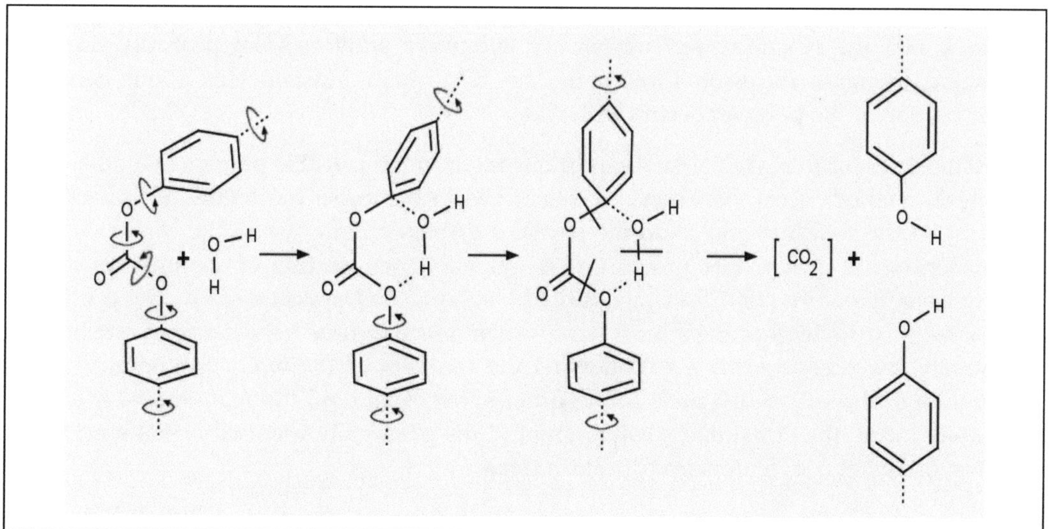

Fig. 7. Hydrolysis mechanism.

Of course the radical reactions path becomes more important at higher temperatures (and of course in practical absence of water). In our conditions water becomes a powerful reactant. It has also to be pointed out that water approaching critical conditions has an increased solubility. Moreover, in this range of temperatures (>230-250 °C), the swelling of the molten polymer offers a large increase of diffusion coefficient into the polymer phase. Formally the homogeneous reaction can be schematized as shown in figure 7.

6. Process parameters

Now it is useful to observe Table 5. In this table are resumed the conditions of the different processes used for recovering BPA from waste PC by using alcoholysis or hydrolysis.

Authors	Reactant	BPA yield	T [°C]	P [MPa]	Time [min]	Activation Energy [kJ/mol]
Piñero et al., 2005	$CH_3OH/NaOH$	80-90%	70-180	2-25	45-130	87.58
Hu et al., 1998	CH_3OH/C_7H_8	7-96%	40-60	0.1	15-330	107.5
Dongpil et al., 2009	CH_3OH	90-96%	160-220	n.a.	40-180	79.5
Jie et al., 2006	C_2H_5OH	20-90%	240-290	n.a.	10-60	97.2
Chen et al., 2004	CH_3OH	20-90%	230-265	7.8-10	10-30	75.72
Bozzano et al., 2010	H_2O	80-90%	240-290	4-8	20-240	84

Table 5. Comparison among obtained results in recovering BPA from PC by alcoholysis or hydrolysis (n.a. = not available, BPA yield=kg BPa/kgPC).

The hydrolysis or methanolysis process can be performed at a temperature level that avoid pyrolysis. It can be observed that, with exception of the paper of Hu et al., 1998, were toluene is used, all the activation energies are very close. This means that the elementary processes and the reaction mechanisms are extremely similar. Most probably they are dominated by mass diffusion limitation. The BPA yield indicate that about complete decomposition of the polymer occurs with water.

Essentially in order to make some observations regarding a possible process that adopts the hydrolysis method, it can be observed that it has to comprise the following aspects: the chopping of the solid materials and the particles dispersion into water; the compression of the suspension till the proper pressure (e.g. 4-8 MPa); the heating of the mixture till the maximum temperature (280-290°C), possibly by superheated pressure steam; the cooling of the reacted mixture by flashing after the convenient reaction time (recovering the steam), till a relatively low pressure (0.5 MPa) allowing the recovery of the molten monomer for the decantation of the suspended inert solid particles; the clean liquid monomer (at 140-160°C) is separated from the remaining small amount of the practically immiscible water and then sent to subsequent conventional purification steps.

Some processes have been already patented by important societies. A list is following: General Electric (Fox D.W. & Peters E.N., 1989; Shaffer S.J., 1994; Caruso A.J. & Lee J.L., 1996; Eijsbouts P. et al., 1997), Bayer AG (Buysch H.J. & Schoen N., 1993; Buysch H.J. et al., 1994; Buysch H.J., 1995), Teijin Chemicals LTD (Kiyoshije K. & Matsumoto K., 1994; Suzuki M. & Matsumoto K., 1995; Suzuki M., 1995, Ogasawara K. & Matsuura T., 2004, Takemoto H., 2005), Mitsubishi Heavy Ind. LTD (Hishihara N. et al., 1999; Hishihara et al., 2001), Victor Company of Japan (Takahashi T. et al., 2001; Tsujita K. et al., 2001a; Tsujita K. et al., 2001b; Tsujita K. et al., 2001c; Tsujita K. et al., 2001d; Kawai N., 2002; Tsujita K. & Kawai N., 2004; Tsujita K. & Kawai N., 2005), Kansai Res. Inst. (Kawashima F. et al., 2003).

7. Conclusions

In this chapter the experimental results of the hydrolysis of poly(bisphenol A carbonate) with water have been presented together with a description of existing methods for recycling PC. Both pure PC and CDs waste have been treated. The adopted temperature levels are subcritical and pressures inside the reactor allow to hydrolyze with liquid water. The tests have shown the feasibility of the process that in these conditions is mainly based on real molecular concerted path reactions. Secondary reactions, requiring a radical path and leading to parasitic byproducts, are characterized by quite higher apparent activation energy and require higher temperature levels than those here adopted. For this reason in the presented process they are negligible.

8. References

Ali S.; Garforth A.A.; Haris D.H.; Rawlence D.J. & Uemichi Y. (2002). Polymer Waste Recycling Over "Used" Catalysts, *Catalysis Today*, vol. 75, pp. 247-255, ISSN 0920-5861

Bozzano G.; Dente M. & Del Rosso R. (2010). Integral Recovery Of Bisphenol From Poly-Bisphenol A Carbonate Via Hydrolysis, proceedings of *20th European Symposium on Computer Aided Process Engineering, ESCAPE 20,* Computer -Aided Chemical Engineering, vol. 28 pp. 1-6, ISBN 978-0-444-53569-6, Ischia, Italy, June, 2010

Buysch H.J. & Schoen N. (1993). Hydrolysis of Polycarbonate(s) to REcover Bisphenol(s) - by Reaction with Water at in the Presence of Organo-Tin Cpds as Catalyst, *DE Patent 4.131.586*

Buysch H.J.; Schoen N. & Kuehling S. (1994). Cleavage of Polycarbonate (s) to Give Bisphenol(s) and Di-aryl Carbonate(s) - by Catalytic Transesterification with Phenol, Isolation of Bisphenol, Flash Distn. of Residue and Fractionation of Distillate. *DE Patent No. 4.220.412*

Buysch H.J. (1995). Process fort he Preparation of Bisphenol A from Polycarbonate, *DE Patent No. 4.324.778*

Caruso A.J. & Lee J.L. (1996). Method for Recovering Phenol and Xanthene Values from Waste Polycarbonate. *US Patent No.5.567.829*

Chen L.; Wu Y.; Ni Y.; Huang K. & Zhu Z. (2004). Degradation Kinetics Of Polycarbonate In Supercritical Methanol. *Journal of Chemical Industry and Engineering*, vol. 55 (11), pp. 1787-1792, ISSN 0438-1157

Chiu S.J.; Chen S.H. & Tsai C.T. (2006). Effect Of Metal Chlorides On Thermal Degradation Of (Waste) Polycarbonate, *Waste Management*, vol. 26, pp. 252-259, ISSN 0956-053X

Dongpil K.; Bo-kyiung K., Youngmin C.; Myungwan H. & Beom-Sik K. (2009). Kinetics Of Polycarbonate Methanolysis By A Consecutive Reaction Model, *Industrial & Engineering Chemistry Research,* vol. 48, No. 14, 6591-6599, ISSN 1520-5045

Day M.; Cooney J.D.; Barrette T. & Sheehan S.E. (1999). Pyrolysis Of Mixed Plastics Used In Electronics Industry, *Journal of Analytical and Applied Pyrolysis*, vol. 52, pp. 199-224, ISSN 0165-2370

Davis A. & J. H. Golden (1968). Thermal Degradation of Polycarbonate, *Journal of the Chemical Society (B) Physical Organic*, pp. 45-47.

Eijsbouts P.; De Heer J.; Hoogland G.; Nanguneri S. & DE Wit G. (1997). Process for Recovery of Bisphenol-A from Thermoplastic Polymer Containing Dihydric Phenol Units. *US Patent No. 5.675.044*

Elmaghor F.; Zhang L.; Fan R. & Li H. (2004). Recycling of Polycarbonate by Blending with Maleic Anhydride Grafted ABS, *Polymer,* vol. 45, pp. 6719-6724, ISSN 0141-3910

Fox D. W. & Peters E.N. (1989). Method for recovering a Dihydric Phenol from a Scrap Polyester. *US patent No. 4.885.407*

Fox D. W. & Schenectady N.Y. (1964). Aromatic Carbonate Resins and Preparation Thereof. *US Patent No. 3.153.008*

Fox D. W. (1964). Polycarbonates of Dihydroxy-Aryl Ethers. *US Patent No. 3.148.172*

Goto M. (2009). Chemical Recycling of Plastics Using Sub- and Supercritical Fluids, *The Journal of Supercritical Fluids*, vol. 47, pp. 500-507, ISSN 0896-8446

Grause G., Sugawara K., Mizoguchi T. & Yoshioka T. (2009). Pyrolytic Hydrolysis of Polycarbonate in the Presence of Earth-Alkali Oxides and Hydroxides. *Polymer Degradation and Stability*, vol. 94, pp. 1119-1124, ISSN 0141-3910

Hishihara N.; Kawamura W.; Kishi M. & Okazaki K. (1999). Decomposition Polyester and Device for Converting Polyester into Monomers. *JP Patent No. 11.100.336*

Hishihara N.; Kawamura W.; Kishi M. & Okazaki K. (2001). Polyester Decomposition Process and Polyester Monomerization System, *US Patent No. 6.214.893*

Hu L.C.; Oku A. & Yamada E. (1998). Alkali-Catalysed Methanolysis of Polycarbonate. A Study on Recycling of Bisphenol A and Dimethyl Carbonate, *Polymer*, vol. 39, No. 16, 3841-3845, ISSN 0032-3861

Jang B.N. & Wilkie C.A. (2005). A TGA/FTIR and Mass Spectral Study on the Thermal Degradation of Bisphenol A Polycarbonate, *Polymer Degradation and Stability*, vol. 86, pp.419-430, ISSN 0141-3910

Jie H.; Ke H.; Qing Z.; Lei Chen, Yongqiang W. & Zibin Z. (2006). Study on Depolymerization of Polycarbonate in Supercritical Ethanol. *Polymer degradation and Stability*, vol. 91, pp. 2307-2314, ISSN 0141-3910

Kawai N. (2002). Method for Useful Substance Recovery from Waste Plastic Consisting Essentially of Polycarbonate Resin, *JP Patent No 2002.212.335*

Kawashima F.; Kitagawa H.; Den K. & Sakai A. (2003). Recycling Method for Polycarbonate Resin, *JP Patent No. 2003.041.049*

Khuwijiitjaru P.; Fujii T.; Adachi S.; Kimura Y.& Matsuno R. (2004). Kinetics on the Hydrolysis of Fatty Acid Esters in Subcritical Water. *Chemical Engineering Journal*, vol. 99, pp. 1-4, ISSN 1385-8947

Kiyoshije K. & Matsumoto K. (1994). Recycling of Aromatic Polycarbanate Resin. *JP Patent No. 6.287.295*

Kitahara M.; Masumi H.; Ban T. & Sawaki T. (2009). Method of Decomposing a Polycarbonate, *US patent* No. 7.585.930 B2

Kruse A. & Dinjius E. (2007). Hot Compressed Water as Reaction Medium and Reactant. 2. Degradation Reactions. *Journal of Supercritical Fluids*, vol. 41, pp. 361-379, , ISSN 0896-8446

Lee L. H. (1964). Mechanism of Thermal Degradation of Phenolic Condensation Polymers. I. Studies on the Thermal Stability of Polycarbonate. *Journal of Polymer Science*, part A vol. 2 (6), pp. 2859- 2873, ISSN 1099-0518

McNeill I. C. & Rincon A. (1991), Degradation Studies of Some Polyesters and Polycarbonates - 8. Bisphenol A Polycarbonate, *Polymer Degradation and Stability*, vol. 31 (3), pp. 163-180, ISSN 0141-3910

McNeill I. C. & Basan S. (1993). Thermal Degradation of Blends of PVC with Bisphenol A Polycarbonate, *Polymer Degradation and Stability*, vol. 39 (3), pp. 145-149, ISSN 0141-3910

Mitan N.M.M.; Brebu M.; Bhaskar T.; Muto A.; Sakata Y. & Kaji M. (2007). Co-Processing of DVDs and CDs with Vegetable Cooking Oil by Thermal Degradation, *Journal of Material Cycles Waste Management*, vol. 9, pp. 62-68, ISSN 1611-8227

Montaudo G. & Puglisi C. (1992). Thermal-Decomposition Processes in Bisphenol-A Polycarbonate, *Polymer Degradation Stability*, vol. 37(3), pp. 91-96 , ISSN 0141-3910

Montaudo G.; Carroccio S. & Puglisi C. (2002). Thermal and Themoxidative Degradation Processes in Poly(Bisphenol A Carbonate), *Journal of Analytical and Applied Pyrolysis*, vol. 64, pp. 229-247, ISSN 0165-2370

Oba K.; Ishida Y.; Ito Y.; Ohtani H. & Tsuge S. (2000). Characterization of Branching and/or Cross-Linking Structures in Polycarbonate by Reactive Pyrolysis Gas Chromatography in The Presence of Organic Alkali, *Macromolecules*, vol.33 (22), pp. 8173-8183, ISSN 0024-9297

Ogasawara K. & Matsuura T. (2004). Method for Obtaining Aromatic Dihydroxy Compoud and Dialkyl Carbonate from Aromatic Polycarbonate. *JP Patent No. 2004.277.396*

Pan Z.Y.; Zhen B. & Chen Y.X. (2006). Depolymerization of Poly(Bisphenol A Carbonate) in Subcritical and Supercritical Toluene, *Chinese Chemical Letters*, Vol. 17, pp. 545-548, ISSN 1001-8417

Piñero R.J.G. & Cocero M.J. (2005). Chemical Recycling of Polycarbonate in a Semi-Continuous Lab-Plant. A Green Route with Methanol and Methanol-Water Mixtures. *Green Chemistry*, vol. 7, pp. 380-387, ISSN 1463-9262

Pinto J.S.S. & Lanças F.M (2006). Hydrolysis of Corn Oil Using Subcritical Water. *Journal of Brazilian Chemical Society*, vol. 17, No.1, pp.85-89, ISSN 0103-5053

Puglisi C.; Stradale L. & Montaudo G. (1999). Thermal Decomposition Process in Aromatic Polycarbonates Investigated by Mass Spectrometry, *Macromolecules*, vol. 32 (7), pp. 2194-2203, ISSN 0024-9297

Shaffer S.J. (1994). Method for Recovering Bis Hydroxy Aromatic Organic Values and Bis Aryl Carbonate Values from Scrap Aromatic Polycarbonate. *US patent No. 5.336.814*

Schnell H.; Bottenbruch L.& Krimm H. (1964). Thermoplastic Aromatic Polycarbonates and their Manufacture. *US Patent No. 3.028.365*

Suzuki M. & Matsumoto K. (1995). Decomposition Method for Aromatic Polycarbonate Resins, *JP Patent No. 7.205.153*

Suzuki M. (1995). Method for Recycling Waste Aromatic Polycarbonate Resin, *JP Patent No. 7.316.280*

Takahashi T.; Tsujita K. & Kawai N. (2001). Method for Decomposing Waste Plastic, *JP Patent No. 2001.139.723*

Tagaya H.; Katoh K.; Kadokawa J. & Chiba K. (1999). Decomposition of Polycarbonate in Subcritical and Supercritical Water, *Polymer Degradation and Stability*, vol .64, pp. 289-292, ISSN 0141-3910

Takemoto H. (2005). Method of Obtaining Aromatic Dihydroxy Compound from Waste Aromatic Polycarbonate Resin, *JP Patent 2005.179.267*

Tsujita K.; Kawai N. & Takahashi T. (2001a). Method for Treating Waste Optical Recording Medium, *JP Patent No. 2001.160.243*

Tsujita K.; Kawai N. & Takahashi T. (2001b). Method for Recovering Useful Material from Waste Plastic Mainly Composed of Polycarbonate Resin, *JP Patent No. 2001.270.961*

Tsujita K.; Kawai N. & Takahashi T. (2001c). Method for Recovering Useful Substance from Waste Plastic Containing Substantially Polycarbonate Resin, *JP Patent No. 2001.302.573*

Tsujita K.; Kawai N. & Takahashi T. (2001d). Method for Recovering Useful Material from Waste Plastic Mainly Composed of Polycarbonate Resin Alloy as Principal Ingredient, *JP Patent No. 2001.302.844*

Tsujita K. & Kawai N. (2004). Method for Recovering Useful Prodict from Waste Plastic, *JP Patent No. 2004.051.620*

Tsujita K. & Kawai N. (2005). Method for Recovering Bisphenol A from Waste Plastic, *JP Patent No. 2005.112.781*

Yoshiki S.; Yasuhiko K.; Koji T. & Noboru K. (2005). Degradation behaviour and recovery of bisphenol-A from epoxy resin and polycarbonate resin by liquid-phase chemical recycling, *Polymer Degradation and Stability*, vol. 89, pp. 317-326, ISSN 0141-3910

5

PVB Sheet Recycling and Degradation

Michael Tupý[1], Dagmar Měřínská[1,2] and Věra Kašpárková[2,3]
[1]Department of Polymer Engineering, Faculty of Technology,
Tomas Bata University in Zlin, Zlin,
[2]Centre of Polymer Systems, University Institute, Tomas Bata University in Zlin, Zlin,
[3]Department of Fat, Surfactant and Cosmetics Technology, Faculty of Technology,
Tomas Bata University in Zlin, Zlin,
Czech Republic

1. Introduction

Increasing growth of raw material prices, environmental aspects and still growing landfill fees bring about the increasing interest encountered with the plastics waste recycling. Globally, the problem has been solved for the common plastics such as polyolefins, poly(ethylene terephtalate) and poly(styrene). Though extensively used for the glass lamination, poly(vinyl butyral) (PVB) does not belong to this group. It is generally known that during the glass lamination process, large volume of PVB trim is formed.

The PVB polymer is mostly used in the form of plasticized PVB sheet for preparation of laminated safety glass (LSG). At present time, Solutia, DuPont, Seki sui and Kuraray are worldwide PVB manufacturers. The PVB is the material which can stick together float glasses with holding excellent optical and mechanical properties of the glass laminate (Ivanov, 2006; v. Elasticity, mechanical strength, toughness, high light transmission and the adhesion to glass are the most significant PVB properties (Tupý, Měřínská, et al, 2010). Toughness of PVB sheet is based on high molecular weight of PVB chain. Certainly, the PVB has to be plasticized for achieving high material elasticity (Iwasaki, et al, 2006; Keller, Mortelmans, 1999; Svoboda, Balazs., et al, 1988). Admittedly, the plasticizer must not reduce light transmittance through the sheet, PVB adhesion to glass, generate haze and yellowness, and migrate out of the polymer matrix (Wade, D'Errico, et al, 2004). In addition, plasticizer must have perfect compatibility with the polymer and low evaporability during processing conditions.

2. Sources for recycling of PVB sheet

Worldwide, 65% of all PVB sheets are used in automotive applications (Dhaliwal, Hay, 2002). According to data from (OCIA, 2007), the worldwide car's production is estimated around 60 millions cars per a year. Assuming that one windshield contains approx. 1 kg of PVB sheet, total amount gives between 60-70 million kg of PVB sheets per year. In addition, by-products from PVB sheets manufacturing (5%) and trimmings (< 10%) from windshield production must be added (Goroghovski, Escapante-Garcia, et al, 2005). It represents total amount of 80 million kg of automotive PVB sheet waste annually. Total worldwide amount of produced PVB sheets for automotive and architectural industry is estimated around 120 million kg per year.

2.1 By-product from sheet manufacturing process

This kind of PVB sheet has the best quality for following recycling process. By-product sheets are not contaminated by powder fragments but there may be present some un-homogeneous parts like plasticizer, light and heat stabilizers, adhesion modifiers, pigments and other elements of the system. The reasons why the sheet has to be re-processed are some specified defects in manufacturing technology (thickness, sheet roughness, edges). In every case, sheet edges must be cut approx. 10-20 cm from the edge beginning. The PVB sheet with a width around 2-3 meters has very high material shrinkage in this place. The edges can be replaced back to the extrusion process for re-extruding (Zvoníček, 1999).

2.2 Trim

The first step in LSG manufacturing is the lamination of PVB sheet between two glasses. It this part, probable defects on laminate edges must be reduced. It is performed by layering of PVB sheet between glasses with a larger sheet surface than the glass sizes are. Prepared "sandwich" is fixed by nipp-roll pre-lamination process and consequently it is possible to trim off oversized PVB sheet. While the sheet oversize would not be used, the laminate would be produced with visible defects (it does not meet the quality specifications) (Svoboda, Balazs., et al, 1988; Zvoníček, 1999).

The quantity of a trim generation depends on the geometry of produced windshield and the geometry of used PVB sheet surface. The width of trim obtained from the pre-laminating process is around 1–20 cm, based on glass geometry. Therefore, the trim quantity is between 5-10% from total amount of processed PVB sheet and the worldwide PVB trim capacity obtained from the windshield laminating process is approx. 4-6 million kg (Svoboda, Balazs., et al, 1988; Zvoníček, 1999).

The trim is created also at manufacturing of architectural LSG. Nevertheless, the ratio of a collected architectural trim is lower than amount of automotive trim. It is influenced by a higher glass powder contamination of this trim (different trimming technology) (Zvoníček, 1999).

2.3 PVB sheet from windshield

Some specialized companies deal with the re-application possibility of PVB sheet obtained from recycled windshield. All present recycling processes produce a good quality of glass scrap which is fully re-used in the glass batch. However, the separated PVB waste is not recyclable due to high amount of glass, water content in the sheet, parts of color PVB sheets, mixed of various PVBs (after blending a haze is created) and foreign plastic matters (Zvoníček, 1999; Plaček, 2006; Recycling .., 2007; Tupý, Měřínská, 2011). Nevertheless, if somebody would develop the recycling technology ensures high-grade of glass separation, obtained PVB sheet may be re-processable to new PVB sheet without optical defects (Tupý, Měřínská, 2011).

Moreover, it is necessary to remain that the laminated (interglassed) PVB is not essentially exposed to UV radiation (up to 320 nm), mechanical stress, elevated temperature, oxygen and any various substances. Thus, the interglassed PVB sheets should keep very similar physical properties as extruded material before the lamination process and it can be re-extruded and re-laminated to new PVB sheet again.

3. PVB sheet composition

3.1 Poly(vinyl butyral)

The way assigned for PVB production is not easy. Firstly, it is necessary to produce poly(vinyl acetate) (PVAc) by the radical vinyl acetate polymerization. Consecutive hydrolysis at acidic or basic ambient creates poly(vinyl alcohol) (PVAl) which provides poly(vinyl butyral) by acetalization with butyraldehyde at acidic environment. The final structure of high-molecular PVB is used for LSG manufacturing and it is compounded from atactic copolymers 80% of vinyl butyral, 18-23% of vinyl alcohol and up to 1% of vinyl acetate [Wade, D'Errico, et al, 2004; Dhaliwal, Hay, 2002; D'Errico, Jemmott, et al, 1996; Nghuen, Berg, 2004; Svoboda, et al., 1970). This chemical structure, viewed in Fig.1, is the same for every manufacturer today. Nevertheless, exact consequence and properties of every PVB sheet depends on every PVB type, manufacturer and PVB sheet composition. The sheet is mostly stabilized by antioxidants and thermal stabilizers in many times (Saflex, 1993).

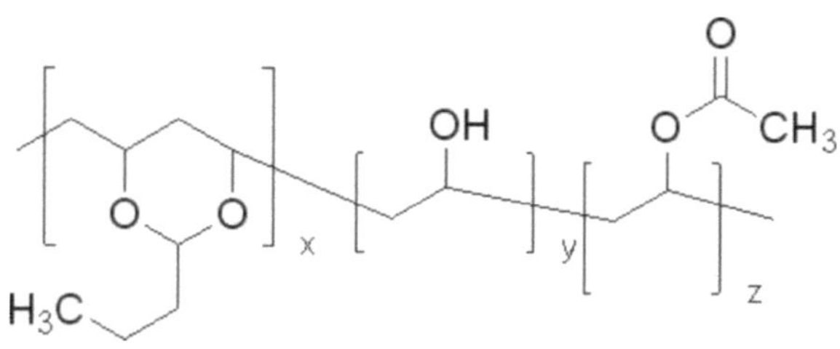

Fig. 1. PVB chain structure with composition of functional groups: vinyl butyral 81%, vinyl alcohol 18-23%; vinyl acetate <1%.

Final PVB properties are assigned by polymerization degree of input PVAc, distribution curve of molecular weight, PVAl hydrolysis degree, catalyst acid strength, reaction temperature and PVAl conversion degree to PVB. The last one is crucial for final polarity of produced PVB. The PVB polymer is white powder, dissolvable in ethanol, THF, ketones and other semi-polar dissolvent (Mrkvičková, Daňhelka, 1984). The PVB solubility depends on -OH group content in the polymer chain and PVB molecular weight (Physical prop.SekiSui, 2001).

3.2 PVB sheet

Certainly, achievement of high PVB toughness must be proved by polymer plasticizing. However, the plasticizer must not considerably reduce light transmittance through the sheet, increase haze and yellowness, reduce PVB adhesion to glass and migrate out of the polymer matrix (Svoboda, Balazs, et al, 1988; Zvoníček, 1999). The plasticizer also must have perfect compatibility with polymer matrix and its evaporability during the processing is prohibited (Zvoníček, 1999). Different plasticizers, used in beyond, were for example triethylenglycol-di-2-ethylbutyrate, dibuthylsebacate, tetraethylenglycol-di-heptanoate and dihexyladipate (Dhaliwal, Hay, 2001; Zvoníček, 1999). Mentioned plasticizers have different molecular polarity. There through, due to this it was indispensable to produce PVBs with different amount of hydroxyl groups in PVB chain (Dhaliwal, Hay, 2001; D'Errico, Jemmott, et al, 1996; Phillips, 2005). The final polymer system has a different absorbability of the plasticizer and water (Mrkvičková, Daňhelka, 1984). At present time, produced PVB sheets assigned for the glass lamination are plasticized with 28% of applied plasticizer; mostly triethylenglycole-bis(2-ethylhexanoete), (labeled 3GO) (Wade, D'Erricco, 2004; Phillips, 2005; Mister, Bianchi, et al, 2007; Smith, Rymer, et al, 2008).

Due to mentioned sheet physical properties, the most important characteristics of windshield are high mechanical strength and absorbability of kinetic energy during car-crash (Keller, Mortelmans, 1999; Svoboda, Balazs, 1988). However, PVB interlayer must keep glass particles on its surface [2, 3]. All these described properties provide an exactly adjusted adhesion grade of PVB to glass. Because the PVB has exceedingly high adhesion degree it cannot be used for automotive glass lamination. Thus, high adhesion degree it is necessary to reduce to ½ of the original adhesion value (Keller, Mortelmans, 1999; Wade, D'Erricco, 2004; Dhaliwal, Hay, 2004; Smith, Rymer, et al, 2008). The virgin adhesion is reduced by an addition of organic salts of alkali metals or alkaline-earth metals during an extrusion process of plasticized PVB sheet. This is described in several patents (Smith, Rymer, et al, 2008; Aoshima, Shohi, 2000; D'Errico, 1995; D'Erroco., 1997; Fowkes, 1987 Herman, Fabian, et al, 1984; Shichiri, Miyai, et al, 2002). The mostly used substances modifying the adhesion are organic salts of Na^+, K^+ and Mg^{2+}. Ion ratio and its total amount are strictly specified (Smith, Rymer, et al, 2008; Aoshima, Shohi, 2000; D'Errico, 1995; D'Errico., 1997; Fowkes, 1987 Herman, Fabian, et al, 1984; Shichiri, Miyai, et al, 2002). Moreover, the water content in PVB sheet must be in range 0.3-0.5% which is necessary for a maintaining the required adhesion grade (Keller, Mortelmans, 1999; Wade, D'Errico, 2004; D'Errico, Jemmot, 1995; Shichiri, Miyai, 2002).

4. Extrusion of PVB sheet

The formation of the PVB trim is the consequence of the production technology, in which laminated glass without defects having irregular shape are manufactured. The PVB assigned for re-processing is usually recycled together with the waste sheets originating from the PVB manufacturing (Tupý, Zvoníček, et al, 2008). However, to find the ideal PVB re-processing conditions is not easy.

Due to its composition, PVB is very sensitive to the degradation and the migration of plasticizer. It was found, that at the atmospheric pressure, plasticizer migrates at the temperature of 260 °C (Dhaliwal, Hay, 2002). During repeated processing of plasticized PVB, both water and plasticizer were reported to be extracted from polymer by vacuum. However, the loss of plasticizer at the common processing temperatures (up to 200 °C) is minimal (Svoboda, Balazs, 1998).

Based on several authors (Keler, Mortelmans, 1999, Svoboda, Balazs, 1998; Nagai, 2001; Neher, 1936; Svoboda, 1987), all PVB sheet manufacturing technologies are based on similar principle. Melt plasticized PVB is extruded by sheet extrusion die at temperature 160-220°C into water. The screw placement is evacuated in order to adjust the water content in the sheet. The PVB melt retains its shape, thickness and specific sheet surface which are necessary for de-aeration at pre-lamination process (Svoboda, Balazs, 1998; Zvoníček, 1999). Plasticizer and additives addition runs at mixing equipments before the extruder in many times.

The PVB sheet is mostly manufactured at the thickness 0.38 mm (LSG for architectonical industry) and 0.76 mm (LSG for automotive and architectonical use). Special applications require thicknesses 1.14 and 1.52 mm. The width of PVB sheet can be up to 3.5 meters. Manufactured PVB sheet is rolled and either separated by thin patterned polyethylene sheet or rolled under-cooled (Tg = 15°C) because the PVB sheet must not be stuck for following use (Svoboda, Balazs, 1998; Zvoníček, 1999; Saflex, 1973).

Shear and thermo oxidative degradation of polymer represents more serious problem observed during reprocessing. Both degradation types induce the cleavage of polymer chains, albeit the degradation mechanism is not the same. Generally, the shortening of polymer chains negatively influences mechanical properties of PVB, resulting in an undesirable lowering of safety characteristics of the produced sheet. The deterioration of mechanical properties of PVB consequently decreases its ability to absorb the mechanical energy (when it is used for the safety car glass) in the case of an accident (Tupý, Zvoníček, 2008; Tupý, Měřínská, 2010). On the other hand, a decreasing of molecular weight and a decreasing of viscosity caused by the degradation can favorably influence rheological properties of PVB melt during the extrusion on the flat die (Měřínská, Tupý, 2010; Grachev, Klimenko, et al, 1974).

In order to decrease power consumption during re-processing, PVB hygroscopicity can be utilized. Because water contained in PVB matrix can act as an additional plasticizer and lower the rigidity of the material, processing of "wet" PVB can be advantageous (Mrkvičková, Daňhelka, 1984). However, during the re-processing of material containing high moisture content (8 %), water can react with butyric groups, which induces the change of the polymer structure. As the consequence, hydrolysis occurs significantly changing the final properties of the re-processed PVB (Dhaliwal, Hay, 2002; Měřínská, Tupý, 2010).

The aim of the following test was to determine the degradation of PVB sheet at different kneading conditions and to estimate an influence of temperature, air oxygen content and mechanical stress on the course of degradation process. The work is also focused on the possibility to find optimal re-processing conditions of PVB whereat the mechanical and thermal degradation as well as yellowness of the re-processed polymer are minimal.

5. PVB sheet degradation by thermogravimetric analysis

First of all, the relative thermal stability of commercial PVB sample was measured by thermogravimetric analysis, from mass loss against temperature plots. As the Fig.2 shows, weight loss occurred in two distinct regions between 175–325 and 325–500 °C and corresponding to about 27-28 and 65-70% mass loss. No marked differences were observed between the various commercial grades. The first part of temperature range is the plasticizer evaporation. The final weight loss was the same for all the samples and a brown residue, approximately 5% of the original mass (Tupý, Měřínská, 2011). The evolved volatiles were analyzed by mass spectrometer as a function of time and temperature at fixed m/z ratios.

These corresponded to the top m/e ratio for acetic acid, butenal, butyraldehyde, benzene and toluene the expected products from the thermal decomposition of PVB (Wade, D'Errico, 2004).

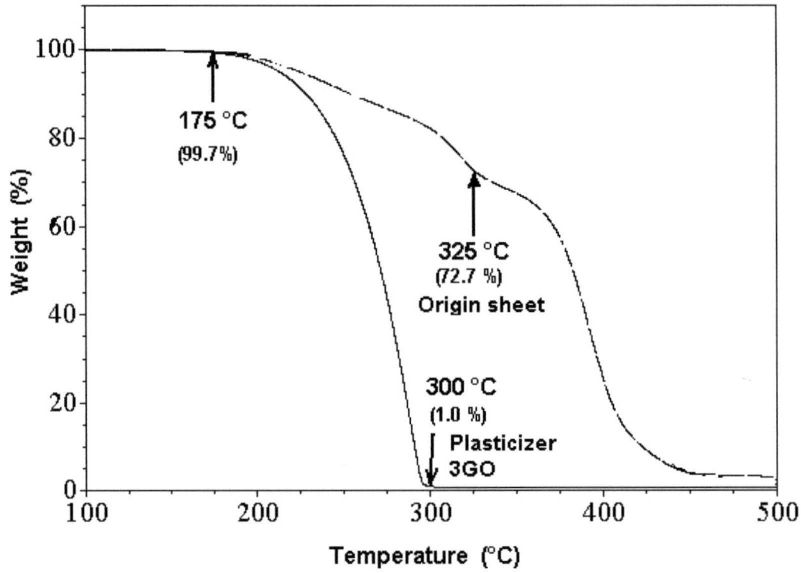

Fig. 2. TGA evaluation of plasticized Butacite PVB sheet.

Little or no degradation products were observed below 250 °C although the PVB samples had lost about 10–12% of mass under these experimental conditions were initially. The major products of the decomposition were observed above 260 °C. Acetic acid was a minor component of the volatile degradation products. Aromatic species, such as benzene and toluene, were also observed. These have been attributed to the break down of the polyene products produced by the elimination reactions.

From the relative % mass loss and the absence of volatiles detected by the mass spectrometer it was deduced that the PVB was primarily losing plasticizer in the temperature volatilization process between 200 and 260 °C. The loss of additives from a polymer is a complex process involving diffusion, transport and evaporation from the surface of the polymer.

6. Degradation by kneading

6.1 Samples preparation

By the reason of finding degradation mechanism of plasticized PVB, this material was reprocessed by kneading, rolling and pressing. Both dry (0.5 % water) and wet (8% water) sheets were tested. Increasing moisture content was reached by the soaking of "dry" PVB sheet in water for 14 days (Grachev, Klimenko, 1974).

Samples stressed by kneading were prepared in the Brabender kneader with two blunders W50 at the friction of 2:3. Volume of the heated chamber was of 55 cm³. Constant amount of

40 g PVB was placed in the chamber and processed for 10 minutes at different temperatures (100, 130, 160, 190, 220 °C) and rotation speeds (40, 60, 80 rpm). The chamber of kneader was filled only to the ¾ of the volume in order to have the sufficient amount of oxygen in order to study thermo-oxidative degradation. During kneading, both thermo-oxidative and shear degradation are assumed to take place.

In order to simulate solely shear degradation with absence of thermal stress, PVB sheets were re-processed by rolling at the temperature of 78°C in the presence of air. Laboratory double-roller was used. Rollers were preheated to 60-70°C in order to allow the PVB calendaring corresponding to processing of the rubber. After the initial preheating, the roller temperature was kept only by the energy dissipation. After 10 minutes the temperature reached 78°C and this value remained almost unchanged.

Pure thermal degradation with low shear stress was simulated by pressing. PVB was placed between two PET sheets preventing the contact with air and thus oxidative degradation. Then, the material was pressed at 1 MPa at temperature of 160, 190 a 220°C for 10 minutes.

Dry PVB was tested at all the above presented conditions; wet one was tested at all temperatures but only with 60 rpm.

6.2 Analysis and methods

Mechanical properties of the stressed samples were determined using a T 2000 Tensile tester *(Alpha Technologies)* with the displacement rate of 500 mm/min at room temperature. For testing, material was pressed onto the plates with the thickness of 1.0 mm at the temperature of 130°C and the standard testing specimens were prepared. Tensile strengths and strain were determined.

Rheological properties of re-processed samples were tested in terms of MFI measurements using the extruding plastometer M201 (Haake) according to EN ISO 1133. Samples were conditioned at 25% relative humidity and then extruded at 150 °C through the 2 mm capillary using the load of 100 N. The MFI correlates to the polymer mass passing through a standard capillary in an interval of 10 minutes, at a given load.

Quantification of water content was carried out by the Karl Fischer method *(Metrohm AG)*. The method is based on the conductometric determination of water evaporated from the sheet into the iodine solution and sulphur dioxide in methanol.

Yellowness was evaluated using the CIE Lab. colour scale. Handy Color *(BYK Gardner)* instrument was applied and calibrated with the white and black standards. Measurement was carried out against the white background at the angle of 10°. Illumination type of D65 corresponding to daylight was applied. Yellowness $\underline{Y_{ID}}$, was calculated from the measurements of spectroscopic values $\underline{L}$, $\underline{a}$ and $\underline{b}$. The obtained value was converted to the value corresponding to the PVB sheet with the standard thickness of 0.76 mm, which is typical for applications in automotive industry and in architecture.

Thermo-gravimetric analysis (TGA) was determined by thermogravimetric analyzer TGA Q500 *(TA Instruments, New Castle, USA)* in open platinum crucibles and weighed-in. Amount of PVB sample for thermal analysis was approx. 8 mg and measurements were taken in temperature interval 20-500 °C, dT/dt = 10 °C min^{-1} in protective nitrogen atmosphere (150 mL min^{-1}).

GPC analyses were conducted using a PLGPC-50 (*Polymer Laboratories*) equipped with a PL differential refractometer (DRI) and on-line viscometer detectors (VIS). Analyses were performed with a PL gel Mixed-C column *(7.8 x 300 mm; Polymer Laboratories)* at 30 °C with the mobile phase flow rate of 1 mL/min. Tetrahydrofurane was used as the mobile phase. The column was calibrated using narrow molecular weight polystyrene standards *(Polymer Laboratories Ltd, Church Stretton, UK)* with molecular weights ranging from 580 to 451 000 g.mol^{-1} (given by supplier). A 100 µL injection loop was used for all measurements. For the determination of molecular weight, universal calibration was applied. Data processing was performed with Cirrus GPC, Multi Detector Software. The concentration was of about 0.2 g /100 ml and samples were dissolved at room temperature for 20 hours under stirring. The combination of both types of detectors enabled to exactly determine molecular weight as well as detect the PVB aggregation.

6.3 The influence of kneading conditions on the change of plasticized PVB sheet properties

From the theory and practice it is confirmed that the PVB re-processing brings about the shortening of macromolecular chains, which induces the change of its mechanical properties. The results show that increasing of the re-processing temperature brought the lowering of melt rigidity (measured as MFI), lowering of tensile strength and strain. These changes are visualized in Figs. 3-4.

The MFI values are shown in Fig.3. To sum up, the MFI increases with the increasing of re-processing temperature and the increasing of rotation speed systematically up to 60 rpm.

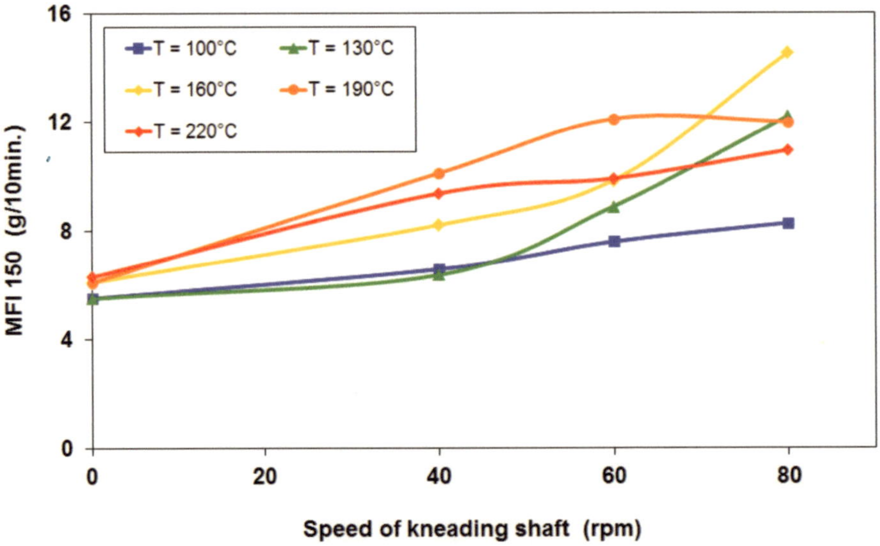

Fig. 3. The change of MFI at the different conditions of the kneading of PVB sheet with water content 0.5%.

It is consequence of thermo-oxidative degradation, which causes reduction of polymer molecular weight. However, at the rotation speed of 80 rpm, the MFI values behave differently. Degradation of PVB macromolecular chains (expressed as increasing of MFI) is reduced at the temperatures above 190 °C, which is indicated by no rising or even slight drop of MFI (see Fig.3). The lowering of the degradation at the higher rotation speed (above 80 rpm) is possible to explain by the sliding of polymer chains in the stressed melt resulting in the lower effect of kneading.

The results obtained from the measurement of tensile strength of the "dry" PVB samples (0.5 % moisture content) in dependence on speed of the kneading shaft are shown in Figure 4. At lower temperatures, degradation of "dry" PVB is proportional to increase of the rotation speed. For example, during processing at 100 °C degradation increased, which can be concluded from the lowering of stress at break values (shape of the curve in concave). On the other hand, increasing of temperature caused straightening of this dependence and for the samples processed above 160 °C the curves exhibit the convex curving. Minima on the curves observed at rotation speed of 60 rpm and temperatures 190 and 220 °C indicate, hence, the highest degradation of PVB. Samples re-processed by pressing were used as a background for the kneaded samples at the same temperatures. Slight increase of tensile strength, strain at break, MFI and yellowness were observed for the samples pressed for 10 minutes at the all tested temperatures.

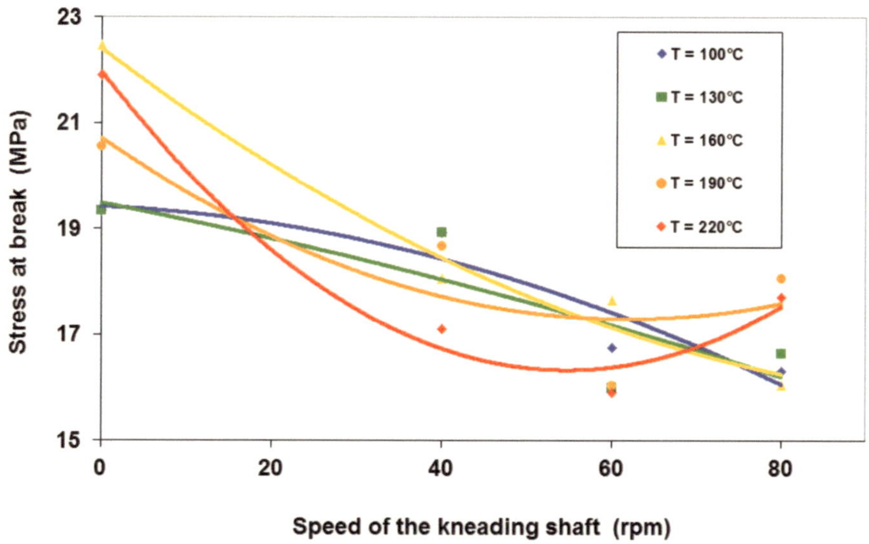

Fig. 4. Tensile strength of the re-processed PVB sheet at the different conditions of the kneading of PVB sheet with water content 0.5%.

6.4 The transformation of the process energy into heat

During the PVB re-processing on the Brabender kneader the temperature of the kneading chamber was measured. The chamber was tempered on the required temperature, but with on-going process of kneading, the temperature slightly increased. The course of temperature changes is summarized in Figure 5. The more noticeable energy transformation was observed at the lower processing temperatures (100 and 130 °C). This effect is clearly correlated to the higher rigidity of the processed material. It is also demonstrated that the evolution of dissipation heat depends on the rotation speed and kneading time. With the higher rotation speed, the amount of dissipated heat rises significantly. Above 130 °C the heat was formed only at the beginning of the kneading, when the material was still rigid enough. At 220 °C, due to the low material rigidity, the evolution of the transformation heat is minimal.

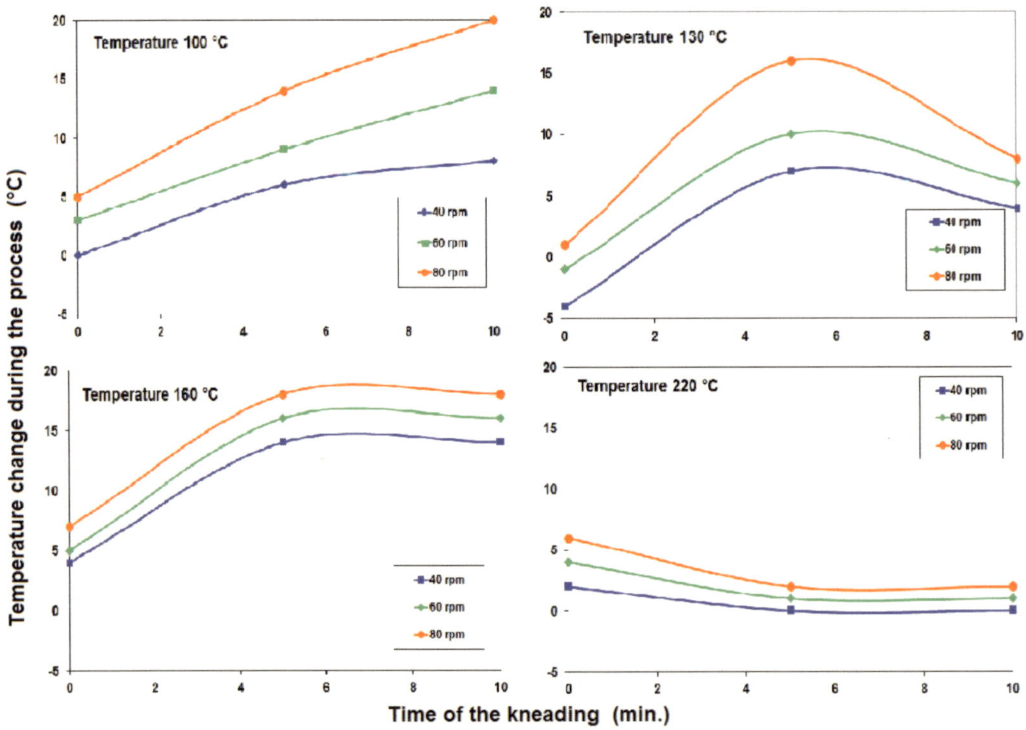

Fig. 5. Temperature change during the kneading process at different temperatures.

7. The influence of water on the change of mechanical properties

In order to lower the energy consumption during the PVB re-processing, its hygroscopicity was employed. As water presented in the PVB matrix can act as an additional plasticizer, it can decrease PVB rigidity (Tupý, Měřínská, 2010). It is supposed that lower material rigidity can decrease the energetic intensity of the re-processing. The comparison of MFI values measured for dry (0.5 % water) a wet samples (8 % water) shows that MFI increases proportionally to the water content (see Fig. 6). On the contrary, the tensile strength decreased. The change of the mechanical properties of "wet" PVB was caused by higher polymer plasticity and the reduction of intermolecular forces.

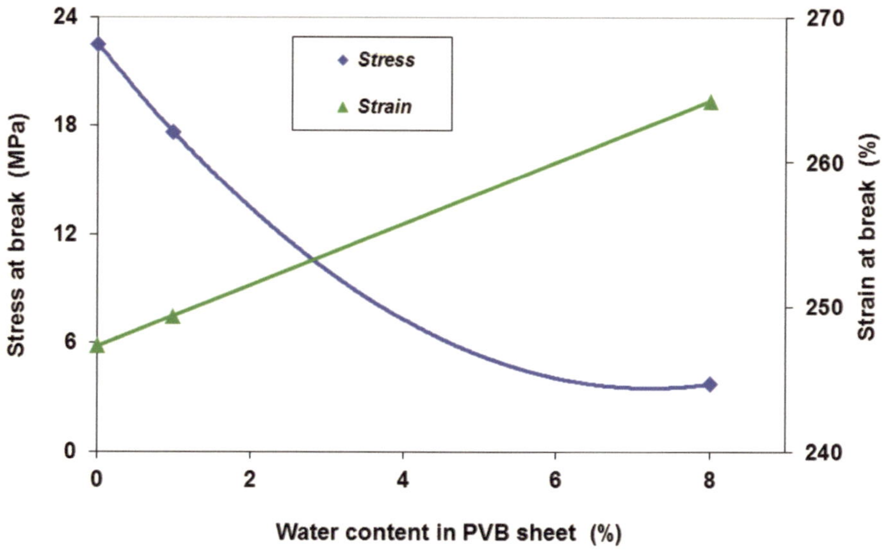

Fig. 6. The influence of water content on the change of mechanical properties of no re-processed PVB sheet.

During the re-processing of wet PVB material at the higher temperature, hydrolysis and elimination of butyric group can occur. This process results in the formation of hydroxyl groups and consequently conjugated double bond, which brings the change of PVB chain structure (Wade, D'Errico, 2004; Remsen, 1991). Thus, the hydrolysis causes considerable changes of the final properties of re-processed PVB. Due to this fact, an effort was done to find the optimal conditions for PVB re-processing with as low hydrolysis as possible. Hydrolysis was qualitatively estimated from the changes of molecular weight and increasing of the sheet yellowness.

Although water present in PVB evaporates very quickly at the beginning of the process, it influences the results of all the tests. The values of MFI for re-processed "wet" PVB show the significant increase in the dependence on temperature in the comparison with the "dry" PVB (see Fig. 7). A notable increasing of MFI values observed for wet PVB is caused by the degradation, which is induced by thermo oxidative reactions and better diffusion of gases into the PVB melt.

The comparison of mechanical properties of "dry" and "wet" PVB presents Fig. 8. The figure shows tensile strength and strain of both PVB types kneaded at different temperatures at the constant rotation speed of 60 rpm. Arrows denote the values measured for the original "dry" PVB. Optimal processing temperature for "dry" PVB, where the degradation was the lowest, is determined at 150 °C as the maximum of the curve. This maximum, with the highest values of tensile strength, corresponds to the minimum degradation of PVB. Below and above 150 °C tensile strength decrease; this can be caused by the lowering of the molecular weight induced by the degradation. Regarding the degradation mechanism, the scission of the "dry" PVB chains less than 150 °C is prevailingly caused by shear stress, whilst at higher temperatures thermo-oxidative degradation takes place.

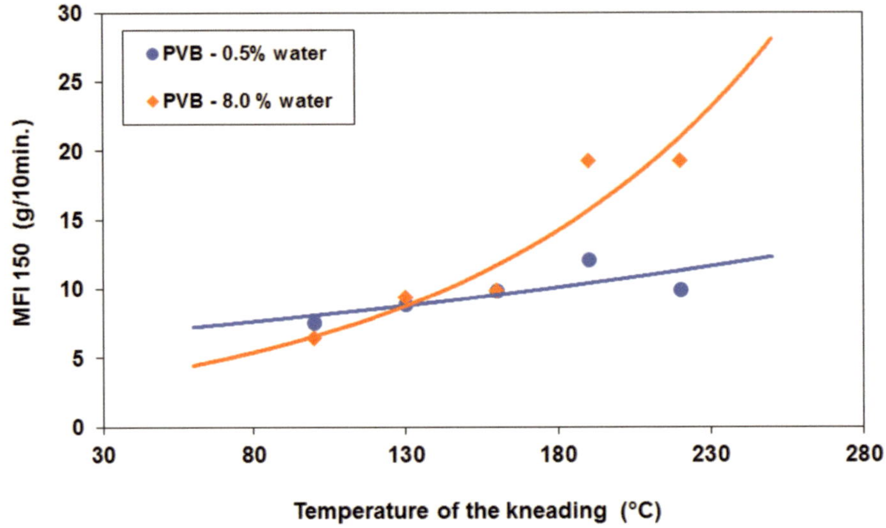

Fig. 7. MFI of kneaded PVB samples with different water content (0.5 % and 8 %) at 60 rpm during ten minutes.

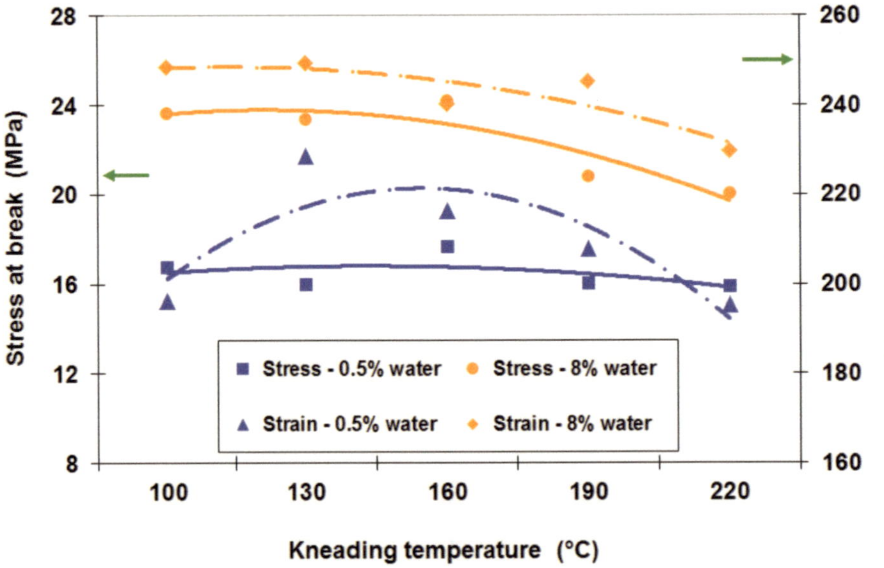

Fig. 8. Tensile strength of kneaded PVB samples with different water content (0.5 % and 8 %) at 60 rpm; arrows point out to values of virgin PVB sheet.

For the wet PVB, the values of stress at break and strain are systematically higher compared to those measured for dry samples. Moreover, dependence stress and strain *vs* temperature is monotonously decreasing not showing maxima or minima. The increase can be explained by the intramolecular crosslinking formed by the hydrogen bonds.

The comparison of "dry" and "wet" PVB indicates that water acts as a plasticizer and "wet" material is less stressed during re-processing and. Hence, the more plasticized "wet" PVB is not significantly stressed by shear and is mostly degraded by thermo oxidative degradation.

7.1 The influence of water on the yellowness

The PVB degradation was the most markedly reflected through the changes of PVB yellowness. Visually and also instrumentally, the yellowness (sometimes even brownness) of re-processed PVB samples was noticeable. Yellowness increased significantly with increasing of re-processing temperature. In the case of „dry" PVB, significant color change was observed above 130 °C. Color of "wet" PVB was significantly changed above 160 °C (Fig. 9). Measurements demonstrated that during kneading at the temperatures below 160 °C yellowness was almost unchanged. This can be explained by the stabilizing function of higher moisture content and consequently higher grade of PVB plasticization. At temperatures lower than 130 °C, the change of the yellowness is insignificantly irrespective of water content. With the increasing temperature, the yellowness grew markedly. This can be explained by thermo oxidative reactions between oxygen and PVB accompanied by better gas diffusion as well as by water induced hydrolysis. It was reported that during hydrolysis, conjugated double bonds are formed [35]. These are more reactive and bring more intensive lowering of molecular weight. The results from the yellowness measurements corresponds the results from determination of MFI (see Fig. 7).

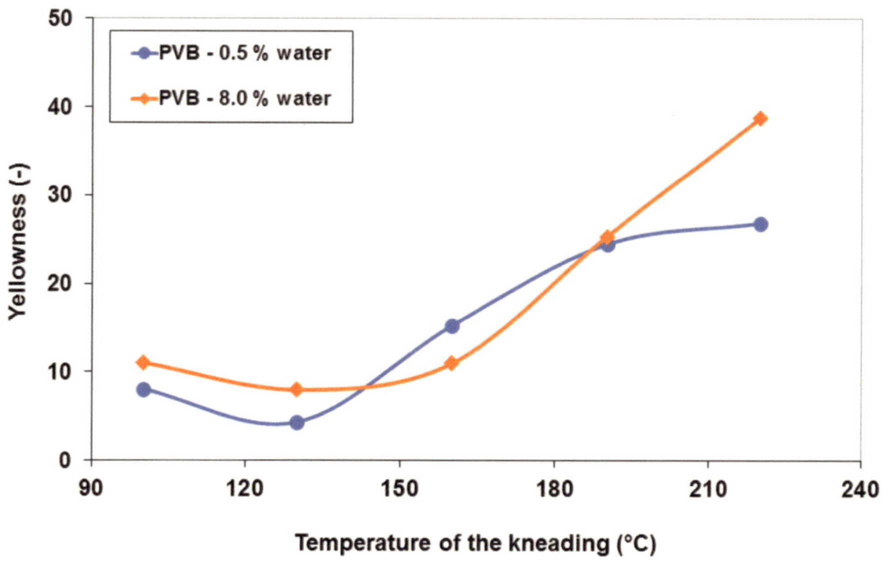

Fig. 9. Yellowness of kneaded PVB samples with different water content (0.5 % and 8 %) at 60 rpm during ten minutes.

8. The change of solution properties influenced by the degradation

Changes in molecular weight and molecular weight distribution of virgin and processed material were followed by gel permeation chromatography. Differential distribution curves of virgin and processed samples (kneading, 100°C, dry) are compared in Fig. 10. From figure it is obvious that the entire distribution of the processed sample compared to virgin one is shifted to lower molecular weight region, which indicates degradation. Moreover the processed sample contains small but distinct peak with molecular weights higher that $2x10^6$ $g.mol^{-1}$ (labeled with arrow). This peak was observed for all the processed samples irrespective temperature and type of processing and its presence indicates that diluted solutions of processed samples contain structures with high molecular weight - aggregates. The aggregation of PVB solution and difficulties with polymer dissolution, even in thermodynamically good solvents, has been reported by several authors (Měřínská, Tupý, 2009; Remsen, 1991).

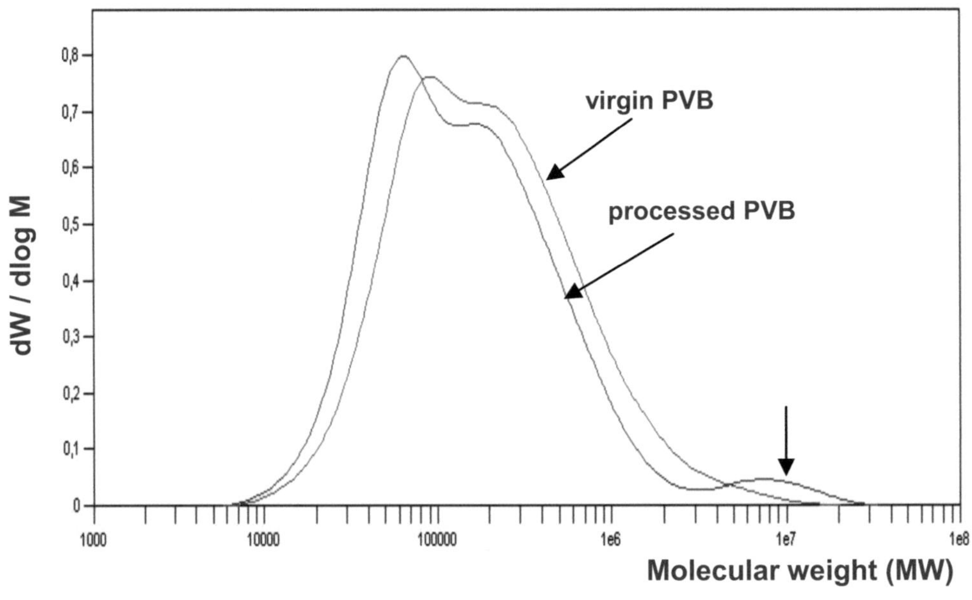

Fig. 10. Comparison of molecular weight distribution curves of virgin and processed PVB (100 °C, dry).

Changes of molecular weight in terms of M_w as a function of increased kneading temperature are for wet and dry samples depicted in Fig. 11. For dry sample, the lowest M_w values (weight average of molecular weight) were measured on samples processed below 150 °C. Under these conditions, predominantly shear degradation takes place resulting in chain scission caused by mechanical stress, Temperature region between 150 °C and 180 °C seems to be favorable for reprocessing of dry PVB. Here, only minor changes in the sample are observed and molecular weights stay almost unchanged. For wet PVB, which posses at lower temperatures low stiffness, molecular weight tends to decrease with increasing processing temperature. At temperatures above 190 °C, molecular weight of dry PVB is comparable to that measured for wet sample. Hence, it can be assumed that degradation mechanism in this temperature region is similar. From Fig.11 it is also obvious that molecular weights of the wet PVB samples, with the exception of the sample processed at 220 °C, were systematically higher compared to dry ones.

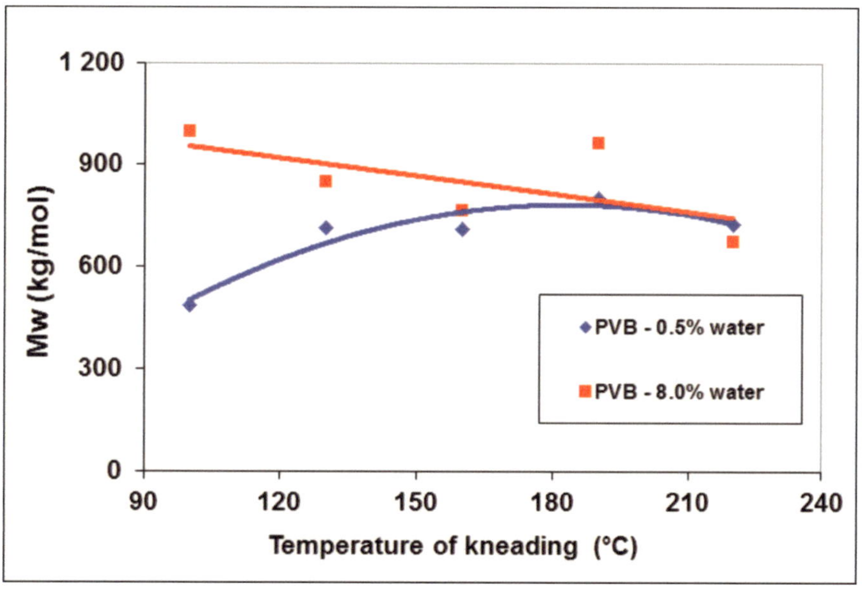

Fig. 11. Influence of kneading temperature on changes of weight average of molecular weight (M_w) recorded for (0.5% of water) dry and wet (8.0 % of water) PVB sheet.

9. Conclusions

In the presented work, conditions for re-processing of plasticized PVB sheets were investigated and influence of temperature, air oxygen content and mechanical stress on the course of degradation was studied. In order to find the possibility for reduction of energy consumption during re-processing, effect of moisture content in PVB sheets on processing parameters and degradation was examined. The obtained results show that, based on the evaluation of MFI and mechanical properties, the optimal conditions for PVB re-processing by kneading occur at the temperature of about 150°C and rotation speed of kneader lower than 60 rpm. These conclusions are in the good agreement with the measurement of PVB yellowness. Below 150 °C yellowness remained almost unchanged and increased significantly above this temperature. GPC measurements corroborate the above conclusions showing minimal changes of PVB molecular weigh for this temperature. Increased amount of water in PVB sheet can act as and additional plasticizer improving workability of polymer melt and decreasing this energy consumption. However the "wet" samples are more susceptible to hydrolytic degradation and compromise decision has to be taken to find the balance between these two effects.

10. Acknowledgement

This article was written with support of Operational Program Research and Development for Innovations co-funded by the European Regional Development Fund (ERDF) and national budget of Czech Republic, within the framework of project Centre of Polymer Systems (reg. number: CZ.1.05/2.1.00/03.0111).

11. References

Aoshima Y., Shohi H., (Jun 6, 2000), Intermediate film for laminated glass and laminated glass, *European Patent EN1227070A1,*

D'erric J.J., Jemmott B.A., et al, (Dec 29, 1995), Plasticized polyvinyl butyral and sheet, *United States Patent 6,559,212*

D'errico J.J., Jemmott B.A., et al, (Dec 24, 1996), Plasticized polyvinyl butyral and sheet, *European Patent EP 0,877,665*

D'errico J.J., (Nov 12, 1997), Control of adhesion of polyvinyl butyral sheet to glass, *European Patent EP 0,938,519B1*

Dhaliwal A.K., Hay J.N., (2002), The characterization of poly(vinyl butyral) by thermal analysis. *Journal Thermochemica acta,* Vol. 391, pp. 245-255

E.E. Remsen, Determination of molecular weight for PVB using size exclusion chromatography / low-angle laser light scattering (SEC/LALLS) in Hexafluoroisopropanol, (1991), *Journal of applied polymer science,* Vol.42, pp.503-510

Fowkes F.M., (1987), Role of acid-base interfacial bonding in adhesion, *Journal of Adhesion Science and Technology,* Vol. 1, pp. 7-27

Gorokhovski A.V., Escapante-Garcia et al, (2005), Composite materials based on wastes of flat glass processing, *Waste Management,* Vol. 25, pp. 733-736

Grachev V.I., Klimenko I.B., et al, (1974), A spectroscopic study of the kinetics of thermal-oxidative degradation of PVB, *Vysokomol Soysed.,* Vol. 16, No. 2, pp.367-373

Hermann H.D., Fabian, et al, (Jan 31, 1984), Polyvinylbutyral films which contain plasticizer and have a reduced adhesive power to glass, *United States Patent 4,533,601*

Ivanov I.V., (2006), Analysis, modeling, and optimalization of laminated glasses as plane beam, *International Journal of Solids and Structures,* Vol. 43, pp. 6887-6907

Iwasaki R., Sato C., (2006), The influence of strain rate on the interfacial fracture toughness between PVB and laminated glass, *Journal de Physique IV*, Vol.134, pp. 1153-1158

Keller U., Mortelmans H., (1999), Adhesion in laminated safety glass–what makes it work, *Glass Processing Days*, Vol. 8, Jun 13–16, pp. 353-56

Měřínská D., Tupý M., (2009), Degradation of plasticized PVB during re-processing by kneading, *Accepted for publication in Macromolecular symposia*, Vol. 286, pp. 107-115

Mister R.E., Bianchi E., et al, (Jan 21-26, 2007), *31st International Cocoa Beach Conference of the American Ceramic Society*, Daytona Brach

Mrkvičková L., Daňhelka J., et al, (1984), Characterization of commercial PVB by gel permeation chromatography, *Journal of Applied Polymer Science*, Vol. 29, pp. 803-808

Nagai K., (Jul 6, 2001), Laminated glass, *United States Patent 6,686,032*

Neher H.T., (Mar 30, 1936), Laminated glass process preparing, *United States Patent 2,032,663*

Nghuen F.N., BERG J.C,. (2004), The effect of vinyl alcohol content on adhesion performance in poly(vinyl butyral)/glass systems, *Journal of Adhesion Science and Technology*, Vol. 18(9), pp. 1011-1026

Ocia, The International Organization of Motor Vehicle Manufacturers, (2007), [june 21, 2011] – Available from: http://www.oica.net/category/production-statistics/

Phillips T.R., (Apr 5, 2005), Process for the aqueous plasticization of polyvinyl acetal resins, *United States Patent 7,285,594,*

Physical Properties of PVB sheet SEKI-SUI, The Sekisui Chemical Co., Copyright 2008, [june 21, 2011] – Available from: http://www.sekisui.co.jp/cs/eng/products/type/slecbk/tech/1183758_5127.html

Plaček T., (2006), Study of impurities separation and PVB trim waste re-processing, *Sindat engineering*

Recycling of windshields glasses in fired bricks industry, (1997), *Key engineering materials*, Switzerland, Vols.132-136

Saflex laminated guide – physical properties and laboratory methods, Monsanto, (1973)

Shichiri T., Miyai J., et al, (Mar 28, 2002), Interlayer for laminated glass and laminated glass, *United States Patent 7,074,487*

Smith R.L., Rymer D.L., et al, (Nov 13, 2008), Polyvinylbutyral interlayer sheet with improved adhesion to glass and a process for preparing same, *United States Patent Application US 20080277045*

Svoboda J., et al., (1970), Research message No.895, New ways for PVB processing, *VÚGPT Gottwaldov*

Svoboda J., (1987), Candidate thesis, VUT in Brno, Faculty of technology - placed in Gottwaldov

Svoboda J., Balazs M., et al, (1988), Industry glass. *Glass union concern*: Research and developing institute of industrial glass in Teplice, Issue C(1)

Tupý M., Zvoníček J., et al, (2008), Problems of recycling PVB sheet used for production laminated safety glass, Tomas Bata University, Faculty of Technology: Plastics and Rubbers, Vol.45, pp. 208-211

Tupý M., Měřínská D., et al, (2010), Influence of water and magnesium ion on the optical properties in various plasticized PVB sheets, *Accepted for publication in Journal of Applied Polymer Science,* Vol. 118, pp. 2100-2108

Tupý M., Měřínská D., (2010), Water and acid-base reactants impact to adhesives properties of various plasticized PVB sheets, *The Paper submitted to Journal of Applied Polymer Science*

Tupý M., Měřínská D., et al, (2001), Windshield recycling focused on effective separation of PVB sheet, The Paper submitted to Resources, Conservation and Recycling

Wade B.E., D'errico J., et al, (Mar 16, 2004), Polymer sheets and methods of controlling adhesion of polymer sheets to glass, *United States Patent Application US 20050208315*

Zvoníček J., (1999), Collection of findings close to PVB extrusion process, *Non-publish message, Retrim-CZ,* Zlín

6

Materials and Methods for the Chemical Catalytic Cracking of Plastic Waste

Luis Noreña, Julia Aguilar, Violeta Mugica,
Mirella Gutiérrez and Miguel Torres
*Applied Chemistry Research Group, Departamento de Ciencias Básicas,
Universidad Autónoma Metropolitana, Azcapotzalco, México D. F.
Mexico*

1. Introduction

Nowadays, plastics play a fundamental role in modern life, they are included in all productive chains. Plastics frequently replace more traditional materials such as wood, metal, glass, leather, paper and rubber because they can be lighter, stronger, corrosion resistant, acid and base resistant, durable and a better insulators. Plastics are polymers with high molecular weight and usually synthesized from low molecular weight compounds, although they can be obtained also through the chemical modification of high molecular weight natural materials such as cellulose (Gervet, 2007).

Plastics can be divided into two major groups, according to their thermal behavior: thermosets and thermoplastics. Thermoplastics soften when they are exposed to heat, and they can be molded and shaped, this heating process can be repeated many times. These plastics contribute to the total plastic consumption by roughly 80% and are used as containers, packaging, trash bags and other non-durable goods (Al-Salem et al., 2009). Some examples are high and low density polyethylene (HDPE, LDP), polystyrene (PS), polypropylene (PP) and polyvinyl chloride (PVC). In contrast, thermosets solidify irreversibly when heated, since an irreversible network of cross-linked covalent bonds is formed, giving a hard, durable, strength and heat resistant products. Such is the case of unsaturated polyurethane (PU), unsaturated polyesters and, alkyd, phenolic and epoxy resins. For that reason, they are used primarily in automobiles, construction adhesives, furniture, kitchenware, inks, and coatings. A third group of plastics, rubber-type, are named elastomers, formed by slightly cross-linked polymer chains; in less proportion than thermosets, giving to these materials elastic properties and relatively good resistance (Morton-Jones, 1993; Aguado & Serrano, 2007, Scheirs & Kaminsky, 2006).

In the last thirty years, plastic industry has raised very quickly, growing around 500%. In 2008, the global plastic production was 245 Mt; the European Community accounts for around 25% of world production, whereas the United States by around 13%; China alone accounts for 15%. Polyethylene has the highest share of production of any polymer type and the packaging and construction sectors represent more than 50% of plastic demand (EC, 2011; USEPA, 2008).

As a consequence of the widespread use of plastics, they represent between 10 to 13% of the municipal solid wastes (MSW) generation in the whole world. In 2008, the total generation of plastic waste in the EU-27, Norway and Switzerland was 24.9 Mt while in the United States was 30.05 Mt. Packaging is by far the largest contributor to plastic waste by 63%. In the United States, the recycling rate for different types of plastic varies greatly, resulting in an overall plastics recycling rate of only 7%. In 2008, however, the recycling rate of some plastics became much higher, for example in 2008, 28% of PET bottles and jars and 29% of HDPE bottles were recycled (USEPA, 2008).

Plastic waste generation imposes negative environmental effects, since these materials are usually non-biodegradable and, therefore, they can remain as waste in the environment for a very long time; they may pose risks to human health as well as to the environment; and they can be difficult to reuse and/or recycle in large-scale practice. An issue of particular concern is that giant masses of plastic waste have been discovered in the North Atlantic and Pacific Oceans, the full environmental impacts of which are not yet fully understood but which may cause severe damage to seabirds, marine mammals and fish (Derraik, 2002, Lavender et al., 2010). Plastic waste in the ocean causes the death of up to one million seabirds, 100 000 marine mammals and countless fish every year (UNEP, 2006).

Once the material enters the waste stream, recycling is the process of using recovered material to manufacture a new product (Hopewell et al., 2009). Recycling, being one of the strategies for minimization of waste, offers three benefits (Edwards, 1999): (i) reduces the demand upon new resources, (ii) cuts down on transport and production energy costs and (iii) uses waste which would otherwise be lost to landfill sites. Plastic wastes are mainly found in MSW mixed with other classes of residues, consequently, their recycling is limited and landfills are the primary destination of these wastes. Recycling plastic waste provides important environmental advantages such as:

1. Saving space in landfills. The need to create new landfills and to put more trash in the earth is ever increasing, and recycling is the only sustainable solution to drastically decrease the waste deposited into landfills. In addition, as common plastics are not degradable, they can remain for centuries without space liberation. One ton of plastic bottles free more than 7 cubic yards of landfills, then, recycling plastic means liberating a lot of space unnecessarily used.
2. Energy conservation. The recycled material is used as a resource, although most of the times, every recycled piece of plastic changes into something completely different after the process. Recycling can use two-thirds the energy to manufacture from recycled products. One pound of recycled PET can save as much as twelve thousand BTU´s energy (Scheirs, 2001).
3. Reduction of air pollution and greenhouse gases. An average net reduction in greenhouse gas emissions (GHG) of around 1.5 ton of CO_2-eq of recycled plastic has been estimated (Department of Environment and Conservation (NSW), 2005). The reduction in pollutant emissions and GHG is mainly due to the substitution of virgin polymer production, decreasing fuel burning.
4. Oil conservation. The increase of the price of oil is actually an important economic issue around the whole world; every ton of recycled plastic saves almost 2000 pounds of oil. Savings can be due to the reduction of raw refinery materials in the manufacturing process, to the use of plastics as a source of combustion fuel in incinerators, as well as fuel production from plastics by catalytic desintegration (Aguado & Serrano, 2007).

5. Saving marine life. It has been reported that a plastic raft (bottle caps, toys, bottles, etc.) floats on the Pacific Ocean, trapping and killing marine life, as well as sea birds and turtles, unable to distinguish plastics from food, dying by malnutrition or asphyxia (Derraik, 2002).

Several end-of-life options exist to deal with plastic wastes. Four categories can be considered for the treatment and recycling processes of plastic wastes (Table 1), depending of requirements of every locality or industry; each method presents advantages and disadvantages. Although primary and secondary recycling schemes are well established and widely applied, it has been concluded that many of the PSW tertiary and quaternary treatment schemes appear to be robust and worth of additional investigation (Al-Salem et al., 2009; Hopewell et al., 2009).

Method	Description	Advantage	Disadvantage
R-extrusion (Primary)	Involves the re-introduction of clean scrap of single polymer to the extrusion cycle in order to produce products of the similar material.	Re-using plastic is preferable to recycling as it requires less energy and fewer resources. Usually they feed the secondary process.	Limited, since rarely possess the required quality. Sorting must be attempted within a short time. Removing paints is necessary.
Mechanical Recycling (Secondary)	Involves reprocessing of plastic by melting, shredding or granulation. Separation, washing and preparation of wastes are essential.	It is an economic and viable route for plastic wastes recovery. It is used mainly to manufacture fibers for carpets, apparel and bottles.	Mechanical recycling of PSW can only be performed on single polymer plastic, e.g. PE, PP, PS, etc. Requires intense energy consumption.
Feedstock or Chemical Recycling (Tertiary)	Refers to techniques used to breakdown plastic polymers into their monomers or small molecules. It is the most sustainable method.	Can be used in petrochemical and chemical production plants; wastes are converted into valuable feedstock chemicals, useful as fuels or raw materials.	Costs of a feedstock plant are still high in comparison with oil plants. Still lacks the proper design and kinetic background to target certain products and chemicals
Incineration and energy recovery (Quaternary)	Implies burning waste to produce energy in the form of heat, steam and electricity.	Produce heat, power and/or gaseous fuels. Plastic waste results in a volume reduction of 90–99%, which reduces the reliability on landfilling	Health and environmental concerns due to the production of large amounts of air pollutants.

Table 1. Methods for treatment and recycling of plastic wastes. Modified from (Scheirs & Kaminsky, 2006; Al-Salem et al., 2009; Hopewell et al., 2009) and based on the ASTM classification.

Among the feedstock (tertiary) recycling methods, we should consider:

- Pyrolysis. Thermal decomposition of polymer chains.
- Gasification. Decomposition under oxygen or stream conditions to yield synthesis gas.
- Hydrogenation. Chain breaking with hydrogen.
- Catalytic cracking. Polymer chain breaking through the action of a catalyst.

Scheme 1 shows an overview of the thermal and chemical recycling processes that can be applied to plastic waste (Kumar et al., 2011) We should point out that methanolyis can be included among alcoholysis methods and hydrolisis is also another chemolysis method.

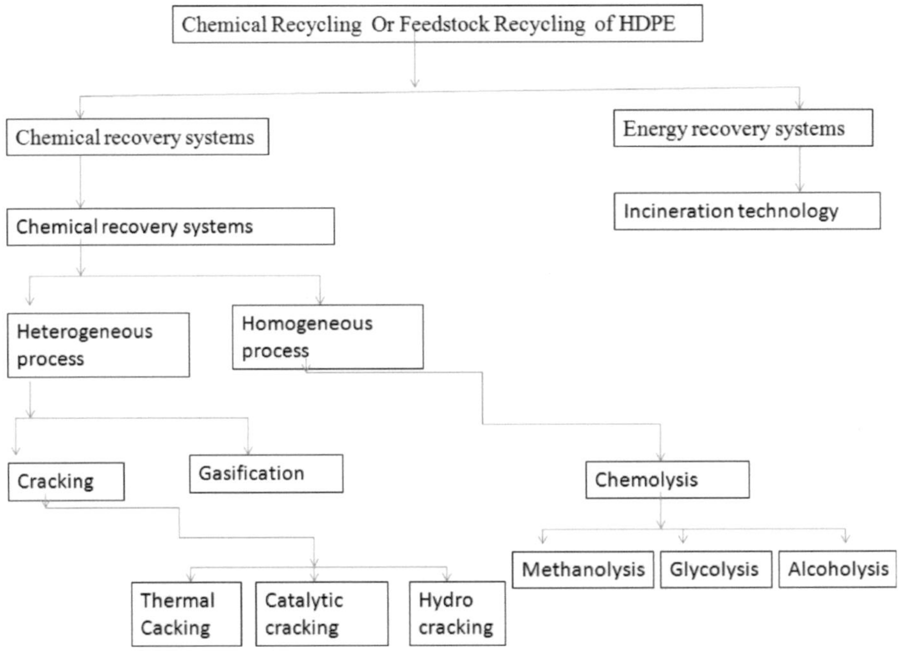

Scheme 1. Feedstock and thermal recycling of polyethylene (Kumar et al., 2011).

Catalytic polymer cracking is one of the actual alternatives of feedstock recycling since very valuable products can be obtained. These technologies can be applied to HDPE, LDPE and PP, which represent around 60% in plastic wastes, although certain polyamides can be efficiently depolymerized. Moreover, in principle, any kind of plastic may be recycled, providing the right catalyst, the right reaction system and the right operation conditions. The industrial application of the catalytic polymer recycling has been limited more because economic considerations rather than technical considerations. Due to the high price of crude oil, these technologies are receiving renewed attention, (Schiers, 2001).

2. Catalytic materials

Nowadays, molecular sieves are the most studied and employed materials for the chemical decomposition of plastic waste. Other catalyst systems may be effective for breaking polymer chains, such as the previously used Friedel-Crafts catalysts, however, they present corrosion and environmental problems (Clark,1999).

Molecular sieves are a wide range of solid materials of porous microstructure. Many of them possess acid or basic properties, very useful in catalytic reactions. There are several types of molecular sieves, depending mainly, on their chemical composition and their pore system. The catalytic decomposition of plastic waste polymer chains takes place following the same pathways as the hydrocarbon catalytic cracking reactions used in petroleum refinery processes and, catalysts employed by this industry are also useful for polymer decomposition. Alumina, silica, amorphous silica-alumina and crystalline aluminosilicates (mainly zeolites) are widely used in petroleum processes. Hydrocarbon cracking reactions may take place through hydride remotion, promoted by Lewis acid sites, or through a carbocation intermediate, from the carbon protonation by Brönsted acid sites. Fluid Catalytic Cracking (FCC) refinery plants around the world employ faujasite zeolite (FAU), having strong Brönsted acid sites, as the principal component of FCC catalyts. Therefore, our research group, as other researchers, have employed commercial FCC catalyts for the catalytic cracking of plastic waste (Cardona & Corma, 2000, De la Puente et al., 2002, Lin et al. 2010, Ortega et al., 2006, Sanchez et al., 2003). A very promising option, from an economical point of view, is to use waste FCC catalysts (also called *equilibrium catalyts*) disposed from refinery plants. In addition to faujasite zeolite (or Y zeolite), other zeolites such as mordenite, clinoptilolite, X zeolite or ZSM-5, have also been employed for the catalytic decomposition of plastic waste (Clark, 1999, Huang et al., 2010). Natural zeolites are other very promising option, from an economical point of view, since they are a lot cheaper than synthetic zeolites.

Research attention has also been focused on new molecular sieves with pore sizes larger than those of zeolites, such as MCM-41 or SBA-15, that allow larger molecules inside the pore channels. According to the IUPAC (International Union of Pure and Applied Chemistry), zeolites are considered microporous materials, with pore sizes smaller than 2 nm, whereas MCM-41 and SBA-15 are considered mesoporous materials, with pore sizes falling within the range between 2 and 50 nm. These materials have mostly weak acid properties as such, but they can be functionalized in many different ways. Their acid properties can be greatly enhanced by the introduction of for instance, aluminum, gallium, iron or zirconium atoms during the synthesis (Chen et al., 2007, Chen et al., 2006, Diaz-Garcia et al., 2010), or by the immobilization of sulfonic or heteropolyacid strong acid groups onto the pore channels, after the synthesis (Boveri et al., 2005, Hernández et al., 2010, Schacht et al., 2010, Wang et al., 2009, Wang et al., 2008, Yang et al., 2009). Recent studies have also employed hybrid micro/mesoporous materials (Serrano et al., 2010).

2.1 Microporous materials: Natural and synthetic zeolites

Nanostructured microporous materials have been the focus of much attention because their unique properties, such as a large specific surface by volume unit (many physical and chemical interactions take place on surfaces), electrical and thermal conductivity, ductility or mechanical resistance. Zeolites are a much relevant class of microporous materials that have found application on diverse fields, such as ion exchangers (i.e. water treatment), adsorbents (i.e. for nuclear plants waste), catalysts, or even as food supplement for farm animals (Mravec et al., 2005). Zeolites have become cornerstone and common household in research centers, industry and domestic environments (Masters & Maschmeyer, 2011).

Zeolites are crystalline aluminosilicates of tridimensional structure and with cavities and channels between 0.3 and 1 nm width. The inner structure is very porous and cavities and channels may be occupied by mobile cations and water molecules. The basic structure is formed by TO_4 tetrahedra (T = Si, Al) with oxygen atoms connecting neighboring tetrahedra. Zeolites composition may be described by three components (Auerbach et al., 2005):

$$\mathbf{M}^{m+}_{n/m} \quad \cdot \quad [\mathbf{Si}_{1-n}\,\mathbf{Al}_{n}\,\mathbf{O}_2] \quad \cdot \quad n\mathbf{H_2O}$$

Extra-framework cations framework adsorbed phase

And their chemical formula by (Dyer, 1988):

$$M_{x/n}\,\mathbf{Al}_x\,\mathbf{Si}_y\,\mathbf{O}_{2(x+y)}\,.w\mathbf{H_2O}$$

Zeolites have an inner surface much larger than their external surface. Zeolites microporosity is open and the structure allows matter transfer between the inter-crystalline space and the surrounding media. This transference of matter is limited by the pore diameter and form, allowing the in and out movement of molecules of a specific critic size (ACS Monograph, 1976).

ZEOLITE	NUMBER OF TETRAHEDRA	PORE (θ) DIAMETER (nm)	EXAMPLES
Small pore	8	$0.3 < \theta < 0.5$	Erionite, A
Medium pore	10	$0.5 < \theta < 0.6$	ZSM-5, ZSM-11
Large pore	12	$0.6 < \theta < 0.9$	Y, β, Ω
Extralarge pore	18	$0.9 < \theta$	MCM-9, VPI-5

Table 2. Zeolite classification.

There are natural and synthetic zeolites. The structure of a synthetic zeolite is exactly the same as its corresponding natural zeolite. Nowadays, there are a lot more synthetic zeolites than natural zeolites. Most natural zeolites are obtained from volcanic-origin geological regions (Dyer, 1988). The catalytic activity of natural zeolites is limited by impurities and a reduced specific surface (Auerbach et al., 2003).

Even if there are more than 10,000 registered patents dealing with the synthesis of zeolites, most synthetic zeolites have been prepared by simple modifications of the Barrer and Milton method, varying only two parameters: the cation and the SiO_2/Al_2O_3 ratio (Mravec et al., 2005).

As previously mentioned, at the beginning of the Catalytic Materials Section, several zeolites have been employed for the catalytic decomposition of plastic waste, and our research group has experience with synthetic zeolites (Torres et al., 2011, Torres et al., 2008). However, considering that natural zeolites are a lot cheaper and ready available in Mexico, we selected natural Mexican mordenite (MOR), Figure 1, and natural Mexican clinoptilolite (HEU) for the catalytic decomposition of low-density polyethylene. Zeolites can be identified by their characteristic XRD pattern and this technique also allows assessing the purity of natural zeolites (Figure 2).

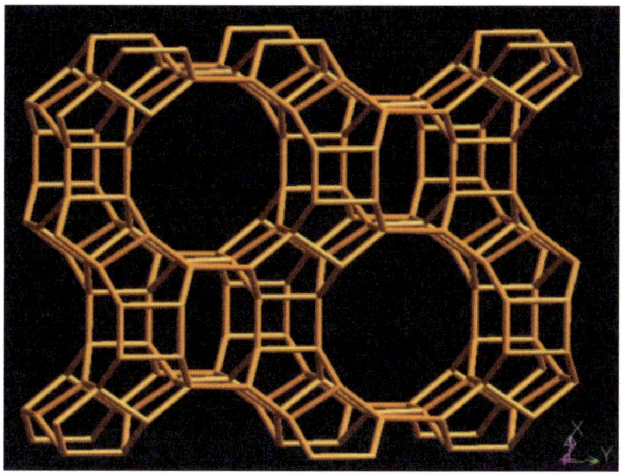

Fig. 1. Mordenite.

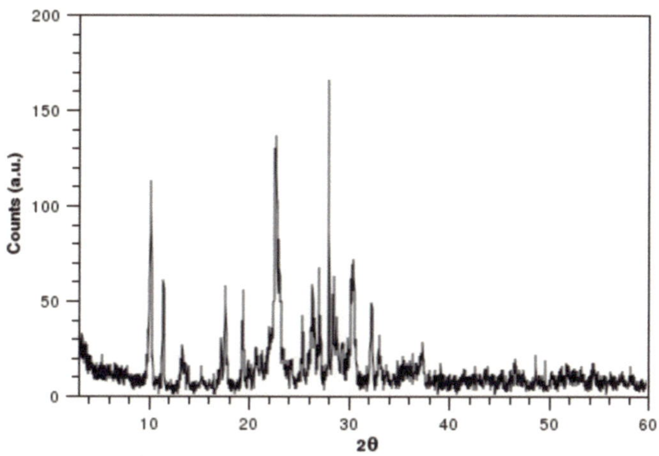

Fig. 2. X-Ray Diffraction pattern of natural clinoptilolite (Hernandez, 2011).

2.2 Mesoporous materials: MCM-41

Among the newly developed mesoporous synthetic materials (MCM, SBA, HMS, MSU or FDU labels), MCM-41 has received special attention for catalytic cracking reactions. Nano-structured MCM-41 mesoporous materials are constituted by an ordered network of hexagonal channels and they have promising applications in adsorption, ion exchange, water treatment and catalysis. These mesoporous materials are synthesized through a templating mechanism employing ammonium surfactants. The ammonium surfactants can form lyotropic liquid crystals with lamellar, cubic or hexagonal mesophases, and, at the right precursors concentration, the mesoporous MCM materials can attain the same geometries (Figure 3). The mesopores have a typical length of 3 nm but it can be varied by using surfactant molecules of different size. MCM-41 materials are essentially silicates but the composition can be modified by the introduction of many other different elements that confer these materials specific properties. The materials can be modified during the synthesis procedure or after the synthesis procedure.

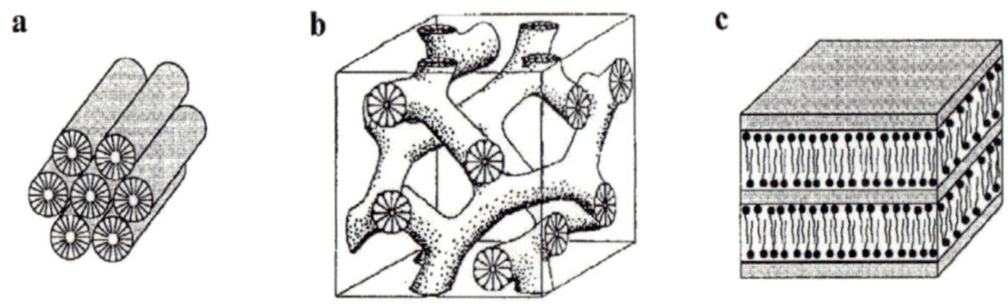

Fig. 3. MCM-41 of hexagonal microstructure (a), MCM-48 of cubic microstructure (b), MCM-50 of lamellar microstructure (c).

The MCM-41 hexagonal channel system has a characteristic XRD pattern (Figure 5) and, it can actually be seen by transmission electron microscopy (Figure 4).

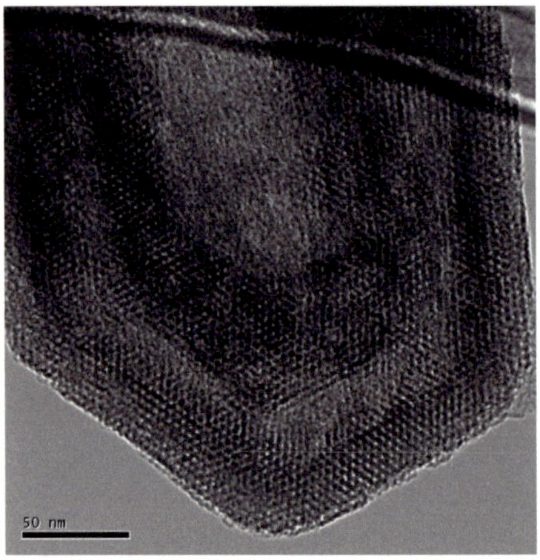

Fig. 4. Transmission electron microscopy image of MCM-41 (Wang et al. 2009).

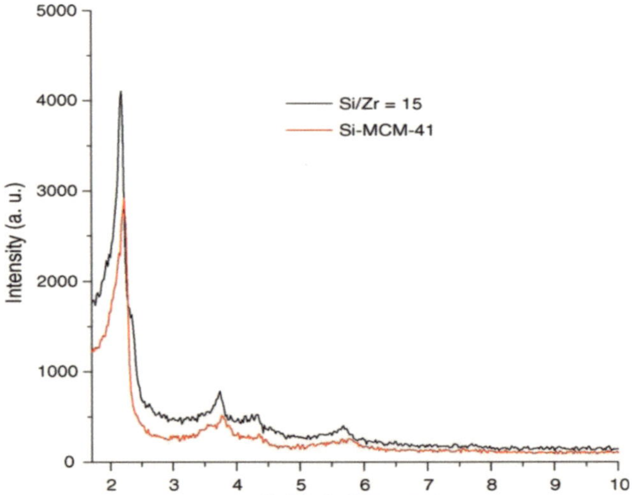

Fig. 5. XRD of siliceous MCM-41 and zirconium-modified MCM-41 (Chen et al., 2006).

Table 3 shows the Si/Al molar ratio and the textural properties of several catalysts we have employed for the catalytic decomposition of low-density polyethylene, which include commercial fresh and waste FCC refinery plant catalyst (with faujasite zeolite as main cracking agent), natural mordenite zeolite and Al-MCM-41 and Ga-MCM-41 mesoporous materials. FCC catalysts were provided by the Mexican Petroleum Institute (IMP).

CATALYST	Si/Al Ratio	Unit Cell Parameter[1] (nm)	Specific Area[2] (m^2/g)
FCC-Fresh	24.5 *	24.30	300.34
FCC-Equilibrium	25 *	24.46	166.15
Al-MCM-41	9	41.75	800
Ga-MCM-41	25 **	46.18	900
Natural Mordenite	5.5	18.1	280

[1] By XRD, [2] By BET method
* Of the zeolite, ** Si/Ga Ratio

Table 3. Composition and textural properties of catalysts employed for the catalytic cracking of polyethylene (Ortega et al., 2006).

We have studied Al-MCM-41 and Ga-MCM-41 nanostructured mesoporous materials for the catalytic cracking of polymers, because the introduction of Al and Ga within the material's framework during the synthesis procedure, results in enhanced Brönsted acidity. We have also immobilized tungstophosphoric acid, $H_3WPW_{12}0_{40}$, onto the MCM-41 surface (after the synthesis) for the same reason.

The pore size, pore volume, pore size distribution and specific surface area of most molecular sieves is often calculated from nitrogen (or other gases) adsorption-desorption isotherms. Such isotherm is shown in Figure 6, for a tungstophosphoric acid-MCM-41

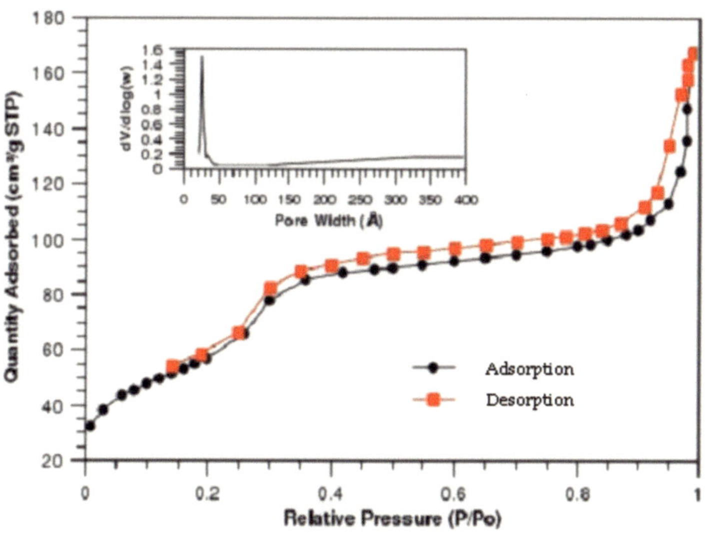

Fig. 6. Nitrogen adsorption-desorption isotherm and pore-size distribution (insert) of the tungstophosphoric acid-MCM-41 material employed in the catalytic cracking of polyethylene (Hernandez et al., 2010).

material. The shape of the isotherm corresponds to a type IV isotherm, according to the Brunauer-Emmet-Teller (BET) classification, typical of mesoporous materials. The pore size distribution (insert) confirms, from the main peak, that the pores are of approximately 3 nm, within the mesoporous range.

The introduction of the heteropolyacid reduces the surface area and the pore volume of the support, but strongly enhances the Brönsted acidity of the material (Table 4)

Sample	BET Surface area (m^2/g)	Pore volume (cm3/g)	Pore size (Å)	Acidity		
				*B	**L	Total
Si-MCM-41	420.0731	0.445895	22	0	820	820
HPA25	340.6419	0.3332	27	41	535	576
HPA50	191.4937	0.2373	23	74	521	595
HPA70	204.0907	0.1923	26	94	339	433

*Brönsted acidity, **Lewis acidity
25, 50 and 70 denote the w/w proportion of $H_3PW_{12}0_{40}$ with respect to MCM-41

Table 4. Textural and acidic properties of MCM-41 and tungstophosphoric acid-MCM-41 employed in the catalytic cracking of polyethylene (Hernandez et al., 2010).

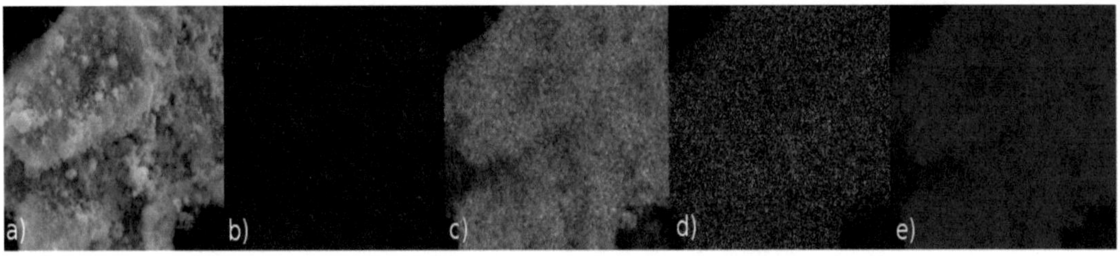

Fig. 7. Scanning electron micrographs of a representative $H_3PW_{12}0_{40}$–MCM-41 material (Hernandez et al., 2010).

From Figure 7, image a) describe the morphology and surface features of the sample, image b) belong to the oxygen atoms distribution, image c) belong to the silicon atoms distribution, while images d) and e) correspond to tungsten atoms distribution from K and L transition signals, respectively. These microscopy images indicate a uniform distribution of the heteropolyacid without cluster formations.

Figure 8 shows the FTIR spectra of the heteropolyacid-impregnated MCM-41 materials, exhibiting the characteristic $H_3PW_{12}0_{40}$ absorption bands at 1080,982, 893 and 822 cm^{-1}. The IR absorption bands in Figure are assigned as: 1080 cm^{-1} belonging to P-O vibrational symmetrical stretching, 982 cm^{-1} to W=O_d stretching coordination, 893 cm^{-1} to W-O_b-W bridge stretching mode (inter-bridges between corner-sharing octahedra) and 799 cm^{-1} to W-O_c-W stretching mode (intra-bridges between edge-sharing octahedra) These absorption bands were retained in the HPW/MCM-41 materials, indicating the preservation of the heteropolyacid Keggin structure. The preservation of the Keggin structure of $H_3PW_{12}0_{40}$ was further confirmed by ^{31}P-MAS-NMR (Hernandez et al., 2010).

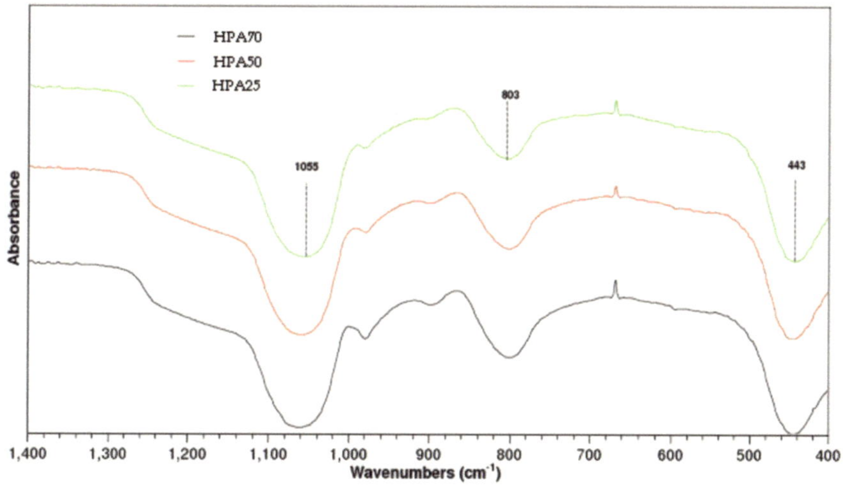

Fig. 8. FTIR spectra of heteropolyacid-MCM-41 materials. 25, 50 and 70 denote the w/w proportion of $H_3PW_{12}O_{40}$ with respect to MCM-41.

2.3 Key catalyst properties

Most catalysts employed for the catalytic polymer cracking are acid micro- or meso-porous solids. Several gas or liquid products are obtained, in different proportions, depending on the catalyst selection. Key factors under consideration are the nature of the acid sites, the acid strenght and textural properties, such as the pore size and the specific surface area. Alumina is widely used in industrial catalytic processes and has Lewis-type acid sites; Lewis sites are also characteristic of several types of silica. Aluminosilicates, such as zeolites or amorphous silica-alumina have a combination of Brönsted and Lewis acid sites. Increasing the aluminum content in the catalyst framework increases the total number of Brönsted acid sites, however, a reduced content of aluminum leads to stronger Brönsted acid sites. The purely siliceous MCM-41 has a medium Lewis acidity and weak Brönsted acidity, but the latter can be greatly enhanced by introducing Al, Ga, Fe, or Zr (among other) atoms within the material's framework, or by anchoring strong acids on the MCM-41 channels, such as heteropolyacids or sulfonic groups. In general terms, strong Brönsted acid sites lead to catalysts with high cracking activity. Strong Brönsted acid sites also tend to produce a larger proportion of gas products (rather than liquid products) and may produce larger amounts of coke.

In order to have relatively more control over the products obtained, it is convenient a catalyst with a regular pore structure and a uniform pore size distribution. Materials with pores of many different sizes tend to produce much more different products and byproducts. In general terms, mesoporous materials favor liquid products, while microporous materials produce a larger proportion of gas products. The catalytic cracking of polymer chains starts at the outer surface of the microporous materials, but once the chain fragments enter the pores, where many active sites are located, small gas molecules come out from them. Zeolites with very small crystal size (of nanometers, instead of micrometers) and large external surface area, may produce less amount of gas molecules while having high cracking activity (Covarrubias et al., 2010). We should point out, that both, gas or liquid products are useful, for chemical industrial processes or as fuels; we may wish to favor certain products depending on the intended application.

3. Reaction mechanism

The reaction mechanism of the catalytic cracking of polymer chains follows similar pathways of those of the hydrocarbon catalytic cracking in petroleum refinery plants. These mechanisms have been studied for a number of years. The catalytic cracking process takes place at temperatures high enough to have parallel thermal cracking reactions (Kumar et al., 2011). Some of the efforts for elucidating the thermal cracking mechanism of polyethylene go back to the end of the 1940's (Oakes and Richards, 1949). The thermal cracking reactions follow a free radical mechanism, by breaking covalent bonds by the action of heat, producing free radical species. Catalytic cracking reactions undergo either by the protonation of carbon atoms in the polymer chain (protons coming from Brönsted acid sites), or by the abstraction of a hydride ion from the polymer chain, by Lewis acid sites.

Protonation:

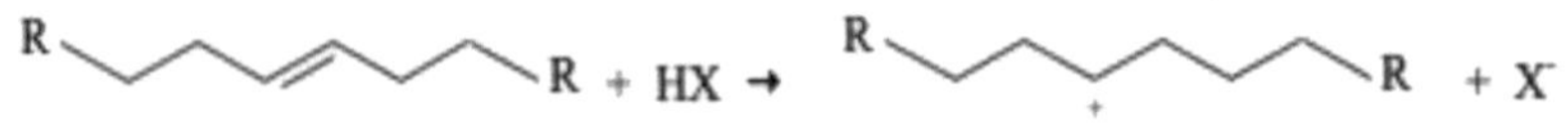

Hydride abstraction:

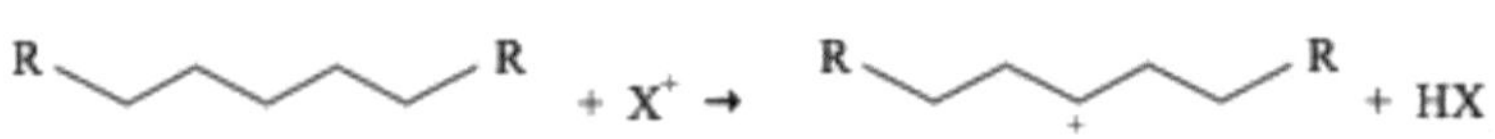

The resulting ions can be stabilized by β-scission, isomerization or hydrogen transfer reactions. Scheme 2 shows different reactions that can take place, each one more or less favored depending on the temperature (Kumar et al., 2011). The surface area and the porous structure of the catalyst also play an important role. The breaking down of polymer chains starts at the external surface of the catalyst. Small enough fragments may get inside the pores, where additional cracking reactions take place, resulting in small-size gas molecules. Unlike thermal cracking, a certain catalyst may promote the selectivity towards specific products (Pinto et al., 1999).

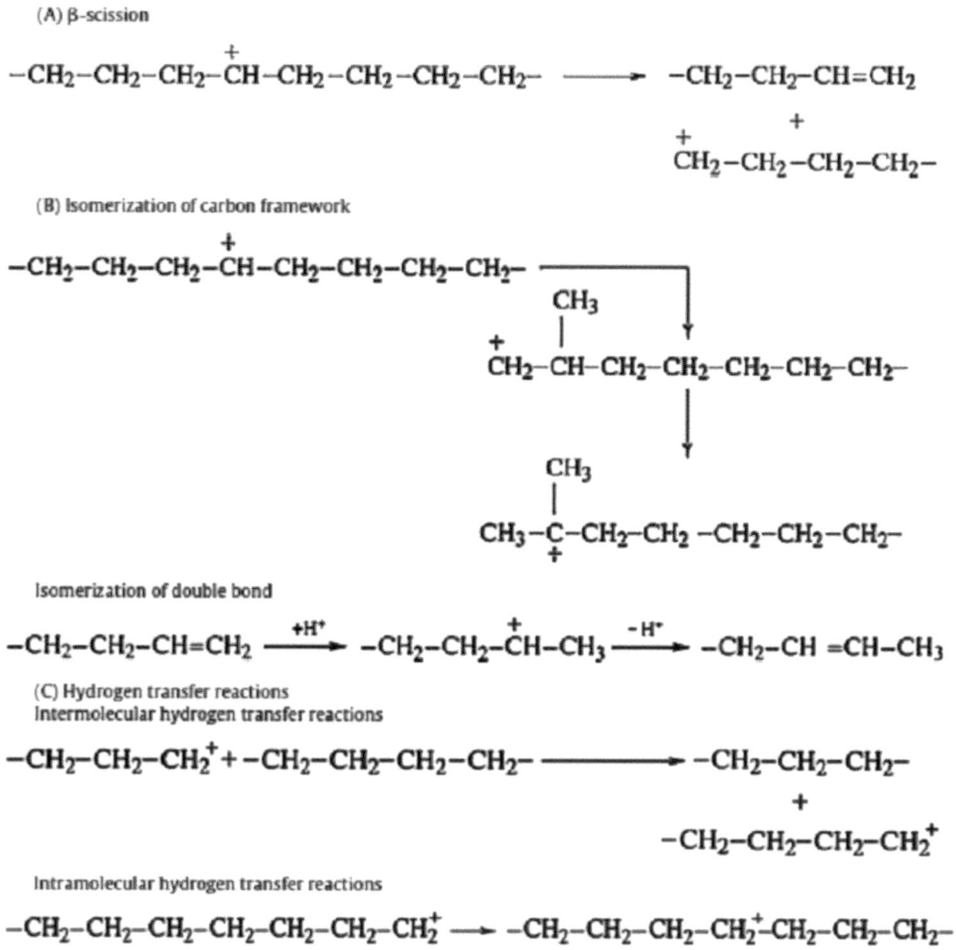

Scheme 2. Reactions involved in the thermal and catalytic cracking of polymer chains (Kumar et al., 2011).

4. Reaction systems

4.1 Reaction system design

For the catalytic cracking reaction we have built a semiautomatic reaction system (Figure 9), initially based on a design reported by Uemichi (Uemichi et al., 1998). The reaction system has (1) a loading section, where the polymer melts, (2) a capillary tube for controlling the polymer feed, employing nitrogen as carrier gas, (3) a fixed bed stainless steel reactor, (4) a condenser and (5), recipients for liquid and gas products. It employs valves (HV), thermopars (TIC) and manometers (PI). The reaction system is heated with electric resists inside a ceramic refractory brick. The temperatures of the 1, 2 and 3 heating zones are automatically controlled *via* Field Point electronic modules from National Instruments; working in an integrated way, through the Labview Professional Development System 6.1 software (Table 5).

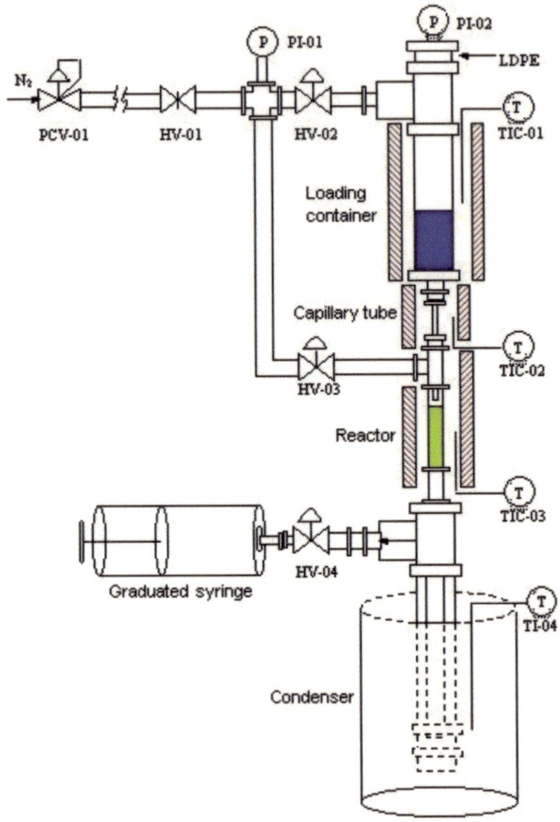

Fig. 9. Reaction system employed in our laboratory (Hernandez et al., 2010, Ortega et al., 2006).

Software	Hardware
Labview professional development system v.6.1	Field Point Modules from National Instruments: • 1FP-1000 RS-232/RS-485, Network interphase.
Field Point v.3.0.2.	• 1FP-TC-120 (temperature reading). • 1FP-PWM-520 (power output).
Toolset Software, NI Developer Suite	
	3 Solid-state Relevators, 40 A, 24-240 V AC. 1 Power Source, 24 V. 1Thermal Fuse, 20 A. 4 Thermopars, K type. 1 Interphase, RS-232.

Table 5. Description of the reactor instrumentation.

We have recently improved the reaction system through a stronger built reactor, better ceramic refractory thermal insulation and better instrumentation, employing National Instruments NI9211 thermocouple differential input modules, NI9472 sourcing digital output modules and a CDAQ-9174 CompactDAQ chassis.

Besides a fixed-bed reactor, several other reactor types have been employed for the thermal and catalytic cracking of polymer waste, such as fluidized-bed, batch or semi-batch reactors, or screw kilns. Main problems for operating reactors are the low thermal conductivity and the high viscosity of plastics. Figure 10 shows a fluidized-bed reactor aimed at limiting the contact between primary volatile products and the polymer/catalyst mixture (Lin et al., 2010).

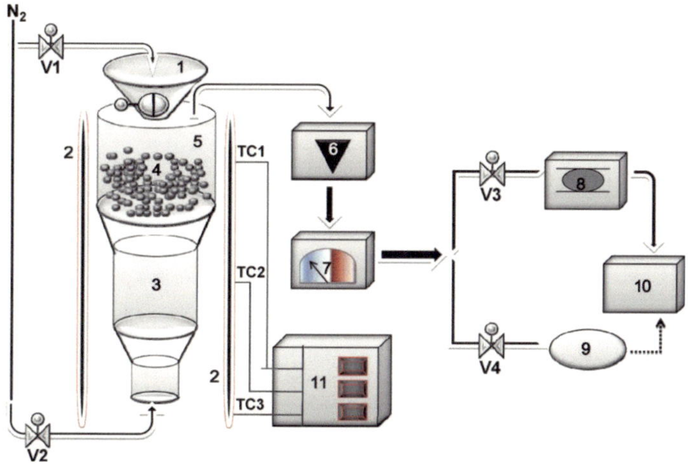

Fig. 10. Fluidized-bed reactor (Lin et al., 2010) (1) feeder, (2) furnace, (3) sintered distributor, (4) fluidized catalyst, (5) reactor, (6) condenser, (7) deionized water trap, (8) 16-loop automated sample system, (9) gas bag, (10) GC, and (11) digital controller for three-zone furnace.

Figure 11 shows the diagram of a continuous spouted bed reactor, aimed at handling sticky, viscous polymer/catalyst mixtures, also looking for reducing residence time, which would favor the formation of light olefins, instead of secondary products such as methane, aromatics or coke (Elordi et al., 2011).

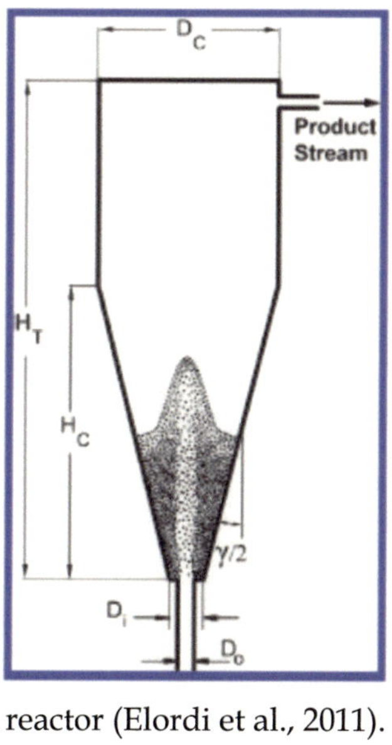

Fig. 11. Continuous spouted bed reactor (Elordi et al., 2011).

4.2 Reaction conditions

Prior to carrying out the reaction, we activated the cataysts at 400 °C for 1 h inside our reactor. The reactor had fitted glass fiber in order to keep the catalyst in place. For every test, we employed 0.9 g of polymer (LDPE). The polymer was melt at 290 C and then introduced into the reactor (0.04 cc/min) The temperature of the capillary tube was set at 330 C and the reactor temperature at 450 C (Table 6).

Polymer (LDPE) feed	0.9 g
Pressure	15 psi
Loading section temperature	290 (°C)
Capillary tube section temperature	330 (°C)
Reactor temperature	450 (°C)
Condenser temperature	-2 (°C)
Reaction time	30 min
N_2 flow (carrier gas)	10 (ml/min)

Table 6. Catalytic cracking reaction conditions (Hernandez et al., 2010, Ortega et al., 2006).

4.3 Product composition determination

Reaction products can be classified as gas products (C1-C4) and liquid products (C5-C44). Gas products were analyzed by gas chromatography. Identification of some specific compounds can be carried out by coupling gas chromatography to mass spectrometry. Liquid products were analyzed by simulated distillation (ASTM method D-2887), corresponding to gasoline, turbosine (or jet fuel or naphtha), kerosene, gas oil and fuel oil fractions, as described in Table 7.

REFINERY FRACTION	NUMBER OF CARBONS	BOILING POINT (°C)
Gasoline	C5-C12	39-220
Turbosine (naphtha, jet fuel)	C13-C14	221-254
Kerosene	C15-C17	255-300
Gas oil	C18-C28	301-431
Fuel oil	C29-C44	432-545

Table 7. Liquid products corresponding to refinery fractions.

5. Results and discussion

We present results we have obtained for the catalytic cracking of low-density polyethylene. Our research group has established a collaboration project with the industrial group ALFA (*via* its IDDEA Office) for the decomposition of several other polymers. ALFA's chemical division is based in Monterrey, Mexico, and has several polymer-precursor industrial plants in different countries. Those results are not presented here.

5.1 Gas products

Several compounds can be obtained among the gas products, such as methane, ethane, ethylene, propane, propylene, n-butane, i-butane, 1-butene or iso-buthylene. The composition of the gas products strongly depends on the catalyst. The gas products distribution obtained with HPA-modified MCM-41 materials can be seen in Figure 12. The gas product consisted of mainly a mixture of ethane, propane and pentane, with very small amounts of ethylene and propylene. The main gas products obtained with Al-MCM-41 and Ga-MCM-41 at the same reaction conditions, were isobutene and propylene (Ortega et al., 2006, Hernandez et al., 2010). We consider that the strong acid sites provided by the HPA resulted into obtaining the smaller ethane molecules. Pentane was propelled into the gas stream by the heated carrier gas.

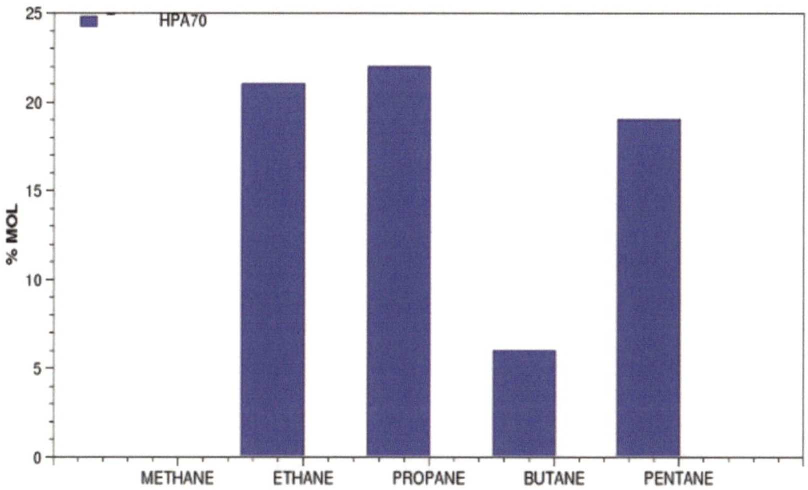

Fig. 12. Gas products distribution obtained from the catalytic cracking of low-density polyethylene employing heteropolyacid-MCM-41 materials (Hernandez et al., 2010).

5.2 Liquid products

Figure 13 shows the relative proportion of liquid products obtained from the catalytic cracking of low-density polyethylene, corresponding to refinery oil fractions (ASTM D-2887 method), employing modified mesoporous MCM-41 (Al-MCM-41 and Ga-MCM-41), Mexican Natural Mordenite and used and fresh commercial FCC (FCC-Eq and FCC-F) catalysts (Ortega et al., 2006) FCC catalysts were provided by the Mexican Petroleum Institute (IMP). Unlike the thermal cracking of polyethylene (pyrolisis), the catalytic cracking of polyethylene produced a high proportion of liquid products, for the several catalysts tested. A relatively higher yield of heavier liquid fractions was obtained with the Al-MCM-41 and Ga-MCM-41 mesoporous materials, whereas zeolite microporous materials produced a larger amount of lighter gasoline fractions, in agreement with discussion in Section 2.3. The smaller proportion of heavier fractions together with the larger proportion of the gasoline fraction was obtained with the fresh commercial FCC catalyst, however, this catalyst also produced a high proportion of gas products. As we pointed out before, from an economical perspective, promising options are the equilibrium (disposal) FCC catalyst and natural mordenite.

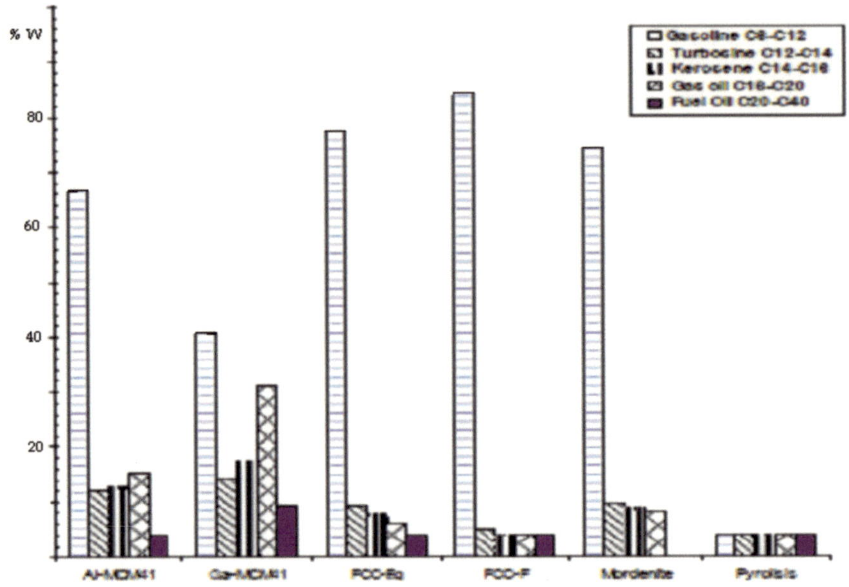

Fig. 13. Liquid products distribution obtained from the catalytic cracking of low-density polyethylene employing different porous solid catalysts (Ortega et al., 2006).

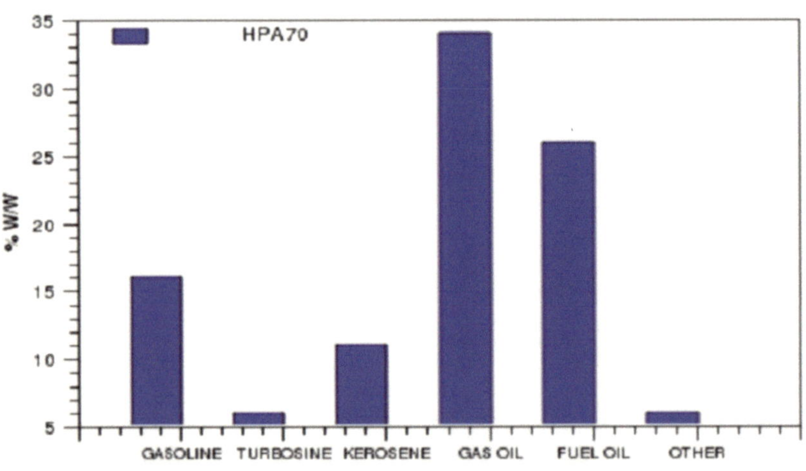

Fig. 14. Liquid products distribution obtained from the catalytic cracking of low-density polyethylene employing heteropolyacid-MCM-41 materials (Hernandez et al., 2010).

While Al-MCM-41 and Ga-MCM-41 produced a relatively large gasoline yield (Figure 13), the HPA-MCM-41 catalyst at the same reaction conditions, produced a large proportion of heavier hydrocarbon fractions, as gas oil, followed by fuel oil and then, gasoline and kerosene (Figure 14). We consider that even if the HPA70 has a large number of initial

Brönsted acid sites, the high proportion of relatively large hydrocarbon molecules obtained (C18-C44) is due to the impaired, formerly strong acidity of the heteropolyacid, at the considerably high reaction temperature, and to the reduced surface area of the material; therefore a small part of active sites can be reached by the polymer chains.

5.3 Solid products

When we have carried out the polyethylene catalytic decomposition, a small proportion of solid products have been obtained, in the form of waxes. For comparison, we run the same polymer decomposition at the same reactor conditions without any catalyst, being in fact, the pyrolysis of the polymer. The pyrolysis test yield a large proportion of solid products, 90.67% w/w (Ortega et al., 2006).

For the particular case of the catalytic cracking of poly(ethylene therepthalate), we have obtained a large proportion of solid therepthalic acid.

5.4 Relationship between reaction conditions and products

The products distribution varies at different reaction temperatures, in general terms (for zeolites as well as for other acid solids) high reaction temperatures increase the yield of gas products and middle boiling point components (C5-C12 gasolines) whereas lower reaction temperatures increase the yield of heavier components (C13-C18) (Gulab et al., 2010).

As the polymer to catalyst mass ratio increases, the system becomes less active (Gulab et al., 2010), however, this effect can be compensated by higher temperatures or larger reaction times. This relative loss of activity also tends to produce a higher yield of liquid products.

It is not uncommon to observe some condensation in gas containers after the reaction, which can be avoided by reducing the flow rate of the carrier gas. Inside the reaction chamber, the heated carrier gas can cause the evaporation of liquid products and also transports them out of the system.

6. Economic considerations

The most important economic issues that influence the viability of primary and secondary (mechanical) plastic recycling are the price of the recycled polymer compared with virgin polymer and the cost of recycling compared with alternative forms of acceptable disposal (Hopewell et al., 2009).

More individuals, organizations, business, and government agencies are collecting materials for recycling than ever before. The number of curbside recycling programs has grown during the last decade and new economic opportunities allow the birth of new markets. Recycling also creates new businesses that haul, process, and broker recovered materials, as well as companies that manufacture and distribute products made with recycled content. By recycling 1 ton of plastic, you can help save the same amount of energy that two people use in one year, or almost 2,000 pounds of oil (USEPA, 2009).

According to the American Chemistry Council, about 1,800 US business handle or reclaim post-consumer plastics. Plastics from MSW are usually collected from curbside recycling bins or drop-off sites. Then, they go to a material recovery facility, where the materials are sorted into broad categories (plastics, paper, glass, etc.). The resulting mixed plastics are sorted by plastic type, baled, and sent to a reclaiming facility. A Mexican study reported than the average cost of one ton of trash is around 350 US Dollars (USD), but if plastics are reduced, the cost would be around 4 USD less (Cortinas, 2009). The investment for recycling of 150 ton could be 300,000 USD producing a monthly utility of 30,000 USD.

The price of virgin plastic is influenced by the price of oil, which is the principle feedstock for plastic production. As the quality of recovered plastic is typically lower than that of virgin plastics, the price of virgin plastic sets the ceiling for prices of recovered plastic. The net heat generation from the use of crude oil in plastic making is roughly $0.4x10^{14}$ kWh from 1939 to 2000. It corresponds to 1.3% of the missing heat and contributes to 0.5% of the global warming (Gervet, 2007).

The profitability of feedstock, chemical recycling methods depends on three key factors: the degree of separation required in the raw wastes, the capital investment involved in the processing facilities and the value of the products obtained (Clark, 1999). For most of feedstock recycling methods, some pretreatment or separation operations are unavoidable. Feedstock recycling methods can be ordered according to the separation steps required (Clark, 1999):

Gasification < thermal cracking < hydrogenation < catalytic cracking < chemical depolymerization

Whereas the value of the products obtained by the feedstock recycling methods follows the opposite order:

Thermal oils < synthesis gas < hydrogenation oils < catalytic olefins and paraffins < monomers

Important benefits of the catalytic cracking of polymer waste with respect to other chemical feedstock recycling methods is the possibility of controlling the selectivity towards desired products and the possibility of reducing energy consumption. The use of cheap catalysts is central under the actual circumstances and some of the current best options are natural materials and waste catalysts from other industries. To use existing industrial facilities is a way for much cost reduction. Since the catalysts and the reaction mechanism of the polymer catalytic cracking are about the same as for the hydrocarbon fluid catalytic cracking (FCC), it is possible to incorporate plastic waste into the FCC refinery feed, with the added benefit that plastic waste has almost no sulfur content and no heavy metals content.

7. Concluding remarks

Active plastic waste catalytic cracking materials involve Brönsted acid sites, present in zeolite catalysts and which we introduced in MCM-41 mesoporous materials by (i) the incorporation of Al and Ga and (ii), by impregnation of the MCM-41 surface with tungstophosphoric acid. The several solid acid catalysts we tested yield gas and liquid products from the LDPE cracking reaction. The gas products consist of a mixture of ethane, propane, butane and pentane, all of them of interest for petrochemical industries or as domestic energy source. Liquid products consist of gasoline, turbosine, kerosene, gas oil, and fuel oil, corresponding to fuel fractions obtained in industrial petroleum refinery plants. The selectivity towards particular products depends mainly on the choice of catalyst and reaction conditions. The study of the catalytic cracking of plastic waste has led to relevant scientific knowledge and to the development of innovative technologies.

The large-scale application of these processes has been limited by economical and profitability reasons. Previous stages involve plastic separation from municipal waste and, since different mixtures of products are obtained, end stages involve product separation processes. Up to now, it is hard to compete with the still cheap option of producing fuels from natural gas and crude oil and to produce plastics from new raw materials. However, worldwide growing concerns about preserving our environment give plenty of room for imaginative ideas on how to scale up these processes to industry level.

8. References

ACS Monograph (1976). *Zeolite Chemistry and Catalysis,* Rabo, J. Ed., American Chemical Society, Washington D.C., 171.

Aguado, J., Serrano, D. P. & San Miguel, G. (2007). European trends in the feedstock recycling of plastic wastes. Global NEST J. 9, pp 12–19.

Al-Salem, S. M., Lettieri, P. & Baeyens, J. (2009). Recycling and recovery routes of plastic solid waste (PSW): A review. *Waste Management,* Vol. 29, pp 2625–2643.

Auerbach, S. M., Carrado, K. A. & Dutta, P. K. (2003). *Handbook of Zeolite Science and Technology,* Ed. Marcel Dekker, New York.

Boveri, M., Aguilar-Pliego, J., Pérez-Pariente, J. & Sastre, E. (2005). Optimization of the preparation method of HSO_3-functionalized MCM-41 solid catalysts. *Catalysis Today,* 107-108, pp 868-873.

Cardona, S. C. & Corma, A. (2000). Tertiary recycling of polypropylene by catalytic cracking in a semibatch stirred reactor, Use of spent equilibrium FCC commercial catalyst. *Applied Catalysis B,* Vol. 25, pp 151-162.

Chen, L. F., Zhou, X., Noreña, L. E., Yu, G., Li, C. & Wang, J. A. (2007). Framework modification and acidity enhancement of zirconium-containing mesoporous materials. *Studies in Surface Science and Catalysis,* Vol. 165, *"Recent Progress in Mesostructured Materials"*, pp 199-202.

Chen, L. F., Noreña, L. E., Zhou, X. L., Wang, J. A., Navarrete, J., Salas, P. & Montoya, A. (2006). Comparative studies of mesoporous Zr-MCM-41 and Zr-MCM-48: synthesis and physicochemical properties. *Applied Surface Science,* Vol. 253, No. 5, pp 2443-51.

Chen, L. F., Noreña, L. E., Navarrete, J. & Wang, J. A. (2006). Improvement of surface acidity and structural regularity of Zr-modified mesoporous MCM-41. *Materials Chemistry and Physics*, Vol. 97, Issues 2-3, pp 236-242.

Clark, J. H. (1999) *Feedstock Recycling of Plastic Wastes,* Royal Society of Chemistry Clean Technology Monographs, London.

Cortinas, C. (2009). Reciclaje de Plásticos, en el Contexto del Desarrollo Sustentable y Humano. SEMARNAT, Mexico.

Covarrubias, C., Gracia, F. & Palza, H. (2010). Catalytic degradation of polyethylene using nanosized ZSM-2 zeolite. *Applied Catalysis A: General,* Vol. 384, pp 186–191.

De la Puente, G., Klocker, C. & Serdan, U. (2002) Conversion of waste plastics into fuels recycling polyethylene in FCC. *Applied Catalysis B: Environmental*, Vol. 36, pp 279-285.

Department of Environment and Conservation (NSW) (2005). Benefits of recycling. Australia: Parramatta.

Derraik, J. G. (2002). The pollution of the marine environment by plastic debris: a review. *Marine Pollution Bulletin,* Vol. 44, pp 842-852.

Díaz-García, M., Aguilar-Pliego, J., Herrera-Pérez, G., Guzmán, L., Schachat, P., Noreña-Franco ,L., Aguilar-Elguezábal, A. & Gutiérrez-Arzaluz, M. (2010). Isomerization of pinene with Al- and Ga- modified MCM-41 mesoporous materials. *Advanced Materials Research*, Vol. 132, pp 162-173.

Dyer, A. (1988). *Zeolite Molecular Sieves: an introduction*, John Wiley & Sons,New York.

EC, DGENV. European Commission, DG Environment. (2011). Plastic Waste in the Environment, Final report. Specific Contract 7.0307/2009/545281/ETU/G2 under Framework contract ENV.G.4/FRA/2008/0112.

Edwards, E. (1999). *Sustainable architecture: European directives and building design*. 2nd ed. Architectural Press, Oxford. U. K.

Elordi, G., Olazar, M., Lopez, G., Artetxe, M. & Bilbao, J. (2011). Continuous Polyolefin Cracking on an HZSM-5 Zeolite Catalyst in a Conical Spouted Bed Reactor. *Industrial and Engineering Chemistry Research*, Vol. 50, pp 6061-6070.

Gervet, B. (2007). The use of crude oil in plastic making contributes to global warming. Renewable Energy Research Group, Division of Architecture and Infrastructure. Luleå University of Technology SE-97187. Luleå, Sweden.

Gulab, H., Jan, M. R., Shah, J. & Manos, G. (2010). Plastic catalytic pyrolysis to fuels as tertiary polymer recycling method: Effect of process conditions. *Journal of Environmental Science and Health*, Part A, Vol. 45, No. 7, pp 908-915.

Hernández, A., Noreña, L., Chen, L. F., Wang, J. A. & Aguilar, J. (2010). Refinery Oil Fraction Fuels Obtained From Polyethylene Catalytic Cracking Employing Heteropolyacid-MCM-41 Materials. *Advanced Materials Research*, Vol. 132, pp 236-245.

Hernández, A. (2011). MSc. Thesis, Universidad Autónoma Metropolitana, Azcapotzalco.

Hopewell, J., Dvorak, R. & Kosior, E. (2009). Plastics recycling: challenges and opportunities. *Phil. Trans. R. Soc. B.*, Vol. 364, pp 2115–2126.

Huang, W. C., Huang, M. S., Huang, C. F., Chen, C. C. & Ou, K. L. (2010).
Thermochemical conversion of polymer wastes into hydrocarbon fuels over various fluidizing cracking catalysts. *Fuel*, Vol. 89, pp 2305–2316.

Kumar, S., Panda, A. K. & Singh, R. K. (2011). A review on tertiary recycling of high-density polyethylene to fuel. *Resources, Conservation and Recycling*, Vol. 55, pp 893– 910.

Lavender-Law, K., Morét-Ferguson, S., Maximenko, N. A., Proskurowski, G., Peacock, E. E., Hafner, J. & Reddy, C. M. (2010). Plastic Accumulation in the North Atlantic Subtropical Gyre. *Science*, DOI: 10.1126/science.1192321. Available at: www.sciencemag.org/cgi/content/abstract/science.

Lin, Y. H., Yang, M. H., Wei, T. T., Hsu, C. T., Wua, K. J. & Lee, S. L. (2010). Acid-catalyzed conversion of chlorinated plastic waste into valuable hydrocarbons over post-use commercial FCC catalysts. *J. Anal. Appl. Pyrolysis*, Vol. 87, pp 154–162.

Masters, A. F. & Maschmeyer, T. (2011). Zeolites–From curiosity to cornerstone. *Microporous and Mesoporous Materials,* Vol. 142, pp 423-438.

Mravec, D., Hudec, J. & Janotka, I. (2005) Some possibilities of catalytic and noncatalytic utilization of zeolites. *Chem. Pap.*, Vol. 59, pp 62-69.

Morton-Jones, D. H. Procesamiento de plásticos (1993). México: Limusa Noriega Editores.

Oakes, W. G. & Richards, R. B. (1949). *J. Chem. Soc.*, pp 2929-2935.

Ortega, D., Noreña, L., Aguilar, J., Hernández, I. & Ramírez, V. (2006). Recycling of plastic materials employing zeolite and MCM-41 materials. *Revista Mexicana de Ingeniería Química*, Vol. 5, No. 3, pp 189-196.

Pinto, F., Costa, P., Gulyurhu, I. & Cabrite, J. (1999). *J. Anal. Appl. Pyrolysis*, Vol. 51, pp 57-71.

Sánchez, M., Aguilar, J., Hernández, F., Rodríguez, A., Noreña, L. & Hernández, I. *Recycling of plastic materials employing FCC catalysts from a refinery plant*, 18th North American Catalysis Society Meeting, Cancun, Mexico, June 1-6, 2003, P-430.

Schacht, P., Aguilar-Pliego, J., Ramírez-Garnica, M., Ramírez, S., Abu, I. & Noreña-Franco, L. (2010). Effect of CoMo/HSO_3^-functionalized MCM-41 over heavy oil. *Journal of The Mexican Chemical Society*, Vol. 54, No. 4, pp 194-200.

Scheirs, J. (2001). *Polymer Recycling*. John Wiley and Sons: Chichester, UK.

Scheirs, J. & Kaminsky, W. (2006). *Feedstock recycling and pyrolysis of wasteplastics.* John Wiley and Sons. Chichester, U. K.

Serrano, D.P., Aguado, J., Escola, J. M., Rodriguez, J. M. & Peral, A. (2010). Catalytic properties in polyolefin cracking of hierarchical nanocrystalline HZSM-5 samples prepared according to different strategies. *Journal of Catalysis*, Vol. 276, pp 152-160.

Torres, M., Gutiérrez, M., Mugica, V., Romero, M. & López, L. (2011). Oligomerization of isobutene with a beta-zeolite membrane: Effect of the acid properties of the catalytic membrane", *Catalysis Today*, Vol.166, pp 205-208.

Torres, M., Gutiérrez, M., López, L., Mugica, V., Gomez, R. & Montoya, A. (2008). Controlled crystal growth of zeolite films on alumina supports. *Materials Letters*, Vol. 62, pp 1071.

Uemichi, Y., Hattor, M., Itoh, Y., Nakamura, J. & Sugioka, M. (1998). Deactivation behaviors of zeolite and silica-alumina catalysts in the degradation of polyethylene. *Industrial and Engineering Chemistry Research*, Vol. 37, pp 867-872.

UNEP (2006). Ecosystems and Biodiversity in Deep Waters and High Seas. In UNEP Regional Seas Reports and Studies No.178., UNEP/ IUCN, Switzerland. Available at: www.unep.org/pdf/EcosystemBiodiversity_DeepWaters_20060616.pdf.

USEPA, U.S Environmental Protection Agency. (2009). Municipal Solid Waste Generation, Recycling, and Disposal in the United States: Facts and Figures.

Wang, J. A., Chen, L. F., Noreña, L. E. & Navarrete, J. (2009). Spectroscopic study and catalytic evaluation of mesostructured Al-MCM-41 and Pt/$H_3PW_{12}O_{40}$/Al-MCM-41 catalysts. *Applied Catalysis A: General*, Vol. 357, pp 223-235.

Wang, J. A., Zhou, X. L., Chen, L. F., Noreña, L. E., Yu, G. X. & Li, C. L. (2009) *Hydroisomerization of n-heptane on Pt/$H_3PW_{12}O_{40}$/Zr-MCM-41 catalysts.* Journal of Molecular Catalysis A: Chemical, Vol. 299, pp 68-76.

Wang, J. A., Chen, L. F., Noreña, L. E., Navarrete, J. & Llanos, M. E. (2008). Mesoporous structure, surface acidity and catalytic properties of Pt/Zr-MCM-41 catalysts promoted with 12-tungstophosphoric acid. *Microporous & Mesoporous Materials*, Vol. 112, pp 61-76.

WRAP (Waste and Resource Action Plan) (2006). Environmental Benefits of Recycling. Final report.

Yang, X. K., Chen, L. F., Wang, J. A., Noreña, L. E. & Novaro, O. (2009). Study of the Keggin structure and catalytic properties of Pt-promoted heteropolycompound/Al-MCM-41 hybrid catalysts. *Catalysis Today*, Vol. 148, pp 160-168.

7

Pyrolysis of Waste Polystyrene and High-Density Polyethylene

Kyong-Hwan Lee
Korea Institute of Energy Research
South Korea

1. Introduction

As the rate of consumption of plastic materials in the world is greatly expanded, more waste plastics are generated. In recent years, their generation amount in Korea becomes about four million tons per year, according to data from the National Institute of Environmental Research. The disposal of waste plastic is mostly achieved by conventional ways such as landfill or incineration. However, these methods have a problem of a social resistance due to the air pollution, soil contamination, and the economical resistance caused by an increase of a space and a disposal cost. Thus, the recycling of plastic wastes as a cheap source of raw materials has become a predominant subject over all countries. The development of technologies acceptable from the environmental and economical fields is one of the most important key factors.

Generally, the recycling methods are classified as the material recycling and chemical recycling. The former is one of the most conventional methods but is limited by difficulties in maintaining the high quality and adequate price of final products, in particular, for the mixture of plastic waste. Thus, application of other procedures such as chemical recycling and energy recovery is required (Al-Salem et al., 2009).

The chemical recycling, referred to as an advanced recycling technology, is included in a tertiary recycling. The process is converted from plastic wastes into smaller molecules corresponding to chemical intermediates through the use of heat and chemical treatment, such as liquids, gases and waxes. These chemical intermediates can be used as the fuel oil and feed stocks of petrochemicals processes, etc. The chemical recycling is described by the routes as follows (Kumar et al., 2011).

The chemical recycling can be mainly explained by the chemical recovery systems, which are classified as a heterogeneous and a homogeneous process. The chemolysis methods as homogeneous process utilize chemical agents as catalysts for depolymerization of polymers to obtain the products with low molecular wieghts. Chemolysis includes the processes such as glycolysis, hydrolysis, methanolysis and alcoholysis. On the other hand, heterogeneous processes are greatly described by gasification and pyrolysis. Gasification as partial oxidation (using oxygen or steam) can generate a mixture of hydrocarbons and synthesis gas (CO and H_2), which are dependent on the type of polymer, biomass, coal and co-mixture, and on quantity of and quality of resulting product.

<table>
<tr><td rowspan="7">Chemical Recovery Systems</td><td rowspan="4">Heterogeneous Process</td><td rowspan="3">Cracking</td><td>Thermal Cracking</td></tr>
<tr><td>Catalytic Cracking</td></tr>
<tr><td>Hydro Cracking</td></tr>
<tr><td colspan="2">Gasification</td></tr>
<tr><td rowspan="3">Homogeneous Process</td><td rowspan="3">Chemolysis</td><td>Methanolysis</td></tr>
<tr><td>Glycolysis</td></tr>
<tr><td>Alcoholysis</td></tr>
<tr><td>Energy Recovery Systems</td><td colspan="3">Incineration Technology</td></tr>
</table>

Fig. 1. Schemes of chemical recycling.

The pyrolysis involves the degradation of the polymeric materials by heating in the absence of oxygen. The method has the routes as the thermal cracking, catalytic cracking and hydro cracking. The recycling of waste plastics by thermal and catalytic degradation processes can be an important source producing alternative fuel oil from the view point of an economical aspect and contributing to the environmental protection from the view point of an environmental aspect (Demirbas, 2004). The method of pyrolysis takes advantage over the incineration and landfill methods because it is based on relatively simplicity into the oil for all thermoplastic mixtures without using the separation treatment for plastic type in the mixture and to lower the environment resistance for air pollutant and soil contamination.

In the pyrolysis, thermal degradation is a simple method for upgrading plastic waste into liquid product at medium temperature (400-600 ^{o}C) in the absence of oxygen. However, this process requires relatively high energy consumption, due to a low thermal conductivity of waste plastic and to an endothermic reaction by degradation of waste plastic. Moreover, the oil obtained by pyrolysis of plastic wastes has a wide molecular weight distribution with poor economical value, which does not have a sufficient quality to use as alternative fuel oils (Marcilla et al., 2009). The pyrolysis of polyethylene with high proportion in mixed plastic produces much more unstable heavy compounds with high viscosity as low grade product (Marcilla et al., 2009; Lee & Shin, 2007). The characteristics of these products depend on the nature of plastic waste and process conditions.

The catalytic degradation process, based on the addition of catalyst, can be conducted at low temperatures and high quality products are obtained in a comparison with thermal degradation process (Miskolczi et al., 2004). The most commonly used catalysts are (1) solid acid catalysts such as zeolite, silica-alumina, FCC catalyst and MCM-41, etc [Miskolczi et al., 2004; Lee et al., 2002; Garcia et al., 2005; Seddegi et al., 2002; Achilias et al., 2007; Miskolczi et al., 2006; Marcilla et al., 2005; Lin & Yang, 2007] and (2) bifunctional catalysts (Buekens & Huang, 1998). In the degradation of the polymer chain using acidic catalyst, the molecular weight of polymer chain could be rapidly reduced through cracking reaction and then carbonium ion intermediates would be rearranged by hydrogen and carbon atoms shifts with producing the isomers of high quality. In the case of bifunctional catalyst consisting of both acidic and metal material as reforming catalyst, the metallic sites catalyze hydrogenation/ dehydrogenation, while the acidic sites on the support catalyze the isomerization reaction. These reactions would improve the octane numbers of light hydrocarbons. Also, hydro-cracking involves the reaction with hydrogen over a bimetallic catalyst at moderate temperatures and pressures, which is focused on obtaining a high quality hydrocarbon product. These catalysts used in refinery hydro-cracking reaction for heavy hydrocarbons incorporate both cracking and hydrogenation.

With regards to the reactant used in this chapter, high-density polyethylene has a linear structure with no or little branching, while polystyrene is cyclic structure with relatively low degradation temperature. Polyolefinic and polystyrene polymers that have above 70% fraction in plastic waste are the major polymeric materials in a municipal plastic waste stream. In case of western Europe, polyethylene plastics make up over 40% of the total plastic content of municipal solid waste (Onwudili et al., 2009). These polymers consisting of mainly hydrogen and carbon atoms are so close to crude oil that the plastic waste would be processed by the reaction methods such as the thermal and catalytic cracking. In the pyrolysis process, polystyrene can be thermally degraded to the corresponding monomer or aromatics with its high selectivity at lower temperatures, whilst thermal degradation of polyolefinic polymers occurs at higher temperatures and lead to a complex mixture of aliphatic hydrocarbons.

This chapter presents the pyrolysis of both polystyrene and high-density polyethylene with different physiochemical properties and also the upgrading of low-grade oil product obtained by thermal degradation. Moreover, the effect of mixing of two plastics and catalyst addition for the pyrolysis would be explained by the yield for gas, liquid, solid products and the composition of liquid components, etc.

2. Basic pyrolysis

2.1 Reaction mechanism of high-density polyethylene and polystyrene

The pyrolysis is basically degraded for large hydrocarbons into smaller ones. From this process, the polymer is converted into paraffins and olefins, etc., with low molecular weights. Thermal degradation is accompanied with a free radical chain reaction. When free radicals react with hydrocarbons, new hydrocarbons and new free radicals are produced. Also, free radicals can decompose into olefins and new radicals. In the reaction mechanism by polymer type (Scheirs & Kaminsky, 2006), high-density polyethylene consisting of straight long carbon chains is pyrolyzed through the random-chain scission, which is broken up randomly into smaller molecules with various chain lengths. This product is obtained with a wide distribution of molecular weight, including hydrocarbons with high boiling point and/or low valuable products like wax. Thus, this means that the addition of catalyst in the pyrolysis can be a more efficient method to produce high valuable products with mainly gasoline range components. On the other hand, pyrolysis of polystyrene with cyclic structure is occurred by both end-chain and random-chain scissions. This polymer is broken up from the end groups successively yielding the corresponding monomers, as well as its breakage randomly into smaller molecules of one or more benzene-ring structures. This product is monomer recovery with a high fraction.

2.2 Thermal and catalytic degradation (Scheirs & Kaminsky, 2006)

(a) Thermal and (b) catalytic degradation of heavy hydrocarbons can be comparatively described with the following items

(a) Thermal degradation

1. High production of C_1s and C_2s in the gas product.
2. Olefins less branched.
3. Some diolefins made at high temperature
4. Wide distribution of molecular weight in the liquid product (poor gasoline selectivity)
5. High fraction of gas and coke products
6. Relatively slow reactions.

(b) Catalytic degradation

1. Short in the reaction time and low in degradation temperature
2. High production of C_3s and C_4s in the gas product
3. Olefins as the primary products and more branched by isomerization
4. More C_5-C_{10} products in the liquid product (high gasoline selectivity)
5. Aromatics produced by olefin cyclization
6. More reactive for larger molecules
7. No reaction for pure aromatics
8. Paraffins produced by H_2 transfer
9. Product distribution controlled by the selection of a catalyst

2.3 Mass balance

To demonstrate the mass balance, it is essential to determine the product yield for gas, liquid and residue, and also the composition of liquid products at different conditions of the various operating parameters such as temperature, residence time and pressure. From this, it is required to mention the economical aspect. Raw materials in a pyrolysis process contain nonproductive constituents, such as moisture, inorganic material, etc. These loss factors have to take into a consideration for the establishment of mass balance. Generally, mass balance is established by input and output amount, based on 100% of feeding amount. In the pyrolysis process, the important operating point is controlled by the maximum of valuable products and minimum of sludge amount. Thus, the operating margin must reach a reasonable level for mass balance in the economic aspect.

3. Pyrolysis of pure waste high-density polyethylene and polystyrene

Although the catalytic degradation of polyethylene over a wide variety of catalysts have been tested, zeolites have proven effective by many researchers [[Miskolczi et al., 2004; Lee et al., 2002; Garcia et al., 2005; Seddegi et al., 2002; Achilias et al., 2007; Miskolczi et al., 2006; Marcilla et al., 2005; Lin & Yang, 2007; Buekens & Hunang, 1998]. Seo et al (Seo et al.,2003) reports that the product characteristics for both thermal and catalytic degradation of waste HDPE using various zeolites are relatively compared as the yields of gas, liquid and residue, and carbon number distribution of liquid products, as shown in Table 1. Yields of liquid were over 70% using all zeolites, with the exception of ZSM-5, as well as thermal degradation. However, the catalytic degradation was produced much more light hydrocarbons (C6-C12) than that of thermal degradation, and moreover ZSM-5 and zeolite Y were more effective than mordenite. ZSM-5 and zeolite Y have a unique three-dimensional micropore structure as well as a strongly acidic property, whereas mordenite has only a parallel one-dimensional pore structure with a restricted diffusion of reactant. Especially, ZSM-5 with a smaller pore size, rather than that of zeolite Y was more cracked into light hydrocarbons such as C_6-C_{12} components and gas products. Since the initially degraded materials on the external surface of catalyst can be dispersed into the smaller internal cavities of catalyst, they can be further degraded to the smaller size of gaseous hydrocarbons. These findings mean that the pore properties of catalyst are important factor in the degradation of heavy hydrocarbons.

Catalysts	Yield			Liquid fraction*		
	Liquid	Gas	Coke	C_6-C_{12}	C_{13}-C_{23}	$\geq C_{24}$
Thermal cracking	84.00%	13.00%	3.00%	56.55%	37.79%	5.66%
ZSM-5 (powder)	35.00%	63.50%	1.50%	99.92%	0.08%	0%
Zeolite Y (powder)	71.50%	27.00%	1.50%	96.99%	3.01%	0%
Zeolite Y (pellet)	81.00%	17.50%	1.50%	86.07%	11.59%	2.34%
Mordenite (pellet)	78.50%	18.50%	3.00%	71.06%	28.67%	0.27%

* wt% were determined by GC/MS

Table 1. Yields of liquid, gas and coke produced from thermal and catalytic degradation of waste HDPE with various catalysts at 450°C (Seo et al.,2003).

In the characteristics of oil product, paraffin, olefin, naphthene and aromatic (PONA) distribution is one of the important factors which can determine the quality of oil product, as shown in Table 2. Oil product from thermal degradation of HDPE consists of 40.47wt% paraffins, 39.93wt% olefin, 18.50wt% naphthenes and a trace amount of aromatics. Relative to thermal degradation of HDPE, catalytic degradation is known to occur at a faster reaction rate and lead to subsequent reactions including isomerization and aromatization, as well as cracking reaction (Vento & Habib, 1979). Subsequent reactions proceeding through carbenium ion-type intermediate generated by acidic catalysts contribute to the greater formation of olefins and aromatics, as shown in Table 2.

Catalyst	Total-Paraffin	(Total-Paraffin)		Total-Olefin	Naphthene	Aromatics	Others*
		n-paraffin	i-paraffin				
Thermal cracking	40.75	40.47	0.28	39.93	18.50	0.68	0.14
ZSM-5(powder)	1.63	1.51	0.12	16.08	23.55	58.75	0.01
ZeoliteY(powder)	5.39	0.00	5.39	79.92	7.68	7.01	0.00
Zeolite Y(pellet)	25.10	20.68	4.42	49.28	12.05	8.43	5.14
Mordenite(pellet)	31.07	30.89	0.18	57.07	11.51	0.13	0.22

*Others mean hydrocarbons containing oxygen or unidentified organic compounds.

Table 2. Weight fraction of each PONA Group in oil products from thermal and catalytic degradation of HDPE with various catalysts at 450°C (Seo et al.,2003).

Catalytic degradation using ZSM-5 with small size increases aromatic hydrocarbons up to 59wt%, as a shape selectivity of catalyst, which is mainly consisting of the alkyl-aromatics with one-benzene ring structure. ZSM-5 is superior to zeolite Y in terms of aromatic formation. Also, the hydrogen atoms in ZSM-5 catalytic degradation contribute to the formation of naphthenes with largely C_6-C_8 hydrocarbons. Paraffins and olefins contain mostly lighter hydrocarbons.

It has been demonstrated that rare earth exchanged zeolite Y is more active than silica-alumina as cracking catalyst (Lin&Yang, 2007; Onwudili et al., 2009), because zeolite can provide a greater acidic site density. Since zeolite Y has more favorable shape selectivity for aromatic formation than non-zeolite catalyst, some intermediate carbenium ion formed by acidic zeolite will choose a pathway to aromatic formation, and some will be left over as olefin. Thus, the oil product from zeolite Y was mostly consisted of C_6-C_9 molecules which would be produced as largely light olefins and some cyclics. Zeolite Y improved the formation of branched isomers by the isomerization of light olefins and in cyclic products, naphthenes and aromatics by cyclization were mostly consisted of C_6 and C_7-C_{10} molecules, respectively.

The oil product over mordenite, among zeolites, appeared differently from other zeolites. This product distribution was similarly shown with that of thermal degradation, rather than other zeolites. This contrasting result of both mordenite and other zeolites seems to be correlated with the crystalline pore structure. Since this physical property is adopted for greater diffusion, mordenite with large pore size of one-dimensional pore structure can provide a greater initial activity than zeolite Y, but it would tend to lose activity more rapidly with time on stream. Coke formation in mordenite is known to be significant in a literature(Chen et al., 1989). As the result, the lighter molecules were less formed in mordenite.

4. Pyrolysis of mixture of waste HDPE and PS

When the pyrolysis is conducted to obtain the oil product, the effects of the mixing of HDPE and PS are described in this section. For the catalytic degradation of two polymers with a different mixing proportions, the cumulative amount distributions of liquid products as a function of reaction lapsed time are shown in Fig. 2. The experiments were performed with a stirred semi-batch reactor at a catalyst amount of 9.1 wt % and at a temperature of 400 °C with the same reaction temperature programming.

The cumulative amount distributions of liquid products clearly increase with an increase in the mixing proportions of PS against HDPE. These results are due to the fact that the increase of PS content in HDPE and PS mixture has much high yield of liquid product and high degradation rate. This means that pyrolysis of PS is predominant over the pyrolysis of HDPE in the mixture. According to the previous result (Lee et al.,2002), waste PS showed higher liquid yield and higher initial degradation rate in the catalytic degradation than waste PP and PE, because PS is mainly converted into stable aromatic components as liquid phase and also the low degradation temperature.

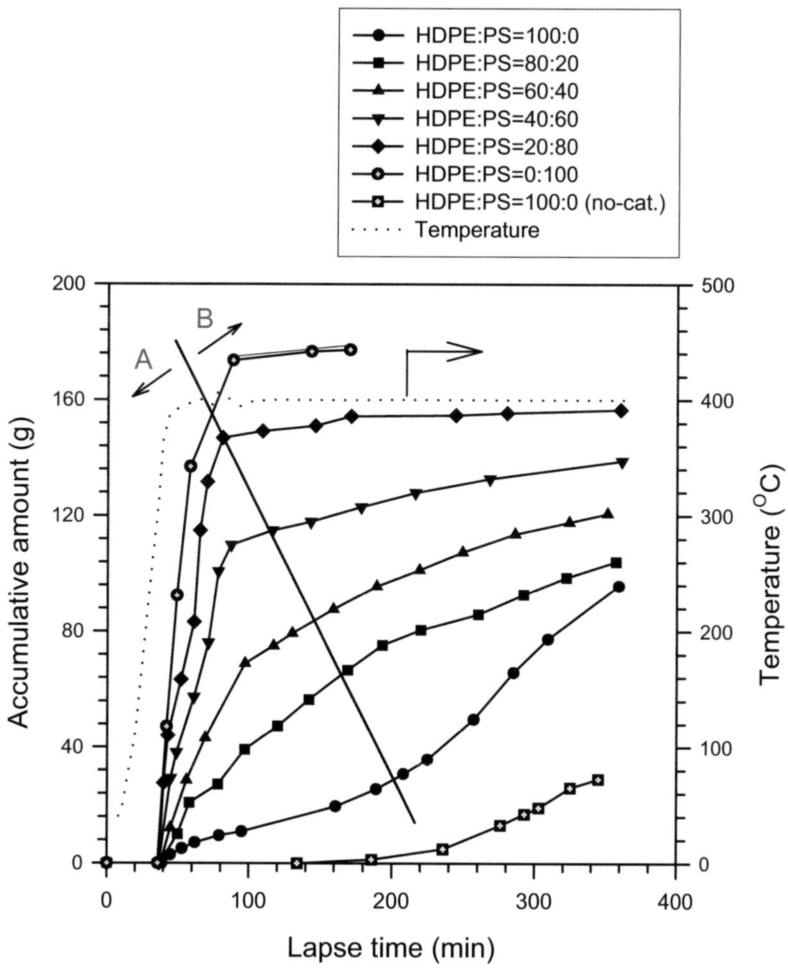

Fig. 2. Cumulative amount distributions of liquid products for catalytic degradation of waste HDPE and PS mixture using spent FCC catalyst at 400°C. (A: Initial degradation region, B: Final degradation region) (Lee et al., 2004).

The slope of the cumulative amount of liquid product versus reaction lapsed time represents as the degradation rate of HDPE and PS mixtures which is needed to obtain liquid products. The initial liquid product was obtained at around 400 °C of reaction temperature. These can be classified as two region of initial (A; initial degradation region) and final (B; final degradation region) lapse time in the reaction time and were appeared as initial and final degradation rate with a function of PS content, as shown in Fig. 3. The initial degradation rates are exponentially increased with increasing PS content in the mixture, while the final degradation rates were also suddenly decreased with increasing PS content, due to the influence of HDPE in the mixture. These results show that the polymers studied do not react independently, but some interaction between samples was observed.

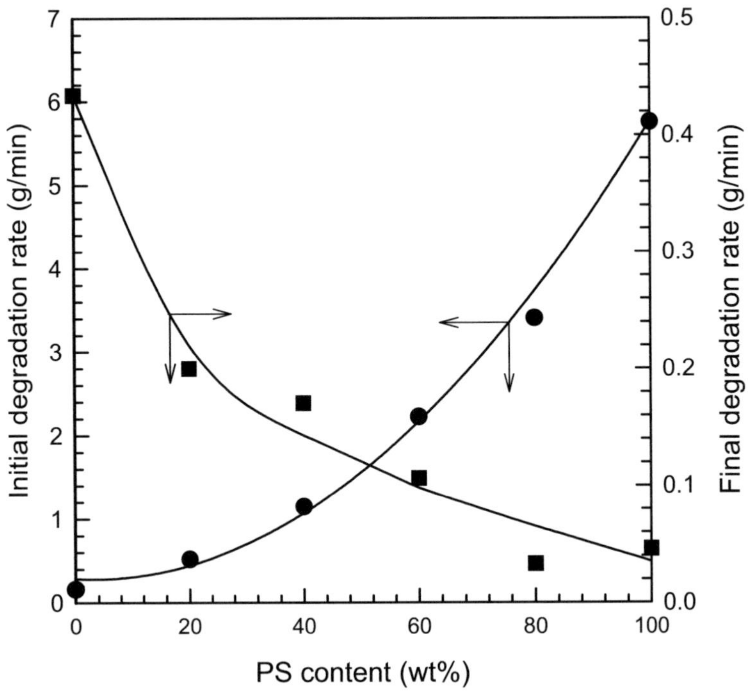

Fig. 3. Initial and final degradation rate as a function of PS content for catalytic degradation of waste HDPE and PS mixture using spent FCC catalyst at 400°C (Lee et al., 2004).

The commercial pyrolysis process yields the pyrolytic oil from the reactor at short contact time of 1-2 hours. It is necessary to know the characteristics of product oil in initial degradation region of Fig. 2. For these results, the distribution of liquid paraffin, olefin, naphthene and aromatic (PONA) products is presented in Fig. 4. Hydrocarbon group compositions of degraded products are strongly dependent on chemical properties of plastic type in plastic waste. As PS is included in the mixture, even though it is low or high, the pyrolysis of this mixture greatly improves the formation of aromatics, whereas the olefins produced by pyrolysis of polyolefins mainly has a much low fraction. This can be explained by the fact that the acceleration of aromatic products stems from the aromatic fragments of PS degradation as well as the cyclization of paraffinic and olefinic intermediates in FCC catalyst containing zeolite. Both the degradation of plastic mixture and the characteristics of oil product obtained are significantly influenced by plastic type in the mixture, as well as zeolite type in the catalyst.

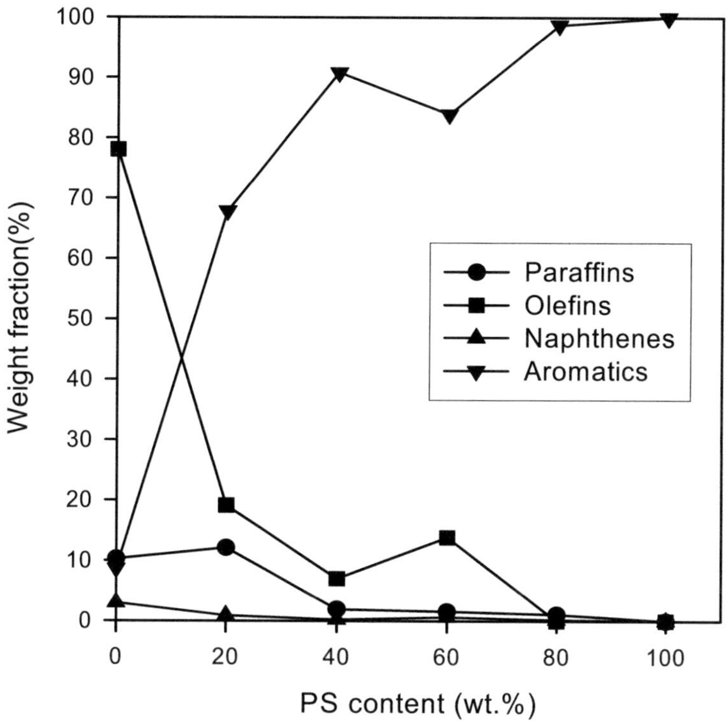

Fig. 4. The distribution of liquid paraffin, olefin, naphthene and aromatic (PONA) products for catalytic degradation of waste HDPE and PS mixture using spent FCC catalyst in the initial degradation time is presented with a function of PS content (Lee et al., 2004).

5. Pyrolysis of municipal plastic waste

Pyrolysis is a suitable process for thermoplastics like polyethylene and polystyrene. For a small mixture of polyvinyl chloride (PVC) and polyethylene terephthalate (PET) included in municipal plastic waste (MPW), an issue of environment and operation problems occurs in pyrolysis process. Thus, the removal of PVC and PET in MPW may be conducted by separation methods such as water separation, because of relatively high density of PVC and PET in a comparison for polyethylene and polystyrene with specific gravity 1.2 or less. Also, after the pretreatment of MPW, the inorganic materials contained with very low content are deposited in solid carbon residue during the pyrolysis. The MPWs are classified as low MPW(<1.0), medium MPW(1.0-1.1) and high MPW(1.1-1.2), based on the specific gravity 1.2 or less. The pyrolysis corresponding to three type MPWs is conducted.

Table 3 shows the yields of liquid, gas and residue products obtained by the pyrolysis of three different MPWs at a stirred batch reactor of 1.1 liter volume size, under the same experimental conditions. From these results, the product distribution is clearly different over various samples of MPW. Basically, the yield of liquid products in all samples is 75% or over. Note the corresponding liquid yields is in the following order; medium MPW >> low MPW > high MPW. Especially, the medium MPW shows highest liquid yield with about 90%. On the contrary, the order of gas and residue yield shows reverse relationship. It can be explained by the result that the plastic type contained in each MPW separated by a difference of specific gravity is an important key on the product distribution obtained. Lee et al. (Lee et al., 2002) have reported the influence of plastic type on liquid, gas and residue yield for pyrolysis of plastic wastes. The pyrolysis of polystyrene, due to the structure of stable benzene-ring, shows higher liquid yield and lower gas yield than that of polyolefinic polymer (PE, PP) with a straight hydrocarbon structure. Polystyrene is less cracked to gas product of 5 carbon numbers or less. Hence, the product distribution is strongly dependent on the plastic type including in municipal plastic wastes.

Sample (S.G..)	Liquid yield (wt%)	Gas yield(wt%)	Residue (wt%)
Low MPW (<1.0)	80.9	11.1	8.0
Medium MPW(1.0-1.1)	89.8	2.9	7.3
High MPW (1.1-1.2)	76.0	9.7	14.3

Table 3. Product yields obtained from pyrolysis of various MPWs at 400°C (Lee, 2007).

As the characteristics of liquid product, the paraffin, olefin, naphthene and aromatic (PONA) components, etc are compared over three different MPWs, as shown in Table 4. Also, their carbon number distributions are plotted in Fig. 5, respectively. These showed a peculiar product distribution, due to the chemical nature and structure of plastic type in MPW.

Product Composition	Sample (S.G.)		
	Low MPW (<1.0)	Medium MPW (1.0-1.1)	High MPW (1.1-1.2)
Paraffins	4.61	1.82	0.06
Olefins	75.93	0.02	0.10
Naphthenes	6.08	0	0.46
Aromatics	11.97	97.19	22.24
Phenols	0.25	-	17.05
Nitro-aromatics	0.15	0.97	1.88
Aldehydes	1.01	-	-
Methylesters	0	-	58.21
$<C_{13}$	92.89	92.05	99.13
C_{13}-C_{24}	7.01	7.95	0.87
$>C_{24}$	0.10	0	0

Table 4. Liquid product composition obtained from pyrolysis of various MPWs at 400°C (Lee, 2007).

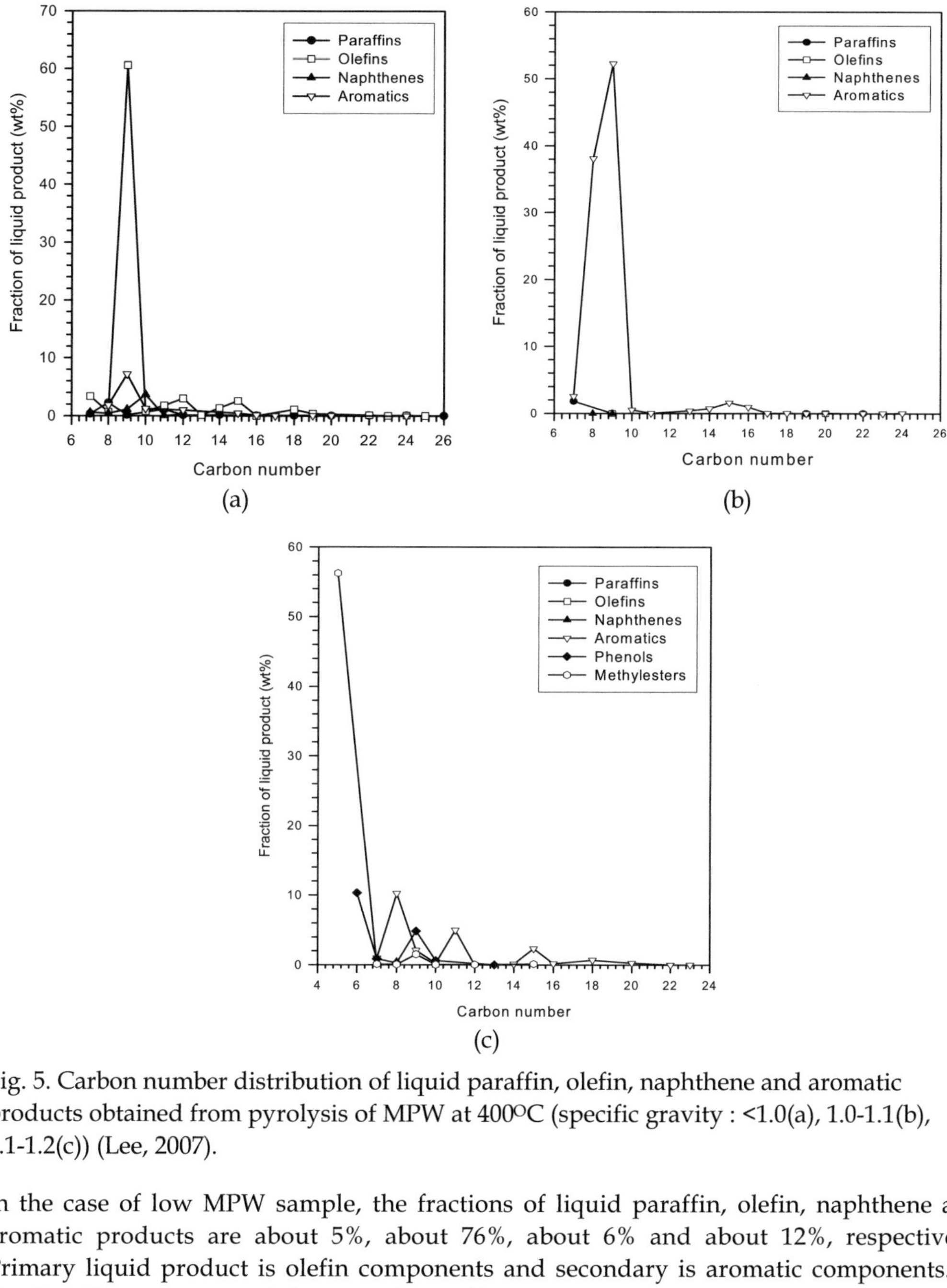

Fig. 5. Carbon number distribution of liquid paraffin, olefin, naphthene and aromatic products obtained from pyrolysis of MPW at 400°C (specific gravity : <1.0(a), 1.0-1.1(b), 1.1-1.2(c)) (Lee, 2007).

In the case of low MPW sample, the fractions of liquid paraffin, olefin, naphthene and aromatic products are about 5%, about 76%, about 6% and about 12%, respectively. Primary liquid product is olefin components and secondary is aromatic components. In the liquid product, naphthene and paraffin components are produced with a little amount. This means that low MPW mainly consists of PP polymer, among polyolefinic polymers, with relatively low degradation temperatures and high olefin fraction of liquid product in pyrolysis process (Lee et al., 2002). Also, aromatic products show 10% or more, because of including a little PS in low MPW. These results were demonstrated by the carbon number distribution of liquid PONA products over the case of low MPW, as shown in Fig. 4(a). Main liquid product was light olefin component with 9 of carbon number. This result was consistent with that of Sakata et al.(Sakata et al., 1999), who produced much more light hydrocarbon with 9 of carbon number from thermal degradation of PP at relatively low degradation temperature.

On the other hand, medium MPW showed highest fraction of liquid product with about 90% and lowest fraction of gas product, among three samples. In liquid product, aromatic components showed about 97% fraction and the rest was less than 2% fraction, respectively, as shown in Table 4. Moreover, phenol, aldehyde and methylester components in liquid products were not appeared and only nitro-aromatic products showed less than 2% fraction. It can be explained by the results that plastic type contained in medium MPW is mostly consisting of polymers with benzene-ring structures, especially PS among these polymers. This result can be reflected by carbon number distribution of liquid product, as shown in Fig. 4(b). Here, carbon number distribution was very short, mainly ranged from 8 to 9, as aromatic components. This result show a similar tendency in a comparison with that of Demirbas study (Demirbas, 2004), which is mainly consisting of 50-60% fraction of styrene and 10-20% fraction of C_5-C_8 hydrocarbons.

For pyrolysis of high MPW sample, the distribution of liquid products shows about 58% fraction in methylmethacrylate component, about 22% fraction in aromatic components and about 17% fraction in phenol components, as a main liquid product. However, straight hydrocarbon and naphthene components mainly obtained from pyrolysis of polyolefinic polymers are produced with very little amount (Lee et al., 2004). This result indicates that high MPW sample did not almost contain polyolefinic polymer type, and was mostly consisting of PMMA and then a little PS. This is demonstrated by the carbon number distribution of liquid products, as shown in Fig. 4 (c). Note the main product is methylmethacrylate monomer, producing from the pyrolysis of PMMA (Smolders & Baeyens, 2004).

6. Upgrading of pyrolysis oil with low quality

6.1 Constant stirred tank reactor

Pyrolytic oil is mainly composed of heavy hydrocarbons with low quality, as well as light hydrocarbons. Heavy hydrocarbons must be cracked to light hydrocarbons, in order to use as the fuel oil. The degradation experiment of pyrolytic oil is conducted by a heating rate of 10°C/min up to 420°C/min and then holding time of 5 hours at that temperature. The effects of degradation temperature and holding time at high degradation temperature on pyrolysis process are investigated. When the pyrolytic oil is degraded in a stirred semi-batch reactor with a bench scale under the condition of degradation temperature programming, the low degradation temperature (at 420 °C or below, short lapse time) only distills each hydrocarbon with the corresponding boiling point within the pyrolytic oil, while at high degradation temperature and long lapse time heavy hydrocarbons are much more decomposed into light hydrocarbons like gasoline ranged components and also a little middle hydrocarbons., as a pattern of GC (gas chromatography) peaks shown in Fig. 6.

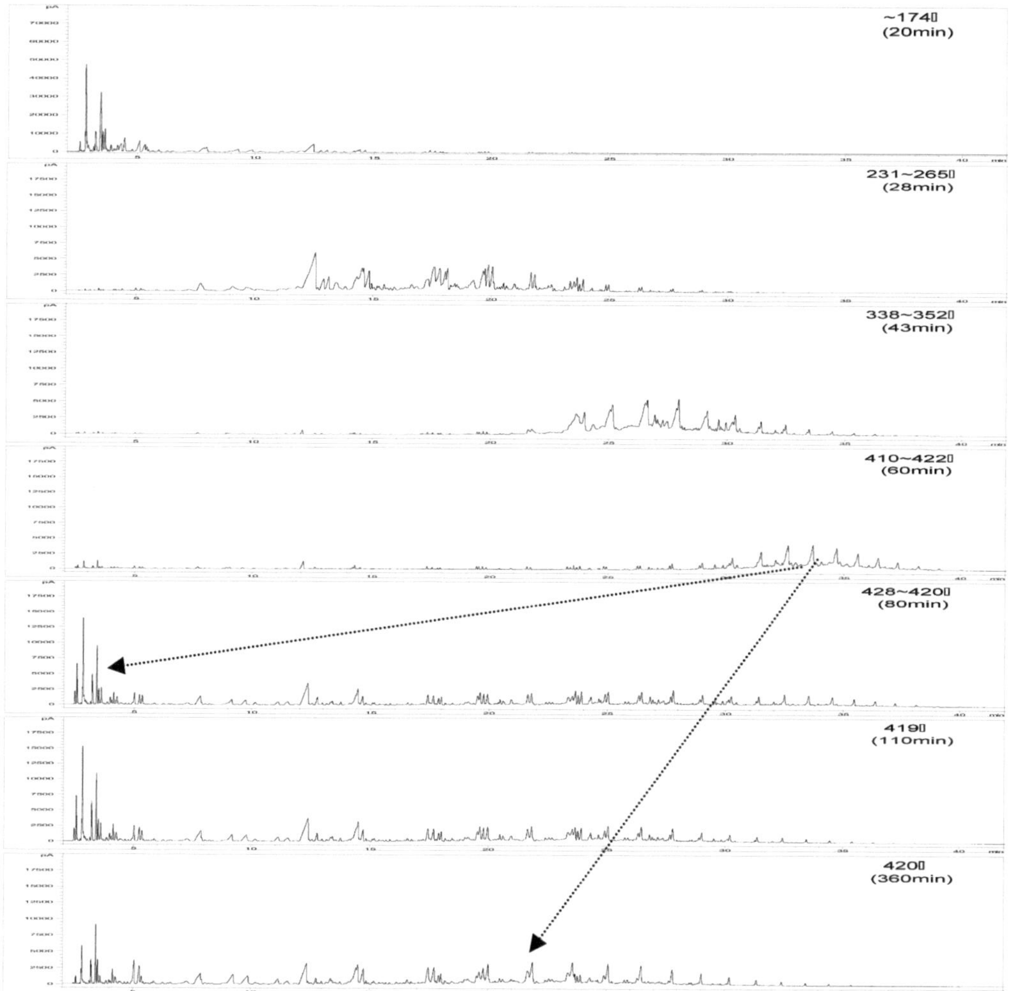

Fig. 6. GC peaks of product oils for thermal degradation of raw pyrolytic oil under degradation temperature programming (Lee, 2009).

Also, the catalytic degradation of pyrolytic oil using powder type FCC catalyst as a commercial cracking catalyst is investigated by a stirred tank reactor. The purpose of the catalytic degradation is to identify the possibility for utility of spent FCC catalyst as a waste catalyst, as well as the application of fresh FCC catalyst. A simple pyrolysis and catalytic degradation using spent or fresh FCC catalyst are compared by cumulative amount distribution of liquid product as a function of lapse time of reaction, as shown in Fig. 7. When a little catalyst (10%) is quickly loaded in the reactor at 420°C, the cumulative yield of liquid product is improved by the effects of catalyst, due to more cracking of heavy hydrocarbons into liquid product. Also, the cumulative yield distribution from catalytic degradation using both spent and fresh FCC catalysts is slightly deviated. This shows that spent FCC catalyst, compared to fresh FCC catalyst, has an effective result on the pyrolysis process.

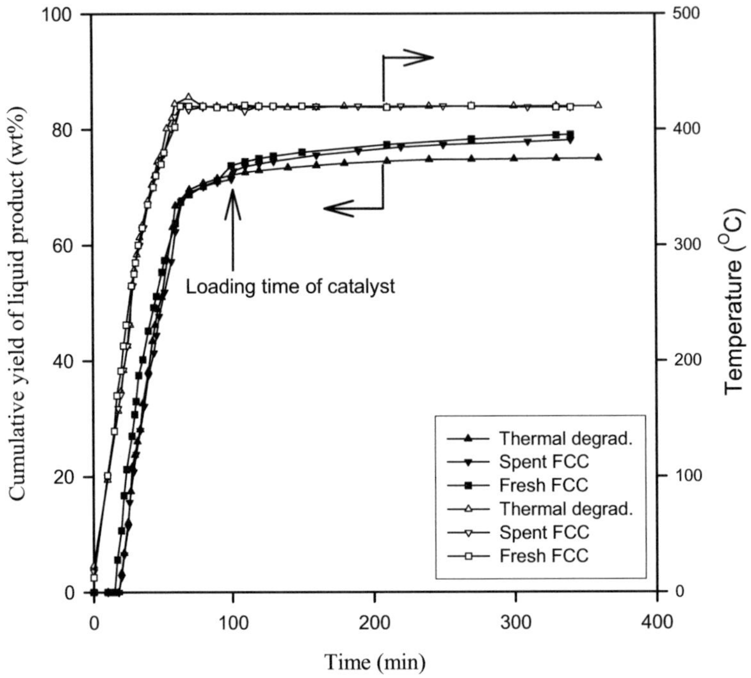

Fig. 7. Cumulative yield distribution of oil product for thermal and catalytic degradation of pyrolytic oil (Lee, 2009).

6.2 Fixed bed reactor

In chemical recycling, the pyrolysis process of plastic wastes by the use of commercial rotary kiln reactor can be taken into a consideration as an efficient method. From this system, municipal plastic waste as a reactant is converted into gas product, oil product (liquid product+wax) and residue. Among pyrolytic oil, the wax oil with a high proportion has a low value for a practical use in industrial companies and moreover difficulty to supply it as an alternative fuel oil, due to its high viscosity and low quality, etc. The SIMDIST (simulated distillation) curves, as the boiling point distribution, over the pyrolysis wax oil and the commercial oils (gasoline, kerosene and diesel) are shown in Fig. 8. The pyrolysis wax oil has much higher boiling point distribution, ranging from approximately 300°C to 550°C, compared to those of commercial oils. It mainly consists of paraffin components and a very wide carbon number distribution ranging from an approximate carbon number of 10 to a carbon number of nearly 40 (Lee, 2012).

Thus, the catalytic upgrading of low-grade pyrolysis wax oil is conducted by a fixed bed reactor, as a continuous gas-phase reaction using zeolite. The distribution of liquid product, gas product and coke over several types of commercial zeolite catalysts is listed in Table 5. Five commercial zeolite catalysts (ZSM-5, zeolite Y and mordenite, with or without clay or alumina as a supporter) used have their unique and different physicochemical properties.

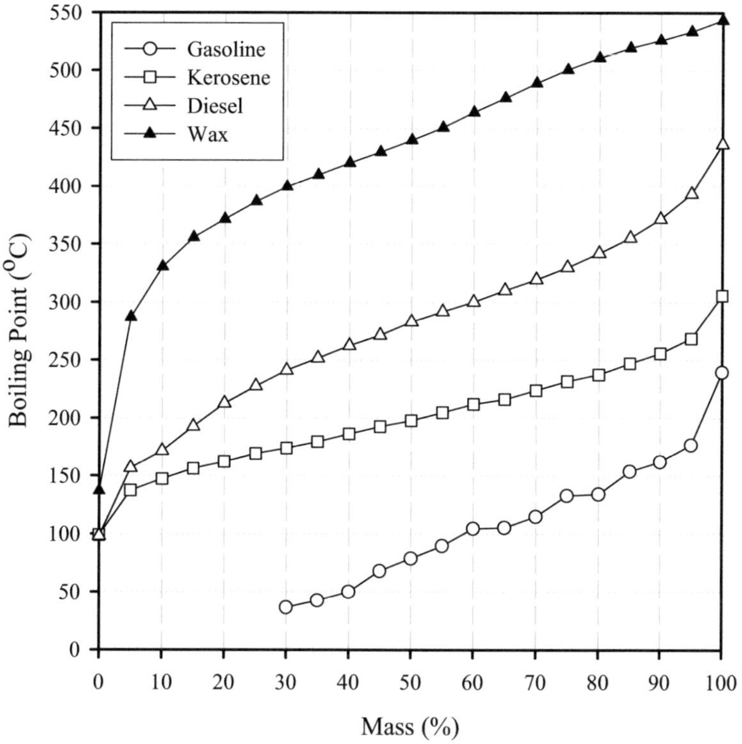

Fig. 8. Boiling point distribution as a function of mass fraction for pyrolysis wax oil and commercial oils (Lee, 2012).

Items	Liquid(wt%)	Gas(wt%)	Coke (wt%)
ZSM-5 (HZ30)	47.18	51.04	1.78
Zeolite Y (HY80)	66.98	28.95	4.08
Zeolite Y(80%)+Clay(20%) (HYC8)	74.12	22.95	2.93
Mordenite(80%)+Clay(20%) (HMC11)	92.12	7.72	0.16
Mordenite(80%)+Alumina(20%) (HMA6)	82.59	15.11	2.3

Table 5. Product distribution for catalytic upgrading of pyrolysis wax oil at 450°C, 1hr (Lee, 2012).

The order of the zeolite for the catalytic degradation of pyrolysis wax oil to gas products is ZSM-5 > zeolite Y > mordenite. ZSM-5 catalyst with a three-dimension pore structure shows the highest activity to gas product at nearly 50%. On the other hand, the catalyst containing mordenite with a one-dimensional pore structure shows the lowest conversion of heavy hydrocarbons into gas product. This indicates that the catalyst of zeolite type plays an important role in the catalytic degradation. As the effect of supporter, the distribution of gas products and coke with both pure zeolite Y and zeolite Y(80%)+clay(20%) shows a slight difference. The catalyst containing clay has low fraction of gas product and coke, compared to pure zeolite Y. The case of mordenite with a different supporter (clay or alumina) also shows a slight difference in the product distribution. It is concluded that the adequate design of both zeolite as a role of main reaction and supporter is a major key to determine the product distribution.

This result can be also sufficiently illustrated by the distribution of the liquid paraffin (normal/iso), olefin (normal/iso), naphthene and aromatic (PONA) products according to zeolite catalysts, as shown in Table 6. Raw pyrolysis wax oil is composed of predominantly normal paraffins and small amount of normal olefins. Among zeolites, ZSM-5 shows the

Items	Paraffins (wt%)			Olefins (wt%)			Naphthenes (wt%)	Aromatics (wt%)
	N-	Iso-	Total	N-	Iso-	Total		
Raw wax	70.7	-	70.7	8.0	-	8.0	-	-
HZ30	2.4	0.5	2.9	-	-	-	17.7	76.8
HY80	12.5	28.5	41.0	6.4	7.5	13.9	9.4	30.2
HYC8	12.2	25.6	37.9	5.5	8.4	13.9	11.9	28.7
HMC11	71.8	-	71.8	8.2	1.6	9.8	0.6	1.8
HMA6	61.3	-	61.3	7.0	10.2	17.2	2.2	8.7

Table 6. Composition of liquid paraffin, olefin, naphthene and aromatic products for catalytic upgrading of pyrolysis wax oil at 450°C, 1hr (Lee, 2012).

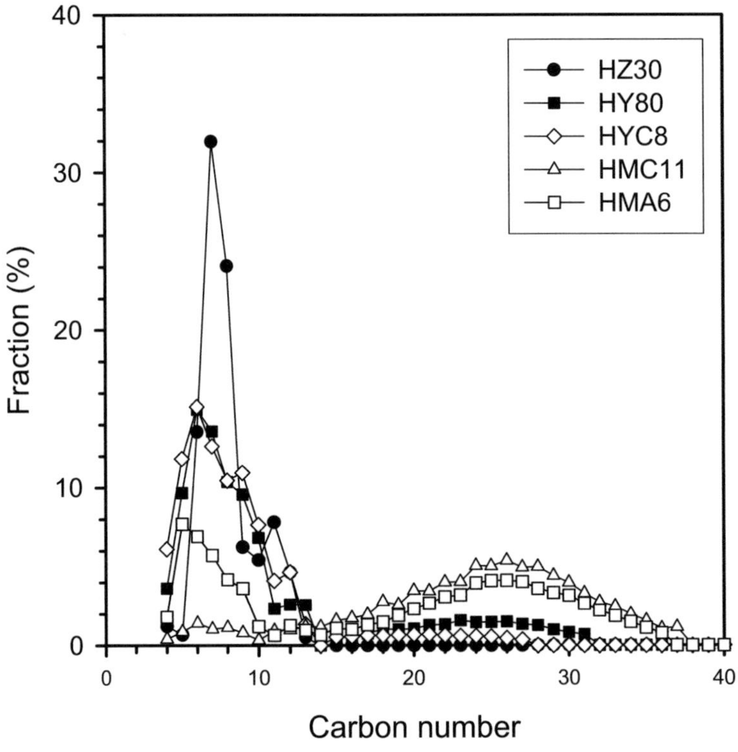

Fig. 9. Carbon number distributions of product oil for catalytic upgrading of raw pyrolysis wax oil at 450°C (Lee, 2012).

highest fraction of aromatic products through the cyclization of light paraffins and olefins. Aromatic products are mainly C_6, C_7 and C_8 components of benzene, toluene, xylene and ethylbenzene, as shown in Fig. 9, due to shape selectivity of ZSM-5. The case of zeolite Y mainly used in a commercial cracking process like FCC process and the hydrocracking process shows also different PONA pattern. Zeolite Y has the highest fraction of branched hydrocarbons with high octane number and also high fraction of aromatic products in the liquid products, which mainly produces gasoline ranged components in liquid product. However, the catalysts containing mordenite with a one-dimension pore structure shows a PONA distribution similar to raw pyrolysis wax oil and also has a wide carbon number distribution ranging from approximately 10 to 40, as shown in Fig. 9. The catalysts contained mordenite do not sufficiently crack for pyrolysis wax oil into light hydrocarbons, as its relatively low activity.

7. References

A. Demirbas, Pyrolysis of municipal plastic wastes for recovery of gasoline-range hydrocarbons, J. Anal. Appl. Pyrolysis, 72 (2004) 97-102

A. Marcilla, J. C. Garcia-Quesada, S. Sanchez, R. Ruiz, Study of the catalytic pyrolysis behavior of polyethylene-polypropylene mixtures. J Anal Appl Pyrolysis 74 (2005) 387-392.

A. Marcilla, M. I. Beltran, R. Navarro, Thermal and catalytic pyrolysis of polyethylene over HZSM-5 and HUSY zeolites in a batch reactor under dynamic conditions, Applied Catalysis B : Environ., 86(2009) 168-176

D. S. Achilias, C. Roupakias, P. Megalokonomos, A. A. Lappas, E. V. Antonakou, Chemical recycling of plastic wastes from polyethylene and polypropylene. J Hazard Materials, 149 (2007) 536-42.

G. Buekens, H. Huang, Catalytic plastic cracking for recovery of gasoline-range hydrocarbons from municipal plastic wastes, Resources, Conversion and Recycling, 23(1998) 163-181

J. A. Onwudili, N. Insura, P. T. Williams, Composition of products from the pyrolysis of polyethylene and polystyrene in a closed batch reactor: Effects of temperature and residence time, J. Anal. Appl. Pyrolysis, 86(2009) 293-303

J. Scheirs, W. Kaminsky, Feedstock recycling and pyrolysis of waste plastics: Converting waste plastics into diesel and other fuels, John Wiley & Sons, Ltd(2006)

K.-H. Lee, N.-S Noh, D.-H. Shin, Y.-H Seo, Comparison of plastic types for catalytic degradation of waste plastics into liquid product with spent FCC catalyst, Polymer Degradation and Stability, 78(2002) 539-544

K.-H. Lee, D.-H. Shin, Y.-H. Seo, Liquid-phase catalytic degradation of mixtures of waste high-density polyethylene and polystyrene over spent FCC catalyst.-Effect of mixing proportions of reactants, Polymer Degradation and Stability, 84(2004) 123-127

K.-H. Lee, D.-H. Shin, Influence of plastic type on pyrolysis of waste thermoplastics into oil recovery, J. Korea Soc. Waste Management, 21 (2004) 646-651

K.-H. Lee, Pyrolysis of municipal plastic wastes separated by difference of specific gravity, J. Anal. Appl. Pyrolysis, 79(2007) 362-367

K.-H. Lee, D.-H. Shin, Characteristics of liquid product from the pyrolysis of waste plastic mixture at low and high temperatures: Influence of lapse time of reaction, Waste Management 27(2007) 168-176

K.-H. Lee, Thermal and catalytic degradation of pyrolytic oil from pyrolysis of municipal plastic wastes, J. Anal. Appl. Pyrolysis, 85 (2009) 372-379

K.-H. Lee, Effect of zeolite type on catalytic upgrading of pyrolysis wax oil, submitted to J. Anal. Appl. Pyrolysis 94 (2012) 209-214

K. Smolders, J. Baeyens, Thermal degradation of PMMA in fluidised beds, Waste Management, 24 (2004) 849-857

N. Y. Chen, W. E. Garwood, F. G. Dwyer, Shape-selectivity catalysis in industrial applications, Marcel Dekker Inc, New York, 1989

N. Miskolczi, L. Bartha, G. Deak, B. Jover, D. Kallo, Thermal and thermo-catalytic degradation of high-density polyethylene waste, , J. Anal. Appl. Pyrolysis, 72 (2004) 235-242

N. Miskolczi, L. Bartha, G. Y. Deak, Thermal degradation of polyethylene and polystyrene from the packaging industry over different catalysts. Polym Degrad Stab 91 (2006) 517-26.

P. B. Vento, E. T. Habib, Fluid catalytic cracking with zeolite catalysts, Marcel Dekker Inc, New York, 1979

R. A. Garcia, D. P. Serrano, D. Otero. Catalytic cracking of HDPE over hybrid zeolitic-mesoporous materials. J Anal Appl Pyrolysis 74 (2005) 379-86.

S. Y. Lee, J. H. Yoon, J. R. Kim, D. W. Park. Degradation of polystyrene using clinoptilolite catalysts. J Anal Appl Pyrolysis 64 (2002) 71-83.

S. M. Al-Salem, P. Lettieri, J. Baeyens, Recycling and recovery routes of plastic solid waste (PSW) : A review, Waste Management, 29(2009) 2625-2643

S. Kumar, A. K. Panda, R. K. Singh, A review on tertiary recycling of high-density polyethylene to fuel, Resources, Conversation and Recycling, 2011,

Y. Sakata, M. A. Uddin, A. Muto, Degradation of polyethylene and polypropylene into fuel oil by using solid acid and non-acid catalysts, J. Anal. Appl. Pyrolysis, 51 (1999) 135-155

Y.-H. Seo, K.-H. Lee, D.-H. Shin, Investigation of catalytic degradation of high-density polyethylene by hydrocarbon group type analysis, J. Anal. Appl. Pyrolysis 70(2003) 383-398

Y. H. Lin, M. H. Yang, Catalytic pyrolysis of polyolefin waste into valuable hydrocarbons over reused catalyst from refinery FCC units, Appl Catal A: General 328 (2007) 132-139.

Z. S. Seddegi, U. Budrthumal, A. A. Al-Arfaj, A. M. Al-Amer, S. A. I. Barri. Catalytic cracking of polyethylene over all-silica MCM-41 molecular sieve. Appl Catal A: General 2002; 225:167-76.

Part 2

Recycling of Tires, Pharmaceutical Packaging and Hardwood Kraft Pulp

8

Recycling of Scrap Tires

Ahmet Turer
Middle East Technical University, Civil Engineering Dept.
Turkey

1. Introduction

As Rachel Louise Carson (1907-1964) successfully noted in her phrase "The human race is challenged more than ever before to demonstrate our mastery - not over nature but of ourselves", we are challenged to find ways to produce more energy, reduce our waste production while minimizing use of limited natural resources. Although recycling of materials has a history going back to the times of Plato BC400 and collecting scrap bronze & metals in Europe in pre-industrial times (Wikipedia, 2011), the demand roar for raw materials in the 19th and 20th centuries with industrial development caused cheaper alternative of reusing scrap material rather than mining them out. Interestingly, 21st century's major driving force has additional items on top of the existing reasons of using recycled material, such as reducing consumption of limited natural resources and lowering carbon dioxide emissions against the greenhouse effect. The increasing demand for energy production and dealing with larger amounts of waste contaminating the nature, forces mankind to find innovative ways to deal with the produced pollutant waste, emit lesser amounts of CO_2, and generate more energy. Recycling of scrap tires turns out to be a perfect match for the recent requirements of the 21st century. This chapter discusses various ways of recycling scrap tires and how they relate to the recent energy, material, and nature needs of our times.

Recycling of scrap tires until the 1960's in the US can be taken as an example; about half of the manufactured automobile tires used to be recycled since only synthetic or natural rubber was used in the tire manufacturing process and tires could have been directly used without major processing. Recycling of used tires was further encouraged by the fact that these materials were also expensive. The increasing use of the synthetic rubber, however, lowered the manufacturing costs and reduced need for recycling. Moreover, the development of steel belted tires in the late 1960's was almost the end of tire recycling since additional processing of tires was needed. Consequently, by 1995, the rate of rubber recycling fell to only 2% [Reschner].

Highway construction industry is a big alternative market for recycling scrap tires. Many studies have been carried out on crumb rubber modified asphalt. In 1995, it was required by all federal states in the U.S. to fund paving projects with tire modified asphalt. After that, the consumption rate of wasted tires in modified asphalt projects was increased, and in some states a maximum recycling rate of 20% was reached [Sheehan]. Other methods to gain the raw material and energy available inside scrap tires are further discussed under each heading below. The outcomes of scrap tire recycling are not only limited by easy access to cost-efficient material such as rubber and steel, but also have positive effects on the environment: Recycling of scrap tires on a global scale can drastically reduce waste yards, soil and atmospheric contamination caused by dump yards and large scale tire fires.

Fig. 1. Scrap tire storage areas and fires.

2. Brief history and production technology of tires

Automotive tires are made of synthetic rubber which is obtained from petroleum. The development of tires was based on improving the performance of natural rubber which is obtained from the liquid latex secreted by certain plants. At the beginning, natural rubber was used to produce waterproof fabrics and to make balls, containers and shoes by Pre-Colombian people in South and Central America. Until the 18th century, Europeans did not make use of rubber except that they utilized it for manufacturing elastic bands and pencil erasers. Joseph Priestley, a founder of the modern study of chemistry, named the substance "rubber" for its use as an eraser (Oven, 2004).

During the 19th century, Charles Goodyear studied on making rubber more resistant to various chemicals. He started his working by mixing rubber with various dry powders, and aimed to find a way to make natural rubber stickier. In 1839, he achieved to obtain the best product by applying steam heat under pressure, for four to six hours at 132 Celsius (270 Fahrenheit) degrees (Goodyear, 2011).

Following the discovery of vulcanization, manufacturers began producing tires from solid rubber which yielded a strong material to resist cuts and abrasions. Although this was a great progress, the tires were too heavy and rigid. In order to decrease vibration and improve traction, Robert W. Thomson, first produced the pneumatic rubber tire which consisted of rubber filled in with air. His idea could not a commercialized since it was introduced too early for its time. John Boyd Dunlop from Ireland, who did not know about Thomson's earlier invention, once more introduced the pneumatic tire to the market in 1888. This time, pneumatic tire caught the public's attention because bicycles were becoming extremely popular and the lighter tire provided a much better ride (Rubberis, 2011).

In early 19th century, manufacturers started producing vehicle tires comprising two parts, i.e., an inner part and an outer part. The inner part, called the inner tube, contained compressed air and the outer part was a casing protecting the inner tube and providing the tire with a better grip. An important element of the outer part were the layers called plies which were made of rubberized fabric cords embedded in the rubber and they were strengthening the casing. They were known as bias-ply tires because the cords in a single ply run diagonally from the beads on one inner rim to the beads on the other rim. The orientation of the cords change from ply to ply so that the cords crisscross each other (Rubbentire, 2011).

The steel-belted radial tires were first produced in 1948 by the Europeans. In those first tires, the ply cords radiate at a 90 degree angle from the wheel rim. Together with this, a belt of steel fabric that wrapped the circumference of the tire reinforced its casing. Radial tire ply cords are composed of nylon, rayon or polyester. The advantages of radial tires include longer tread life, better steering and less rolling resistance. On the other hand, radials have a harder riding quality, and are about twice as expensive as the tires without radials. The production sequence of steel-belted radial tires is briefly illustrated in Fig. 2.

3. Scrap tire disposal related problems

Massive disposal sites of scrap tires is common in many cities of modern times as about 1 scrap tire is produced per person every year. The stored used tires slowly degrade under the effects of solar radiation as well as rusting of steel takes place. Degraded material would slowly contaminate soil and underground water over years. The disposal sites waiting under the sun for extended periods of time might catch on fire either by accident or because of bottles or broken glass focusing sunlight. Tires burn with thick black smoke and heat, quickly spreads over the whole disposal area, and leaves oily residue contaminating the soil. Such fires are difficult to put off and generates significant amount of air pollution.

One of the overseen problems of scrap tire disposal yards is that these areas become breeding places for rodents and mosquitoes. Stagnant water that collects inside tires is a suitable breeding place for mosquitoes. Elimination of scrap tire disposal sites by proper recycling would also have secondary advantages of eliminating disposal related problems.

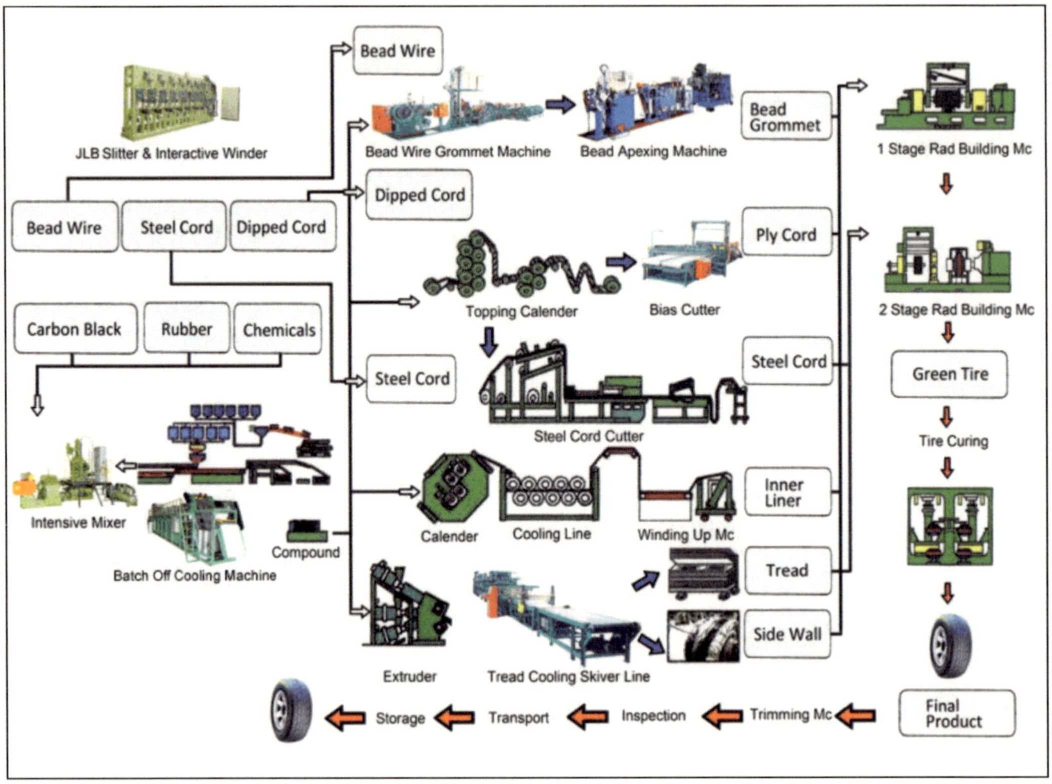

Fig. 2. Tire production line (Courtesy of PHT, Pam Hsiang Trading Co., Ltd.).

4. Major methods and reasons of recycling tires

The recycling of scrap tires may be defined under two different categories: i) using the scrap tires as whole or mechanically modified shapes (in crumps or shredded), and ii) chemical decomposition or separation of scrap tire contents into different materials.

Recycling as-is or after mechanical process has the advantages of directly using scrap tires without major investment. For example, scrap tires can be directly used as boat bumpers at marinas to protect ships from scratching or hitting at the side of wharf (Fig. 3). Similarly, old tires can be placed side by side in half tire shifted pattern for slope stability or under roads for improved stability (Mechanical Concrete®). Ripped tire pieces in large chunks can be directly used as light weight infill material at embankments. Smaller scrap tire pieces (Fig. 4) can be used as mixture in concrete as gravel substitute to improve tensile capacity or in asphalt paved roads for better traction. Smaller crumbs can be bonded together to generate walking or running mats or soft surfaces for playgrounds. Drainage around building foundations, erosion control for rainwater runoff barriers, wetland establishment, crash barriers at sides of race tracks are other uses of scrap tires without much modification.

Fig. 3. Innovative uses of scrap tire (a,b,c) road sublayer stability, (d) tire pieces as fill material, (e) slope stability, (f) ship bumper at warf.

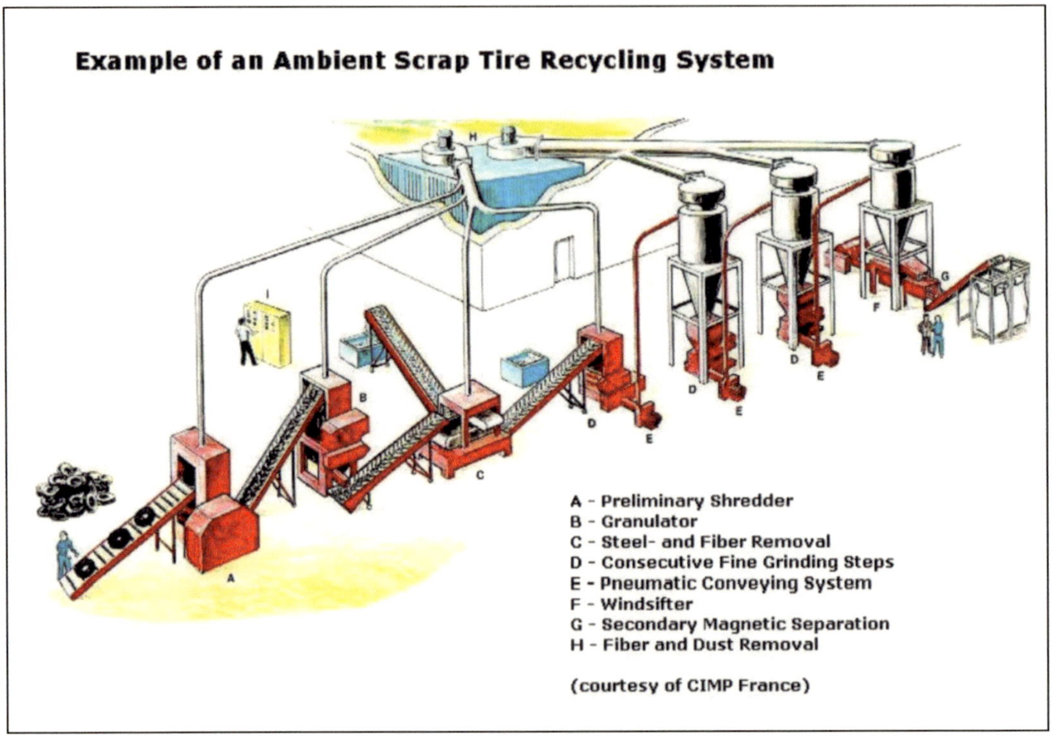

Fig. 4. Example scrap tire recycling system (courtesy of CIMP France).

Recycling of scrap tires at element level that includes some form of chemical decomposition or transformation is different than the mechanical process. Chemical recycling has additional advantages of obtaining well defined building blocks of a tire separately (such as steel wires, natural gas, oil, carbon black, charcoal etc.). The process in a way reverses the manufacturing process and obtains the elements forming a tire backwards. The materials then can be directly sold or used for energy in factories or diesel cars. Alternatively, burning scrap tires may also be included as a chemical process since long chained carbon based molecules are divided into smaller molecules and carbon forms new bounds with oxygen generating heat and carbon dioxide (CO_2). Additionally, hydrogen in the molecules also forms bounds with oxygen forming water (H_2O). Further details of the chemical process can be found in the literature [Murugan et., al., Wikipedia cement kiln].

4.1 Separation of scrap tire contents by thermo-chemical decomposition (Pyrolysis)

Pyrolysis is the common name used for decomposing organic material at elevated temperatures in the absence of oxygen. The oxygen needs to be absent otherwise organic material may burn. Typically the process takes place under pressure and operating temperatures above 430 °C (800 °F). The word is originated from Greek based words "pyr" and "lysis" meaning "fire" and "separating", respectively.

Initial studies on pyrolysis of scrap tires have shown that tire-derived activated carbon, carbon black, boudouard carbon, and fuel gas are obtained. Considering recycling of scrap tires in the road industry couldn't pass much beyond 2% of available scrap tire production; therefore, pyrolysis of scrap tires have enough resources to keep the system running. Gas obtained from the decomposition of scrap tires can directly be used in the pyrolysis process itself; therefore, the production can support the process for energy saving and sustainability. Economical evaluation of the pyrolysis have shown that when tipping fee for collecting scrap tires (F), revenue received from sale of products (R), processing cost for operating the facility (C), cost for transportation of tires (T), cost of tire shredding (S), cost of disposal of waste products (D) are considered with the assumption of 35% char, 20% gas, 45% oils, and using 50% of char burn-off during activation, net profit (P) is found to be USD 1.5/tire (1996 prices) with about 6 million USD/year gross income with investment payback of about 3.3 years (Marek, 1996).

$$P = F + R - C - T - S - D \tag{1}$$

Recent evaluation of scrap tires pyrolysis by Rubber Manufacturers Association in 2009 contains some disheartening comments. Even after the increase in oil prices reaching USD 150 per barrel, the market did not support this technology. Carbon black, charcoal, and waste oils demand would determine if the operation is viable. Although methane gas is produced during the process and can be used to operate the pyrolysis facility, the manufactured amount is not large volumes enough to sell economically. The excessive gas is usually flared off. Pyrolysis produces pyrolytic carbon char, often confused as carbon black.

Although pyrolytic carbon char has a high carbon content, it is dissimilar to carbon black, which is a highly engineered product. Pyrolytic carbon char is said to have limited market as a filler in some materials and as a colouring agent for some plastics after extensive refining and cannot be easily sold in carbon black markets where there is a lot of competition. The liquid hydrocarbon material obtained from pyrolysis unfortunately contains some contamination and may not be suitable to be directly used as diesel fuel or in home heating; it should be either used as waste oil or further refined. As a result, pyrolysis technology today could not reach its intended target yet. If it comes to the choice between either dumping the scrap tires to large storage areas as housing to rodents and mosquitoes every often catching fire and polluting air, soil, water or pyrolysis to melt down the scrap tire stocks while obtaining less than perfect charcoal, gas, and oil to be further refined is a relatively easy choice. It would be easier if the process can become environmentally friendly and profitable without government subvention.

4.2 Burning scrap tires for energy

Another chemical process on scrap tires is burning in high temperature ovens for energy. The burning is usually carried out at thermoelectric power plants and cement production in kiln with clinkers. Although burning a tire usually produces a dark heavy smoke, burning at high temperature furnaces with proper chimney filtering achieves a complete burning without similar smoke. Using scrap tires as fuel is referred as TDF (tire derived fuel) by Scrap Tire Management Council, which was established in 1990 by the North American tire manufacturers.

Cement is produced in high temperature kilns as the raw materials are placed in cement kiln and heated to a temperature range of 1455 to 1510 °C (2650 to 2750 °F). At this temperature the formation of tricalcium silicate (ALITE), the principal compound of portland cement clinker, occurs. A flame temperature of 1925°C (3500°F) is necessary to arrive at this temperature. Scrap tires (TDF) can be completely destroyed in cement kilns since the temperatures are extremely high along with a positive oxygen atmosphere and relatively long periods of 4 to 12 seconds at the elevated temperatures ensures the complete combustion of the scrap tire; therefore, incomplete combustion (PICs) or black smoke or odors release is prevented.

When tires are burned in the cement production, the production rates may increase in preheater kilns. This is made possible as the preheater calcination rate is increased in the preheater when burning tires compared to the normal calcination rate when burning coal only. Calcination rates were reported to be increased from 45% to 56% when burning tires instead of burning coal. The carbon dioxide transported by the kiln is reduced when scrap tires are burned in kilns; in this way, additional oxygen be used in the kiln, which allows for the burning of additional clinker (Scrap Tire Management Council, 1992).

Burning scrap tires raises some concerns from environmental point of view since tires include up to 17 heavy metals (e.g., lead, chromium, cadmium, and mercury) in addition to natural rubber from rubber trees, synthetic rubber made from petrochemical feedstocks, carbon black, extender oils, steel wire, other petrochemicals and chlorine. Synthetic rubber often contains the organic chemicals styrene and butadiene. Styrene, a benzene derivative, is a suspected human carcinogen. Butadiene is known to cause cancer in laboratory animals and is a suspected human carcinogen. Studies show a strong association between leukemia and butadiene. Extender oils contain benzene based compounds which cause cancer in laboratory animals but totally burnt at high temperatures. A coal and tire chlorine content comparison showed that tires may contain as much as 2 to 5 times the chlorine level of coal. The coal averaged a chlorine weight of 0.04% and tires showed a weight range of 0.07% to 0.2%. (CIWMBA, 1992). Most of the mentioned toxic material are in low percentages and remain in the burnt wastes or bound inside the cement. Factories and power plants that burn tires must therefore have proper filtering at chimneys in case the pollutants remain in the ashes or emitted gasses (Page, 1980). The non-condensable gases are filtered (using a demister filter) and are passed through a wet scrubbing system to remove acid components by NaOH (4%) injection (Sharma et. al.).

Typical composition of fuel derived from tyres have Sulphur <1.8%, Chlorine 0.07, Mercury <2mg/kg, Cadmium and thallium <79mg/kg, Antimony, arsenic, chromium, cobalt, copper, lead, manganese, nickel, tin and vanadium <640mg/kg while control limits are <2%, <0.2%, <10mg/kg, <80mg/kg, <1200mg/kg, respectively (European Commission 2003, Castle Cement 1996 reported by EA 2001a).

4.3 Use of scrap tires as a whole or after mechanical processing

Scrap tires can be utilized by making use of their sturdy nature and steel reinforcement inside the rubber. The steel wires are usually protected inside the rubber if the rubber is not severely cracked or eroded. Therefore, the tires can survive for long periods of time even under harsh environments such as as a boat bumper in salty sea water, under a paved road with heavy traffic, or constant pressure of unstable slopes (Fig. 3). The tires can be used at the edge of sloped soil to maintain the soil at the edge from washing away with rainwaters. The reduced erosion at the edges help the soil to maintain its slope and integrity for extended periods of time. When tires are placed side by side and connected each other by clamps or wires, they help to keep the soil together as shear locks. The soil or gravel fill are trapped inside the tires, cannot expand due to high strength steel wires inside the tires and stabilizes the medium.

When tires are shredded or crumb size divided, they can be used as mixture to concrete for additional tension material and making lightweight concrete. Also, can be mixed with asphalt for extra traction and tensile capabilities. Smaller size crumbs are used to make children play grounds and running track surface finishing.

4.4 Strengthening structures using scrap tires, structural engineering applications

In a recent study of using scrap tires as confinement material for concrete columns (Abdulmoula and Saatcioglu, 2009), the tires were used as peripheral material to confine concrete. When concrete is axially loaded, it tends to expand defined by Poisson's ratio. If the lateral expansion is prevented, axial compressive strength of concrete is significantly increased. In the mentioned study, a series of concrete columns were cast inside scrap tires which were placed on top of each other. The cylindrical shape formed by carefully and centrally alined pile of scrap tires have also formed a natural form to be easily filled by concrete. Following the strength gain of concrete in 28 days, the steel wires and rims inside the scrap tires served as horizontal confinement for the columns. It was shown both experimentally and analytically that steel-belted tires can be used effectively to confine concrete in reinforced concrete columns. The exterior scrap tire also protects the column and steel reinforcement inside the column from corrosion.

4.4.1 STP scrap tire pads for seismic base isolation

Seismic base isolation is the placement of a laterally flexible system between the footing (ground) and upper structure to isolate earthquake induced seismic forces. The natural vibration periods of the suspended building or structure shifts towards larger values in the response spectrum causing reduction in the forces and accelerations in the suspended building. The accelerations that correspond to the natural period of the structure decrease, therefore the demand of the earthquake on the structure reduces. Inter-storey drifts decrease considerably and the superstructure on isolation system behaves similarly to a rigid body during earthquake motions.

Seismic base isolation systems can be studied in two main groups: elastomer-based and sliding-based systems. Elastomeric bearings are the most common isolators used in the design of seismically isolated structures. Among the most common types of elastomeric isolators are the low-damping rubber with damping ratio (ξ) about 2% to 3%, high-damping rubber (ξ=10%-20%), lead-plug, and the fibre-reinforced elastomeric bearings. The high vertical stiffness of elastomeric bearings is provided by horizontal steel or fiber reinforcement whereas the low horizontal stiffness is provided by flexible laminated rubber layers. Elastomeric-based isolators may be mimicked using pads made out of scrap tires which are called Scrap Tire Pads (STPs).

Since automobile tires are produced by vulcanizing steel mesh and cords with the rubber, when the part that touches the ground is removed from the sidewalls of the tire and piled on top of each other as rectangular rubber sheets, they form an STP. Steel cords inside tire layers have similar effect as the steel layers inside an elastomeric isolator. The terminology used for STP includes; disposed scrap tires (Fig. 6a); a tire ring is the tread part of a tire that touches the ground and is obtained after cutting off the sidewalls of the tire (Fig. 6b). Tire band is the same part after cutting the ring in transverse direction (Fig. 6c). Tire layers are about 0.20m long pieces of scrap tire bands (Fig. 6d). The scrap tire pad, i.e., the STP, is formed when a set of scrap tire layers are placed on top of each other (Fig. 6e).

Although layers forming an STP can be glued together using epoxy, the friction between tire layers is large enough to keep STP layers intact and working together. The mechanical and dynamic properties of various brand STP samples used in this study were obtained using axial compression, static shear, dynamic (impact), and shaking table experiments.

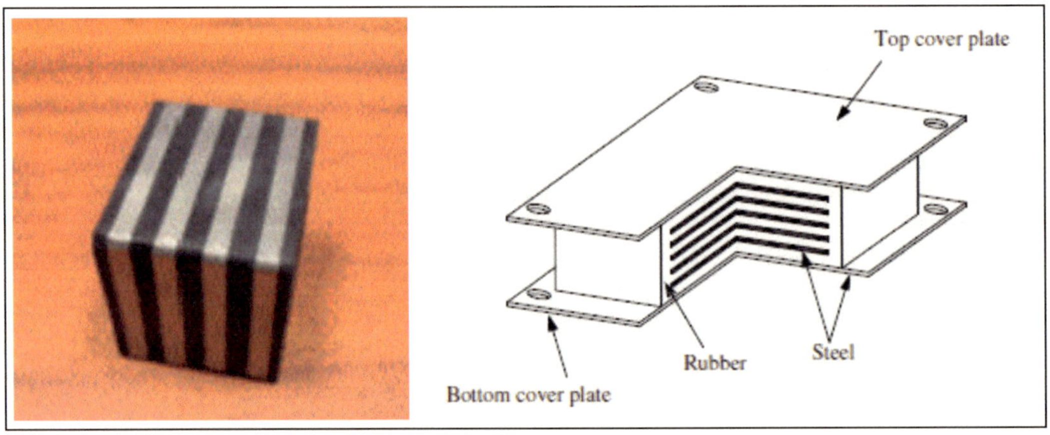

Fig. 5. Seismic isolator, laminated rubber bearing (LRB).

Lateral dynamic tests were conducted on different height of STP specimens and horizontal stiffness values of STPs were found to be linearly decreasing as the number of STP tire layers increased (Fig. 7). The reduction in stiffness with the increase in number of layers was similar to the behaviour of common elastomeric isolators as indicated by Equation 2.

$$K = G \cdot \frac{A}{t_r} \quad (2)$$

where, K is the stiffness of isolator, G is the shear modulus of isolator material, A is the isolator contact surface area, and t_r is the thickness (height) of the isolator. Nonlinearities in Fig. 7 are observed as more than 8 layers of tire were used, and therefore accepted as an indication of a stability problem. The linear relationship between horizontal stiffness and number of layers (up to 10 layers) implies that the stiffness of STPs can easily be adjusted by changing the number of tire layers. The transverse and the longitudinal direction stiffness graphs are parallel to each other (up to the 10-layer mark) and decline linearly as the

(a)

(b)

(c)

(d)

(e)

Fig. 6. Scrap a) tire, b) ring, c) band, d) layer, and e) scrap tire pad (STP).

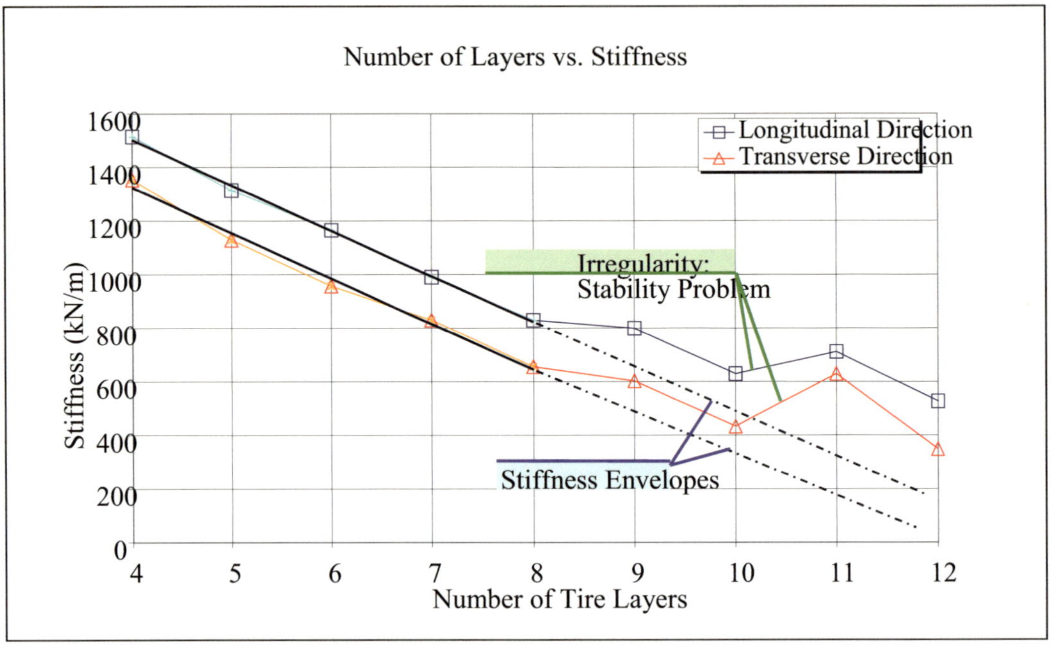

Fig. 7. Stability graph of scrap tire pads (STP).

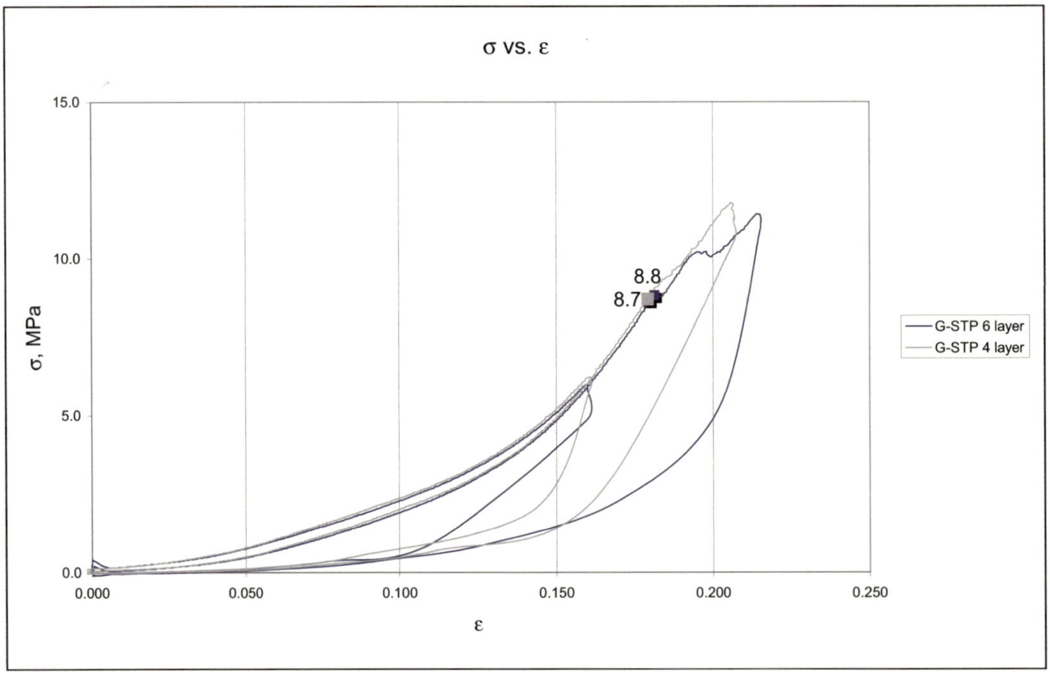

Fig. 8. Stress-strain graph of scrap tire pads (STP) under axial loading.

number of tire layers is increased. The constant 200 kN/m difference between the lateral stiffness terms in the two principle directions is an interesting result and deemed to be a shape factor since the width and length were 0.18m and 0.20m, respectively (Turer, 2008). The axial load capacity of STPs were obtained to be around 8 MPa (Fig. 8), which is relatively low compared to commercially available laminated rubber bearings.

Axial compression tests revealed that the non-linear compressive behavior of STP specimens is close to the compressive behavior of common steel reinforced elastomeric isolators (SREI). The axial compression tests showed that an allowable vertical stress level of 4 MPa for STP specimens can be obtained if a safety factor of 2 is accepted. The free vibration test results showed that the damping values of various STP specimens change between 7% and 14%. For design purposes, taking into account the displacement safety margins, the minimum value of 7% should be accepted as the damping value for the STP specimens. The free vibration tests also showed that, the lateral stiffness of STP specimens can be adjusted by changing the number of tire layers composing the STPs. However, the stability should also be satisfied for higher numbers of tire layers; i.e., larger than 8 layers for 0.18m×0.20m sized STP. The horizontal behavior of STP specimens were determined by conducting static shear tests. Shear modulus values of STP specimens were calculated to be between 0.9 MPa and 1.85 MPa. Relatively high levels of shear modulus values and their dependence on the brand of the tire present difficulty for the design of STPs.

The experimental and analytical program including shaking table tests have shown that STP based base isolation is possible within certain constraints. Softer type of scrap tires, such as winter tires, may be used with additional recycled steel plates placed between each layer would increase the vertical load capacity while maintaining a relatively low horizontal stiffness.

STP based seismic base isolation can be used for rural bridge supports as a low cost and practical material while recycling and reducing pollutants. The STPs would also serve as temperature compensation devices in bridges.

4.4.2 Post-tensioned elastic walls using scrap tires

Post-tensioning is a well-known technique used in modern civil engineering such as light poles and bridge girders spanning relatively larger gaps. The theory is based on applying a compressive stress field on usually a brittle material (such as concrete), which has weak properties under tensile forces. The compressive strength being about 10 times the tensile stress, the structure highly benefits from the even compression field generated by post-tensioning.

In the case of scrap tire based post-tensioning, poor housing in the seismically active zones were targeted. Those houses are usually made of masonry and occupants generally have low income and undereducated. The poor economic and social background of the residents also means that masonry constructions do not receive any engineering services and, therefore, are susceptible to heavy damage or total collapse during earthquakes. Earthquake-induced forces cause masonry houses to collapse in a sudden (brittle) manner. In other words, the disintegration of masonry constructions built from adobe, brick, or stone is very quick and it leads to a total collapse of the roof which is traditionally composed of very heavy earth (usually up to 1 meter (3 feet) thick). The sudden disintegration of walls and collapse of the heavy roof of masonry houses kill the residents instantly not leaving any "life pockets" which might be formed during collapse of reinforced concrete houses. The economically constrained residents living in such houses do not have sufficient resources to build their houses from reinforced concrete. Masonry construction in rural areas is traditional and same inferior construction is repeated since losses during earthquakes are perceived as an act of God. In developing countries, the problem of finding an efficient solution to strengthening masonry houses is further exacerbated by the fact majority of the building stock is generally masonry type.

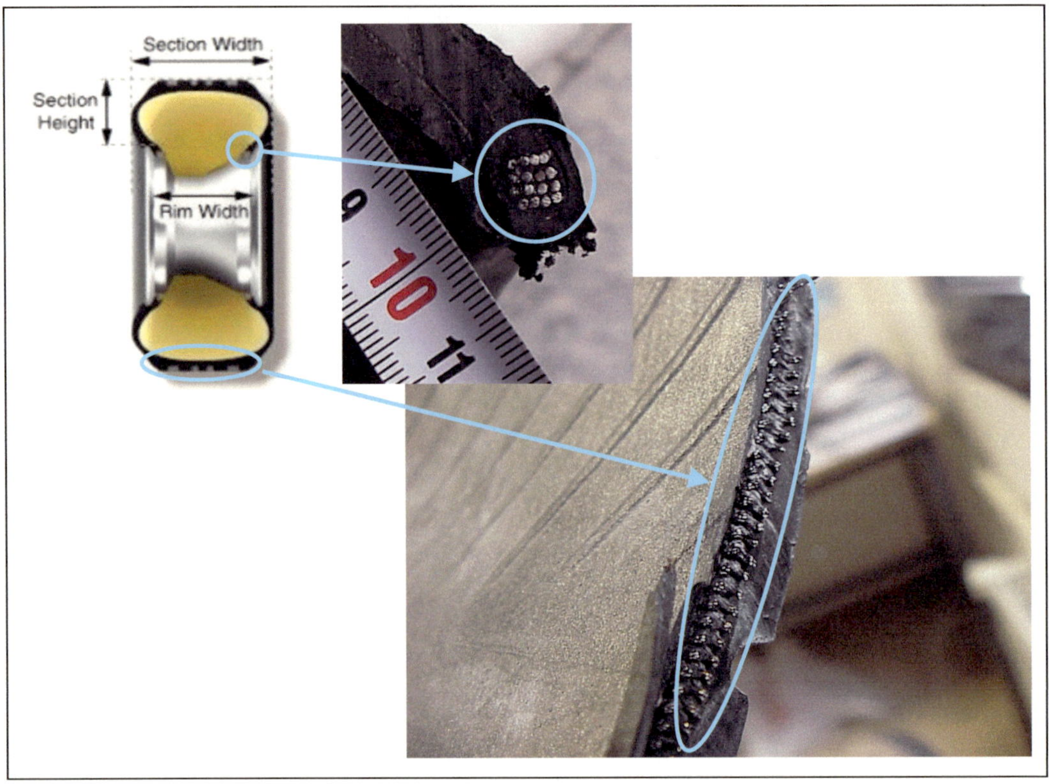

Fig. 9. Typical cross section of a scrap tire.

Implementation of the method is economically affordable and environment-friendly due to the following reasons. First, scrap tires have steel mesh inside with high tensile strength that makes them suitable reinforcement material. Second, except for the low cost of transportation, scrap tires can be obtained free of charge rendering them as low-cost (strengthening) materials. Third, tires can be prepared using simple tools (e.g., a utility knife). Recycling tires has additional advantages preventing waste yards. Finally, the application of scrap tires on walls is simple and easy, and does not require complicated tools and practices. It is believed that the owners of masonry houses in poor countries would be able to implement the strengthening work by themselves since many of them have already built their own houses. Consequently, this will eliminate any workmanship costs thus contributing to the overall affordability and applicability of the strengthening project.

The scrap tire ring test results are obtained in terms of load-displacement curves (Fig. 10). The results indicate that the mean and standard deviation of the ultimate tensile load capacities of scrap tire rings (STRs) were calculated as 133 kN and 32.1 kN. Assuming average weight of a passenger car to be around 12 kN, each single STR can carry more than the weight of 10 cars, which is an amazing performance.

The STRs are placed in the form of chains and tied around the brick walls to generate compression stresses on the masonry wall for post-tensioning. Bolted connections to tie STRs together was used to shorten the distance between tires, in turn, applying force on the system (Fig. 11). The tensile forces generated on the STRs would be in balance with the forces acting on the masonry walls. The walls were tested before and after the post-tensioning application using STRs and performance of the walls were compared for before and after conditions.

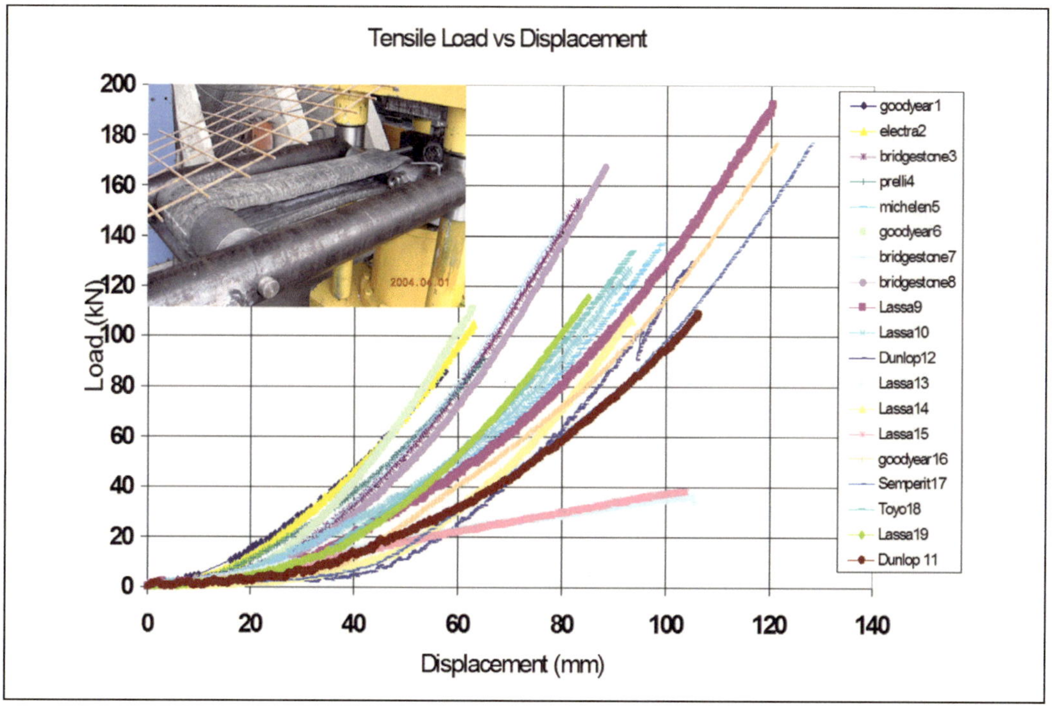

Fig. 10. Load - deflection graph of scrap tire rings (STRs) under axial tension.

The experimental studies showed that the nominal lateral load capacities of the brick walls in out-of-plane direction can be improved up to about 10 times by applying 100 kN (per 0.885m of length) axial post-tensioning force using STR chain and about 16 times using hybrid system. The better improvement ratio of the wall post-tensioned with the hybrid system in the third test could have resulted from higher stiffness associated with the STR chain and/or possibility of having a relatively high unintended initial (nominal) strength of the hybrid test's masonry wall.

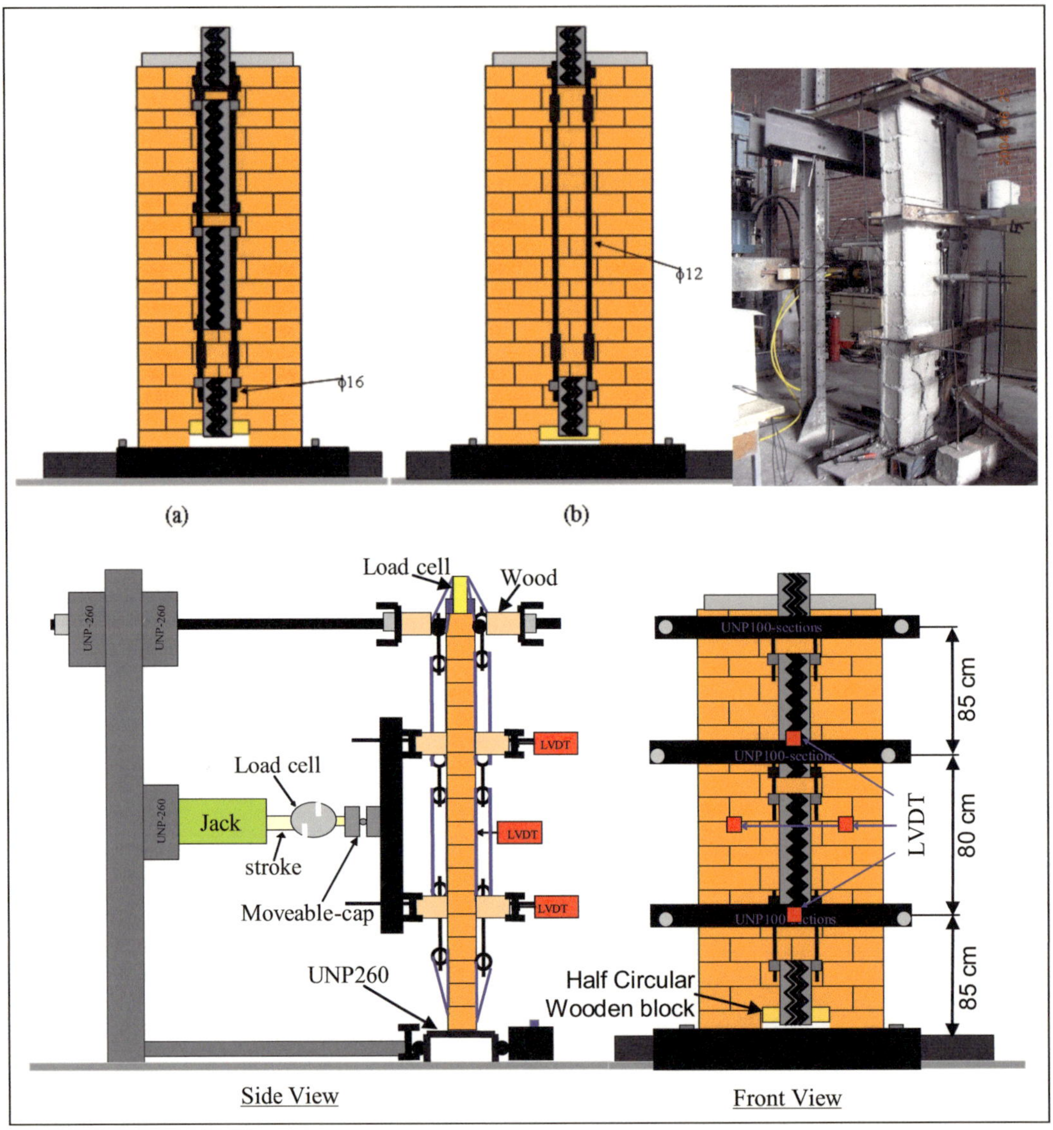

Fig. 11. Application of post tensioning by (a) STR chains, (b) hybrid system, (c) test setup.

5. Conclusions

This chapter illustrated various uses of scrap tires through recycling as a whole, in parts, or after chemically decomposition of materials inside scrap tires. Industrial development brought luxury of cars to our modern lives that produces scrap tires in an increasing rate. As in the cases of other natural resources in the world, we need to learn using less of natural resources while recycling readily available tires by finding ways not to pollute the environment. All cars in the world constantly generating about one scrap tire per person every year causes scrap tires generation in the order of billions on a global scale. The ideal solution would have been recycling each scrap tire to a brand new tire, since when someone throws away a used tire has to buy a new tire.

Using tires on slope stability and land fill, inside asphalt and concrete is not adequately spread enough and in right quantities to use all manufactured tires. Structural uses of scrap tires remain to be at limited instances either enforced by government such as in the case of roads and pavements or experimentally sparse and rare mostly applied by good intentioned environmentalists or low-budgeted projects. Chemical decomposition using pyrolysis is a highly promising approach; however, could not quite reach its full potential yet. On the other hand, burning scrap tires at high temperature furnaces at cement producing kilns and thermo electric power plants as fuel is quite efficient and widely used. Provided that chimney filtering is defined by regulations and rules are properly enforced from toxic material emissions, scrap tire burning seems to be a good source of recycling and transforming otherwise useless and harmful discarded material into energy.

6. Acknowledgment

This study was made possible by World Bank DM2003 SPIM-1451 as well as MAG(İÇTAG)-I599/01 (104I011) projects. Author acknowledges contribution from research assistants Mr. Mustafa Golalmis and Mr. Bayezid Ozden.

7. References

A. Turer, B. Ozden. "Seismic base isolation using low-cost Scrap Tire Pads (STP)", Materials and Structures, 41:891-908 (2008).

Abdulmoula, B.A., Saatcioglu, M., "Concrete columns confined with scrap tires", Masters Abstracts International, Volume: 39-01, page: 0255, 2009.

CIWMB, California Integrated Waste Management Board, Tires as a Fuel Supplement: Feasibility Study. Sacramento, CA, 1992.

European Commission - Directorate, General Environment Refuse Derived Fuel, Current Practice and Perspectives (B4-3040/2000/306517/Mar/E3), Final Report; July 2003. http://ec.europa.eu/environment/waste/studies/pdf/rdf.pdf

Goodyear web page, "History: The Strange Story of Rubber", 2011.

Kurt Reschner, "Scrap Tire Recycling; A Summary of Prevalent Disposal and Recycling Methods", 2008; cited on 2011 at http://www.entire-engineering.de/Scrap_Tire_Recycling.pdf

Mojtowicz, M. A., Serio, M. A., Pyrolysis of scrap tires: Can it be profitable ?, Available from http://www.dnr.state.oh.us/recycling/awareness/facts/tires/rubberis.htm (2005)

Owen A. Rosenboom and Mervyn J. Kowalsky, "Reversed In-Plane Cyclic Behavior of Post-tensioned Clay Brick Masonry Walls", "the Journal of Structural Engineering ASCE, Vol. 130, No. 5, May , 2004, pp-787-798.

Page, A. et al. "Other Trace Metals" Impact of Heavy Metal Pollution on Plants. Volume 1: Effects of Trace Metals on Plant Function. N. Lepp, ed. London: Applied Science Publishers. 1980.

Ruben Tire and Auto Service, Available from http://www.rabentire.com/?PageData=37134 2011.

S.Murugan, M.C. Ramaswamy, G.Nagarajan, "The Use of Tyre Pyrolysis Oil in Diesel Engines", Waste Management, Volume 28, Issue 12, December 2008, Pages 2743-2749.

Scrap Tire Management Council, "The use of scrap tires in cement rotary kilns", 1992. Available from http://www.rma.org/scrap_tires/scrap_tire_markets/cement_ kiln_report.pdf

Sharma et. al., "Disposal of waste tyres - review", Elsevier, Energy Convers. Mgmt Vol. 39, No. 5/6, pp, 511-528, 1998.

Turer, A., Ozden, B., Golalmis, M., World Bank DM2003 "Seismic Performance Improvement of Masonry Houses" (SPIM-1451) project final report, 2005.

Wikipedia, Cement kiln, cited on 2011 at http://en.wikipedia.org/wiki/Cement_kiln

Wikipedia, recycling Available from http://en.wikipedia.org/wiki/Recycling, 2011.

William Sheehan, "Tires an d Glass Markets for Rural Georgia", 1995; cited on 2011 at http://www.p2pays.org/ref/24/23747.pdf

9

Waste Tire Pyrolysis Recycling with Steaming: Heat-Mass Balances & Engineering Solutions for By-Products Quality

Uladzimir Kalitko
Heat-Mass Transfer Institute, HMTI,
Belarus National Academy of Science, Minsk,
Belarus

1. Introduction

Waste tires pyrolysis is well known method for their thermal recycling by heating at near 500°C with purpose of liquid oil and carbon black by-production as near 50% and 35% yield correspondingly, including about 10% combustible off-gas residual after oil condensing and 5% wire steel cord in rest (all relatively to tire mass). There are many patents claimed in the world such as [1-20] and others, as well as many research papers published in this field such as [21-42] and others since the 1980s mainly. Not considering such simplest as batch-battery type and such complicated as fluidized bed one and some others, the drum-kiln and screw-auger type of pyrolysis reactor should be noticed as most preferable for commercial use, being both of them operated continually with tire shreds as 2-3 inch size. The tire material is gasifying in a sealed pyrolysis reactor and volatile hydrocarbons (pyrolysis gases or simply pyrogas,) are piping from reactor to condenser for the liquid pyrolysis oil. A few of the oil as 5-10% is burning for heating the reactor, provided with all the off-gas fuel after condensing the oil is afterburning too. As a solid rest the tire carbon char is continually discharging from reactor for its powdering, separating off steel wires and producing the carbon black.

But even in 2000s with reference to [32] it has been concluded as there was not an operating commercial plant in the world that could be recognized as operated successfully with a high commercial productivity and quality of both by-products. Particularly, for the carbon black could be used commercially in the rubber industry again, its quality must be as 1-2 % tire oil volatile matters residual content. In recent years the tire pyrolysis plants of rotary-batch type are many referenced in the net as an alternative and widely used in China, Malaysia, Taiwan, etc., operating simply with a whole tire bulk and producing so way the carbon black of low quality as 5-6% and more of the residual content above, operating a priory with a low productivity as the batch-type.

In connection with the carbon black quality and with reference to [1, 2] the vacuum tire pyrolysis method should be mentioned as claimed in the 1980s, being performed under the low-pressure and resulted in 4% and less of oil residual content. It should be noticed as well corresponded to the theoretical solution on such of dependence with pressure and pyrolysis temp condition considered just in the next part of the article. And specially concerning the steam use for tire pyrolysis in connection with that and with reference to U.S. Patent 866 758, it had been first claimed even in 1907, resulting in the pure carbon char yield with heating directly by steam at 315°C that was well enough for such of recycling (reclaiming in original), provided with the rubber particles intensively piping by the same steam flow. Much more late with reference to [9, 10, 33] the direct heating pyrolysis method by super-heated steam feeding into reactor at 500-600°C has been claimed and published as a new idea of that in the 2000s. In accordance with that a multi-batch pyrolysis tunnel system has been elaborated in HMTI for recycling the whole tires in cartridges which continually moving through the long tunnel heating by this way. The idea was realized in a large scale commercial plant per 120 t/day in Lithuania, 2004-2005 but it was not effective as both low productivity and quality of carbon black because of low operating conditions as for heating the bulk of whole tires in cartridge, as for heating the large-long tunnel by steam feeding at all.

Independently the reactor type and heating system with references to [25, 28, 30, 31] some of catalytic pyrolysis methods could be noticed as proposed for enhancing the oil productivity and provided with such of solid or liquid additives as Na2CO3, AlCl3, KOH, Y-, USY- or ZSN- zeolites etc. Being restricted in detail consideration on that for sake of the present article, it can be shortly characterized as no radical catalytic dependence was obtained for commercial use so as a little of addition could be used for its action without contamination of the oil or carbon black with rest of the same addition in kind of solid or gas. Proposing the steam actually is not a catalyst, but simply a carrier gas, nevertheless with reference to [24] it is interesting to notice that the oil yield with steam pyrolysis of the oily shale (in a laboratory scale) was increased by 34%, comprising that with nitrogen at the same operating and thermal conditions. In contrast, no real difference for oil yield rate was obtained with olefins or tires pyrolysis in [22, 27], but excluding only a high steam reactivity with tire char for its next purification with carbon black by-production.

As for the steam pyrolysis in the present article, it is obviously proposed that oil residual content in the carbon black is objectively corresponded to the pyrogas concentration in pyrolysis reactor where the gas is saturating all inside, including the carbon black porous structure too as considered next in the present article. Even the tire pyrolysis would be first performed ideally as 100% oil volatile matters gasified entirely, next pyrogas inside the carbon porous structure will be cooling and condensing there in kind of the same oil residual matters after the carbon discharge from reactor. To replace the pyrogas from the carbon and so way to clean that simply, it is also well known as an inert gas (e.g. nitrogen as most available) could be feeding into pyrolysis reactor finally, provided the gas feeding rate to be corresponded to the tire pyrolysis rate. If no such of inert gas blowing up the reactor theoretically it is about 3% oil residual matters in the carbon by-product by this way. But really it is not commercial solution because of high cost for any inert gas supply relatively to price of the carbon black by-product. A second thermal processing (firing) of the carbon black at 750-800°C is required after pyrolysis so to purify that off the oil residual for next treatment with commercial use, or even for its storage to be clean off the specific smell which is steady appeared at the oil residual content 5-6% and more.

In this connection with reference to [33-38] a new pyrolysis system for waste tire commercial recycling in the reactor of double-screw type with steam has been elaborated in HMTI (Belarus) for ENRESTEC Co. (Taiwan) in 2006, being next installed and applied by author's guidance in 2007-2008. The plant has been designed as a double-line pyrolysis system per 1 t/hr every (Pict.1, Fig.1) with carbon black quality as 1-2% oil residual matters, including more proof-explosion safety inside the reactor due to diluting the pyrolysis gases (pyrogas) with steam, as well as more sealing the reactor with steam feeding against air penetration into reactor under the low-operating pressure as required. Steam is self-producing in a second-heat boiler after heating the reactor provided by own pyrolysis oil-fuel combustion in a special furnace apart of reactor, including the pyrolysis off-gas afterburning therein too. For processing the steam is super-heating up to pyrolysis operating temp 400-450°C, being that performed in a steam coil tube around and along with reactor heating together, and feeding into reactor as shown and considered next bellow. With reference to [36] for all of unit numbers to be observed, the general design overview of the plant is shown in Pict.2 where only the units above are noticed in caption.

Pict. 1. ENRESTEC thermal units for waste tire pyrolysis with steam: 1 - oil fuel burner, 2 - furnace, 3 - pyrolysis reactor heating box, 4 - double-screw reactor in the box, 5 - second heat steam boiler.

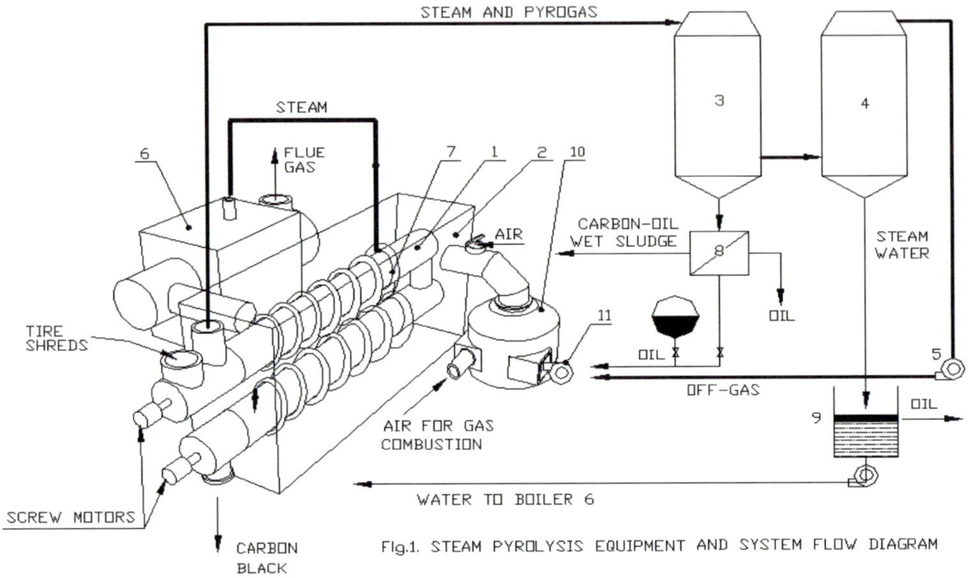

Fig. 1. Steam pyrolysis equipment and system flow diagram: 1 – pyrolysis reactor, 2 - reactor heating box, 3 - oil condenser, 4 - steam condenser, 5 - off-gas fan, 6 - second heat steam boiler, 7 - steam super-heating coil, 8 - oil Laval-separator, 9 - oil-water gravity separating tank, 10 - gas-oil furnace.

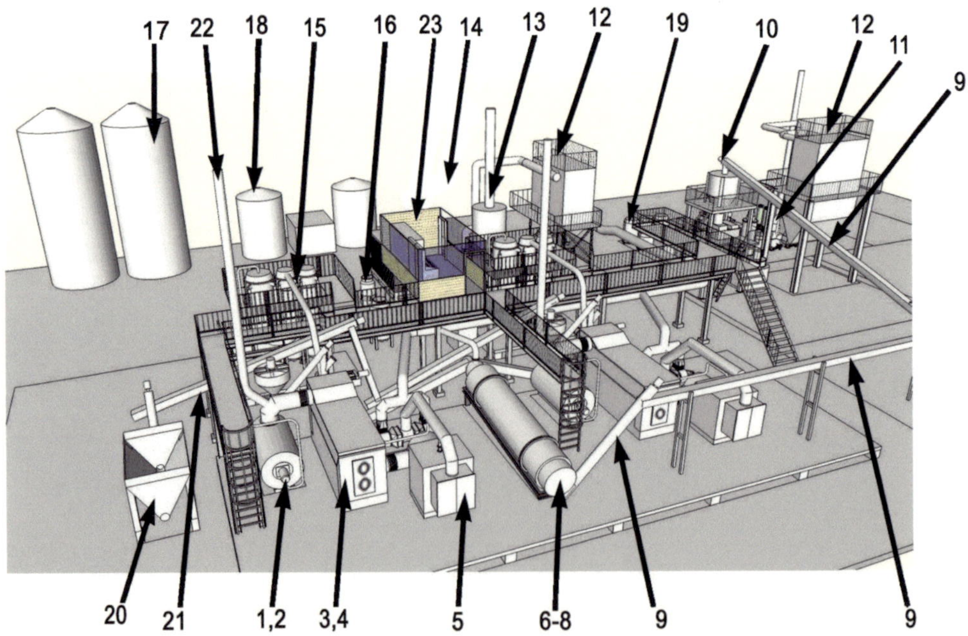

Pict. 2. Design overview of the ENRESTEC plant: 1,2 - both furnaces with oil fuel burners at front; 3, 4 - both heating boxes with every of double-screw reactor inside; 5 - heat utilizing steam boilers.

As for some history, the first steam use for rubber waste pyrolysis by U.S. Patent 866758 had been tested even in 1907, being concluded that the temp 600 F (315°C) is well enough for rubber vulcanized particles to be full pyrolized in the steam pneumatic flow condition at this temp. Comparing that to pyrolysis at 450–500°C as required without steam now, that is good evidence of steam effect by its diffusion with heat penetration inside every of the particles even at the lower temp. Being not so porous initially, with pyrolysis process in time the rubber is destructing and becoming as the carbon black of its fine porous structure that finally is well suitable for its cleaning by steam diffusion inside. In this connection it is all reason first to consider and evaluate even theoretically why and what is a limit on the carbon black quality by scrap tire pyrolysis recycling without steam (in terms of residual oil matters of its content).

2. Theoretical limit on carbon black quality without steam

By usual pyrolysis way as without steam, even all the tire volatile matters are proposed to be initially gasified, next there is to be objectively some of tire oil matters residual in carbon black (CB) because of its secondary contamination with the same volatized matters that re-condensed in the CB porous structure after its cooling with discharge from reactor. Being some idealized and simplified, such of contamination can be theoretically formulated and estimated as bellow here. Let use the ideal gas law equation for the tire pyrolysis volatile matters (hydrocarbon vapors and inherent gases or simply pyrogas) which are proposed to be heated finally up to the temp $T = 450°C$ (723 K) which is rather above all of pyrolysis liquid-gas phase transition points, and so allows the pyrogas can be considered as a super-heated one similar to an ideal gas, which molecular weight is proposed to be equal to that of tire pyrolysis oil condensed from the pyrogas as the same. With reference to [1.2] it is about $\mu = 210$ and we have the next pyrogas mean-total density in tire pyrolysis reactor at the temp 450°C and near the normal pressure operating conditions ($p = 10^5$ Pa):

$$pV = \frac{m}{\mu} RT , \tag{1}$$

$$\rho = \frac{m}{V} = \frac{p\mu}{RT} = \frac{10^5 \cdot 210}{8314 \cdot 723} \cong 3.4 \text{ kg/m3}. \tag{2}$$

Pyrogas of the density above is saturating all inside the reactor operating volume, including the bulk between CB particles and inside every of the particles too, being these of the inner porous structure. By this consideration we have the next pyrogas quantity discharged with anyone particle of CB porous product which will be next condensed therein as the oil residual matters:

$$m_{oil(i)} = V_i \rho \cdot \sigma , \tag{3}$$

where V_i is volume of the particle, σ is factor of the particle structure porosity which is objectively proposed as not bellow $\sigma = 90\%$, and the pure mass of the particle solid-porous structure is a light as conveniently proposed by density not above $\rho_{cb} = 100$ kg/m3:

$$m_{cb(i)} = V_i \rho_{cb} , \tag{4}$$

by which relation the next simple estimation is obtained:

$$\frac{m_{oil(i)}}{m_{cb(i)}} = \sigma \frac{\rho}{\rho_{cb}} = 0.9 \frac{3.4}{100} \cong 3\%. \tag{5}$$

Concerning the theoretical limit on steam dilution with pyrogas as for its proof-explosive condition in reactor, with author's reference to [34] it is above 5 kg steam per 1 kg tire required so to provide such of condition with air. It is too much for real observation, comparing that to available steam self-producing rate by second heat after heating the reactor. And it is the question for analytical consideration and formulation bellow as it was realized in the operating process with steam in Taiwan.

3. Heat-mass balance analysis on waste tire pyrolysis with steam

Referencing namely to [35], the first version of the thermal units and flow diagram of the tire-steam pyrolysis process in Taiwan in 2007-2008 is shown in Fig.1, where steam is self-producing in the second heat boiler 6 with flue gas flow after heating the double-screw reactor 1 in the hot gas box 2 connected with the furnace 10 for oil combustion with off-gas after-burning together. Steam is super-heating in the tube coil 7 inserted as along-around the reactor in the hot box too, and next steam is feeding into the reactor for steaming the tire pyrolysis as considered above. The pyrogas with used steam flow together is piping to the oil and steam condenser 3-4 in line correspondingly, being provided by suction performance of the gas fan 5 that pipes the combustible pyrolysis off-gas residual after oil-steam condensing into the furnace. Flue gas from the furnace is piping for heating both reactor and steam boiler in line, and next to exhaust scrubber. The real view on these thermal units (furnace, reactor, steam boiler) is shown in Pict.1.

The tire scrap is moving and mixing by screw along-inside the reactor by usual way of such processing, being pyrolyzed and discharged to outside with the carbon black to be next cooling, screening and crashing for magnetic separation against some of steel cord wire residual. Pyrolysis oil is separating by Laval centrifugal unit 8 so to produce the own light oil for burning in the furnace and so heating the reactor and boiler. Ideally the pyrolysis oil and steam were proposed to be condensing in 3-4 separately and next steam water to be cycling and pumping to the boiler again simply as it is shown in Fig.1. Really it was not of such ideal proposition and steam was partially condensing with oil together, as well as all different benzene's and low-temp aromatic fractions of the pyrolysis oil were condensable and soluble with steam water too. Considering that especially for development next bellow in p.6, here we formulate and calculate simply the heat-mass balance of reactor and other thermal equipment above, not including the condenser as not involved with pyrolysis process. Due to the reactor heating is based on the off-gas afterburning, being the latter well available and corresponded to the pyrolysis rate, as well as the steam for the process is self-produced after heating the reactor and next super-heated along with reactor heating too, all of that is evidently depended on each other and so we can formulate analytically and calculate numerically the oil fuel specific consumption per 1 kg tire additively to the off-gas burning. It is a novel question on the tire pyrolysis recycling because even without steam there is not such of general analysis in this field until now. With all references to [35, 36] the question on the oil fuel quantity to be combusted with pyrolysis off-gas together for heating the process with steam feeding and self-producing at the same time, it was answered by the heat-mass balance solution on that as following:

$$X = \frac{G_{oil}}{G_t} = \frac{q_t - \left(Q_{gas}E_r E_f - \alpha_2 A\right)\cdot GAS}{Q_{oil}E_r E_f - \alpha_1 A}, \tag{6}$$

$$A = q_{ss}\frac{q_g g_a}{q_s}, \tag{6'}$$

$$Y = \frac{G_s}{G_t} = \frac{c_{p(g)}(T_{g2} - T_{g3})g_a}{c_{p(w)}(T_s - T_a) + h_s}(\alpha_1 X + \alpha_2 GAS) = \frac{q_g}{q_s}\left(\alpha_1 X + \alpha_2 \frac{GAS}{100}\right), \tag{7}$$

where $g_a \approx 15$ kg/kg is air stoichiometry index per 1 kg of liquid fuel, $\alpha_1 = 1.3$–1.35 is air excessive supply index for fuel combustion, $\alpha_2 = 1.05$–1.1 is the index for gas fuel combustion, $E_f = E_r$ = 0.95–0.97 is the furnace and reactor thermal efficiency by every heat emission to outside 3–5% proposed, including the supplementary specific calculations as follows:

$q_t = c_{p(t)}(T_p - T_a) + h_t$ is a specific heat capacity per 1 kg pyrolysis,

$q_{ss} = c_{p(s)}(T_p - T_s)$ is a specific heat per 1 kg steam super-heating,

$q_s = c_{p(w)}(T_s - T_a) + h_s$ is a specific heat per 1 kg steam producing,

$q_g = c_{p(g)}(T_{g2} - T_{g3})$ is an enthalpy per 1 kg furnace gas flow.

With the same references to [35, 36] the question what is the off-gas burning rate to be for heating the pyrolysis reactor without oil fuel, it has been simply obtained from (7) with proposition $X = 0$ which numerical solution at the pyrolysis conditions above is $GAS_{\min} \cong 10\%$:

$$GAS_{\min} = \frac{q_t}{Q_{gas}E_r E_f - \alpha_2 A}, \tag{8}$$

The testing-operating data on the process with max 10% approximation are presented in the Table 1, including the standard analysis data on the pyrolysis oil and carbon black products quality referenced to [36]. It should be noted initially that testing the commercial process as considered above, as well as any other of such thermal processes with heat-mass balance calculation too, it is performed as 5–10 % discrepancy usually to be allowed. At the same time, concerning the carbon black recycling up to 1–2% of C_nH_m-residual quality as required for the market, it is clear that the latter can not be the subject for modeling, but only testing at the thermal parameters of the process under question. As for the non-calculating parameters which could be important for the carbon black quality, it is also clear that the pyrolysis exposition time and tire chips-shreds size, and more exactly even the scrap thickness size, are both of most important, provided the size of the shreds was used as min 2 inches and some more. Really and simply the low, middle and high temp pyrolysis condition in range 350–450°C had been tested with tire processing rate within 0.5–1 t/hr, provided the reactor length and screw rotary speed resulted in the tire processing time as max 13 min so as carbon black quality to be near the same in specification range 1–2%. As for the scrap tire thickness, it was supplied in range of 5–15 mm.

Pyrolysis oil condensed heat value: 42 MJ/kg Off-gas heat value (without steam): 39 MJ/m^3	Tire pyrolysis rate, kg/hr		
	500	750	1000
Off-gas burning rate (10% tire mass at 55 C), m^3/hr	50	75	100
Light pyrolysis oil mean-daily burning rate, kg/hr	40	30	20
Steam self-producing rate (feeding to reactor), kg/hr	200	250	300
Off-gas-oil furnace max operating temperature, °C	950	1100	1150
Furnace gas temp Tg1 for reactor heating inlet, °C	850	900	950
Furnace gas temp Tg2 for reactor heating outlet, °C	450	480	510
Pyrolysis operating temp Tp inside reactor, °C	350	400	450
Sulfur content in pyrolysis oil, %	-	1.15	-
Carbon content in pyrolysis oil, %	-	1.1	-
CnHm-content in carbon black, %	-	1.5	-
Sulfur content in carbon black, %	-	2.3	-

Table 1. Testing-Operating Data on Tire Pyrolysis Recycling With Steam (Taiwan-2008).

The process calculation by (6)–(8) is presented in Table 2 as carried out with low, middle and high-temp pyrolysis condition at 350, 400 and 450°C inside reactor correspondingly, correlating that to the tire shreds thickness 5, 10 and 15 mm proposed. The subject and result of the calculation is the oil fuel consumption and steam self-producing rate per 1 kg tire which is well corresponded to the testing data at the high-temp pyrolysis condition.

GAS (oil condensing corresponded temperature), %	Low-temps: 5 mm shreds, $Tg1$ = 850°C, $Tg2$ = 450°C, Tp = 350°C		Middle-temps: 10 mm shreds, $Tg1$ = 900°C, $Tg2$ = 500°C, Tp = 400°C		High-temps: 15 mm shreds: $Tg1$ = 950°C, $Tg2$ = 550°C, Tp = 450°C	
	X, kg/kg	Y, kg/kg	X, kg/kg	Y, kg/kg	X, kg/kg	Y, kg/kg
6 (35-40 C)	0.00896	0.164	0.0199	0.236	0.0333	0.330
7 (40-45 C)	0.00261	0.162	0.0111	0.232	0.0245	0.326
8 (45-50 C)	-	0.159	0.0024	0.229	0.0157	0.322
9 (50-55 C)	-	-	-	0.226	0.0069	0.318
10 (55-60 C)	-	-	-	-	-	0.314

Table 2. Calculating Data (6)-(7) on Oil Burning (*X*) And Steam Self-Producing Rate (*Y*) With Variable Pyrolysis Off-Gas Burning Rate at the Different Operating Temperatures .

The factor of steam feeding rate as required for pyrolysis reactor of the screw tubular type has been some tested in connection with the carbon black dusting by an excessive steam flow, being the pyrolysis oil next condensed with much of the carbon sludge after its centrifugal separation from the oil finally. To prevent the carbon dusting the steam feeding rate is appropriated as max 200–250 kg/hr with reactor diameter 0.6 m, or simply 1 t/hr per 1 m^2 of cross-section square of that in specific terms. So way it is enable to provide the steam feeding rate for carbon black purification and so on as considered next in p.6.

With reference to [35, 36] it was a few as 1–2% of wet carbon-oil slurry after its gravity and centrifugal separation from the pyrolysis oil with water. After the separation slurry was well marketable in Taiwan for use as the asphalt component in the road construction as they doing there. Otherwise, the slurry is proposed to be mixing with the scrap tire and recycling with pyrolysis too. The question concerning what is max possible sludge mixing-recycling rate with tire pyrolysis together (in percent relatively to tire), it has been obtained in [35, 36] by the similar heat-mass balance solution as following:

$$SLU_{\max} = \frac{\Delta X}{X} \cdot \frac{\left(Q_{gas} E_r E_f - \alpha_2 A\right) \cdot GAS}{wB}, \tag{9}$$

$$B = c_{p(w)}(100 - T_a) + h_s + + c_{p(s)}(T_p - 100), \tag{9'}$$

where ΔX is an additional oil fuel consumption for sludge pyrolysis together with tire by which calculation with conditions above it is obtained as $SLU_{\max} \cong 6\%$.

4. Process development with oil venturi condenser

As an imperfection of the first plant the steam was condensed with pyrolysis oil together and steam water after its gravity or centrifugal separation from the oil it was contaminated with different benzene and other low-temp aromatic fractions of the tire pyrolysis oil as well soluble with water. Such of contaminated steam water has become a regulation problem for its normal cycling to steam boiler again and so for the new process to be certificated in Taiwan and elsewhere, including some other operating problems considered in [36, 37] and being all the problems resulted simply from the water tube condensers applied and operated with water at 35-40°C. Moreover, operating by this way it was resulted in a low-quality of pyrolysis oil fuel in terms of the flash point temp which was about 40°C correspondingly.

With reference to [36, 37] in project for ALPHA RECYCLAGE FRANCHE COMTE (France) in 2009-2010 the tire-steam pyrolysis system has been developed and modernized so as a new condenser of venturi type with steam too is used (Fig.2), being first and only one referenced as without steam but namely for such of application in [2]. The new steam pyrolysis system is operating with oil condensing at near 100°C, proposing so way its quality in terms of the flash point temp to be high as near 80°C. Steam is not condensing and all piping with residual off-gas to furnace by which way only the oil condenser 4 is required (see Fig.2). And so way the furnace gas flow is rather enhanced with steam for heating reactor and next boiler where steam is acting in a new manner as a heating agent too which analyzed here.

With furnace flue gas together steam is piping to scrubber and condensing therein simply with water, being so way water is far from the oil and nothing of oil-water separating equipment is required.

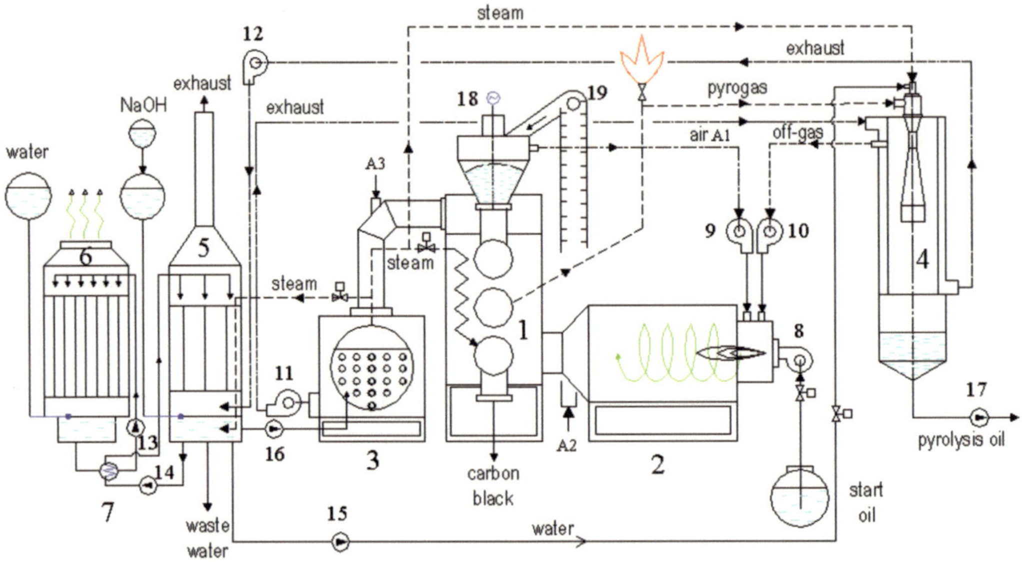

Fig. 2. Diagram of the modernized steam pyrolysis system: 1- reactor, 2 – furnace, 3 – steam boiler, 4 – venturi condenser, 5 – exhaust scrubber, 6 – cooling tower, 7 – water cooler for scrubber , 8 – oil fuel burner, 9-12 – air blowers and gas fans, 13-17 – water and oil pumps, 18-19 – scrap tire feeding system.

Basing and referencing to the data on the tire pyrolysis oil distillation with temp in [1, 2] and with the same reference to [36, 37] it is analytically obtained that the new process is characterized as *OIL* = 40-45% pyrolysis oil to be condensed at 100°C and correspondingly *GAS* = 17-18% incondensable off-gas to be residual (relatively to tire mass), being the latter well enough for the process heating without oil at all. By the analytical heat-mass balance it is resulted in the next formulation on the steam self-producing rate:

$$STEAM = \frac{G_s}{G_t} = \frac{Q_{gas}E_f}{[A_n c_{p(a)} + c_{p(s)}](T_{g1} - T_a)} \frac{GAS}{100}, \tag{10}$$

where the complex A_n is for the math compaction too:

$$A_n = \frac{G_a}{G_s} = \frac{c_{p(w)}(T_s - T_a) - c_{p(s)}(T_{g2} - 200) + h_s}{c_{p(a)}(T_{g2} - 200)}. \tag{10'}$$

By this way it is rather more steam self-producing rate then before in Taiwan and to avoid the off-gas would be diluted with steam too much as no ignition by [34] with reference to [36, 37] the steam self-producing rate is to be limited and reduced by $\Delta STEAM \geq 15\%$ which is formulated and calculated as follows:

$$\Delta STEAM = \frac{\Delta G_s}{G_s} \geq 1 - \left(\frac{Q_{gas}}{Q_{gas(\min)}} - 1 \right) \frac{\rho_s}{\rho_{gas}} \frac{GAS}{STEAM}, \tag{11}$$

provided simply by air injection into boiler (similar to its injection into furnace in Fig.1) that is also obtained by the analytical heat-mass balance as following:

$$\frac{\Delta G_a}{G_a} \geq \frac{T_{g2} - 200}{T_{g3} - T_a} \cdot \frac{0.15}{1 - \dfrac{c_{p(s)}(T_{g2} - 200)}{c_{p(w)}(T_s - T_a) + h_s}}, \tag{12}$$

where the total air injection rate (with combusting air and relatively to tire pyrolysis rate) per 1 kg tire is obtained:

$$AIR = \frac{G_a}{G_t} = \frac{A_n Q_{gas} E_f}{[A_n c_{p(a)} + c_{p(s)}](T_{g1} - T_a)} \frac{GAS}{100}, \tag{13}$$

Due to the more off-gas afterburning rate as GAS = 17-18% instead of around 10% before, including all the steam used and air injected as above, now it is a rather more furnace gas flow available for heating the pyrolysis reactor which results in rather less acting temp difference of the flow between the reactor inlet-outlet as follows:

$$T_{g1} - T_{g2} = \frac{c_{p(t)}(T_p - T_a) + h_t + c_{p(s)}(T_p - T_s)}{E_f E_r \left(c_{p(s)} + c_{p(a)} \dfrac{AIR}{STEAM} \right)}. \tag{14}$$

Calculating data on the modernized process by (10)-(14) with max 1% iterating discrepancy and max 5% calculating accuracy are presented in Table 3, where these are compared to operating data on the first steam pyrolysis system in Taiwan. Shortly it can be concluded as the new system is rather more capable for convective heating the pyrolysis reactor first by factor of the furnace gas flow than by its high temperature (i) in which accordance the reactor is also developed properly as a triple-screw design of a long heating surface (see Fig.2). In second, by more heating and 1.5 times longer process way the carbon black quality is proposed to be surely high as 1% and less of the tire residual matters. In third, steam boiler is correspondingly also heating much more and steam producing rate is proposed to be well enough as for feeding the reactor, as for injecting into the oil Venturi condenser (iii) as it is considered in the next p.5, including the oil quality is to be also high as noted yet above.

The numerical data on steam limitation with air injection by (13) are presented in the Table 3 so as the low-calorific heat value of the off-gas with steam mixture can be well increased as 5.5 MJ/m³ (1300 kcal/m³). In this connection and with the same reference to [36, 37] the special flow-vortex burner for the low-calorific gas fuel condition is proposed to be applied, being that well appropriated namely in the similar steam process with carbon black pyrolysis recycling from coal in Russia in commercial scale in 70-80-ths, where the similar off-gas was also much diluted by 80% with both steam and nitrogen as 3.3–3.8 MJ/m³ (800–900 kcal/m³), i.e. even bellow the critical value above. The burner is operating by the gas pre-mixing and ignition with air in a vortex-flame tunnel just before the furnace (see Fig.3) by which way there is an area in the tunnel where the gas flame is every moment torching and so igniting just near from the furnace.

Pyrolysis oil heat value: 42 MJ/kg Off-gas heat value : 39 MJ/m3	Operating process in Taiwan (1000 kg/hr)	Modernized process (1000 kg/hr)	
		Steam self-producing rate by (10)	Steam producing limited with air injection by (13)
Oil combustion rate, kg/hr	20	-	-
Off-gas burning rate, m3/hr	100	180	180
Combusting air flow, nm3/hr	2650	5000	5000
Furnace heat capacity, MW	1.2	1.95	1.95
Furnace gas temp inlet reactor, °C	900	800	800
Furnace gas temp outlet reactor, °C	500	590	590
Furnace gas temp inlet the boiler, °C	500	590	500
Air injecting rate in boiler, nm3/hr	-	-	1350
Steam self-producing rate, kg/hr	350	1300	1000
Steam residual with off-gas, kg/hr	15	1300	1000

Table 3. Modern Process Calculation in Comparison With Operating Process in Taiwan-2008

5. Oil venturi condenser with water spraying by steam jet

To prevent the condenser cooling surface could be contaminated with carbon soot the pyrogas is proposed to be condensing for oil in the condenser of venturi type which is operating now with cooling-spraying water instead of pyrolysis oil cycling before as it referenced in [2]. Water is well corresponded to the pyrolysis process with steam so as it is spraying and evaporating for steam too while the pyrogas cooling and condensing, including the carbon soot catching with oil droplets at the same time. The new condenser is developed with a steam jet for both spraying water and ejecting pyrogas from reactor (Fig.3).

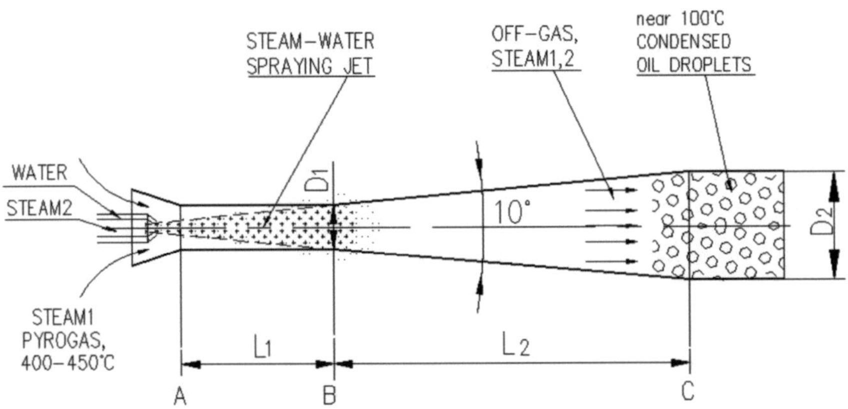

Fig. 3. The flow structure in the oil venturi condenser: pyrogas, steam, water, condensed oil.

With steam the oil condensing temp is well provided as near 100°C for the flash point to be at near 80°C. Moreover, the Venturi tube part is actively cleaning against the carbon soot by the same steam jet for spraying. There are three acting specific points with pyrogas-steam reactor flow mixing with steam-water jet and resulting in condensed oil droplets with incondensable pyrolysis off-gas and all steam residual flow in venturi tube as shown in Fig.3:

Inlet point A:

- pyrogas with used steam from pyrolysis reactor;
- cooling water flow spraying by boiler steam jet;
- steam jet self-wetting max 3% by throttle effect;
- water spraying by steam jet for droplets ≤ 10 μm;

Mixing point B:

- water droplets mixing-heating with pyrogas flow;
- pyrogas flow mixing-cooling with water droplets;
- steam content is passive thermal agent of the flow;

Condensing point C:

- pyrogas is condensed for oil droplets at 100°C;
- water droplets are evaporated at the same temp;
- pyro-carbon soot is captured with oil droplets;
- incondensable pyro-off-gas is residual at 100°C.

Proposing a simple sonic type of steam nozzle for spraying water, initially a self-cooling and wetting effect with steam jet discharged at near the sonic velocity (throttle effect) to be considered, which can be maximally estimated by [40] under the next steam min pressure and temp conditions to be: $p_{\min}$ = 0.4 MPa – steam pressure (abs), T_1 = 142°C = 415 K – steam boiler temp, T_2 – steam temp after the throttle effect,

$$T_2 = T_1\left(\frac{2}{k+1}\right) = 415\cdot\left(\frac{2}{1.4+1}\right) \cong 72\ ^{\circ}\mathrm{C}. \tag{15}$$

Due to steam jet after its discharge is really cooling down only to 100°C and next a few steam is condensing at the same temp and normal pressure, it is acting for steam self-wetting as follows:

$$w_{s(\max)} = \frac{c_{p(s)}(100 - T_2)}{h_s} \cong 3\%, \tag{16}$$

where $c_{p(s)}$ = 0.5 kcal/kg·C is steam specific heat capacity, h_s = 540 kcal/kg is steam specific heat value at the normal (atmospheric) pressure.

5.1 Spraying water specific flow rate

With author's reference to [35-37] the tire (rubber) pyrolysis specific heat required for its thermal destruction at near to steady pyrolysis temp 400-450°C is experienced approximately as $h_t = 640$ kJ/kg. Proposing the oil condensed is max 45% of tire pyrolysis mass, we have the next condenser heat capacity provided with spraying and evaporating water:

$$Q_{oil} = 0.45 G_t h_t \,. \qquad (17)$$

Heat with pyrogas condensing for oil and co-pyrolysis steam from reactor cooling down to Venturi operating temp 100°C is following:

$$Q_1 = Q_{oil} + c_{p(s)} G_{s1} (T_p - 100)\,, \qquad (18)$$

heat for spray water heating-evaporating is following:

$$Q_3 = G_w [c_{p(w)} (100 - T_a) + h_s]\,, \qquad (19)$$

heat for steam wetness evaporating is following:

$$Q_2 = w_s G_{s2} h_s = w_s r G_w h_s \,, \qquad (20)$$

steam jet as ratio of spraying water rate is following:

$$G_{s2} = rW \,, \qquad (21)$$

and condenser total heat-with-mass balance is following:

$$Q_1 = Q_2 + Q_3 \,. \qquad (22)$$

By substitution-solution (10)-(15) we have:

$$\frac{W}{G_t} = \frac{0.45 h_t + \dfrac{G_{s1}}{G_t} c_{p(s)} (T_p - 100)}{c_{p(w)} (100 - T_a) + h_s (1 - w_s r)} \,. \qquad (23)$$

As noted here initially in 4 and with reference to [35, 36] the co-pyrolysis steam for reactor supply is self-producing after heating the latter and it is formulated, calculated and well tested relatively to tire mass in the next terms and rates:

$$\frac{G_{s1}}{G_t} = 25\text{-}30\%$$

With reference to steam common application for liquid spraying (e.g. for heavy oil fuel combustion), the factor of steam-water spraying mass ratio $r = 1$ is well enough. Taking that and all other given condition above into account, by (23) we have the next analytical solution on the water spraying for venturi condenser performance:

$$\frac{W}{G_t} \cong 0.2$$

5.2 Water droplets spraying and evaporating

With all reference to [39] and other fundamentals, the Nukiyma-Tanasawa equation can be applied for calculation on a liquid droplet diameter sprayed with a high-speed gas flow in venturi tube as follows:

$$d_i = \frac{0.585}{\vartheta_g}\sqrt{\frac{\sigma}{\rho}} + 59.7\left(\frac{\mu}{\sqrt{\sigma\rho}}\right)^{0.45} \cdot \left(\frac{G}{G_g}\right)^{1.5}, \tag{24}$$

where ϑ_g is gas flow velocity, ρ is liquid density and σ is liquid surface tension. Being both of the flows presented in terns of the volume flow rate, the liquid flow G is usually much less then that of spray gas G_g and equation is used conveniently as simplified to the first part only:

$$d_i \cong \frac{0.585}{\vartheta_{spr}}\sqrt{\frac{\sigma}{\rho}}. \tag{25}$$

Proposing the steam pressure could be so much as the jet velocity to be near of sonic as not less $\vartheta_s \cong 500$ m/s (water density is $\rho = 1000$ kg/m3 and water surface tension at near 100°C is a few as not above $\sigma \leq 0.05$ N/m), we have the next the water droplets by steam jet spraying way:

$$d_i \leq \frac{0.585}{500} \cdot \sqrt{\frac{0.05}{1000}} \leq 10 \text{ μm}$$

There are many analytical and experimental data and references on that, concerning the liquid fuel combustion and specially considering the sphere droplet lifetime by its quick-transient heating and vaporizing (gasification) at the flame thermal condition, referencing that for example to [41]. And there is another approach which is all of near the same consideration but at quasi-steady thermal conditions – for example, in [42] concerning the wood particle combustion by its gasification too.

To estimate the time under question we use the latter as a method at quasi-steady thermal conditions, being there is not so evidence as weakness of that relatively to transient analyzing method. So way it can be analyzed by the convenient heat-mass transfer analogy, taking into account the heat for evaporating the water droplet of diameter $d_i = 2r$ is provided under the heat transfer criterion condition $Nu_{\min(i)} = 2$ which with reference to [9] is a minimal criterion value of that for spherical particle at a zero-flow velocity condition when the heat is transferred by the gas conductivity only. Due to near the same condition is proposed for a fine water droplet injecting and moving by steam at near the same flow velocity (zero-flow velocity relatively one other), it can be differentially formulated as bellow, beginning from the heat transfer coefficient by the criterion above and so on:

$$\alpha_{\min(i)} = \frac{Nu_{\min(i)} k_s}{d_i} = 2\frac{k_s}{d_i}. \tag{26}$$

Evaporating surface of the water spherical droplet is following:

$$f_i = \pi d_i^2 , \tag{27}$$

droplet mass evaporated with the surface layer dr is following:

$$dm_i = \rho_i f_i dr , \tag{28}$$

heat for the layer dr above to be evaporated is following:

$$dQ_i = dm_i h_s , \tag{29}$$

the same heat to be transferred during in $d\tau$ is following:

$$dQ_i = \alpha_i f_i (T_p - 100) d\tau . \tag{30}$$

By (26)-(30) substitution and integration we have the next solution on the question in title:

$$d\tau = \frac{\rho_i h_s}{k_s (T_p - 100)} r dr , \tag{31}$$

$$\tau = \frac{d_i^2 \rho_i h_s}{8k_s \left(T_p - T_{oil}\right)} \ln \frac{T_p - 100}{T_{oil} - 100} , \tag{32}$$

where $k_s = 0.02$ kcal/m·hr is steam thermal conductivity at 100 C, $T_{oil} = 105\text{-}110°C$ is pyrolysis oil condensing temp to be proposed, and where the current arithmetic temp difference $(T_p - 100)$ by its integrating in (30) is logically resulted in a mean-logarithmic temp difference ΔT as following:

$$\Delta T = \frac{(T_p - 100) - (T_{oil} - 100)}{\ln \dfrac{T_p - 100}{T_{oil} - 100}} = \frac{T_p - T_{oil}}{\ln \dfrac{T_p - 100}{T_{oil} - 100}} . \tag{33}$$

So way we have the next numerical estimation on the droplet 10 μm evaporating time :

$$\tau = 3600 \cdot \frac{10^{-12} \cdot 1000 \cdot 540}{8 \cdot 0.02 \cdot (400 - 105)} \ln \frac{400 - 100}{105 - 100} \cong 0.015 \text{ sec}$$

With venturi condenser or scrubber under consideration, it is about 50 m/s of gas flow velocity as a minimal value of that to be in the narrow part of the venturi tube for its effective performance. Next in the conic part of the tube the flow is extending with a spherical angle about 10° so as the flow velocity dropping down about by one order as for an ordinary gas pipe to be. In particularly, considering the venturi condenser above, the narrow tube for 1 t tire pyrolysis per hour is proposed to be about 4 inches in diameter, proposing so the length of the conic tube to be not less then $L2$ = 1 m (see Fig.3). The minimal exposition time for water droplets evaporation in the venturi tube, even proposing the droplets save the initial flow velocity as 50 m/s above, it is min 0.02 sec which is obtained simply by the conic length divided by the venturi velocity above.

6. Steam: Inner heating, carbon black cleaning & air sealing lock

6.1 Inner heating and increasing the tire pyrolysis rate with steam

The steam counter-feeding effect for the carbon black purification at the end of processing just inside the reactor of screw type is illustrated in Fig.4 where with author's reference to [38] the longitudinal diagram of scrap tire pyrolysis is presented and where a multi-tube reactor is simplified as one line. Steam is well penetrating into the moving-mixing bed of scrap tire simply by its diffusion, as well as into the every porous fragment or particle of that too, acting so for cleaning them off the volatile residue matters at the end of processing, even there is not convective steam flow inside the bed and most of steam is flowing above that. Along with such cleaning there is evidently some of inner heating the scrap tire by steam diffusion into the bed which question is a quite easy for estimation by value of inner specific surface per 1 m3 bulk of scrap tire minimally as $f_i =$ 20 m2/m3, including the standard mean-average thickness of chips $d_i = 10$ mms, bulk density of tire scrap $\rho_{t(b)} = 500$ kg/m3 and Nusselt number for particle heating without convection is $Nu_{\min(i)} =$ 2 to be taken into account:

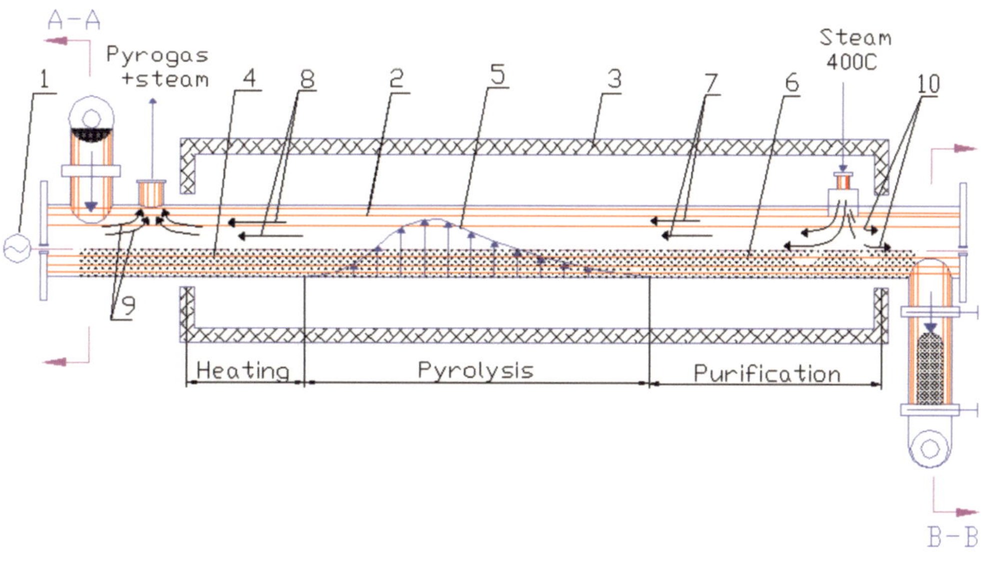

Fig. 4. The longitudinal and linearized diagram of scrap tire pyrolysis recycling in reactor of screw-tubular type with steam (the helix of the screw inside reactor is not shown):
1 – geared motor of screw, 2 – reactor tube shell, 3 – reactor heating box, 4 – scrap tire,
5 – release of tire volatile matters, 6 – carbon black, 7 – cleaning steam flow, 8 – resulting steam flow with pyrolysis gases (pyrogas), 9 – some of possible and allowed air inflow at reactor loading side, 10 – steam counter-flow pulse impact toward the air inflow as a steam seal-lock at the reactor unloading side.

$$\sum F_t = f_t \frac{G_t}{\rho_{t(b)}}, \tag{34}$$

$$\alpha_i = \frac{\mathrm{Nu}_{\min(i)} k_{ss}}{d_i}, \tag{35}$$

$$\Delta T = \frac{T_p - T_a}{ln \frac{T_{ss} - T_a}{T_{ss} - T_p}}, \tag{36}$$

$$\Delta Q_t = \alpha_i \sum F_t \Delta T, \tag{37}$$

where the inner tire heating by steam ΔQ_t to be provided by steam superheating as ΔQ_{ss}:

$$\Delta Q_t = \Delta Q_{ss}, \tag{38}$$

$$\Delta Q_{ss} = c_{p(s)} G_s (T_{ss} - T_s), \tag{39}$$

and the heat for tire pyrolysis is formulated as before too:

$$Q_t = G_t \left(c_{p(t)} (T_p - T_a) + h_t \right). \tag{40}$$

With reference to [38] the numerical data on that are presented in Table 4 where it is calculated by (34)-(40) per 1 t/hr tire rate at the minimal thermal pyrolysis condition as follows: ambient air temperature is $T_a = 20$°C, boiler steam temperature is $T_s = 100$°C, tire pyrolysis temperature is $T_p = 350$°C, steam super-heating temperature is $T_{ss} =$ 400°C, steam feeding rate is $G_s = 300$ kg/hr, tire pyrolysis heat is $h_t = 630$ kJ/kg and others by the nomenclature:

$\sum F_t$, m²	α_i, W/(m²·°C)	ΔT, °C	ΔQ_t, kW	ΔQ_{ss}, kW	Q_t, kW
40	8	162	52	52	350

Table 4. Numerical Data (34)-(40) on Inner Heating with Steam per 1 t/hr Tire Pyrolysis.

The additional tire inner heating (37)-(39) could be objectively corresponded to increasing the pyrolysis rate, being that compared to the similar tire processing without steam and being all other conditions equaled. Taking into account the calculating accuracy of that as 2-3% the obtained result can be concluded as 10-15% which is a theoretical limit for increasing the tire pyrolysis rate with steam at the given conditions:

$$\frac{\Delta Q_{ss}}{Q_t + \Delta Q_t} = \frac{52}{350 + 52} = 13\%. \tag{41}$$

6.2 Steam feeding rate for reactor air-lock sealing

At last, with reference to [38] there is other effect with steam feeding into reactor of screw type which really acts as a hydrodynamic seal-lock preventing any possible air inflow through the reactor unloading system which is usually proposed to be seal but should be taken in mind as possible to be otherwise too. In any case the steam feed forms a local hydrodynamic counter-pressure pulse (steam seal-lock) which precisely keeps air from entering the reactor. With purpose of the uniform steam inlet and sealing impact all over the reactor cross-section in-side, it is feeding into there via the multi-jet deflector as shown in Fig.5. The same is shown by dashed arrows as a steam counter-flow acting for sealing toward the possible air inflow in the apposite direction in Fig.4. The steam pulse above is formulated usually as its dynamic pressure depended on the flow velocity:

$$\Delta p_1 = \frac{\vartheta_{ss}^2}{2}\rho_{ss}, \quad (42)$$

where the velocity is to be calculated by half of reactor cross-section square whose second half is filled with scrap tire initially and carbon black finally (see Fig.5):

$$\vartheta_{ss} = \frac{2G_s}{S_r \rho_{ss}} = \frac{2g_s}{\rho_{ss}}. \quad (43)$$

Proposing the reactor unloading system would be not sealed with a double-gate or double-flap valve etc., it would be a chimney draft effect acting as a static low-pressure by the temp difference between the reactor inside and outside which is additionally depended on the height of the reactor installation as shown in Fig.5:

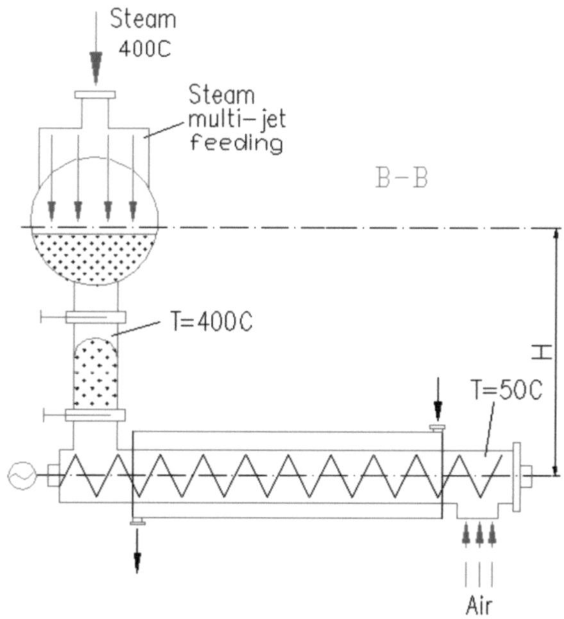

Fig. 5. Reactor unloading system with a water-cooling screw and double-gates as for consideration on the steam sealing effect against the possible air inflow from outside.

$$\Delta p_2 = \rho_a g H \left(1 - \frac{T_a}{T}\right), \tag{44}$$

which is to be equal to the dynamic pressure (42) with steam feeding above and by which substitution the steam feeding rate under question is following:

$$g_{s1} = \frac{G_{s1}}{S_r} = \sqrt{\frac{g\rho_a\rho_{ss}H}{2}\left(1 - \frac{T_a}{T}\right)}. \tag{45}$$

With reference to [38], as well as simply and evidently by equation (45), the steam feeding-sealing effect is to be objectively enhanced with cooling of the carbon discharge from reactor, being the opposite draft effect for air inflow is depended on the cooling temp just by another way. The numerical data on that are presented in Table 5 and theoretically it even would be nothing of sealing required if the carbon discharge temp could be entire equal to outside so as nothing of chimney effect appeared. At the same time in Table 6 there are numerical data on the steam feed required as obtained by (45) at the different ambient temperature which is other factor of the chimney draft effect simply by outside condition of the pyrolysis plant location in different region. With the same reference to [38] it was well experienced practically and particularly in Taiwan at about 30°C where it was no evidence of the air penetration inside with the steam feeding rate appropriated minimally as 200–250 kg/hr for the reactor diameter 0.6 m, or near 1 t/hr per 1 m² of cross-section square of that in specific terms.

carbon cooling temp T, °C	400	300	200	100	50
steam feeding rate by (45), t/(m²·hr)	3.7	3.45	3.0	2.3	1.55

Table 5. Steam Feeding Rate per 1 m2 Reactor Cross-Section Square for Air-Lock Seal With Carbon Product Cooling Temp for Discharge.

ambient air temp T_a, °C	30	20	10	0	-10
steam feeding rate by (45), t/(m²·hr)	1.07	1.55	1.74	1.94	2.12

Table 6. Steam Feeding Rate per 1 m² Reactor Cross-Section Square for Air-Lock Seal With Air Ambient Temperature Outside Reactor.

7. Conclusion: Brief engineering-economy analysis on steam use way

In conclusion, in connection with possibility for steam self-producing along with tire pyrolysis recycling, it is a reason to analyze numerically and economically what is more effective way for steam use: power generation by turbine machine or carbon black production could be more as min 10% of tire rate additionally by inner heating with steam feeding into reactor as it is considered above in 6.1? By first way we have near to the next thermal data for steam generation after heating the pyrolysis reactor per 1000 kg/hr tire processing:

Pyrolysis off-gas combustion heat value (pure-dry gas as without steam) = 39 MJ/m3;
Heat capacity by max 18% off-gas after-burning (relatively to tire mass) = 1.7 MW;
Heat emission to outside with both of furnace and reactor by 5% totally = 0.1 MW;
Heat consumption for tire heating and pyrolysis (process without steam) = 0.35 MW
Heat capacity for 1000 kg/hr steam super-heating to 400°C (for turbine) = 0.175 MW
Heat residual and available for steam generation after all of these above = 1.075 MW;
Exhaust gas flow temperatures inlet/outlet the steam second-heat boiler = 600/200°C

Thermal efficiency of the boiler: $E = \frac{600-200}{600-20} \cong 70\%$;

Heat for steam generation: 0.7x1.075 MW = 0.7525 MW;
Steam rate (by index 1400 kg/hr per 1 MW) = 1050 kg/hr;
Thermal efficiency of small steam turbine (max) = 35%;

By the steam enthalpy operating range in turbine as max h1 = 770 kcal/kg (4.0 MPa, 400°C) and h2 = 570 kcal/kg (0.05 MPa, 40°C), it is the next power per 1000 kg/hr tire processing:

1050 x (770–570) x 0.35 x 1.16 = 0.085 MW

Proposing as max $0.06 price per 1 kW-hr power, we have the next economy if to sale that:

$0.06 x 85 = $5.1

By min 10% waste tire pyrolysis rate to be more by steaming way and with reference to [32] as the carbon black price is min $0.3 per 1 kg, we have the next economy if to sale that more:

Carbon black (per 1000 kg/hr pyrolysis) = 350 kg/hr
Carbon black rate additionally: 0.1 x 350 = 35 kg/hr
Economy for sale the carbon black: $0.3 x 35 = $10.5

8. Acknowledgement

The author wishes sincerely and friendly to thank Mr. Wu Chun Yao (Morgan) and Mr. Horng Jiang (Rhine) for assistance and good co-operation in Taiwan.

9. Nomenclature

w – wetness of components, %
ρ – density of components, kg/m^3;
μ – molecular weight of components;
m – mass of steam or gas components, kg;
p – pressure of steam-gas components, bar;
V – volume of steam-gas components, m^3;
c_p – specific heat of components, J/kg·°C;
T – temperature of components and others, °C;
T_{g1} – furnace gas temp for reactor heating inlet, °C;
T_{g2} – furnace gas temp for reactor heating outlet, °C;

T_{g3} – furnace gas temp for boiler heating inlet, °C;
T_p – pyrolysis reactor operating temp inside, °C;
T_a – ambient air temperature outside, °C;
G – mass flow rate of components, kg/hr;
G_t – tire pyrolysis mass flow rate, kg/hr;
G_{s1} – steam feeding rate into pyrolysis reactor, kg/hr;
G_{s2} – steam injecting rate into Venturi condenser, kg/hr;
G_g – furnace (combusted) gas mass flow rate, kg/hr;
W – spraying water rate for Venturi condenser, kg/hr;
d_i – spraying water droplet diameter, m;
GAS – off-gas flow rate relatively to tire, %;
OIL – oil condensing rate relatively to tire, %;
AIR – air mass flow rate relatively to tire, %;
STEAM – steam mass flow rate relatively to tire, %;
SLU – oil slurry mass rate relatively to tire, %;
Q – heat flow capacity of components, MW
Q_{oil} – condensed oil fuel heat value, MJ/kg;
Q_{gas} – incondensable off-gas heat value, MJ/m³;
Q_f – combustion furnace heat capacity, MW.
Q_t – heat capacity for tire pyrolysis, MW;
Q_s – heat capacity for steam generation, MW;
Q_{ss} – capacity for steam super-heating, MW;
Q_r – pyrolysis reactor total heat capacity, MW;
q – specific heat per 1 kg components, MW/kg
h_t = 640 – tire pyrolysis specific heat, kJ/kg;
h_s = 2260 – steam specific heat value, kJ/kg;
R = 8314 – ideal gas constant value, J/kg·K

10. References

[1] Solbakken A. et al. Process for Recovering Carbon Black & Hydrocarbons from Used Tires. U.S. Patent 4 250 158 (1981).
[2] Solbakken A. et al. Process for Recovering Carbon Black & Hydro-carbons from Used Tires. U.S. Patent 4 284 616 (1981).
[3] Cha Chang Y. et al. Pyrolysis Method with Product Oil Recycling. U.S. Patent 4 983 278 (1991).
[4] Ledford C.D. Process For Conveying Old Rubber Tires Into Oil & Useful Residue. U.S. Patent 5 095 040 (1992).
[5] Wu Arthur C. & Chen Sabrina C. Thermal Conversion Pyrolysis Reactor System. U.S. Patent 5 411 714 (1995).
[6] Kanis D.R. Pyrolysis System and Method of Pyrolyzing. U.S. Patent 5 636 580 (1997).

[7] Avetisian V. et al. Tire Liquefying Process Reactor Discharge System & Method. U.S. Patent 5 705 035 (1998).
[8] Meador W.R. Tire Liquefying Process Reactor Discharge System & Method. U.S. Patent 5 720 232 (1998).
[9] Zhuravsky G.I. et al. Method of Treating Plastic Waste. U.S. Patent 5 771 821 (1998).
[10] Mulyarchik V.M. et al. Processing Waste Rubber by Steam. U.S. Patent 5 780 518 (1998).
[11] Chang Y. Cha et al. Process for Co-Recycling Tires & Oils. U.S. Patent 5 735 948 (1998).
[12] Bouziane R. Pyrolyzing Apparatus. U.S. Patent 5 820 736 (1998).
[13] Denison G.W. Method & System For Recovering Marketable & Products From Waste Rubber. U.S. Patent 5 894 012 (1999).
[14] Takegawa T. et al. Method For Recovering Carbon Black from Waste Rubber Such As Tires & Apparatus Therefore. U.S. Patent 5 961 946 (1999).
[15] Faulkner et al. Pyrolysis Process For Reclaiming Desirable Materials From Vehicle Tires. U.S. Patent 6 221 329 B1 (2001).
[16] Bullok B.P. Hydrocarbon Conversion Apparatus & Method. U.S. Patent 6653517 (2003).
[17] Holden H.H. & Holden H.S. Continuous Temperature Variance Pyrolysis for Extracting Products From Tire Chips. U.S. Patent 6 657 095 B1 (2003).
[18] Massemore B. & Zarrizski R. Process for Pyrolizing Tire Shreds & Tire Pyrolysis System. U.S. Patent 6 736 940 B2 (2004).
[19] Platz G.M. et al. Resource Recovery of Waste Organic Chemicals By Thermal Catalytic Conversion. U.S. Patent 6 683 227 (2004).
[20] Nichols R.E. Low Energy Method of Pyrolysis of Hydrocarbon Materials Such As Rubber.U.S. Patent 6 833 485 (2004).
[21] Day M. et al. Pyrolysis of Automobile Shredder Residue: Analysis of the Products of Commercial Screw Kiln Process. *Journal of Analytical & Applied Pyrolysis.* 1996. Vol. 37, pp. 49-67.
[22] Simon C.M., Kaminsky W. & Schlesselmann B. Pyrolysis of Polyolefins With Steam To Yield Olefins. *Journal of Analytical & Applied Pyrolysis.* 1996. Vol. 38, pp. 75-87.
[23] Cunliffe A.M. et al. Composition of Oils Derived From The Batch Pyrolysis of Tires. *Journal of Analytical & Applied Pyrolysis.* 1998. Vol. 44, pp. 131-152.
[24] El-harfi K., Mokhlisse A. & Ben-Chanaa M. Yield & Composition of Oil Obtained By Isothermal Pyrolysis of the Moroccan Oil Shale With Steam Or Nitrogen as Carrier Gas. *Journal of Analytical & Applied Pyrolysis.* 2000. Vol. 56, pp. 207-218.
[25] Teng H., Lin Yu-Chuan & Hsu Li-Yeh. Production of Activated Carbon From Pyrolysis of Waste Tires Impregnated With Potassium Hydroxide. *Journal of the Air & Waste Management Association.* 2000. Vol.50, No.11, pp. 1940-1946.
[26] Kaminsky W. & Mennerich C. Pyrolysis of Synthetic Tire Rubber in Fluidized Bed Reactor To Yield Butadiene-Styrene Oil & Carbon Black. *Journal of Analytical & Applied Pyrolysis.* 2001. Vol. 58-59, pp. 803-811.
[27] Zabaniotou A.A. & Stavropoulos. Pyrolysis of Used Automobile Tires & Residual Char Utilization. *Journal of Analytical & Applied Pyrolysis.* 2003. Vol. 70, pp. 711-722.
[28] Williams P.T., Brindle A.J. Fluidized Bed Pyrolysis & Catalytic Pyrolysis of Scrap Tires. *Environmental Technologies.* 2003. Vol. 24, No. 7, pp. 921-929.
[29] Barbooti M.M. et al. Optimization of Pyrolysis Conditions of Scrap Tires Under Inert Gas Atmosphere. *Journal of Analytical & Applied Pyrolysis.* 2004. Vol. 72, pp. 165-170.

[30] Shen B., Wu C., Wang R., Guo B. & Liang C. Pyrolysis of Scrap Tires with Zeolite USY. *Journal of Hazardous Materials.* 2006. Vol. 137, No. 2, pp. 1065-1073.

[31] Zhang X., Wang T. & Chang J. Vacuum Pyrolysis of Waste Tires with Basic Additives. *Waste Management.* 2008. Vol. 28, No. 11, pp. 2301-2310.

[32] Kiser J.V. Scrap Tire Pyrolysis: Impossible Dream? *Scrap Magazine.* 2002. Vol. 59, No. 5. pp. 34-41.

[33] Zhuravskii G.I., Matveichuk A.S. & Falyushin P.L. Obtaining of Fuels Based On Products Of Organic Waste Steam Thermolysis. *Journal of Engineering Physics and Thermophysics.* 2005. Vol. 78, No. 4, pp. 684-689.

[34] Kalitko V.A. Steam-Thermal Recycling of Tire Shreds: Calculation on the Explosion-Proof Steam Feeding Rate. *Journal of Engineering Physics and Thermophysics.* 2008. Vol. 81, No 4, pp. 781-786.

[35] Kalitko V. A., Morgan Wu Chun Yao et al. Steam Thermolysis of Discarded Tires: Testing and Analyzing of the Specific Fuel Consumption with Tail Gas Burning, Steam Generation and Secondary Waste Slime Processing. *Journal of Engineering Physics and Thermophysics.* 2009. Vol. 82, No 2, pp. 236-245.

[36] Kalitko U. & Morgan Chun Yao Wu. Tire Scrap Pyrolysis Recycling By Steaming Way: Heat-Mass Balance Solutions and Developments. In: *Pyrolysis: Types, Processes, Industrial Sources and Products.* Edited by W.S. Donahue and J.C. Brandt. Nova Science Publishers. NY, 2009, pp. 79-115.

[37] Kalitko V.A. Tire Shreds Steam-Thermal Recycling Process Modernization and Development by Inherent Gas Burning with Steam. *Journal of Engineering Physics and Thermophysics.* 2010. Vol. 83, No 1, pp. 179-187.

[38] Kalitko V.A. A Thermal-Hydrodynamic Lock Sealing with Steam Feeding for Tire Scrap Pyrolysis in Reactor of Screw Type. *Journal of Engineering Physics and Thermophysics.* 2010. Vol. 83, No 2, pp. 324-330.

[39] Perry's Chemical Engineers' Handbook. 8th ed./prepared by a staff under the editorial direction by D. W. Green, editor-in-chief, R. H. Perry, late editor. Printed in China by Copyright TP151. 2007.

[40] Leonard John H. A Heat Transfer Textbook. Original American edition published by Prentice-Hall Inc., NJ, USA. Published and Reprinted by JWANG JUAN PUBLISHING CO., Taipei, Taiwan, 1981.

[41] Schiller D., Li J. & Sirignano W.A. Transient Heating, Gasification and Oxidation of Energetic Liquid Fuel. *Combustion and Flame.* 1998. Vol.114, Issue 3-4, pp. 349-358.

[42] Ouedraogo A., Mulligan J.C. & Cleland J.G. A Quasi-Steady Shrinking Core Analysis of Wood Combustion. *Combustion and Flame.* 1998. Vol.114, Issue 1-2, pp.1-12.

10

Study on the Feasibility of Hazardous Waste Recycling: The Case of Pharmaceutical Packaging

Vincenzo Gente[1] and Floriana La Marca[2]
[2]*Environmental Engineer, Rome*
[2]*DICMA, Sapienza University of Rome*
Italy

1. Introduction

Hazardous waste management should fulfil the following three main goals: (i) to protect human health and the environment, (ii) to reduce waste while conserving energy and natural resources and (iii) to reduce or eliminate the volume of waste to dispose of. The last two of these goals may derive from recycling, which aims at reducing raw materials and energy consumption and decreasing the volume of waste materials that must be treated and disposed of.

However, recycling must be conducted in a safe way, ensuring human health and environment protection. Recycling activities should be regulated at a different degree on the basis of the risk they cause to human health and the environment. A hazardous waste destined for recycling must be identified by type and recycling process in order to determine its level of regulation (Linninger & Chakraborty, 2001).

Pharmaceutical packaging represents a very small percentage of hazardous waste, but its management can cause problems for the environment, depending on the type of packaging waste is concerned (Sacha et al., 2010). Such waste may include:

- uncontaminated waste (assimilated to domestic waste: paper, cardboard, glass, plastic);
- contaminated waste (paper, cardboard, glass, plastic), e.g. waste that has been in contact with cytotoxic products, blood, blood-derived products or radioactive products.

Waste is created at all stages of the supply-chain: production, distribution and use of a pharmaceutical product. At each step, care therefore needs to be taken, either by the manufacturer or the end-user, to protect the environment (Biniecka et al., 2005; Dillon & Rubinstein, 2005).

In several European countries, pharmaceutical manufacturers must dispose of their waste, or by themselves or by external specialized companies, and are encouraged to recover packaging waste. In both cases, waste management represents a considerable cost for the manufacurers.

The use of environmental-friendly packaging (i.e. recyclable or degradable packaging) has to be considered. Valuable packaging materials, such as aluminium paper, glass and plastic materials, can been extensively recycled if they have not been in contact with toxic or dangerous substances (Bauer, E.J. , 2009).

This chapter is focused on a feasibility study for the management of packaging waste from a pharmaceutical plant, considering the following phases:

- waste materials characterization;
- preliminary tests on waste processing;
- set up of size reduction (comminution) operations.

Experimental tests have been executed on several typologies of packing, as listed:

- primary packaging:
  - bottles in high density polyethylene (HDPE), for suspension to be reconstituted;
  - bottles in poly(ethylene terephthalate) (PET), for syrup;
  - plastic bags and films of varying composition and thickness;
- pharmaceutical waste:
  - flexible multi-layered (plastic and aluminium) sachets containing granular medicine.

2. Experimental

In a first stage, the products under investigation were characterized as derived from the industrial process. The results of the characterization were utilized to set up preliminary tests on waste processing, in particular the comminution operations were evaluated. Finally, an experimental plan was carried out to assess the feasibility of waste recycling.

2.1 Methods for characterization

The following methods were adopted in the experimental set-up for the characterization of products under investigation:

- image analysis, to measure geometric and morphologic characteristics; the results were evaluated by statistical methods;
- dry sieve analysis, to classify the size distribution of particles;
- laser granulometry, to classify size distribution of particles in the interval between 0.1 e 1,000 μm;
- infrared spectroscopy, to recognize chemical composition of polymeric materials under investigation.

2.1.1 Image analysis

The images for characterization were acquired by the stereoscopic microscopy Leica Wild M8 and a by a digital camera Olympus C-5060 Wide Zoom. The image analysis was carried out by the software SigmaScan Pro© Version 5.0.0 (Systat Software Inc., 2007), which provides a complete set of tools to analyse structure and dimension of an object's image.

Firstly, calibration allowed to convert image unit from pixel to millimetre (Figure 1). After calibration, the image quality for data elaboration has been enhanced, increasing the distinction between particles and background, by varying contrast, brightness and colour of the image (Figure 2). In the measurement process, the software automatically recognizes objects on the image (Figure 3) and computes geometric and morphologic parameters, accordingly to operator's choice.

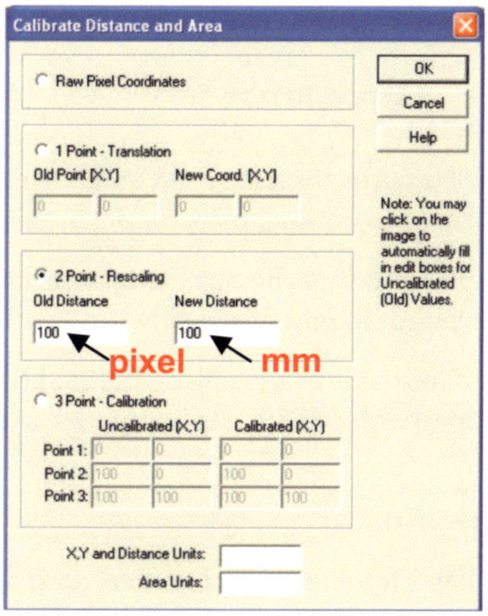

Fig. 1. Calibration process to convert image unit from pixel to millimetre.

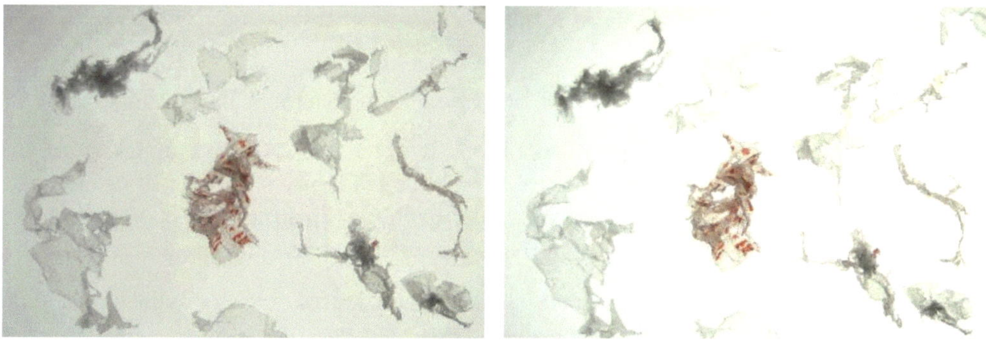

Fig. 2. Variation of contrast, brightness and colour of an image.

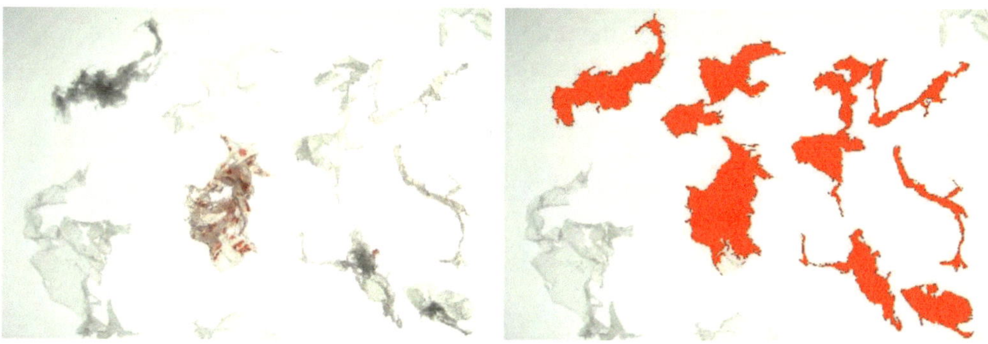

Fig. 3. Automatic recognition of objects on the image.

The following geometric and morphologic measurements were considered.

Area: reports the area in mm2 for the selected object.

Compactness: reports a numeric non-dimensional measurement of the shape of an object. It is defined as the perimeter squared, divided by the area:

$$\text{Compactness} = \frac{\text{perimeter}^2}{\text{area}} \tag{1}$$

The minimum Compactness of a perfectly measured and digitized circle is 4π (about 12.57). As an object tends toward the shape of a line, the Compactness tends towards infinity.

Major Axis Length: calculates major axis of the object (defined by the two most distant points on the object) and reports the length in mm of the axis.

Minor Axis Length: calculates minor axis of the object (defined by the two most distant points on the object that creates a line perpendicular to the major axis) and reports the length in mm of the axis.

Perimeter: returns the perimeter in mm of an object.

Shape Factor: measures the shape (circularity) of a measured object. This non-dimensional measure is defined as 4π times the object's area divided by the perimeter squared:

$$\textit{Shape factor} = \frac{4\pi \cdot area}{perimeter^2} \tag{2}$$

A perfect circle will have a Shape Factor of 1. A line's Shape Factor will approach zero.

Feret Diameter: describes the shape of an object. It gives the diameter of the equivalent circular object that has the same area as the current object. For each object, it calculates the theoretical diameter of the object if it were circular in shape. This measure is often compared with an object's major and minor axes lengths to create new shape parameters.

The results obtained from image analysis were evaluated by considering statistical parameters, as described in the following.

Number of objects: counts the numeric values in the considered set.

Mean: returns, as central tendency, the order of magnitude of the value for each considered measurement. The arithmetic mean $\bar{x}$ is calculated as the sum of all measurements divided by the number of observations in the data set:

$$\bar{x} = \frac{1}{N}\sum_{i=1}^{N} x_i \tag{3}$$

where x_i is the single measurement and N is the total number of measurements.

Minimum: returns the least value of the considered data set.

Maximum: returns the greatest value of the considered data set.

Standard deviation: shows how much variation or dispersion there is from the mean in the data set and it is measured in the same unit of the data. The standard deviation σ is directly derived from the variance (σ^2, which unit is the square unit of the considered data), as its square root:

$$\sigma_x = \sqrt{\frac{\sum_{i=1}^{N}(x_i - \overline{x})}{N}} \tag{4}$$

where $\overline{x}$ is the arithmetic mean.

Standard error: returns an estimation of the standard deviation σ_x of the estimator, giving a valuation of its imprecision. If the estimator is the arithmetic mean of N independent measurements with the same statistical distribution, the standard error is given via the equation:

$$se = \frac{\sigma_x}{N}. \tag{5}$$

Confidence interval: refers to the range of values preceding or following a mean value where it is expected an unknown population parameter (e.g. the true mean) is located. The width of the confidence interval gives an indication about uncertainty of the unknown parameter. If independent samples are taken repeatedly from the same population, and a confidence interval calculated for each sample, then a certain percentage (confidence level) of the intervals will include the unknown parameter. The confidence level is the probability value (1 - α) associated with a confidence interval. For example, say α = 0.05, then the confidence level is equal to 0.95, i.e. a 95% confidence level.

Let's be the true mean the considered unknown parameter; the confidence level is given by:

$$\overline{x} \pm A_{conf}\left(\frac{\sigma_x}{\sqrt{N}}\right) \tag{6}$$

where A_{conf} is area under the normal distribution curve that is equal to the chosen confidence level. In the case under investigation confidence levels of 95% and 99% have considered.

2.1.2 Dry sieve analysis

The dry sieve analysis is a mechanical method to assess the particle size distribution. A set of sieves with wire mesh cloth is stacked in column, so that each lower sieve has smaller openings than the one above; at the base of the column there is a round pan. A representative weighed sample is poured into the top sieve. The column is typically placed in a mechanical shaker, that shakes the column for a fixed amount of time. After the shaking is complete the material on each sieve is weighed and divided by the total weight to gain the percentage retained on each sieve. In this study, certified high-precision sieves Giuliani in stainless steel (ASTM series) were utilized.

2.1.3 Laser granulometry

The laser granulometry analyses of the effect of diffracted light produced by a laser beam passing through a dispersion of particles. The angle of diffraction increases as particle size decreases. After mixing in distilled water (or alcohol), the representative sample is introduced in the measuring cell. The laser beam (wavelength = 632.8 nm; power = 5 mW) passes through the suspension and is deviated by particles accordingly to their particle size. The deviation is then analyzed by detectors. This method can measure particle sizes between 0.1 and 1,000 μm.

The laser granulometer utilized in this investigation was SYMPATEC HELOS/KA.

2.1.4 FTIR spectroscopy

The Fourier transform infrared (FTIR) spectroscopy utilizes the infrared region of the electromagnetic spectrum (between 0.8 and 1,000 μm wavelength) to recognize chemical composition of materials. In the case of plastic materials it allows to identify the structural polymer. The infrared spectrum (by transmittance or absorbance) of a sample is recorded by passing a beam of infrared light through the sample. A data-processing technique called Fourier transform converts raw data into the sample's spectrum. Then the sample's spectrum is compared to reference spectra. The samples were cleaned with water and mild detergent, rinsed with deionized water and then dried in oven with air convection at 450 °C for 24 hours.

The characteristics of the instrument utilized in this study for FTIR spectroscopy are:

- FTIR Perkin-Elmer SpectrumOne;
- equipped with HATR, crystal ZnSe, 45°, (pressure 90) ;
- wavenumber range: 4000-630 cm^{-1};
- resolution: 4 cm^{-1};
- number of scanning: 4.

2.2 Methods for waste processing

Waste processing was carried out at laboratory scale to assess the feasibility of recycling. In particular, for the treatment of the different typologies of investigated pharmaceutical waste, comminution operations were evaluated. According to the composition of products (polymeric materials) cutting mills were employed, which apply shearing to reduce particle size. In this study a cutting mill Retsch - SM 2000 equipped with interchangeable sieves to control particle size in output product was utilized to carry out comminution tests.

2.3 Materials

The bottles in HDPE are utilized for suspension to be reconstituted. The analyzed samples are composed by waste bottles, collected at the end of the production line and manually emptied. The bottles are without labels and caps and may contain residual powder. A synthesis of HDPE bottles characteristics is reported in Table 1.

Material	HDPE
Longitudinal dimension (cm)	11.0
Transversal dimension (cm)	5.5
average weight (g)	17.4
average powder content (mg)	40.0
average powder content (%)	0.05

Table 1. Synthesis of the characteristics of HDPE bottles.

The bottles in PET are utilized for syrup packaging. The analyzed samples are composed by waste bottles, collected at the end of the production line and manually emptied. The bottles may have labels and aluminium caps and may contain varying amount of residual syrup. A synthesis of PET bottles characteristics is reported in Table 2.

Material	PET
Longitudinal dimension (cm)	14.2
Transversal dimension (cm)	5.2
average weight (g)	23.2
average syrup content (mg)	19.0
average syrup content (%)	8.99

Table 2. Synthesis of the characteristics of PET bottles.

Plastic bags and films derive from the packaging of raw materials utilized in production processes. The analyzed samples are of varying composition and thickness, and contain residual powders, whose composition is in relation to the production cycle. Four different typologies of plastic bags and films were identified:

- white and red bags, containing bicarbonate;
- thin-film;
- bags, with printed character "A" in black;
- bags, with printed character "A" in blue.

The samples were analysed by FTIR spectrometry to identify the polymeric composition (Figure 4).

As resulting from the comparison of acquired spectra with the reference one, all 4 types of materials are polyethylene (PE), in particular the recognized polymeric structure is low-density polyethylene (LDPE).

The residual powders were analysed by dry sieve analysis to identify size distribution, characterized by a mode of the distribution equal to 112 μm, while the top-size is lower than 1,000 μm.

The flexible multi-layered (plastic and aluminium) sachets containing granular medicine are wasted at the end of the production line because of incorrect filling. In this case, the sachets are collected and sent to disposal. The number of wasted sachet is $4{\cdot}10^6$ per year on average. Table 3 reports a synthesis of the sachets characteristics.

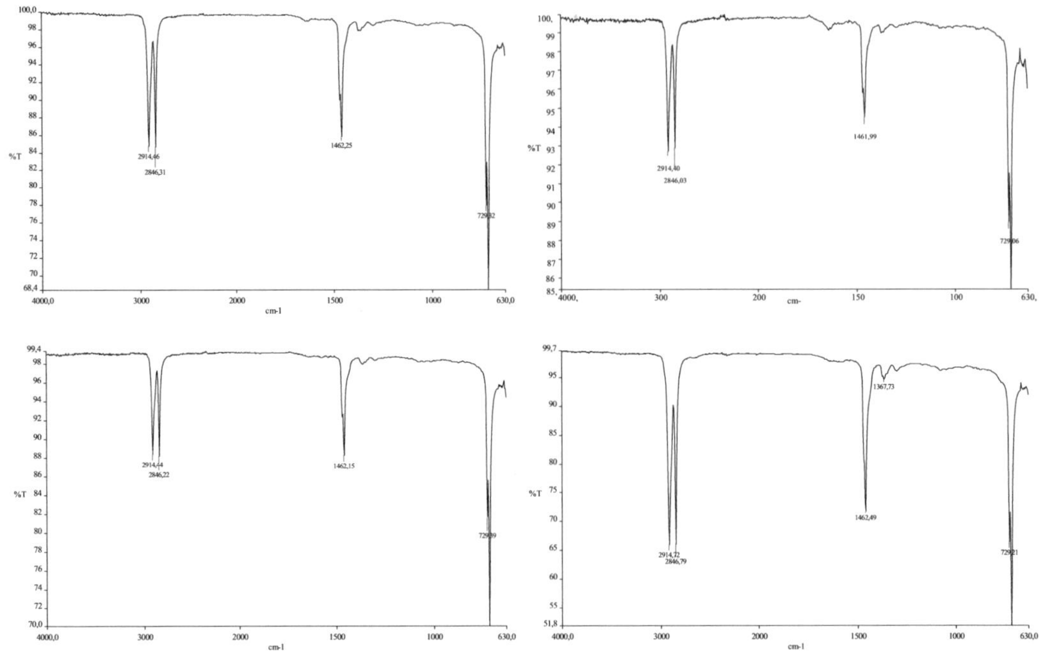

Fig. 4. FTIR spectra of plastic bags and films (y-axis: transmission %; x-axis: wavenumber cm^{-1}) : white and red bags, containing bicarbonate (above left), thin-film (above right), bags, with printed character "A" in black (below left), bags, with printed character "A" in blue (below right).

Material	plastic and aluminium
Longitudinal dimension (cm)	11.8
Transversal dimension (cm)	2.2
average weight of sachet(g)	7.4
average powder content (mg)	3.2
average powder content (%)	0.46

Table 3. Synthesis of the characteristics of flexible multi-layered (plastic and aluminium) sachets.

The granular medicine contained in sachets was analysed by dry sieve analysis and by laser granulometer to identify size distribution. Size distribution was analyzed by dry sieve analysis and laser granulometry (Figure 5), showing different mode and top-size. In particular, laser granulometry shows lower value of both (mode: 40 µm, top size: 100 µm) than dry sieve analysis (mode: 280 µm, top size: 1000 µm). This is probably due to the break-up of aggregated granules during mixing in water.

2.4 Preliminary tests

For the recovery of the waste materials, in order to evaluate the possibility of adopting an industrial shredder installed in the production plant under investigation, preliminary comminution tests have been carried out on the following waste typologies:

- primary packaging:
  - bottles in high density polyethylene (HDPE), for suspension to be reconstituted;
  - bottles in poly(ethylene terephthalate) (PET), for syrup;
  - plastic bags and films of varying composition and thickness;
- pharmaceutical waste:
  - flexible multi-layered (plastic and aluminium) sachets containing granular medicine.

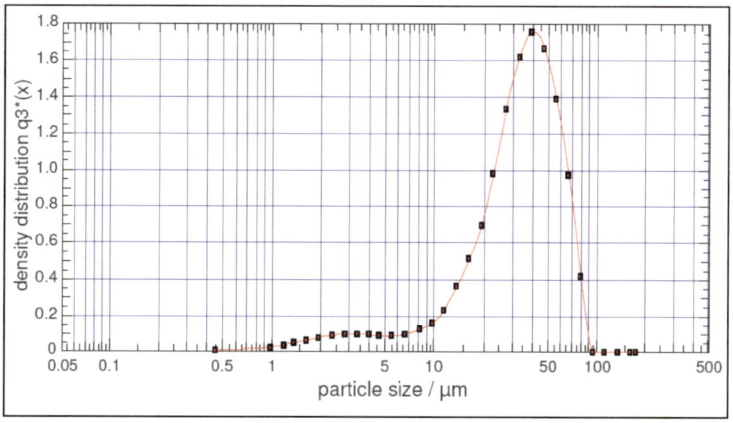

Fig. 5. Size distribution of granular medicine contained in sachets by laser granulometry.

The technical details of the industrial shredder are shown in the Table 4.

Producer	Satrind S.p.A.
Model	F615
Engine power	11 kW
n. shafts	2
Speed shafts	19/10-15/8 rpm
n. blades of 30 mm	19

Table 4. Technical characteristics of industrial shredder installed in the investigated plant.

On the two shafts of the shredder 19 blades are fixed that, thanks mainly to the application of cut stresses, are able to break the waste materials (Figure 6). Some of the material is broken by tear action due to the rotation of the blades. In the preliminary lab tests, the application of cut stresses have been obtained by mean of a blade mill RETSCH - SM 2000, that can be equipped or not with different grids that allow to control the size of the comminuted products. The comminution chamber of the mill is shown in Figure 6.

The preliminary laboratory tests have been conducted in dry conditions adopting two different operational configurations:

- without the grids for the control of the size of the comminuted products;
- with a 20 mm mesh grid.

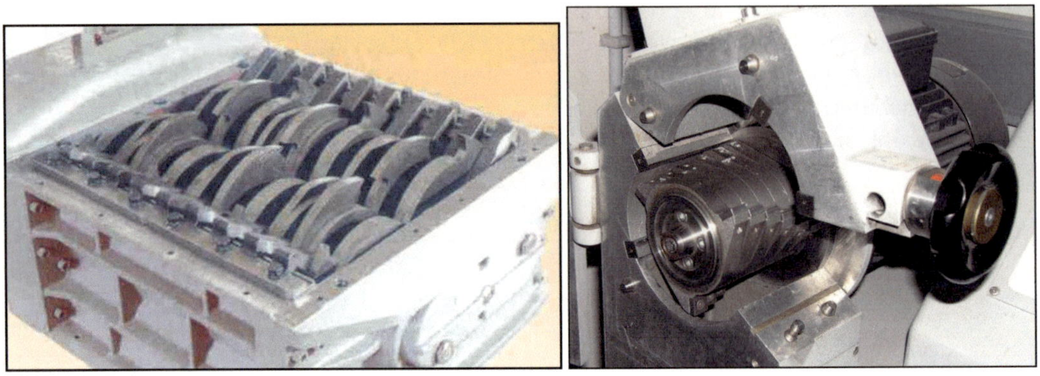

Fig. 6. Industrial shredder blades (left) and comminution chamber of the blade mill adopted in the preliminary tests (right).

2.4.1 Results of the preliminary laboratory tests

The preliminary lab tests have shown the effectiveness of the application of cut stresses to break the considered typologies of pharmaceutical waste materials. Moreover, the comminuted products obtained in the tests are characterised by an average lower size that is above the higher average size of the powder and granular medicine contained in the waste sachet (1.0 mm). The results obtained in the preliminary tests are reported for each considered waste typologies in the following.

- *Bottles in high density polyethylene (HDPE), for suspension to be reconstituted*

The bottles have been divided in two parts in order to reach dimensions suitable for the laboratory blade mill.

The tests carried out without the adoption of the control grids did not produce useful results, as no breakage were observed in the bottles collected in the output.

On the contrary, in the tests carried out with the adoption of a 20 mm mesh control grid it was possible to break the bottles in HDPE and reach comminuted products mainly belonging to the size class +1.0 mm. The comminuted products are shown in Figure 7.

- *Bottles in poly(ethylene terephthalate) (PET), for syrup*

The bottles have been divided in two parts in order to reach dimensions suitable for the laboratory blade mill. The bottles have been divided in two parts in order to reach dimensions suitable for the laboratory blade mill. The tap and aluminium ring have been kept in the sample. The bottles submitted to the tests did not contain syrup. The tests carried out without the adoption of the control grids did not produce useful results, as no breakage were observed in the bottles collected in the output. On the contrary, in the tests carried out with the adoption of a 20 mm mesh control grid it was possible to break the bottles in PET and reach comminuted products mainly belonging to the size class +1.0 mm. The comminuted products are shown in Figure 7.

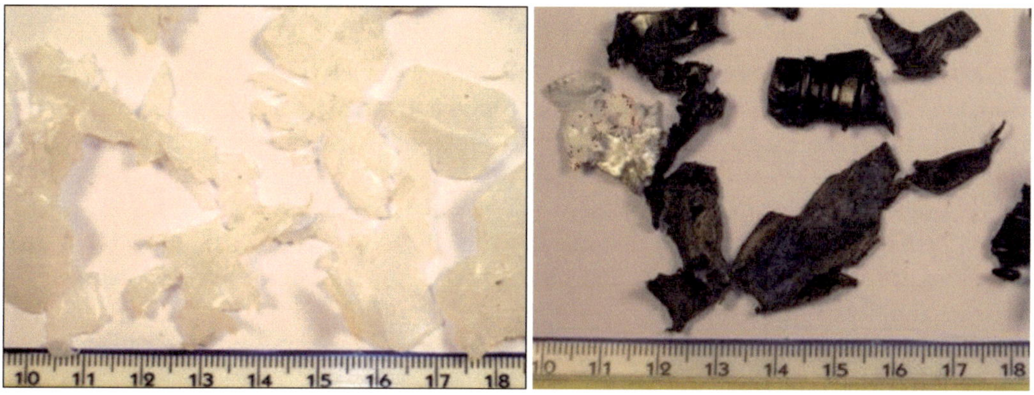

Fig. 7. Comminuted bottles in HDPE (left) and PET (right) obtained in the preliminary tests with the adoption of a 20 mm mesh control grid.

- *Plastic bags and films of varying composition and thickness*

The preliminary comminuted tests have been carried out on samples of plastic bags containing bicarbonate and on samples of films. Both sample typologies are made of LDPE.

The plastic bags and films have been cut in samples of 50×50 mm and 100×100 mm in order to reach dimensions suitable for the laboratory blade mill. The tests carried out without the adoption of the control grids did not produce useful results, as no breakage were observed in the samples of 50×50 mm, and clogging and consequent stoppage of the mill for the samples 100×100 mm took place. On the contrary, in the tests carried out with the adoption of a 20 mm mesh control grid it was possible to break the plastic bags and films, both the 50×50 mm and 100×100 mm samples. The comminuted products of 100×100 mm samples of bags and films are shown in Figure 8. When submitted to sieving classification, the comminuted products presented average size generally above 1.0 mm and, therefore, higher than the higher average size of the powder medicine contained in the bags and films.

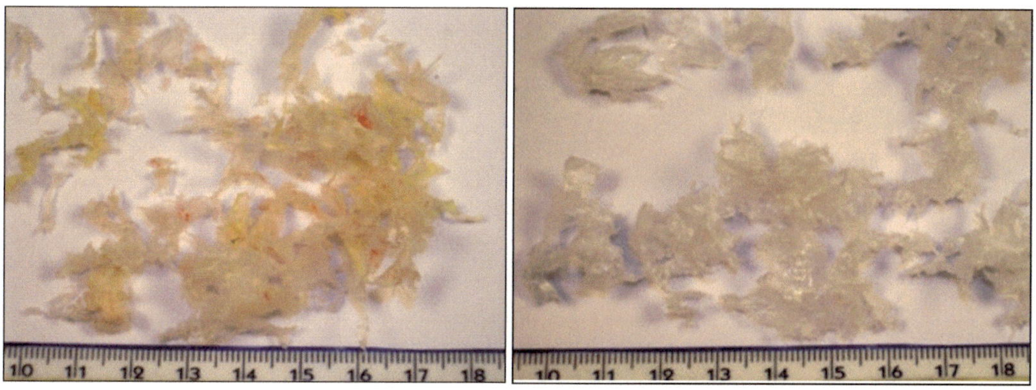

Fig. 8. Comminuted 100×100 mm samples of bags (left) and films (right) obtained in the preliminary tests with the adoption of a 20 mm mesh control grid.

- *Flexible multi-layered (plastic and aluminium) sachets containing granular medicine*

Preliminary tests were conducted on flexible multi-layered sachets containing granular medicine. The comminution tests resulted effective both without and with the adoption of the 20 mm mesh control grid. Notably, due to the lower resident time, the milling operations conducted without the control grid produced particles belonging to size classes greater than 1.0, i.e. greater than the maximum size of the granulate medicine contained in the sachet. Figure 9 shows the comminuted products classified in the size class +1.0 mm, while Figure 10 and Figure 11 show the comminuted products classified in the size classes obtained -1.0 +0.5 mm and -0.5 mm.

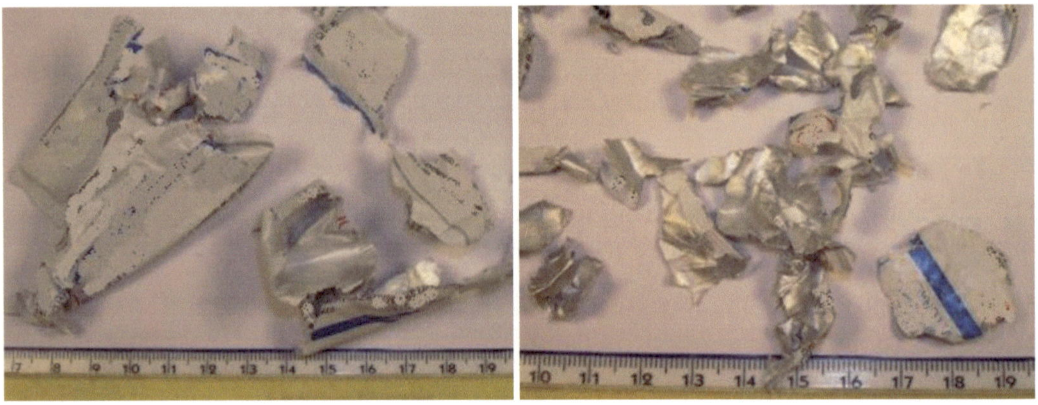

Fig. 9. Comminuted flexible multi-layered sachets containing granular medicine, belonging to the size class +1.0 mm, obtained in the preliminary tests without control grid (left) and with 20 mm mesh control grid (right).

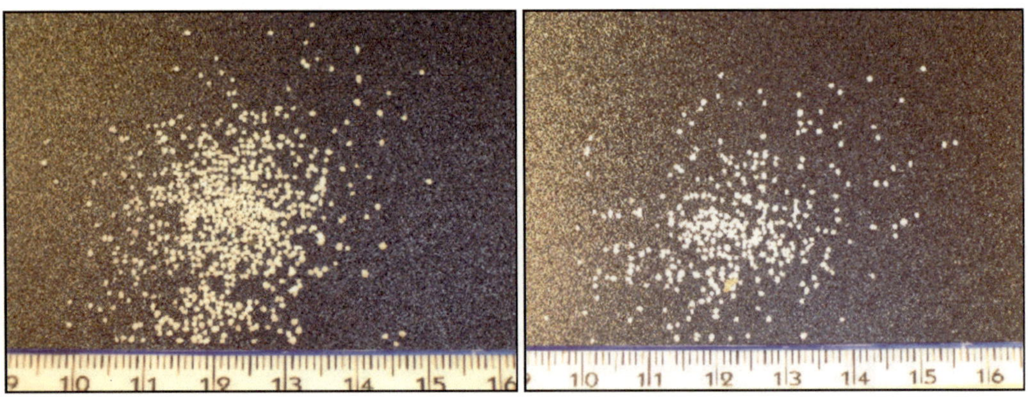

Fig. 10. Comminuted flexible multi-layered sachets containing granular medicine, belonging to the size class -1.0 +0.5 mm, obtained in the preliminary tests without control grid (left) and with 20 mm mesh control grid (right).

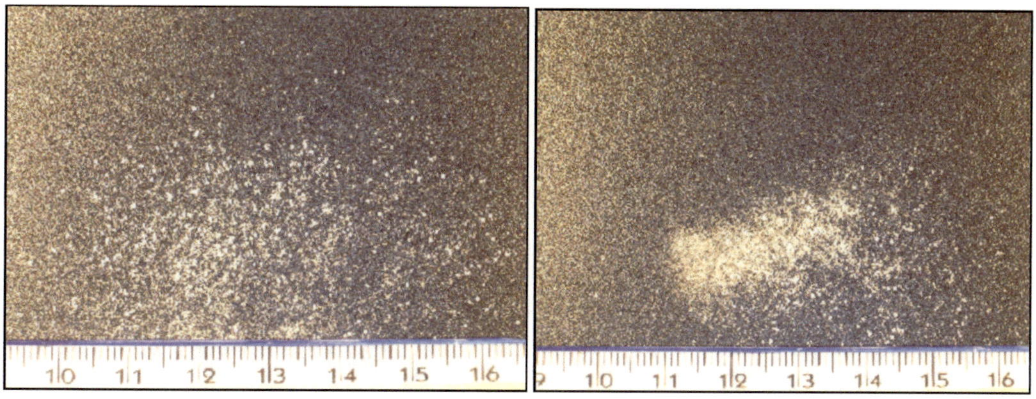

Fig. 11. Comminuted flexible multi-layered sachets containing granular medicine, belonging to the size class -0.5 mm, obtained in the preliminary tests without control grid (left) and with 20 mm mesh control grid (right).

2.5 Waste processing (comminution) tests

On the basis of the results of the preliminary tests, the comminution processes in laboratory scale have been set up. Notably, the tests have been carried out adopting the blade mill Retsch - SM 2000 under two operational conditions:

- dry milling, with 2 cm mesh control grid;
- wet milling, with 2 cm mesh control grid.

Wet drying has been realised feeding the mill with the waste materials together with little quantities of water. In such a way, the operational conditions that could be achieved with the shredder installed in the considered industrial plant have been simulated. For the Flexible multi-layered (plastic and aluminium) sachets, tests have been conducted with the blade mill and with a mini-shredder, in order to evaluate the possibility of recovering the granular medicine they contain. The comminuted products have been analysed by mean of:

- dry sieving;
- laser granulometry;
- imagine analysis.

The comminution processes are described in the following paragraphs, for each considered waste typology.

- *Bottles in high density polyethylene (HDPE), for suspension to be reconstituted*

The HDPE bottles have been cut in their longitudinal axes before feeding them to the blade mill.

In the dry comminution tests, samples made of 5 bottles in HDPE have been adopted. The products of the dry and wet comminution tests have been submitted to dry sieving and image analysis. The sieving tests have been conducted adopting sieves of ASTM series with 2.0 and 1.0 mm mesh. The results of the sieving tests are reported in Figure 12 in terms of cumulative passing for dry and wet comminution tests.

Comparing the results of sieving tests (Figure 13), the dry and wet comminution tests do not show substantial differences in the size distribution of their products.

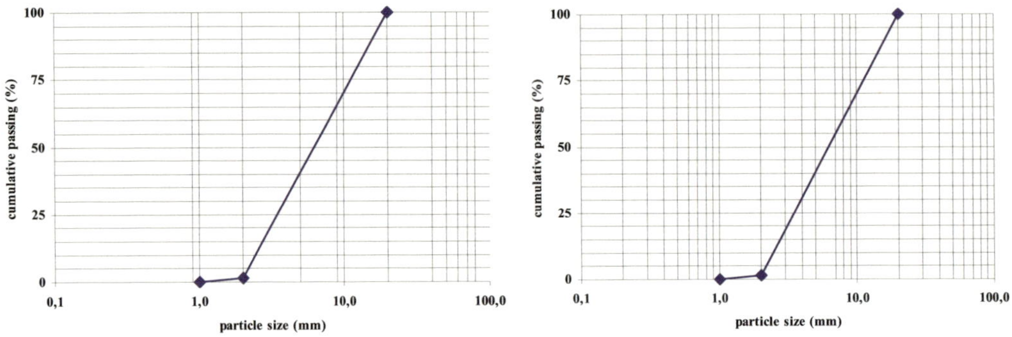

Fig. 12. Bottles in HDPE, 20 mm grid, cumulative passing, dry (left) and wet (right) comminution.

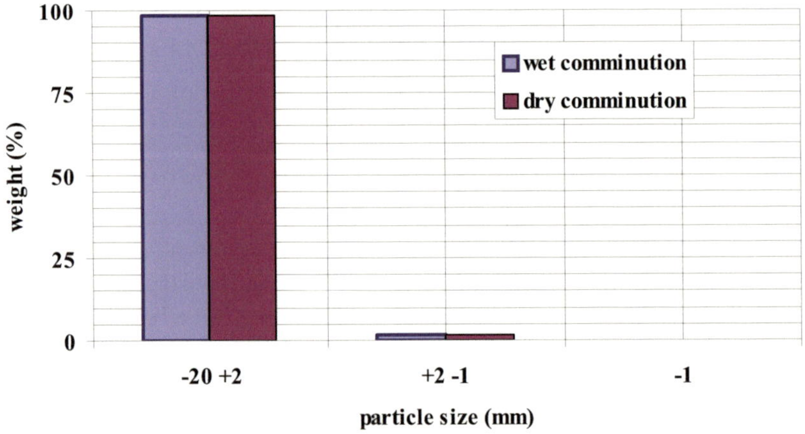

Fig. 13. Bottles in HDPE, comparison of the products of dry and wet comminution tests in terms of size distribution.

After the classification of particles in the size classes +2 mm, -2 mm +1 mm, and -1 mm obtained by sieving, image analysis has been conducted on the products of dry and wet comminution tests.

The results of image analysis for the bottles in HDPE are given in Tables 5-10.

Examples of images of the dry and wet comminution products are shown in Figure 14 and Figure 15 respectively.

Comparing the results of the image analysis for the considered size classes, no major differences can be observed in the particles size of the products obtained in the dry and wet comminution tests in terms of the values of the parameters *Area, Major Axis Length, Minor Axis Length and Feret Diameter.*

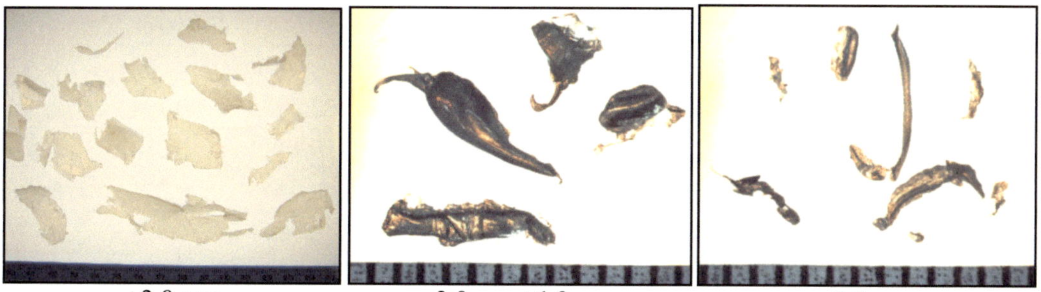

+2.0 mm *-2.0 mm +1.0 mm* *-1.0 mm*

Fig. 14. Bottles in HDPE, images of the products of dry comminution tests.

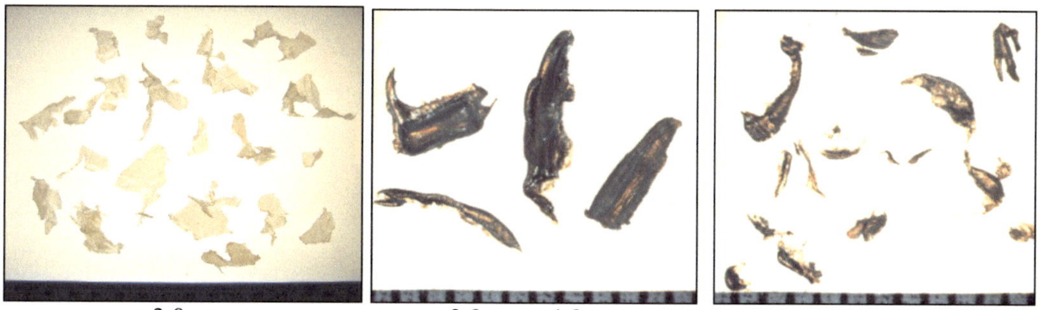

+2.0 mm *-2.0 mm +1.0 mm* *-1.0 mm*

Fig. 15. Bottles in HDPE, images of the products of wet comminution tests.

The high values of the parameter *Compactness*, measured in the products of both dry and wet comminution products, are in relation with the irregular morphology of particles, reasonably due to the cut stresses applied by the blades of the mill.

The values of the parameter Shape Factor describe a shape of elongated particles.

The statistic parameters, notably the standard deviation, show a high variability in the analysed particles, with standard error and confidence intervals substantially constant for both dry and wet comminution products.

	Area	Compact	Maj Len	Min Len	Perim	S Factor	Feret D
# Obj	120	120	120	120	120	120	120
Mean	205.066	188.244	24.942	13.659	173.927	0.124	15.124
Min	7.272	34.974	6.584	2.208	27.283	0.009	3.043
Max	961.603	1437.503	73.747	24.068	753.578	0.359	34.991
Std Dev	152.176	232.241	10.375	5.290	129.450	0.077	5.712
Std Err	13.892	21.201	0.947	0.483	11.817	0.007	0.521
95% Conf	37.879	57.808	2.583	1.317	32.222	0.019	1.422
99% Conf	49.781	75.973	3.394	1.730	42.347	0.025	1.869

Table 5. Bottles in HDPE, dry comminution, size class: +2.0 mm.

	Area	Compact	Maj Len	Min Len	Perim	S Factor	Feret D
# Obj	42	42	42	42	42	42	42
Mean	5.917	107.792	5.397	2.067	22.963	0.194	2.649
Min	0.909	22.001	2.209	0.863	9.138	0.018	1.076
Max	15.461	684.613	11.824	3.209	92.040	0.571	4.437
Std Dev	3.175	110.433	2.439	0.589	14.100	0.122	0.727
Std Err	0.490	17.040	0.376	0.091	2.176	0.019	0.112
95% Conf	0.790	27.488	0.607	0.147	3.510	0.030	0.181
99% Conf	1.039	36.126	0.798	0.193	4.612	0.040	0.238

Table 6. Bottles in HDPE, dry comminution, size class: -2.0 +1.0 mm.

	Area	Compact	Maj Len	Min Len	Perim	S Factor	Feret D
# Obj	48	48	48	48	48	48	48
Mean	1.040	67.968	2.205	0.768	7.582	0.263	1.023
Min	0.011	19.554	0.185	0.054	0.514	0.055	0.118
Max	3.614	227.840	6.093	2.334	21.682	0.643	2.145
Std Dev	0.930	45.895	1.494	0.433	5.192	0.148	0.532
Std Err	0.134	6.624	0.216	0.062	0.749	0.021	0.077
95% Conf	0.232	11.424	0.372	0.108	1.292	0.037	0.132
99% Conf	0.304	15.014	0.489	0.142	1.698	0.048	0.174

Table 7. Bottles in HDPE, dry comminution, size class: -1.0 mm.

	Area	Compact	Maj Len	Min Len	Perim	S Factor	Feret D
# Obj	70	70	70	70	70	70	70
Mean	178.616	286.454	25.949	13.569	203.690	0.084	14.500
Min	22.718	38.559	10.575	3.509	42.665	0.009	5.378
Max	349.484	1430.789	41.267	22.467	558.069	0.326	21.094
Std Dev	94.828	263.140	7.478	4.529	112.930	0.066	4.175
Std Err	11.334	31.451	0.894	0.541	13.498	0.008	0.499
95% Conf	23.604	65.500	1.861	1.127	28.110	0.016	1.039
99% Conf	31.021	86.081	2.446	1.482	36.943	0.022	1.366

Table 8. Bottles in HDPE, wet comminution, size class: +1 mm.

- *Bottles in poly(ethylene terephthalate) (PET), for syrup*

The PET bottles have been fed to the blade mill without any pre-treatment, therefore including the aluminium cap and relative ring, and in some cases also the paper labels. The comminuted tests have been carried out only in dry conditions, as in the wet tests the

	Area	Compact	Maj Len	Min Len	Perim	S Factor	Feret D
# Obj	29	29	29	29	29	29	29
Mean	5.231	91.432	4.927	1.874	20.587	0.176	2.517
Min	0.885	26.783	1.890	0.700	8.621	0.051	1.062
Max	9.565	245.852	10.922	3.359	34.566	0.469	3.490
Std Dev	2.190	45.358	1.892	0.656	6.504	0.100	0.580
Std Err	0.407	8.423	0.351	0.122	1.208	0.018	0.108
95% Conf	0.545	11.290	0.471	0.163	1.619	0.025	0.144
99% Conf	0.717	14.838	0.619	0.215	2.128	0.033	0.190

Table 9. Bottles in HDPE, wet comminution, size class: -2 +1 mm.

	Area	Compact	Maj Len	Min Len	Perim	S Factor	Feret D
# Obj	61	61	61	61	61	61	61
Mean	1.279	71.463	2.384	0.911	8.827	0.223	1.178
Min	0.115	20.377	0.529	0.255	1.529	0.077	0.382
Max	7.051	164.213	7.085	3.252	28.419	0.617	2.996
Std Dev	1.151	34.111	1.293	0.481	4.833	0.115	0.495
Std Err	0.147	4.367	0.166	0.062	0.619	0.015	0.063
95% Conf	0.287	8.491	0.322	0.120	1.203	0.029	0.123
99% Conf	0.377	11.159	0.423	0.157	1.581	0.038	0.162

Table 10. Bottles in HDPE, wet comminution, size class: -1 mm.

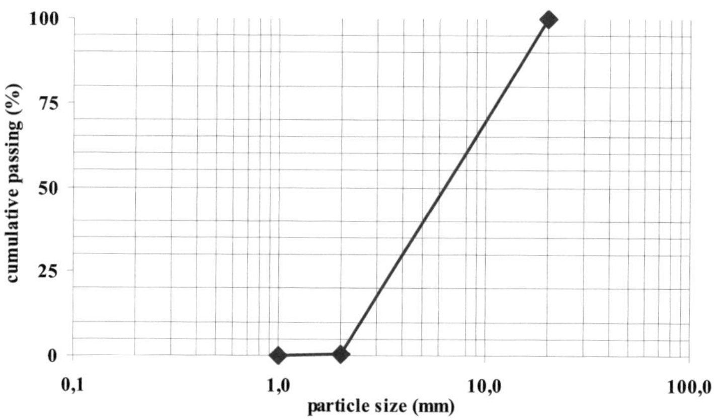

Fig. 16. Bottles in PET, 20 mm grid, cumulative passing, dry comminution.

products of comminution could not be easily extracted from the mill due to the presence of the syrup acting as a bonding agent for the PET particles and the mill surface. The dry comminution tests have been carried out on samples composed of 5 PET bottles. The products of the dry comminution tests have been submitted to dry sieving, laser granulometry and image analysis. The sieving tests have been conducted adopting sieves of ASTM series with 2.0, 1.0 mm and 38 µm mesh. Results of sieving tests and of laser granulometry analysis are reported in Figure 16 and Figure 17, respectively, both as cumulative passing for the dry comminution tests. The size class -38 µm has not been analyzed due to the presence of paper fibres of the labels including fine plastic particles.

After the classification in the particle size classes +2 mm, -2 mm +1 mm, and -1 mm +38 µm obtained by sieving, image analysis have been conducted on the products of dry comminution tests. The results of image analysis for the bottles in PET are given in Tables 11-13. Examples of images of the dry comminution products are shown in Figure 18.

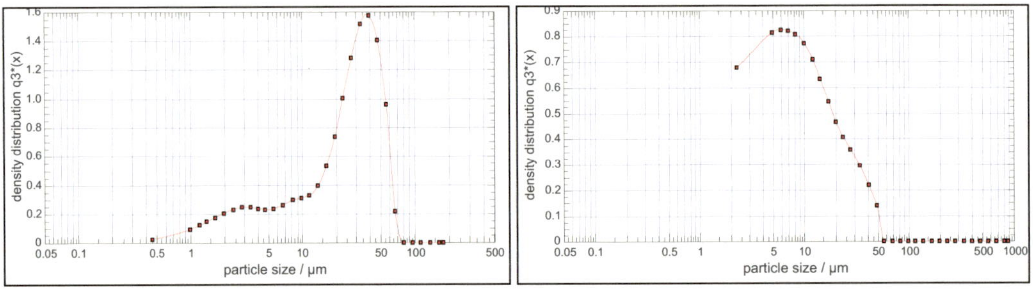

Fig. 17. Bottles in PET, 20 mm grid, size distribution, dry comminution, size class +2.0 – 1.0 mm (left) and -1 mm +38 µm (right), laser granulometry.

+2.0 mm *-2.0 mm +1.0 mm* *-1.0 mm +38.0 µm*

Fig. 18. Bottles in PET, images of the products of dry comminution tests.

The high values of the parameter *Compactness*, measured in the products of both dry comminution products, are in relation with the irregular morphology of particles, reasonably due to the cut stresses applied by the blades of the mill. The values of the parameter Shape Factor describe a shape of elongated particles. The statistic parameters, notably the standard deviation, show a high variability in the analysed particles, in particular for the higher size class (+2 mm), while the intermediate size class (-2 +1 mm) presents more homogeneous values of the morphologic and dimensional parameters.

	Area	Compact	Maj Len	Min Len	Perim	S Factor	Feret D
# Obj	183	183	183	183	183	183	183
Mean	164.865	44.458	21.181	11.656	77.378	0.334	13.151
Min	0.563	16.702	1.171	0.805	4.024	0.083	0.847
Max	571.600	152.309	50.237	27.126	264.501	0.752	26.977
Std Dev	140.469	20.766	9.826	5.912	42.869	0.132	6.097
Std Err	10.384	1.535	0.726	0.437	3.169	0.010	0.451
95% Conf	34.965	5.169	2.446	1.472	10.671	0.033	1.518
99% Conf	45.952	6.793	3.214	1.934	14.024	0.043	1.995

Table 11. Bottles in PET, dry comminution, size class: +2 mm.

	Area	Compact	Maj Len	Min Len	Perim	S Factor	Feret D
# Obj	46	46	46	46	46	46	46
Mean	5.681	47.475	4.676	2.072	15.530	0.327	2.587
Min	0.463	19.162	1.763	0.560	4.587	0.105	0.768
Max	17.147	119.185	11.921	3.325	38.087	0.656	4.672
Std Dev	3.379	24.258	2.045	0.559	7.238	0.142	0.744
Std Err	0.498	3.577	0.302	0.082	1.067	0.021	0.110
95% Conf	0.841	6.038	0.509	0.139	1.802	0.035	0.185
99% Conf	1.106	7.936	0.669	0.183	2.368	0.046	0.243

Table 12. Bottles in PET, dry comminution, size class: -2 +1 mm.

	Area	Compact	Maj Len	Min Len	Perim	S Factor	Feret D
# Obj	8	8	8	8	8	8	8
Mean	4.690	72.620	3.232	1.592	18.006	0.320	2.002
Min	0.111	28.067	0.595	0.334	1.764	0.038	0.376
Max	22.233	330.335	7.219	4.877	85.700	0.448	5.321
Std Dev	7.236	104.285	2.024	1.471	27.651	0.127	1.498
Std Err	2.558	36.870	0.716	0.520	9.776	0.045	0.530
95% Conf	1.801	25.958	0.504	0.366	6.883	0.032	0.373
99% Conf	2.367	34.115	0.662	0.481	9.046	0.042	0.490

Table 13. Bottles in PET, dry comminution, size class: -1 mm +38 µm.

- *Plastic bags and films of varying composition and thickness*

The composition of the samples that contain the 4 typologies of plastic bags and films used in the dry and wet comminution tests are reported in Table 14.

Material typology	Polymer	Weigh (%)	
		dry comminution	wet comminution
white and red bags containing bicarbonate	LDPE	23	23
thin film	LDPE	23	20
bags, with printed character "A" in black	LDPE	24	27
bags, with printed character "A" in blue	LDPE	30	30

Table 14. Composition of the samples of plastic bags and films used in the dry and wet comminution tests.

The plastic bags and films have been cut in samples of 50×50 mm in order to reach dimensions suitable for the laboratory blade mill. Moreover, the samples maintained their content of bulk powder.

The products of the dry and wet comminution tests have been submitted to dry sieving, laser granulometry and image analysis.

The sieving tests have been conducted adopting sieves of ASTM series with 2.0 and 1.0 mm mesh. The results of the sieving tests are reported in Figure 19 in terms of cumulative passing for dry and wet comminution tests.

Comparing the results obtained in the sieving tests (Figure 20), the wet comminution shows less particles belonging to the +2 mm size class than the dry comminution.

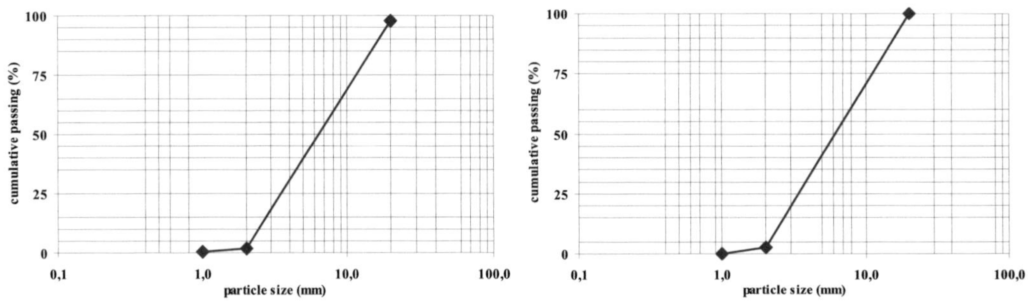

Fig. 19. Plastic bags and films, 20 mm grid, cumulative passing, dry (left) and wet (right) comminution.

The results of laser granulometric analysis are shown in Figure 21 in terms of size distribution for products of dry and wet comminution tests belonging to the -1 mm size class.

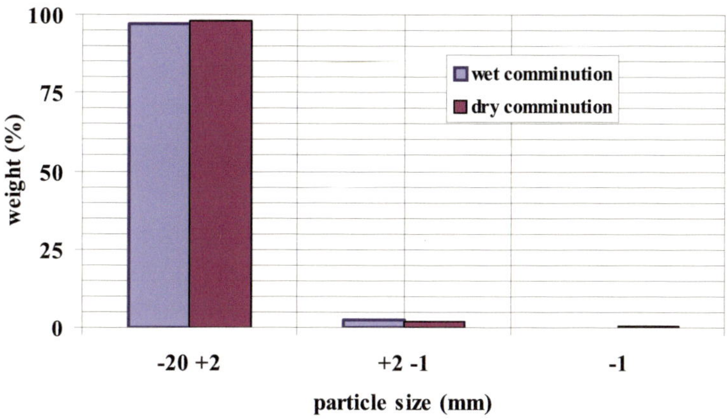

Fig. 20. Plastic bags and films, comparison of the products of dry and wet comminution tests in terms of size distribution.

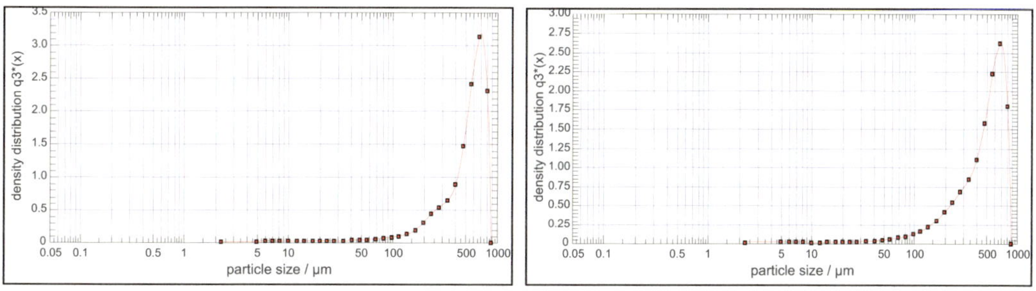

Fig. 21. Plastic bags and films, 20 mm grid, size distribution, size class - 1.0 mm, dry comminution (left) and wet comminution (right), laser granulometry.

Comparing the results obtained in the laser granulometry analysis, the dry and wet comminution tests do not show substantial differences in the size distribution of their products.

After the division in the particle size classes in +2 mm, -2 mm +1 mm, and -1 mm obtained by sieving, image analysis have been conducted on the products of dry and wet comminution tests. The results of image analysis for the plastic bags and films are given in Tables 15-20. Examples of images taken of the dry and wet comminution products are shown in Figure 22 and Figure 23 respectively.

The results of image analysis are reported in the following for all the considered size classes.

Comparing the results of the image analysis for the considered size classes, the difference between dry and wet comminution tests can be observed in the dimensions of the collected particles, measured by the values of *Area, Major Axis Length, Minor Axis Length e Feret Diameter*: the analysed particles generally belong to smaller size classes.

The high values of the parameter *Compactness*, measured in the products of both dry and wet comminution products, are in relation with the irregular morphology of particles, reasonably due to the cut stresses applied by the blades of the mill to very thin material (LDPE).

+2.0 mm *-2.0 mm +1.0 mm* *-1.0 mm*

Fig. 22. Plastic bags and films, images of the products of dry comminution tests.

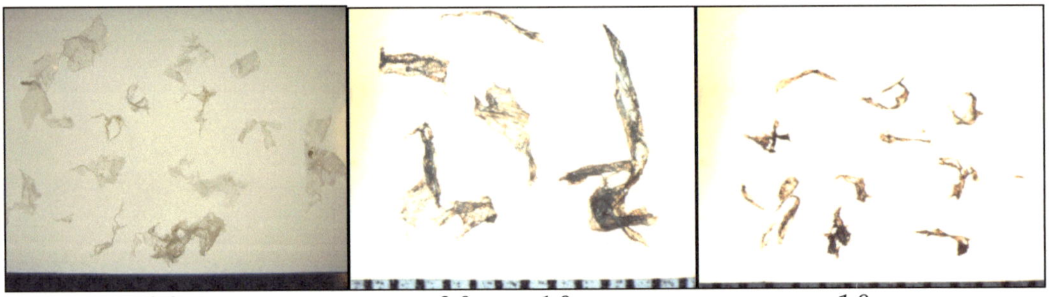

+2.0 mm *-2.0 mm +1.0 mm* *-1.0 mm*

Fig. 23. Plastic bags and films, images of the products of wet comminution tests.

The values of the parameter Shape Factor describe a shape of elongated particles.

The statistic parameters, notably the standard deviation, show a high variability in the analysed particles, with standard error and confidence intervals substantially constant for both dry and wet comminution products.

	Area	Compact	Maj Len	Min Len	Perim	S Factor	Feret D
# Obj	84	84	84	84	84	84	84
Mean	110.169	858.617	22.404	11.392	276.048	0.033	10.525
Min	0.353	45.314	0.943	0.691	3.999	0.003	0.670
Max	460.604	4994.348	54.142	28.985	1175.889	0.277	24.217
Std Dev	101.475	828.363	12.335	6.681	232.520	0.038	5.464
Std Err	11.072	90.382	1.346	0.729	25.370	0.004	0.596
95% Conf	21.700	177.145	2.638	1.429	49.724	0.008	1.168
99% Conf	33.195	270.983	4.035	2.186	76.065	0.012	1.787

Table 15. Plastic bags and films, dry comminution, size class: +2 mm.

	Area	Compact	Maj Len	Min Len	Perim	S Factor	Feret D
# Obj	67	67	67	67	67	67	67
Mean	4.586	147.115	4.563	2.075	23.798	0.151	2.325
Min	0.349	31.033	2.480	0.474	8.716	0.016	0.666
Max	15.881	765.458	13.879	3.402	110.257	0.405	4.497
Std Dev	2.654	143.863	1.931	0.627	17.288	0.101	0.665
Std Err	0.324	17.576	0.236	0.077	2.112	0.012	0.081
95% Conf	0.636	34.448	0.462	0.150	4.140	0.024	0.159
99% Conf	0.868	47.062	0.632	0.205	5.655	0.033	0.218

Table 16. Plastic bags and films, dry comminution, size class: -2 +1 mm.

	Area	Compact	Maj Len	Min Len	Perim	S Factor	Feret D
# Obj	62	62	62	62	62	62	62
Mean	0.886	103.529	2.157	0.847	8.870	0.169	0.974
Min	0.070	23.744	0.483	0.173	1.579	0.052	0.300
Max	3.647	242.053	5.013	2.413	21.939	0.529	2.155
Std Dev	0.725	59.113	1.078	0.479	5.028	0.100	0.428
Std Err	0.092	7.507	0.137	0.061	0.639	0.013	0.054
95% Conf	0.181	14.714	0.268	0.119	1.251	0.025	0.107
99% Conf	0.237	19.338	0.353	0.157	1.645	0.033	0.140

Table 17. Plastic bags and films, dry comminution, size class: -1 mm.

	Area	Compact	Maj Len	Min Len	Perim	S Factor	Feret D
# Obj	55	55	55	55	55	55	55
Mean	55.820	777.032	16.333	8.129	189.172	0.034	7.174
Min	4.040	73.678	4.976	1.902	22.982	0.002	2.268
Max	418.924	6791.312	56.129	31.755	1686.727	0.171	23.095
Std Dev	81.603	939.346	10.465	5.434	250.611	0.033	4.469
Std Err	11.003	126.661	1.411	0.733	33.792	0.004	0.603
95% Conf	20.312	233.818	2.605	1.353	62.381	0.008	1.112
99% Conf	26.695	307.289	3.423	1.778	81.983	0.011	1.462

Table 18. Plastic bags and films, wet comminution, size class: +2 mm.

- *Flexible multi-layered (plastic and aluminium) sachets containing granular medicine.*

The flexible multi-layered sachets have been fed to the blade mill without any pre-treatment, including the granular medicine they contained. For the dry and wet comminution tests, samples made of 25 sachets (equal to 184.87 g) and of 10 sachets (equal to 73.70 g) have been respectively used.

	Area	Compact	Maj Len	Min Len	Perim	S Factor	Feret D
# Obj	55	55	55	55	55	55	55
Mean	3.638	319.329	4.975	1.915	32.679	0.051	2.039
Min	0.354	113.617	1.577	0.451	7.857	0.016	0.671
Max	9.377	776.877	10.951	3.989	83.139	0.111	3.455
Std Dev	2.309	172.573	2.079	0.847	17.157	0.026	0.695
Std Err	0.311	23.270	0.280	0.114	2.313	0.003	0.094
95% Conf	0.575	42.956	0.517	0.211	4.271	0.006	0.173
99% Conf	0.755	56.454	0.680	0.277	5.613	0.008	0.227

Table 19. Plastic bags and films, dry comminution, size class: -2 +1 mm.

	Area	Compact	Maj Len	Min Len	Perim	S Factor	Feret D
# Obj	75	75	75	75	75	75	75
Mean	0.679	123.913	2.000	0.809	8.781	0.163	0.851
Min	0.049	16.139	0.463	0.136	1.124	0.027	0.249
Max	3.544	463.666	5.469	2.341	38.710	0.779	2.124
Std Dev	0.606	86.252	1.026	0.527	6.619	0.127	0.377
Std Err	0.070	9.960	0.118	0.061	0.764	0.015	0.044
95% Conf	0.151	21.470	0.255	0.131	1.648	0.032	0.094
99% Conf	0.198	28.216	0.336	0.172	2.165	0.041	0.123

Table 20. Plastic bags and films, dry comminution, size class: -1 mm.

The products of the dry comminution tests have been submitted to dry sieving, laser granulometry and image analysis.

The sieving tests have been conducted adopting sieves of ASTM series with 1.0 mm, 0.85 mm and 0.50 mm mesh.

The results of the sieving tests are reported in Figure 24 in terms of cumulative passing for the dry comminution tests.

The wet comminution tests have been carried out in order to verify the effect of the use of water on the granular medicine. The results of the tests have been analysed in qualitative terms. In the Figure 25 are shown the images of the products of the comminution test in which 0.4 l of water have been fed to the mill together with the sachets. From the images it can be observed that the sparkling granular medicine has not relevant effects. Moreover, the presence of water allowed reducing the dispersion of powder in the environment during the comminution.

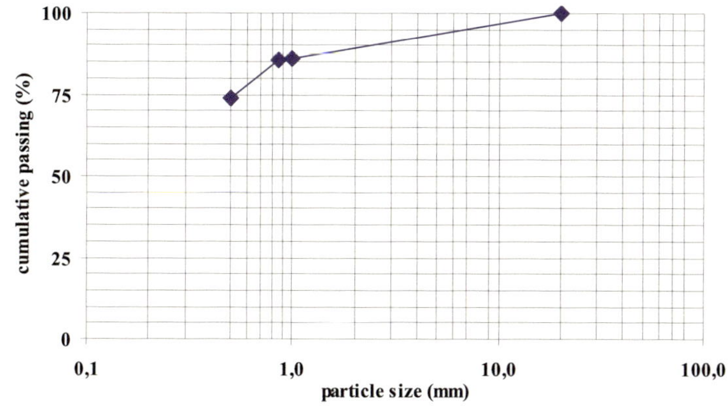

Fig. 24. Flexible multi-layered sachets, 20 mm grid, cumulative passing, dry comminution.

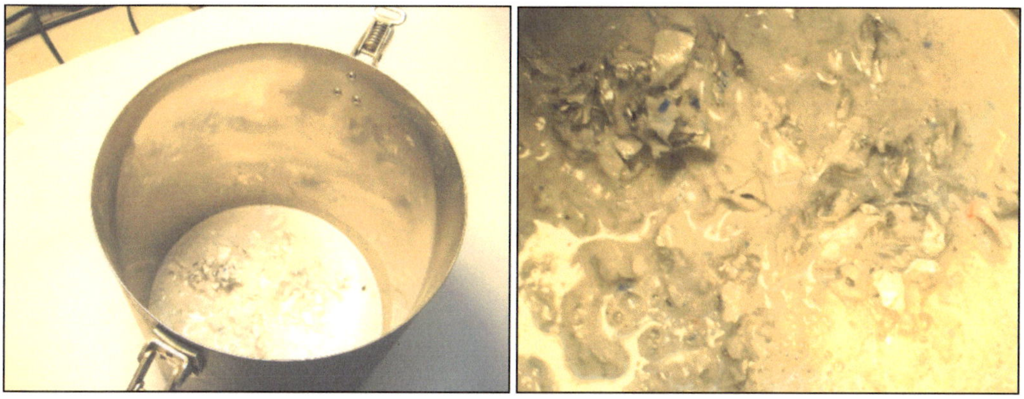

Fig. 25. Flexible multi-layered sachets, 20 mm grid, cumulative passing, wet comminution.

The results of laser granulometry analysis are shown in Figure 26 and Figure 27 in terms of size distribution for products of dry comminution tests belonging to the -1 mm +0.85 mm, -0.85 mm +0.5 mm and -0.5 mm size classes.

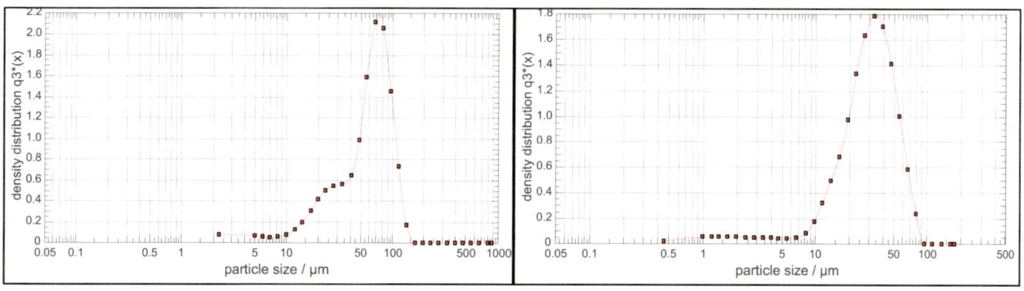

Fig. 26. Flexible multi-layered sachets, 20 mm grid, size distribution, size classes -1 +0.85 mm (left) and -0.85 +0.5 mm (right), dry comminution, laser granulometry.

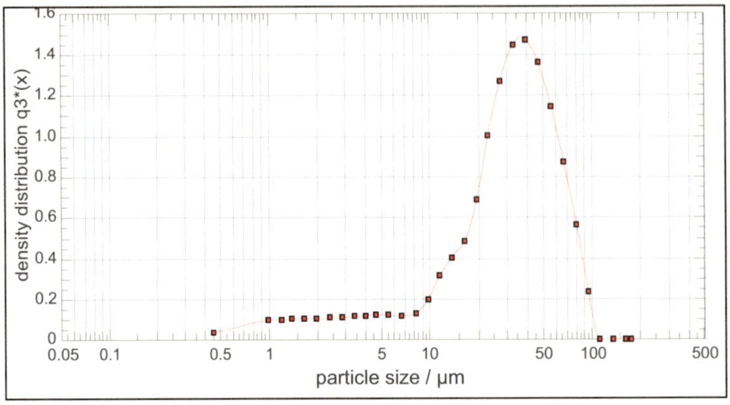

Fig. 27. Flexible multi-layered sachets, 20 mm grid, size distribution, size class -0.5 mm, dry comminution, laser granulometry.

The results of the sieving tests show that the comminuted dry sachets are mostly found in the +1.0 mm and -0.5 mm size classes. In fact, in these classes are respectively collected the multi-layered materials and the granular medicine particles. In the size class –1.0 mm +0.85 mm, the results of laser granulometer analyses show two principal modes, reasonably due to the presence of both multi-layered materials and granular medicine.

After the division in the particle size classes in -1 mm +0.85 mm obtained by sieving, image analysis have been conducted on the products of dry comminution tests. The results of image analysis for the multi-layered sachets are given in Table 21. Examples of images taken of the dry comminution products are shown in Figure 28.

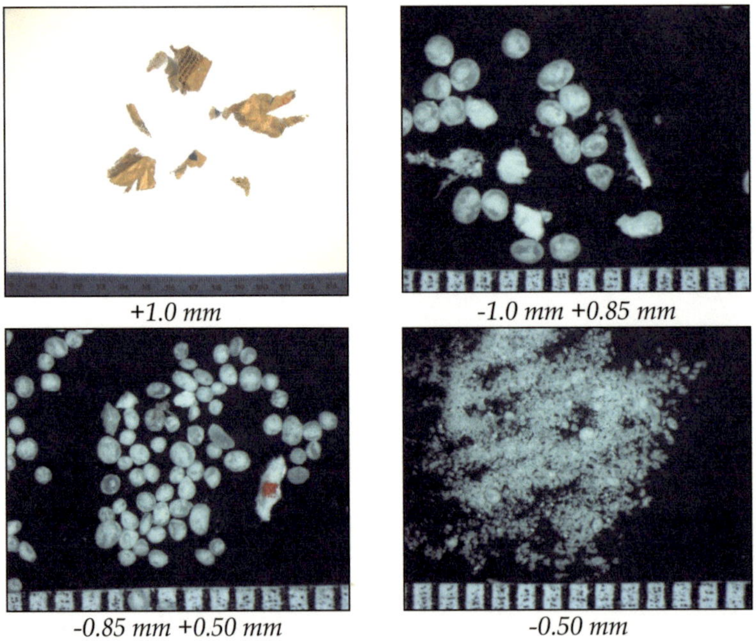

Fig. 28. Flexible multi-layered sachets, images of the products of dry comminution tests.

The high values of the parameter *Compactness*, measured in the products of dry comminution products, are in relation with the irregular morphology of multi-layered particles, reasonably due to the cut stresses applied by the blades of the mill.

The values of the parameter Shape Factor describe a shape of elongated particles.

The statistic parameters, notably the standard deviation, show a high variability in the analysed particles, due to the simultaneous presence of multi-layered and granular particles.

	Area	Compact	Maj Len	Min Len	Perim	S Factor	Feret D
# Obj	11	11	11	11	11	11	11
Mean	253.133	73.910	27.032	15.394	131.359	0.193	16.894
Min	32.653	38.744	10.198	3.748	35.569	0.092	6.448
Max	495.707	136.919	41.900	26.437	227.952	0.324	25.123
Std Dev	154.107	27.727	9.990	7.452	63.551	0.072	6.371
Std Err	46.465	8.360	3.012	2.247	19.161	0.022	1.921
95% Conf	38.360	6.902	2.487	1.855	15.819	0.018	1.586
99% Conf	50.413	9.070	3.268	2.438	20.789	0.024	2.084

Table 21. Flexible multi-layered sachets, dry comminution, size class: -1.0 mm.

3. Conclusions

The results of experimental tests demonstrate the effectiveness of shear stress to comminute primary packaging and waste pharmaceutical product under investigation.

The comminution tests by blade mill RETSCH – SM 2000 show the following outcomes:

- shear stresses on plastic materials determined an irregular and elongated shape on output particles;
- wet and dry conditions are irrelevant on geometric and morphological characteristics of output particles;
- statistical analysis on image analysis data evidenced a high variability in geometric and morphological parameters: this is probably due to plasticity property of materials under investigation and to applied shear stresses;
- size distribution of the plastic particles after comminution is always greater than 1.0 mm and, therefore, greater than powder eventually contained inside packaging (e.g. in pharmaceutical waste).

Considering the outlined results, the comminution process seems to be a feasible treatment for pharmaceutical waste, in order to reduce particle size and to separate packaging materials (mainly plastics) and powder eventually contained.

The wet comminution, even if not influential on geometric and morphological characteristics of output particles, can be adopted to avoid powder dispersion in air.

4. References

Achilias, D.S., Giannoulis, A. & Papageorgiou, G.Z. (2009). Recycling of polymers from plastic packaging materials using the dissolution–reprecipitation technique, *Polymer Bulletin*, Vol.63, (2009), pp. 449–465, ISSN (printed) 0170-0839, ISSN (electronic) 1436-2449

Bauer, E.J. (2009). *Pharmaceutical packaging handbook*, Informa Healthcare, ISBN 978-1-4200-1273-6, London, United Kingdom

Biniecka, M., Campana, P. & Iannilli, I. (2005). The technological and economic management of the environmental variable in the pharmaceutical–chemical industry, *Microchemical Journal*, Vol.79, (2005), pp. 325– 329, ISSN 0026-265X

Dillon, P. & Rubinstein, L. (2005). *Managing Pharmaceutical Waste: Best Management Practices for Plastic Medication Containers from Consumers*, Report of Northeast Recycling Council, Inc., Vermont, USA

Linninger, A.A. & Chakraborty, A. (2001). Pharmaceutical waste management under uncertainty, *Computers and Chemical Engineering*, Vol.25 (2001), pp. 675–681, ISSN 0098-1354

Sacha, G.A., Saffell-Clemmer, W., Abram, K. & Akers, M.J. (2010). Practical fundamentals of glass, rubber, and plastic sterile packaging systems, *Pharmaceutical Development and Technology*, Vol.15, No.1 (January-February 2010), pp. 6-34, ISSN (printed) 1083-7450, ISSN (electronic) 1097-9867

Systat Software Inc. (2007). *SigmaScan Pro© Version 5.0.0 – User guide*, Systat Software Inc., Chicago, Illinois, USA

11

Recycling of the Hardwood Kraft Pulp

Jarmila Geffertová and Anton Geffert
Technical University in Zvolen, Faculty of Wood Sciences and Technology
Slovakia

1. Introduction

According to „Key Statistics 2009 European Pulp and Paper Industry" CEPI member countries produce 21,6 % of world fibre production, while North America 37,4 % and Asia 23,8 % (Key Statistics, 2009).

Pulp made by kraft process is the most widely used raw material for paper and board production.

For paper and board production is not using exclusively fresh pulp fibres. The portion of recycled fibres increases gradually.

While the portion of new and recycled fibres used in CEPI countries was equivalent in 2008, the consumption of recycled fibres (as a waste paper) increases 2.6 mil. tonnes (fig. 1).

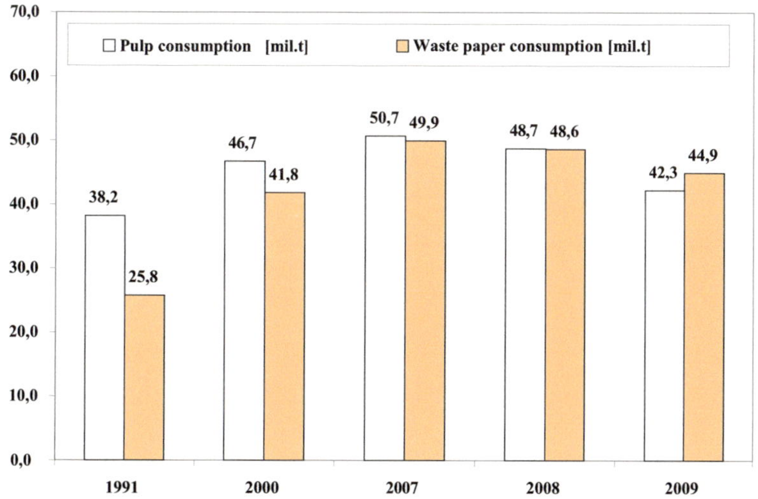

Fig. 1. Evolution of new and recycled fibres.

The year 2009 was the first year when CEPI countries consume more waste paper (44.9 mil. tonnes) than pulp (42.3 mil. tonnes) for production of 88.7 mil. tonnes of paper and board.

The usage level of recycled fibres is evaluated based on predefined parameters - waste paper utilization rate and recycling rate.

The waste paper utilization rate (percentage ratio of waste paper consumption to total paper and board production) in CEPI countries reaches v 50.7 % and recycling rate (percentage ratio of waste paper consumption to paper and board consumption) 72.2 % in 2009. Thereby the obligation of 13 different sectors of paper chain to achieve recycling level 66 % in 2010 was surpassed (www.paperrecovery.eu, 2007).

Figure 2 illustrates the evaluation of waste paper usage rate and recycling level within 1991 and 2009 in CEPI countries (Key Statistics, 2009).

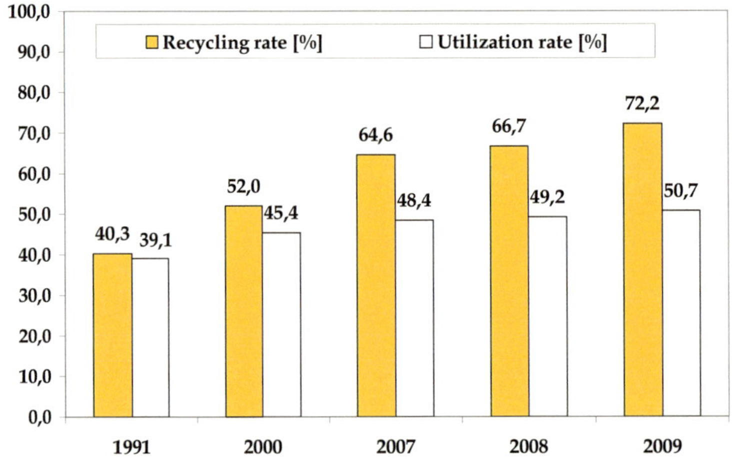

Fig. 2. Evaluation of waste paper usage rate and recycling level in CEPI countries.

Consumption of paper and board per inhabitant is one criteria of industrial and cultural country level. Average world consumption of paper and board was 56.3 kg per habitant in 2004. However, paper and board (P&B) consumption of some countries (USA, Finland) was around 300 kg per inhabitant in 2004 (Lešikár, 2006).

Table 1 indicates actual paper and board consumptions of some world countries. (http://swivel.com/charts/2381-Paper-Consuption-per-capita-by-Country).

Country	Consumption of P&B [kg/inhabitant]	Country	Consumption of P&B [kg/inhabitant]	Country	Consumption of P&B [kg/inhabitant]
Finland	334	Germany	235	Greece	76,2
USA	312	Switzerland	210	Bulgaria	35,8
Canada	277	Great Britain	208	Russia	34,5
Belgium	264	Luxemburg	198	Romania	25,6
Denmark	257	Italy	195	China	33,6
Japan	250	France	183	Vietnam	6,3
Austria	245	Spain	166	Cuba	4,7
Germany	235	Czech Rep.	116	India	4,4
Nederland	227	Hungary	82	Iran	9,8
Sweden	223	Poland	80	Iraq	1,4

Table 1. Paper and board consumption of some world countries in 2009.

Nowadays adverse price evolution in the area of hygienic paper products is examined. Increased demand for basic raw material leads to the increase in price of pulp (about 40 %) and waste paper (about 80 %) (ŠK, 2010). This evolution was influenced by enormous demand of China for fibres and waste paper. Chinese expansion of paper and board production represents 180 % within the years 1999 and 2009 (MD, 2010). Mentioned negative evolution was influenced by earthquake in Chile; floods in USA and Portugal; transporters strike etc. All these factors conduce to the input increase and the price grow of hygienic paper products at about 10 % at average (ŠK, 2010).

1.1 Paper recycling

Recycling is implemented in paper production for centuries (from 105 A.C.). Initially the paper was made from fibres obtained from old cloth. The usage of wood fibres starts from the half of 19th century (Blažej and Krkoška, 1989).

Europe has long tradition in the effort of the old paper re-usage. First notices are from 1695 (Denmark), chemical processing by decolouration was patented in England in the year 1800. Technical importance of recycling begins to increase in 20th century. Nowadays the development is focussed on old paper processing. Old paper utilization is important for national economy because of:

- Natural fibres and wood material savings,
- Decrease of specific fuel and electricity consumption for paper production,
- Water savings,
- Improvement of production economics, decrease of capital and operating costs,
- Conduce to environment protection (Blažej and Krkoška, 1989; Hnětkovský, 1982 and 1983a, b).

Paper and paper products are the most widely used transmission form of written and printed information and packaging material. Paper and board consumption is authoritative and validated parameter of society economic and cultural level. Therefore used paper is the most widespread waste. So paper occurs everywhere where people are living. Besides of very good useful paper properties, paper provides several advantages contrary to competitive material. Firstly the paper is degraded relatively quickly and without negative effect on environment. Then paper represents valuable secondary raw material mainly because of its regeneration ability after usage, collection and separation.

Waste paper occurrence, its quality and amount vary within time periods, seasons and regional relatives. It depends on paper industry production relatives of country.

European paper industry, companies focussed on paper collection and waste paper vendors are persuaded that it could be achieved more in long-term environment protection and that "paper cycle" could be closed whereas primary and secondary fibres complete each other.

Paper production, processing and utilization create waste paper which is an important source of secondary fibres. If this paper is not used, it represents valueless waste and it contaminates the environment. Nowadays waste paper is not considered as waste but it represents valuable fibre raw material which could be multiply used - recycled. Secondary fibres are routinely recycled for 3- to 5-times (Souček, 2009).

Whereas the majority of mixed grades and OCC grades of recovered paper primarily find use in the production of packaging papers and board (84 %), almost all deinking grades (88 %) go to graphic paper production (Putz, 2000).

Waste paper participates in total paper production (60 %) mostly in Nederland and Great Britain. This limit is almost reached in Germany. Whereas Nordic countries that have sufficient wood raw material source are using only 10 - 15 % of waste paper as a secondary fibres (Milichovský, 1994).

1.2 Recycling influence on pulp fibre quality

Raw material source for paper production including usage of secondary fibres as well as ways of its obtaining varies continually. This fact represents basic and more and more difficult issue in paper production research.

Paper is nonhomogeneous network of pulp fibres. Except of fibres real papers contain also fillers, sizing agents, colours and other auxiliary materials. Paper properties are defined by properties of all used materials and technology of paper process. Describing this wide complex of variables is actually unreal. It is possible to partially describe the properties of laboratory testing sheets prepared under standard conditions or eventually of orthotropic paper made from pulp fibres on paper machine.

Description of fibres properties is based on the dimension characteristic and derived numeric parameters. This description is insufficient and only approximative. Dimensional analysis does not express fibres status and vanish the fine portion which influence significantly paper properties (Blažej and Krkoška, 1989).

Pulp fibres dimensions and physic-chemical properties are modified significantly by process of defibering, beating and refining, i.e. via mechanical effect of pulper and beating apparatus in aqueous medium. The interaction between pulp fibres and water is crucial at mentioned operations. The result reached by beating is influenced by fibres parameters (theirs origin, preparation method, drying method and deepness), mill parameters and beating process parameters (e.g. swelling level) (Hnětkovský, 1983a).

The characteristic differences of paper made by secondary fibres and freshly prepared fibres are similar to deviations between properties of paper made by fibres in wet and dry stage. Basic properties of fresh wet fibres change within drying process. Defibering and beating represent regeneration process. However this process is not complete, moreover it introduces additional destructive changes. Similar effect is observed during old paper defibering. Several fibres properties are changed irreversibly. Changes relevancy depends on the cycle count of fibres utilization (regeneration and usage for new paper production) and the way of paper products utilization (ageing destruction). The parameters used for the description of primary fibres paper properties are also suitable for the description of paper properties made by secondary fibres and fibre ageing changes (Blažej and Krkoška, 1989).

The fibres wear irreversibly and change theirs properties within the count of utilization cycles. Defibering and beating induce water absorption, swelling and partial regeneration of original properties. Repeated beating and drying during several production cycles provoke gradual decrease of swelling ability, which determine fibres bonding ability. Moreover a fibre shortening is observed within increasing count of utilization cycles. Mentioned modifications reflect in paper properties (Blažej and Krkoška, 1989).

Experiences gained during old paper utilization prove that these fibres show significantly different properties comparing to freshly prepared fibres (Blechschmidt, 1979; Nordman, 1976; Laivins and Scallan, 1993; Hubbe et al., 2007; Howard, 1990, 1994, 1995; Nazhad and Paszner, 1994; Phipps, 1994; Ackermann et al., 2000; Shao and Hu, 2002; Hubbe and Zhang, 2005; Nazhad, 2005). Fibre re-utilization creates extremely non-homogenous mixture of variously old fibres. Old paper is composed by all types of manufactured paper and board. Additional inhomogenity is caused by the presence of fibres used for several times but unequally (Attwood, 1983).

Pulp fibres are modified within the paper production by beating. Whereas the beating condition optimization is very important because of created fibrillation of fibre surface, release of fine portion and cell wall delamination. Paper rigidity increases during the beating. However recycled fibres repeated beating and drying lead to the decrease of inter-fibre bonding potential (Stürmer and Göttsching, 1979; Peng et al.,1994).

Göttsching (1976) detected by strength measuring at zero-span of jacks that recycled fires resistance does not change practically. Decrease of paper resistance made by recycled fibres could be explained as the consequence of inter-fibre bond strength decrease.

Decrease of bonding ability and strength of recycled fibres bring the improvement of several utility characteristics as increased velocity of dewatering and drying, air permeability, blotting paper ability, improvement of light diffusion, opacity, and dimensional stability of paper (Göttsching and Stürmer, 1975), which is linked with decreased fibres ability of swelling in contact with water (Ackermann et al., 2000).

Fibres swelling in width orientation and increase of wet fibre flexibility prove inner fibrilation caused by beating. Beating process creates submicroscopic areas in lamellar structure of kraft pulp fibre cell wall. Mentioned areas have tendency to close themselves semi-reversibly during drying (Jayme and Büttel ,1968; Paavilainen, 1993).

2. Experimental

Recycling process of kraft pulp fibres was observed on white kraft pulp sample prepared from hardwood mixture.

Pulp sample was processed into sheets of surface weight 800 - 900 g.m^2 and brightness 82.7 % MgO.

Original sample after first processing by defibering, milling and drying represents zero recycling. Following pulp fibres processing simulated recycling. Pulp fibres undergo recycling for 8 times. Simulation could be considered as sufficient because usually pulp fibres are re-utilized for 4 to 5 times in practice.

Pulp was returned back into defibering, milling and drying process. The beating value was chosen as 29 °SR because of achieving sufficient strength of paper sheets without redundant fibre weakening and ensuring the possibility of next recycling.

The drying temperature influence on chosen pulp fibre properties was observed at 80 °C, 100 °C a 120 °C. These chosen temperatures cover the action of usually used temperature (100 °C) and extreme temperature values (80 °C and 120 °C) which were chosen for comparison.

Pulp samples taken from each recycling level were used for sheet preparation. Sheets surface weight was determined as relative value needed for calculation of chosen characteristics.

Standardly used procedures were applied for recycled pulp treatment (dry substance determination, wet laboratory defibering, laboratory beating in PFI mill, laboratory sheet preparation for physical tests, determination of dewatering ability according to Schopper-Rieglera, determination of surface weight).

Chosen properties were observed on the sheets prepared from original pulp, pulp after first processing (zero recycling) and pulp from particular recycling degrees:

- Fibre dimensional properties (fibre length, fibre width, shape factor, local fibre deformation a. o.)
- Mechanical characteristics (breaking length, tearing index)
- Physical properties (porosity, swelling, polymerization degree)
- Optical characteristics (brightness, colourity)

3. Results

3.1 Dimensional characteristics of recycled fibres

Basic fibre wood information is data of length, width, cell wall thickness, eventually lumen width. Dimensional parameters provide first fibre information and could be used for deduction of fibres utilization suitability for specific paper type.

Hardwood has more complicated structure of cell types comparing to coniferous wood. Tracheids occupy 90 % of total wood volume in coniferous wood. Theirs length is moving within 2 to 6 mm and it is 100 to 300 times longer than width. Tracheids are at average 2 to 4 times longer than libriform fibres of hardwood typical for Slovak republic. Their length is moving between 0.3 to 2.2 mm. Libriform fibres constitute approximately 50 % of hardwood volume, vessels represent 20 % and parenchymatic fibres 1/3. Vessels dimensions vary depending on wood type, their location in year circle, post and climatic conditions. Their length could be within 0.2 to 0.8 mm and width could reach 0.4 mm (Požgaj, 1993; Blažej et al., 1975).

Changes of dimensional parameters and physico-chemical characteristics created during native fibre isolation decrease primary importance of dimensional parameters. Blažej and Krkoška (1989) claim fibre shortening during wood isolation by comparing of fibre wood length (Chovanec, 1975). Dimensional analysis allows expressing average fibre length and width, as well as cell wall thickness. However it does not describe the fibre stage.

Repeated pulp beating during recycling causes numerous changes of pulp fibre dimensional characteristics. It means fibre shortening (wherever and under random angle across the fibre), inner and outer fibre fibrilation, fibre delamination and fibre downiness.

Dimensional characteristics were observed by microscope until recently. Nowadays modern apparatus which can measure faster and are able to evaluate fibre dimensions are used.

Dimensional characteristics were monitored by Fiber Tester, which allows the evaluation of app. 20 000 fibres in pulp suspension containing 0.1 g of fibres in 100 ml of suspension in one measure (Karlsson, 2006). Double-dimensional view allows fibre length measuring separately and measuring of pulp fibre deformation during the running between two-glass plates.

Following pulp fibre characteristics in recycling process were monitored during dimensional analysis:

- Average fibre length
- Average fibre width
- Fibre shape factor
- Fibre length distribution – in 75 classed divided by 0.1 mm upto 7.5 mm
- Fibre width distribution – in 50 classes divided by 2 μm upto 100 μm
- Fine ratio – fibres up to length 0.2 mm
- Percentual fibre distribution within configured length and width interval
- Local fibre deformation – angle of fibre deviation and fibre length between particular angles of fibre deviation.

Average values of chosen dimensional characteristics - length, width and final fibre shape factor from multiple measures of fibre suspension are stated in tab. 2.

Dimensional characteristics of pulp fibres									
Pulp sample	Average fibre length [mm]			Average fibre width [μm]			Fibre shape factor [%]		
	Drying temperature [°C]								
	80	100	120	80	100	120	80	100	120
Original sample	0,824			20,8			90,1		
0th recycling	0,801	0,792	0,792	20,3	21,0	20,3	89,7	89,9	89,9
1st recycling	0,774	0,792	0,782	20,2	20,9	20,2	89,9	89,5	90,1
2nd recycling	0,760	0,784	0,789	20,0	20,6	19,8	90,0	89,2	91,2
3rd recycling	0,766	0,777	0,771	19,8	20,5	20,0	89,5	89,5	89,6
4st recycling	0,769	0,770	0,772	20,4	20,4	20,2	87,2	89,6	90,2
5st recycling	0,767	0,769	0,761	19,8	20,3	19,9	86,7	89,7	89,3
6st recycling	0,757	0,768	0,761	19,5	20,3	19,9	90,3	89,5	89,4
7st recycling	0,769	0,762	0,754	19,9	20,2	19,8	89,4	89,5	89,4
8st recycling	0,771	0,762	0,744	19,8	20,2	19,8	89,8	89,7	89,3

Table 2. Dimensional characteristics of pulp fibres.

Gained results prove that recycling process cause decrease of length, width and fibre shape factor. The most significant modifications were observed during first beating treatment. Although pulp fibres milled only at 29 °SR within repeated recycling, displayed shortening and fibrillation beating effect by decrease of length, width and fibre shape factor values.

Average fibre length of original hardwood pulp mixture was 0.824 mm. The highest decrease of average fibre length within whole monitored recycling range (fig. 3) was observed at drying temperature 120 °C. Average length decrease after 8th recycling represents 9.7 % at drying temperature 120 °C, 7.5 % at 100 °C and 6.4 % at 80 °C comparing to original fibre length.

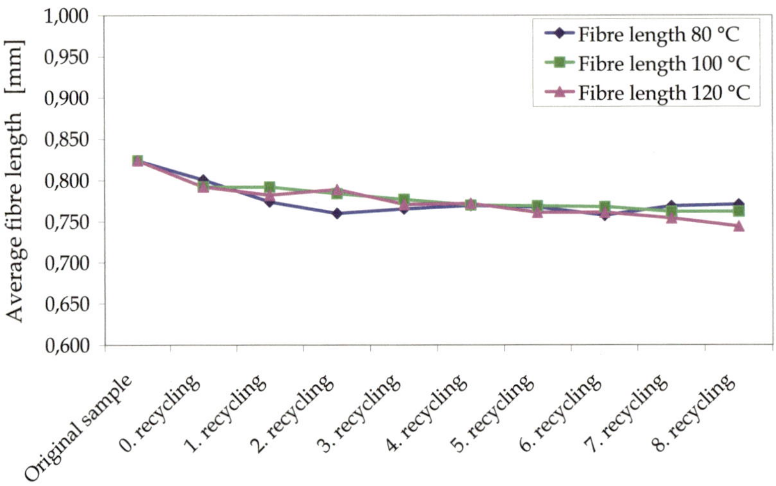

Fig. 3. Average fibre length in recycling process.

Descending trend of average fibre length (tab. 2) during repeated recycling correspond to growing amount of fine ratio (see tab. 3). This effect is clearly visible at drying temperature 100 °C where average fibre length decreases and fine ratio ascend within whole recycling range. The progression was not so explicit at drying temperatures 80 °C and 120 °C. However the correlation between fine ratio content and average fibre length was kept.

Average original pulp fibre width was 20.8 μm and the change of average fibre width after 8th recycling was negligible.

Fibre shape factor is defined as the ratio of join between fibre ends and real fibre length (in %). This factor declines during recycling depending on recycling rate, temperature and fibre beating process (tab. 2).

Relative fibre distribution in length and width interval is more characteristic for pulp than average length and width (fig. 4).

The highest relative distribution of longer fibres was in original pulp sample. Content of longer fibres decreases and of shorter fibres increases because of beating fibre shortening. Although the distribution curve progress at particular temperatures is similar and the curves are almost overlaid, relative distribution of shorter lengths is the highest at drying temperature 120 °C. This fact gives evidence of higher fibre fragility and hornification caused by the impact of higher temperature.

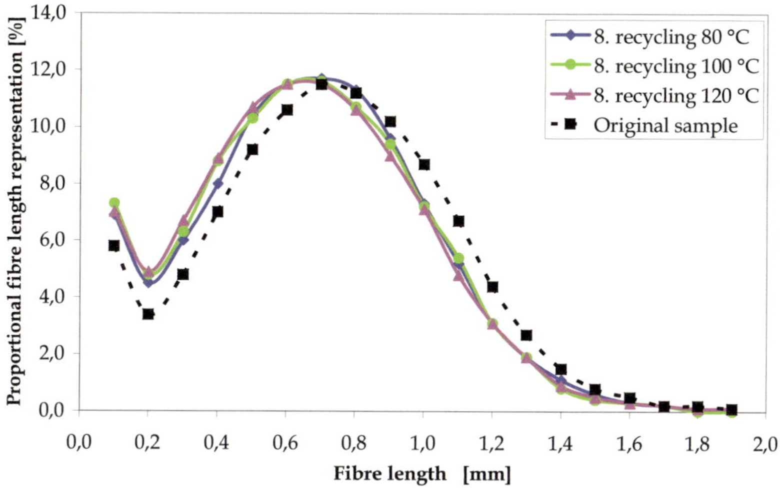

Fig. 4. Relative distribution of pulp fibre length dried at 80, 100 and 120 °C.

Fine ratio content represents in Fiber Tester apparatus record the fibre length from 0 to 0.2 mm. The ratio attained the lowest value in original sample and increased within recycling process even if the fine ratio of mucus and powder was partially eliminated by dewatering.

Table 3 presents average values of fibre fine ratio in original sample and in samples after 8th recycling at monitored temperatures.

Fibre Fine ratio influences negatively the mechanic properties (weak bonding properties) as well as dewatering ability during paper formation in drying machine (Karlsson, 2006).

Fine ratio content [%]						
Drying temperature	80 °C		100 °C		120 °C	
Fraction [mm]	0 - 0,1	0,1 - 0,2	0 - 0,1	0,1 - 0,2	0 - 0,1	0,1 - 0,2
Original sample	5,8	3,3	5,8	3,3	5,8	3,3
8th recycling	6,9	4,6	7,3	4,8	7,3	5,2

Table 3. Fine ratio content of pulp fibres.

All tested samples of recycled pulp contained lower portion of fraction 0.1 to 0.2 mm. The augment of this fraction increased during recycling. The highest values were observed at drying temperature 120 °C.

Relative fibre width distribution varies within recycling.

Distribution curves (fig. 5) prove that relative pulp fibre distribution of the biggest widths were in original pulp sample. Lower width distribution increased after 8th recycling.

Pulp fibres were suspended in water and repeatedly treated in beating system during 8-times recycling. Fibre beating caused fibre deformation - waving as a result of fibre compression, flattening and distortion beside of shortening and fibrillation effect. Fiber Tester measured and evaluated local fibre deformations beside of basic dimensional characteristics. Local fibre deformations were defined as the modification of main fibre line direction. The deviation was recorded if the achieved angle was higher than 20°.

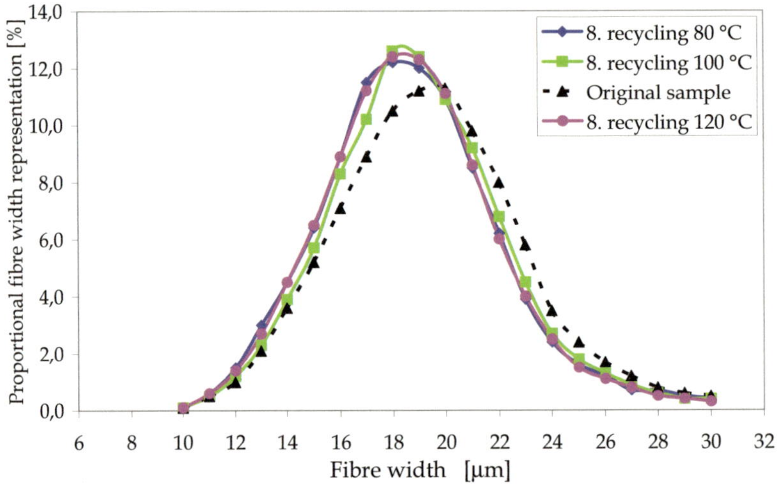

Fig. 5. Relative width distribution of pulp fibres dried 80, 100 and 120 °C.

Number of local fibre deformation informs about fibre weakening which influences strength paper characteristics.

Table 4 shows average values of pulp fibre local deformations in particular recycling levels.

Pulp sample	Average angle of fibre deviation			Deformation count per mm			Deformation count per fibre			Medium index of fibre distortion		
	Drying temperature [°C]											
	80	100	120	80	100	120	80	100	120	80	100	120
Original sample	50			0,60			0,50			1,40		
0th recycling	47	49	48	0,74	0,63	0,67	0,56	0,48	0,51	1,73	1,49	1,58
1st recycling	48	49	48	0,69	0,69	0,67	0,52	0,53	0,51	1,64	1,65	1,59
2nd recycling	48	50	47	0,69	0,72	0,55	0,51	0,55	0,42	1,63	1,76	1,29
3rd recycling	48	50	49	0,76	0,69	0,72	0,56	0,52	0,54	1,79	1,67	1,73
4st recycling	51	50	47	0,98	0,69	0,66	0,74	0,52	0,50	2,42	1,66	1,55
5st recycling	52	50	50	1,06	0,70	0,77	0,79	0,52	0,57	2,63	1,68	1,85
6st recycling	47	46	49	0,69	0,73	0,76	0,51	0,56	0,56	1,63	1,63	1,82
7st recycling	48	50	49	0,76	0,72	0,76	0,57	0,53	0,56	1,80	1,73	1,83
8st recycling	48	49	49	0,69	0,72	0,77	0,51	0,53	0,56	1,65	1,72	1,85

Table 4. Local deformations of pulp fibres.

Average deviation angle is moving within 46 to 52° during recycling. The highest values of deformations per mm, deformations per fibre and fibre distortion index were measured after 8th recycling at drying temperature 120 °C. This fact proves negative impact of higher temperature on pulp fibre deformation.

3.2 Mechanic characteristics of recycled fibres

3.2.1 Breaking length

Breaking length defining paper strength rate was one of observed mechanic characteristics.

Enlargement of fibre active surface due to the outer pulp fibre fibrillation occurred during beating process (Blažej and Krkoška, 1989). This effect induces the increase of fibre bonding, paper strength and finally the increase of breaking length. The fibres are simultaneously shortened during beating which causes deterioration of mechanic properties. Moreover, the negative impact on dewatering ability and bonding properties is amplified by growing ratio of fine fraction (Hubbe and Heitmann, 2007).

Pulp fibre recycling causes the decrease of fibre wall thickness as well as tension strength. Per contra the ratio of lumen diameter and fibre width increases whereas is the direct proportion between tension strength and fibre wall thickness (Okayama, 2001).

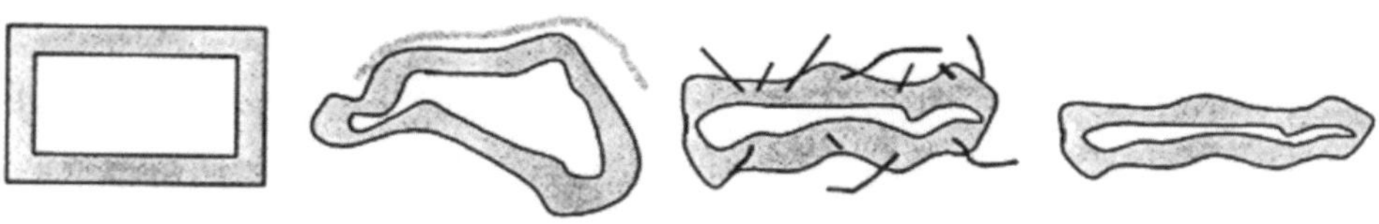

Fig. 6. Different levels of fibre changes (Karlsson, 2006).

Strength properties changes occurred during recycling are closely linked on morphological fibre characteristics (lumen diameter, thinness ratio) (Okayama, 2002).

Tension strength at zero clamping corresponds with fibre strength and increases owing to beating and recycling. Decrease of sheets strength prepared from recycled fibres is caused by the decrease of inter-fibres bonds. Specific strength of inter-fibres bonds is not influenced by recycling fibre treatment. So decreased strength of inter-fibres bonds could be induced only by decrease of inter-fibres bonding surface (Khantayanuwong, 2002).

Figure 7 shows the evolution of pulp sheets breaking length during the process of 8-times recycling.

Intense breaking length increase was caused by the active fibre surface increase after first fibre beating and drying (0th recycling). Whereas the highest value of pulp breaking length was determined at drying temperature 120°C. Consecutive recycling decreases the strength because of pulp fibre shortening and liberation of fibrillated fragments from their surface.

While the values of pulp breaking length dried at 80 °C and 100 °C decrease about 47 %, the breaking length of pulp dried at 120 °C decrease about 68 %.

Based on the evolution of breaking length modification during recycling it could be conclude that the pulp dried at 120 °C reflects the lowest values of breaking length. On the other hand, the highest values of breaking length after 8th recycling were observed on pulp dried at 80 °C. This fact proves the negative impact of higher temperature on strength properties characterized by breaking length.

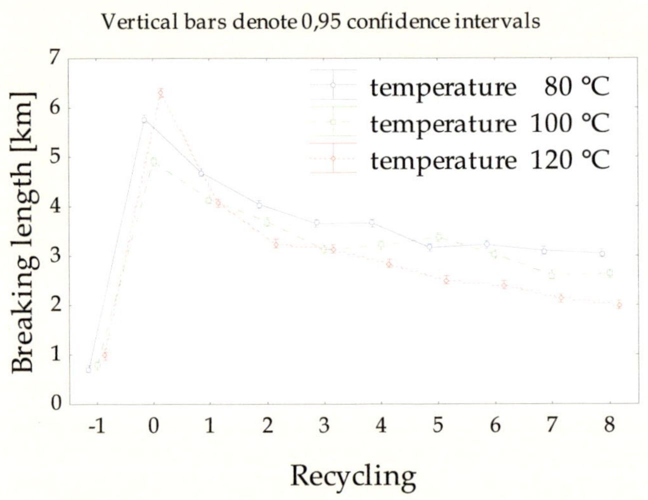

Fig. 7. Pulp breaking length during recycling.

3.2.2 Tearing index

Further monitored parameter characterising strength properties of pulp sheet is tearing index which was established by the method according to Elmendorf.

Tearing strength is characterised by the effort needed for tearing of pre-cut paper sample under established conditions. Tearing strength is then recalculated on square weight and expressed as tearing index.

Tearing strength depends on the character of used fibres and on its treatment level. Higher tearing strengths are observed on kraft pulp and significant decrease of tearing strength is caused by paper bleaching, sizing and intensive drying (Souček, 1977).

Tearing strength is significantly influenced by the fibres amount in tearing area, fibres length, fibres bond density and strength. If the more fibres are longer, the more strength needed for paper tearing divided into bigger surface. Reversely the strength is concentrated on small surface when the fibres are short.

The strength of the paper with small and insufficiently evolved bonding surface is small. It is caused by easy fibre evulsion from the paper structure.

Fibre surface and tearing strength is increasing while fibres beating until the effect of fibres shortening is displayed.

The evolution of tearing index is illustrated by figure 8.

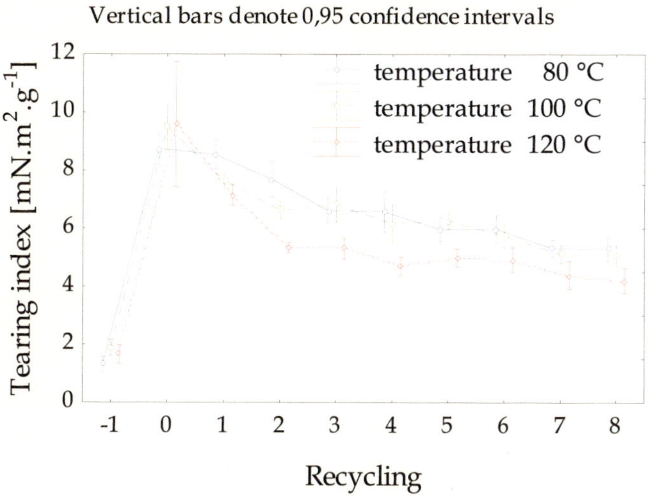

Fig. 8. Tearing index during recycling.

The first beating during 0th recycling shows the highest impact on pulp fibres. Similarly to breaking length the first beating increased significantly tearing index at all observed drying temperatures because of the bonding surface increase. This fact caused by first intensive fibre beating 29 °SR. The highest tearing index value observed on pulp dried at 120 °C, lower at 100 °C and the lowest at drying temperature 80 °C.

Tearing index shows descending character within the whole recycling process. The pulp was beaten at 29 °SR in each recycling level so the fibres were not enormously destroyed.

The pulp dried at 120 °C shows the lowest values of tearing index after 8th recycling. This diminution represents 53 % of maximally achieved value in comparison to 0th recycling. This significant difference is caused by repeated beating and drying at higher temperature. Lower tearing index decrease (46 %) was observed at drying temperature 100 °C and 38 % at 80 °C. Stated tearing index values confirm growing negative impact of increasing drying temperature during whole recycling process.

The negative impact on pulp sheets tearing strength was amplified by growing fine ratio content. The highest fine ratio (0-0.2 mm) content was measured by Fiber Tester after 8th recycling at drying temperature 120 °C and represented 12.5 %, 12.1 % at 100 °C and 11.5 % at 80 °C.

Repeated pulp beating and drying caused fibres surface changes (fig. 6). Enlarged fibre surface (increase of contact surface due to the fibrillation) caused by beating ensures the increase of mechanical characteristics values due to the higher inter-fibres hydrogen bond amount which provides the improvement of fibrils and fibres enlacement during dewatering (Hubbe, 2006 b).

Klofta and Miller (1993) present in their article the importance of liberated fibrils capillary strength and fine ratio on fibre surface during the creation of bleached kraft fibres network.

According to Ackermanna et al. (2000) the fibres shortening influence on fibres mechanic characteristics is not critical since these characteristics are influenced by presence or absence of fine ratio (Ackermann et al., 2000). Moreover, the fibre after repeated recycling is fragile and inclinable to destruction during beating (de Ruvo and Htun, 1983).

The results prove negative impact of drying on mechanic characteristics of recycled fibres. While the evolution of mechanic characteristics changes (fig. 7 and 8) at drying temperatures 80 °C and 100 °C was almost the same, the evolution at drying temperature 120 °C was significantly lower. Fibre hornification and fragility is demonstrated significantly at 120 °C which is proved by increase of fine ratio within recycling process. Matsuda et al. (1994) introduced that typical temperature interval of hornification is 80-120 °C.

3.3 Physical characteristics of pulp fibres

Pulp fibre recycling caused beside of dimensional and mechanic changes, also physical characteristics modifications.

3.3.1 Porosity

Porosity is defined as ability of paper, cartoon or board to permeate the air under established conditions.

Air permeability is given by permeable pores which connect both paper sites and allows air circulation in the direction of hydraulic gradient. Air permeability depends on paper porosity and decrease with increasing beating level, content of fillers and sizes (Souček, 1977).

Pulp sheet samples porosity was observed after particular recycling levels based on ISO 5636/3 according to SCAN P 26:78.

Figure 9 illustrates evolution of porosity changes measured after recycling levels at monitored drying temperatures.

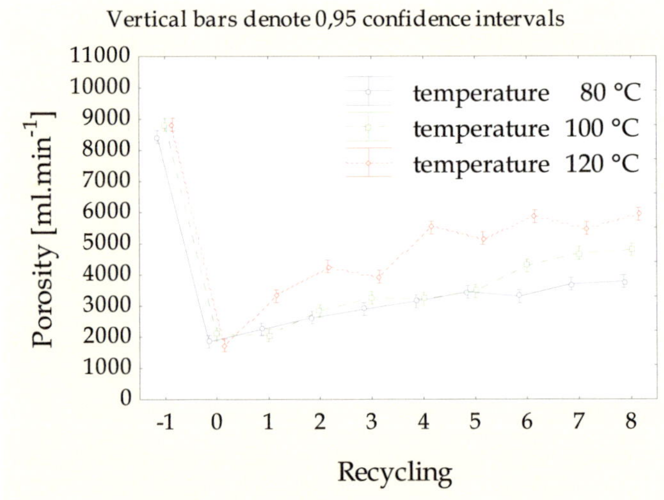

Fig. 9. Evolution of porosity during recycling.

The evolution of porosity values was opposite to the evolution of monitored mechanic characteristics. The highest porosity was observed in original pulp sample where pulp fibres were not affected by beating and air permeability between pulp fibres was high.

First intensive treatment of original fibres by beating increased bonding surface, strength properties (breaking length, tearing index), and the sheet created more consistent fibres network, decreased inter-fibres areas and the porosity plummets.

Beating at 29 °SR in following recycling levels caused fibre shortening, fibrillation and consequent unstuck of cell wall outer primary part and elimination of fine ratio. This fact caused decrease of mechanic characteristics and porosity increase.

The porosity values of pulp sheets dried at 80 °C and 100 °C were much closed until 5th recycling level. The negative impact of higher temperature was demonstrated from next recycling levels. The porosity of sheets dried at 120 °C was higher within all range of repeated recycling.

The increase of sheets porosity between 8th and 0th recycling was 2-times at drying temperature 80 °C, 2.3-times at 100 °C and 3.4-times at 120 °C.

The influence of higher drying temperature was manifested by fibre hornification, theirs shrinkage due to the pore closing in cell wall (Alince, 1997, 2002; Stone and Scallan, 1968), which leads to consequent increase of pulp sheets air permeability (porosity). Electron microscope photos (fig. 10 and 11) shows the surface of recycled kraft fibres and proves this shrinkage.

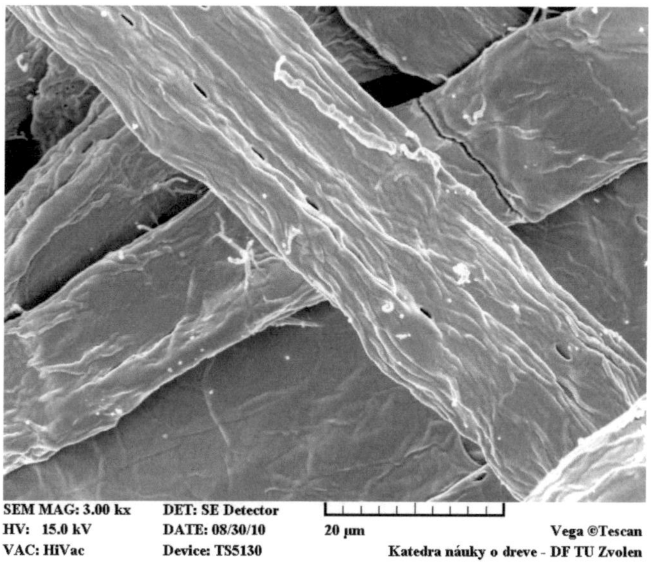

Fig. 10. Surface of dried non-beated pulp fibres.

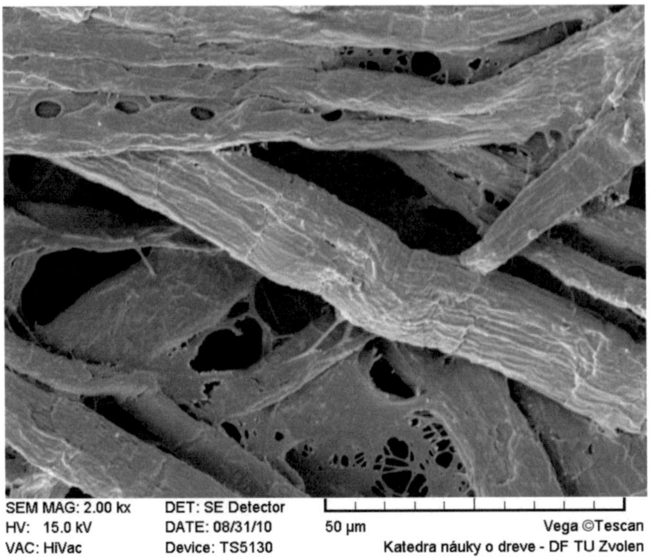

Fig. 11. Surface of pulp fibres after 8th recycling and drying at 120 °C.

3.3.2 Swelling

Swelling is defined as wood ability to increase its linear dimensions, surface or volume via bonded water absorption in range from 0 to the point of fibres saturation. The water comes into the amorphous areas of cellulosic fibrils and enlarges cell wall of particular elements and whole wood. The wood swells during water and water steam absorption until the point of fibres saturation. Swelling does not caused additional increase of water content because the water fill in only lumens or eventually inter-cells cavities (Požgaj et al., 1997).

Swelling ability is given mainly by free -OH groups which are able to bond water via hydrogen bonds. The wood content of free –OH groups is markedly limited because of lignin-saccharide bond blocks. Bleached pulp fibres with almost complete lignin removal contents higher amount of free -OH groups. Swelling is more intensive with regard to speed as well as maximal values.

Figure 6 shows different levels of fibres changes which they pass during recycling process because of repeated beating and drying. Original pulp fibre cell structure is devoid of outer layer parts by beating (P, S1). Moreover the fibres are fibrillated and swell in water medium. Consecutive drying causes fibres collapse which induces decrease of swelling ability.

Kraft pulp with significantly decreased strength during recycling swells less and has lower bonding potential. Recycled fibres are markedly less hydrophilic than original fibres. Water contact angle increases extremely which is related to fibre surface inactivation during recycling known as "irreversible hornification" (Takayuki, 2002).

Plenty of scientists (Stone and Scallan, 1966; De Ruvo and Htun, 1983) concluded that the loss of repeated pulp fibres swelling abilities is caused by the pores closure in pulp walls and pore disability to re-open when watered.

Beating is one of the most important operations for obtaining paper potential of recycled fibres and it could partially reverse recycled pulp fibres hornification. Beating increases fibres flexibility and their swelling ability which improve bonding and strength (Woodward, 1996).

Fibre crystallinity index varies within recycling process. This effect is caused by negligible crystallinity increase of specific amorphous fibre area during recycling. The insufficient lumen re-opening of moist recycled fibres causes that water content bonded in fibre wall influences considerably re-swelling of recycled fibres. The decrease of recycled fibres re-swelling ability is parallel with the decrease of cell wall bonded water content. This event is caused by the diminution of amorphous areas known as sub-morphological changes of fibre wall (Khantayanuwong, 2002).

Modified method of swelling kinetics determination (Solár et al., 2006) was used for experimental monitoring of pulp fibres swelling. The method is based on dimensional changes recording of test pulp sheets swelled in water via sensors and their transfer into electric signals. Swelling is defined as the difference between immediate and initial sample dimensions (in %).

The experimental results of recycled pulp fibres swelling changes depending on the time allow concluding that:

- maximal pulp fibres swelling speed is achieved during first seconds of contact between pulp sheet and used medium - water. Swelling speed descends consequently till zero value
- swelling ability of pulp fibres decreases with recycling
- maximal swelling of pulp fibres was achieved always during 0th recycling (after first beating).

Figure 12 indicates the swelling evolution of pulp fibres dried at 80 °C.

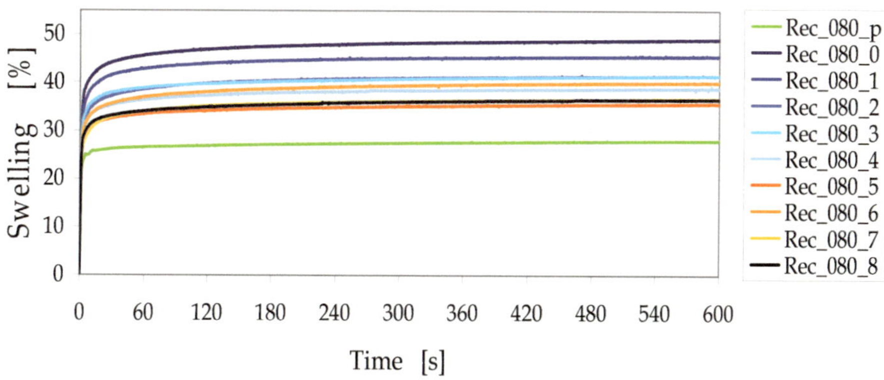

Fig. 12. Swelling evolution of pulp fibres dried at 80 °C.

Swelling of recycled pulp fibres dried at different temperatures after one-second contact with water shown on figure 13.

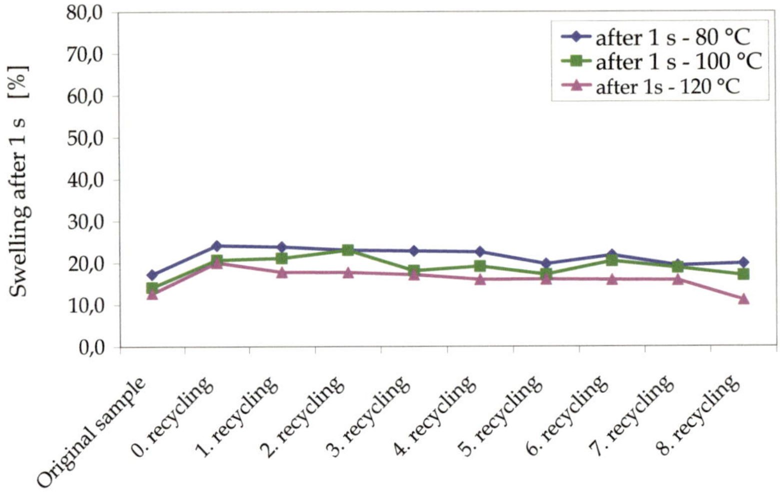

Fig. 13. Swelling of pulp fibres after one second contact with water.

Pulp fibres dried at 80 °C shown the highest swelling value after one second contact with water within whole recycling range.

Figure 14 illustrates swelling values of pulp fibres after 600 seconds contact with water.

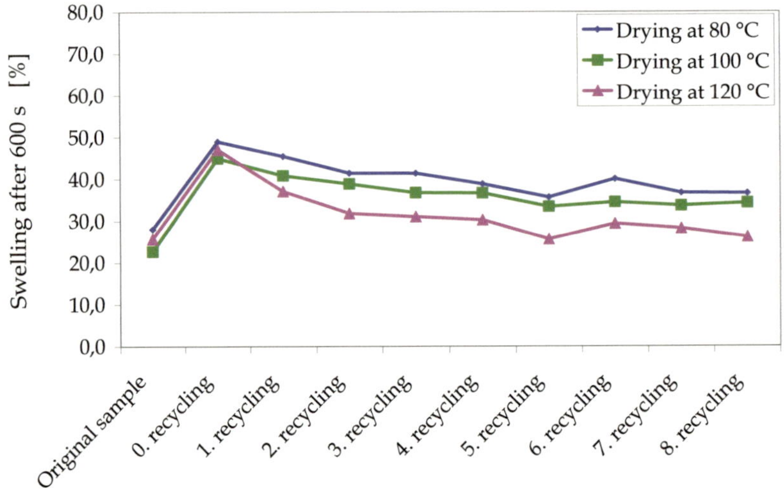

Fig. 14. Swelling of pulp fibres after 600 seconds contact with water.

Swelling ability of pulp fibres decreases with increasing recycling level. Maximal swelling of pulp fibres was achieved during 0th recycling (after first beating).

Difference of swelling abilities at different drying temperatures increases after 600 seconds of pulp fibres swelling.

The fibres swelling decreases because of repeated recycling which covers water plucking, refining and re-drying at monitored temperatures. It is caused by pulp fibres hornification as a result of irreversible pore closing in cell walls. Whereas significant negative influence of higher temperature on swelling values was observed.

3.3.3 Degree of polymerization

Cellulose is natural polymer with variably long chains. Several authors indicate higher degree of polymerization (DP) as 6000-8000 (Albersheim, 1965), 14000-15000 (Hon, 1994, Rowell, 2005), 14000 Marx et al. (1966) for native cellulose form. Wood delignification and cellulose chains degradation occur during kraft cooking and consecutive pulp bleaching. This process continues during recycling and drying process. DP depends on wood species, technological treatment process and cellulose determination (isolation) method. The values are moving within 500-1500 (Kačík and Kačíková, 2007) and 100-10000 Solár (2001). Zugenmaier (2008) mentions the interval of DP within 950-1300 for kraft pulp.

DP of cellulose constituting cell wall skeleton influences several paper characteristics, mainly strength that is given by fibres strength and inter-fibres bond amount.

Humidity and temperature action causes beside of fibres fragility and hornification, also degradation of cellulose chains (Kato and Cameron, 1999; Ackermann et al., 2000; Hubbe et al., 2003; Dupont, 1996) which is manifested by DP decrease. This DP decrease is more significant during the action of higher temperatures on wet pulp fibres than on dry fibres.

Viscosity measure in solvent FeTNa (ISO 5351/2-1981 (E) where the cellulose is relatively stable, is suitable method for monitoring of cellulose degradation. This method allows to establish limit viscosity number (LVN) which is used for DP calculation based on Staudinger equation $DP = [LVN]/8{,}14.10^{-4}$.

Figure 15 shows DP changes of hardwood pulp in recycling levels.

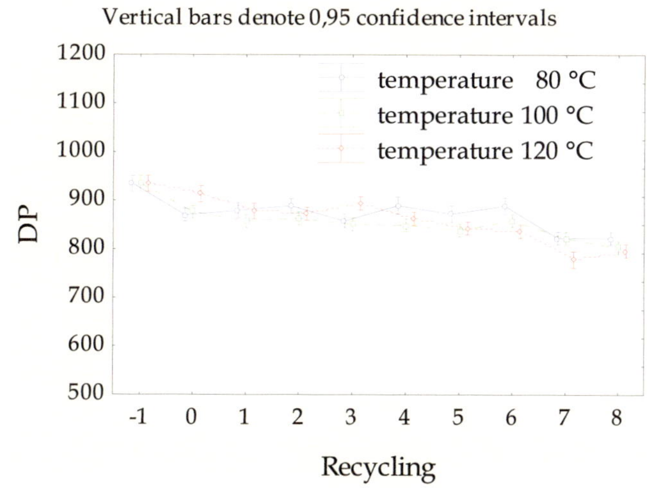

Fig. 15. DP of hardwood pulp during recycling (Čabalová and Geffert, 2009).

Cellulose samples contain cellulose chains of different length. Viscometric determination of DP represents the value characterising average chain length. Gained DP results allow concluding that cellulose chains shorten during recycling because of glucosidic bond fission whereas DP decrease was the slightest at drying temperature 80°C.

The disadvantage of viscomeric measure is that it provides only the information about average values of DP. Therefore, it is better to characterise recycled pulp by gel permeability chromatography (GPC) which allows detecting the distribution of molar weights too (Dupont and Mortha, 2004).

The cellulose in pulp samples was derivatived on tricarbanilates according to Kačíka et al. (2007) for GPC needs. The samples were analysed and evaluated based on the conditions stated by Kačíkom et al. (2009). The main advantage of this method is that derivatization process of complete cellulose tri-substitution runs in one reaction step and without degradation impact on cellulose. GPC method for the analysis of cellulose tricarbonilates was used by several authors (Daňhelka et al., 1976; Kačík et al.,2007; Kučerová and Halajová, 2009; Čabalová et al., 2009).

Measured results (Kučerová and Halajová, 2009) of distribution - log (M) upon taking into account complete tri-substitution (DP = M/519) were used for the evaluation of relative distribution of cellulose DP in pulp dried at monitored temperatures (80 °C, 100 °C a 120 °C) within recycling levels.

Figure 16 illustrates the relative distribution of cellulose DP in pulp after 8th recycling at drying temperatures 80 °C, 100 °C and 120 °C.

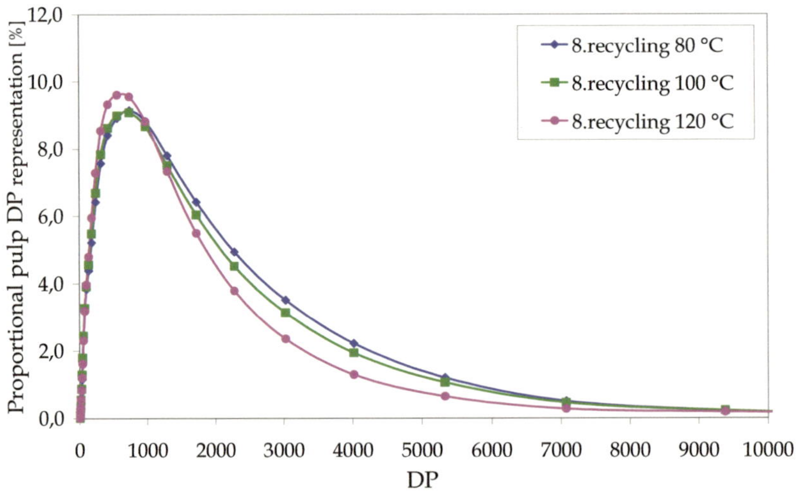

Fig. 16. Relative DP of pulp dried at 80 °C, 100 °C and 120 °C

The comparison of relative cellulose DP distribution after 8th recycling at all monitored temperatures shows that relative distribution of shorter cellulose chains increases with increasing drying temperature.

Maximal DP value of pulp dried at 80 °C and 100 °C was 736, while the maximum DP of pulp dried at 120 °C was 645. Particular deviations of DP decrease were observed during recycling since this process is relatively complicated and influenced by many factors acting positively as well as negatively.

Some authors (Khantayanuwong, 2003; Chen, Wang, Wan and Ma, 2009) realized beside of DP decrease, also crystallinity increase in recycling process. This increase could be relative since shorter hemicellulose chains and amorphous cellulose portion is degraded as first during repeated recycling and drying.

3.4 Optical characteristics of recycled fibres

Many various optical events perceived as diffusion surface reflection appears during light fall on paper. One light part is reflected from paper as emitted light and the other part is absorbed by fibres and pigments. Four optical phenomena are observed - reflection, refraction, absorption and diffraction of light.

Reflection, refraction and diffraction are merged in one term - light diffusion which represents important property in paper technology domain (Pauler, 2002). Fibrillar paper, cartoon and boar structure with fibres and filler particles divided by pores filled by air creates suitable conditions for diffused reflection. Each interface between solid matter and air creates light rays reflection and refraction (Souček, 1977).

3.4.1 Brightness

Brightness is defined as material ability to diffuse light flow upon minimal absorption or penetration in the way that spectral composition of reflected light in visible domain remains the same as falling light. Brightness is expressed in reflectivity percentage of basic brightness standard (MgO) measured at wave $\lambda = 457 \pm 5$ nm (Souček, 1977).

Hardwood pulp sheets were prepared after each recycling level. Experimental sheets were used for measurement and statistics evaluation of brightness. Optical changes depending on drying temperature occur during the process of repeated recycling and drying. Figure 17 shows brightness variation of pulp sheets surface at monitored drying temperatures (80 °C, 100 °C and 120 °C).

Pulp brightness decreases in 0th recycling at all temperatures because of fibres surface changes due to intensive beating (in average from 15 °SR to 29 °SR). First beating caused the increase of fibre surface due to the liberation of primary fibre wall and the significant increase of light diffusion. Brightness increase observed during next recycling when pulp fibres were slightly refined at 29 °SR. This increase was the most significant at 80 °C. This effect could be explained firstly as successive leafing of small fibre primary wall particles and theirs removal from suspension during dewatering and then as decrease of total fibres surface. Deviations of brightness values during recycling are probably linked to other fibres surface changes due to the refining (shortening, fibrillation, delamination).

Statistic evaluation of achieved results reflects significant influence of recycling, temperature and the interaction of both factors on brightness values.

Negative influence of higher sheets drying temperature on pulp brightness is illustrated on graphic dependence of figure 17. Lower brightness of pulp sheets surface due to the higher drying temperature (120 °C) was examined during whole 8-times recycling process. This is the consequence of cellulose and hemicellulose oxidation reaction at higher temperature in the presence of air oxygen. Reactions are running on primary and secondary hydroxyl groups of pyranose circle and create carbonyl and carboxyl groups. This fact causes paper yellowing since arising substances are chromophores and have the ability of visible radiation absorption (Solár, 2001; Margutti et al., 2001).

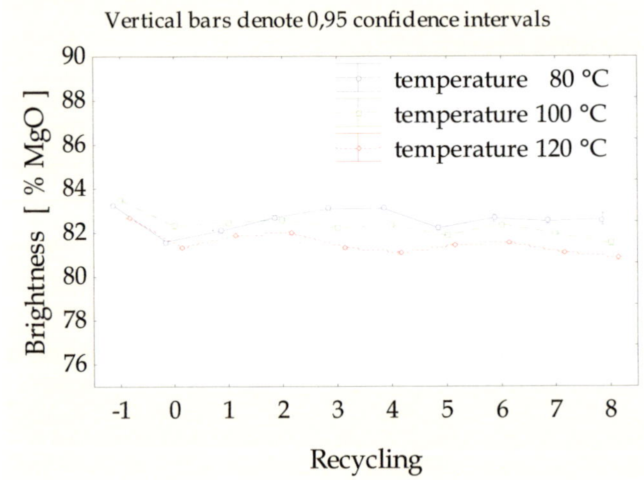

Fig. 17. Brightness of sheets surface during recycling at different drying temperatures.

3.4.2 Colourity

Paper industry has noted important changes of paper optical properties measurement in 70th and 80th years. Due to this development the brightness was not sufficient to paper quality characterization. Spectrophotometers allow to measure and define paper lightness and colour.

World widely used colour system CIE L*, a*, b* which consists of grey scale axis L*, yellow-blue axis a* and green-red axis b* in three-dimensional colour system based on 4basic colours (Pauler, 2002).

The colour of sheets surface was measured by spectrophotometer Minolta CM 2600d and the process of measure and evaluation was realized by software „Spektra Magic".

Sheets surface reflection was measured in whole range of visible spectrum from 360 nm to 740 nm with resolution of 10 nm. Whereas difference spectrum was evaluated from the difference of average sheets reflection values after particular recycling levels at monitored drying temperatures.

Negative influence of higher drying temperature manifests already on sheets surface reflection of original pulp sample (fig. 18). While the reflection within whole measured spectrum at drying temperatures 80 °C and 100 °C was almost overlaid, the reflection values mainly in lower wave lengths were lower at drying temperature 120 °C.

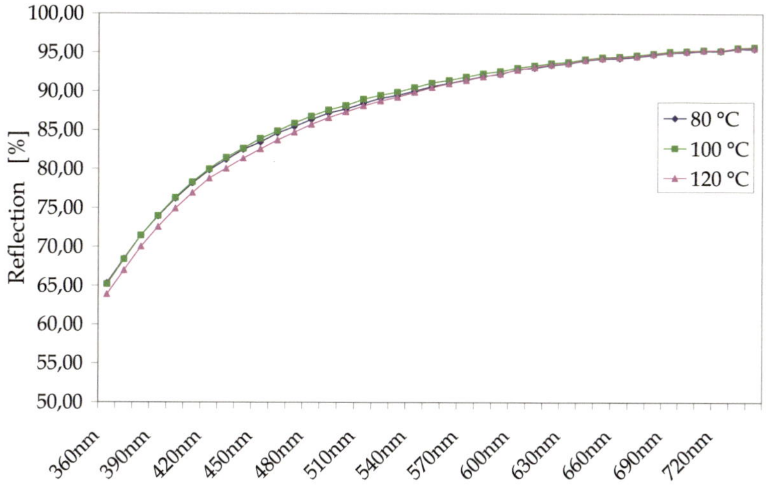

Fig. 18. Reflection of original pulp sample.

The reflection difference of pulp dried at monitored temperatures was the most significant after 8th recycling (fig. 19). The highest reflection in whole measured spectrum was kept only for pulp fibres dried at 80 °C. The lowest reflection was measured for pulp dried at 120 °C. The reflection of pulp dried at 100 °C was moving between the reflection values of 80 °C and 120 °C, at higher wave length it approximates to the reflection values of pulp dried at 80 °C.

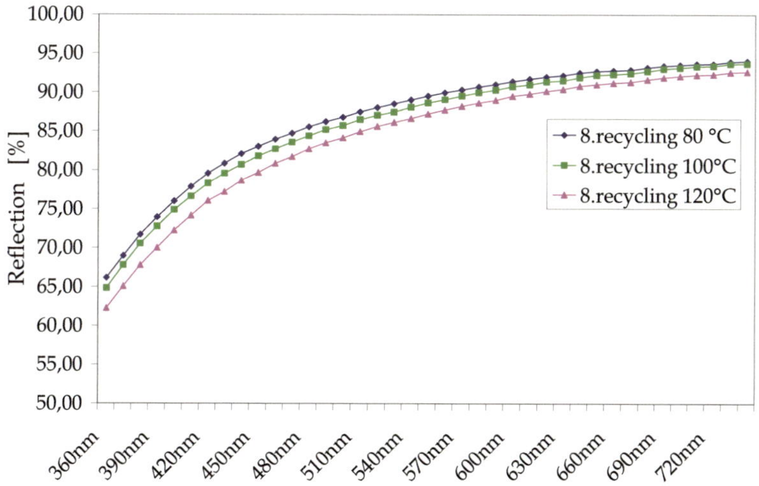

Fig. 19. Pulp reflection after 8th recycling.

Graph of differential spectrum indicates the change of average reflection measured on pulp sheets surface within the wave length range 360 nm to 740 nm after 8th recycling with the regard to average reflection of original sample at monitored temperatures (fig. 20).

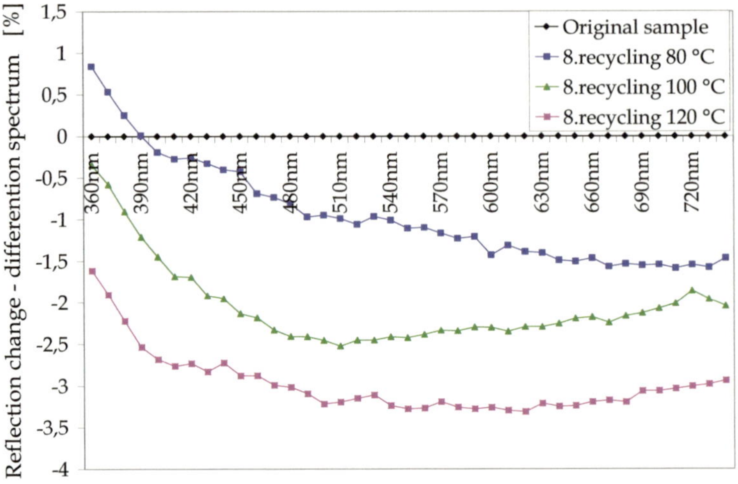

Fig. 20. Differential spectrum at monitored temperatures.

The smallest difference between the sheets surface reflection after 8th recycling and reflection of original pulp was determine at 80 °C and the biggest one at 120 °C.

The evolution of sheets surface differential spectrum dependence after 8th recycling is similar at all monitored drying temperatures. The decrease from 360 nm to 410 nm was linear, the differential spectrum of pulp dried at 80 °C slightly descends in whole wave length range, while for drying temperature 100 °C it was observed slight difference decrease 520 to 740 nm and the decrease was moved to the end of monitored spectrum within 680 to 740 nm for drying temperature 120 °C.

Achieved results indicate that higher drying temperature causes chromophores creation absorbing at whole monitored spectrum range.

Light reflection decrease of sheets surface made by repeatedly recycled and dried pulp fibres indicates the creation of secondary chromophores.

Absorption maximum at 350-368 nm is given by resting lignin chromophores. This effect is caused by Cα-Cβ π bonds conjugated system with aromatic nucleus, phenolic and methoxyl structures which results to increased light absorption. Moreover kraft pulp contains chalcones chromophores with absorption maximum at 368 nm and extremely high absorption maximum at 478 nm is given by p, p´-stilbenchinones present in lignin (Solár, 2009). Absorption spectrum of cellulose materials below 420 nm is influenced mainly by lignin and the absorption of other components is relatively low (Blažej et al., 1975). In addition α-karbonyl conjugated with aromatic nucleus is intensive chromophore in UV spectrum (358-365 nm). In the case of extended system conjugation with phenolate anion in p-position of aromatic nucleus the absorption maximum is moved into the zone of 400 nm. Insatureted γ-karbonyl structures with resting lignin macromolecule are important chromophores that cause the absorption of visible radiation within 340-440 nm (Solár, 2001). Absorption maximum of thermic oxidated cellulose is set at wave length 360 nm (Bos, 1972).

The reason of pulp colouration in visible spectrum could be also the presence of microelements, iron and heavy metals that create colour complex with phenolic structures (Falkehag and Marton, 1961; Blažej and Šutý, 1973).

Colour changes of paper surface was evaluated in colour space CIE L*, a*, b* depending on colour shade shift along axis a* (from green to red) and along axis b* (from blue to yellow). The change of lightness colour component was characterized by the value of specific lightness L*.

Sheets surface colour after particular recycling levels was evaluated based on measured reflection values in visible spectrum area (360 nm to 740 nm), coordinates values in colour CIE **L***, **a*** a **b***.

Specific lightness evolution (L*) of monitored sample surface dried at three different temperatures is shown on figure 21 and in colour space moves from black to white colour.

Recycling caused slight decrease of sheets surface specific lightness at all three drying temperatures. Specific lightness values at 80 °C and 100 °C were very closed in whole recycling range.

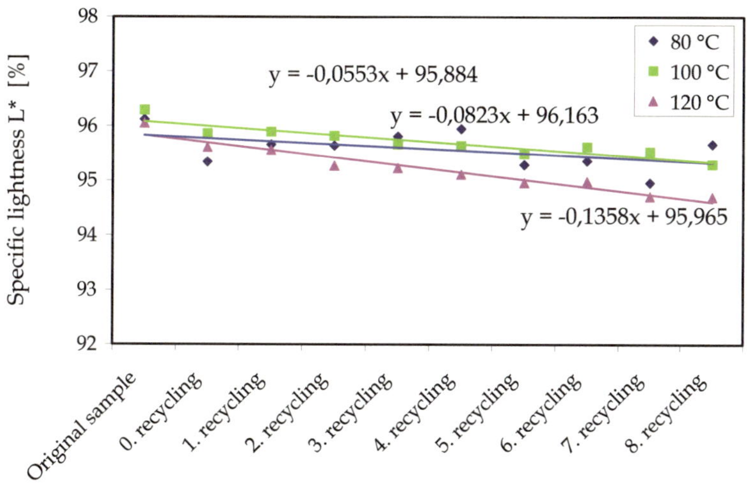

Fig. 21. Specific lightness change L*.

Evolution of values b* in particular recycling levels that characterized the coordinate in colour space from blue to yellow is shown on figure 22 at all three temperatures.

Sheets dried at 120 °C display the highest values of coordinate b* (shift into yellow shape) and the lowest values were determined at 80 °C (closer to blue shape). The values b* of sheets dried at 100 °C are situated between them.

Figure 23 illustrates the shift in plane L* b* of colour space CIE L*a*b* of original pulp sample and pulp after 8th recycling at all three monitored temperatures.

Introduced figure indicates that the colour of pulp sheets moves into dark and yellow shapes with increasing drying temperature.

Variation of coordinate a* during recycling was negligible.

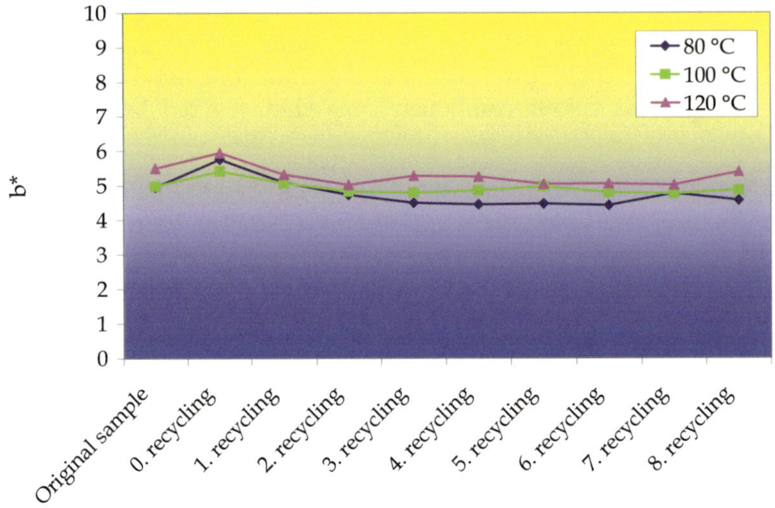

Fig. 22. Change of colour shape coordinate b*.

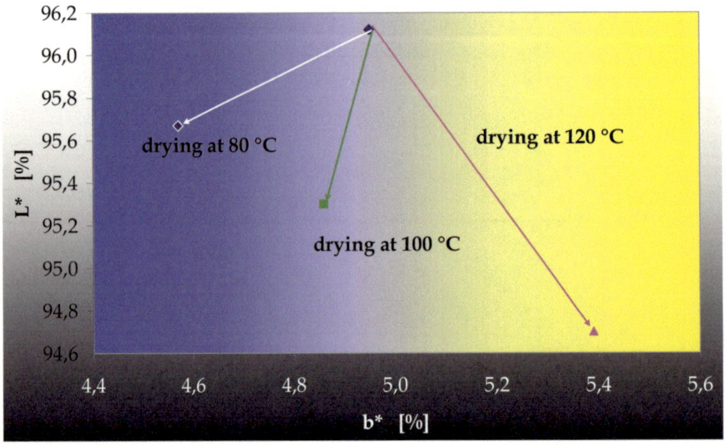

Fig. 23. Shift in plane L*b* of colour space in the beginning and in the end of recycling.

3.5 Additional beating influence on changes of recycled pulp fibres properties

Pulp fibres were refined at 29 °SR within each 8-times recycling level. Monitored chosen characteristics deteriorate depending on recycling level and drying temperature. The biggest variations of qualitative parameters were observed on the fibres of bleached hardwood pulp dried at 120 °C. The aim of 9th recycling – additional beating at 49 °SR and drying at 120 °C was deeper investigation of 8-times recycled fibres influenced by higher loading at beating.

Characteristics	Original sample	After 8th recycling	After refining at 49 °SR
Fibre length [mm]	0,824	0,744	0,729
Fibre width [μm]	20,8	19,8	19,5

Table 5. Dimensional characteristics of fibres refined at 49 °SR after recycling.

Relative distribution of fibres length and width were moved into lower values (fig. 24 and 25) due to the additional beating at 49 °SR.

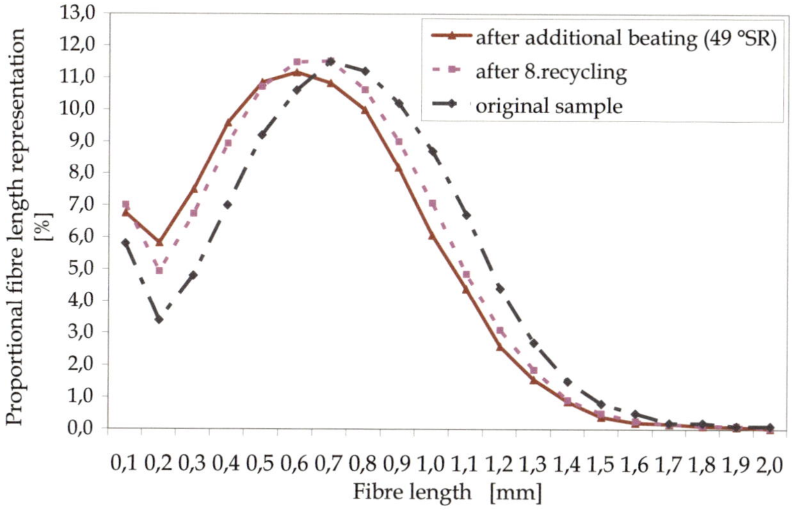

Fig. 24. Relative distribution of pulp fibres length (49 °SR, 120 °C).

Table 5 and figures 24 and 25 indicate that intensive additional beating at 49 °SR after 8-times recycling explicitly decrease the distribution of longer and wider fibres and increase the distribution of shorter and thinner fibres. Higher drying temperature (120 °C) causes that the fibres were more keratinized and fragile. Intensive additional beating at 49 °SR occurred not only to fibrillation, but also to shortening.

Although the values of pulp fibres dimensional characteristics were decreased, increased beating effect influenced positively pulp fibres mechanic properties. Increased fibres fibrillation induced the increase of inter-fibres bonding surface which caused the augmentation of breaking length and tearing index (see tab. 6).

Characteristics		Original sample	After 8th recycling	After refining at 49 °SR
Breaking length	[km]	0,97	2,00	5,22
Tearing index	[mN.m^2.g^{-1}]	1,60	4,51	6,77

Table 6. Mechanic characteristics of fibres refined at 49 °SR after recycling.

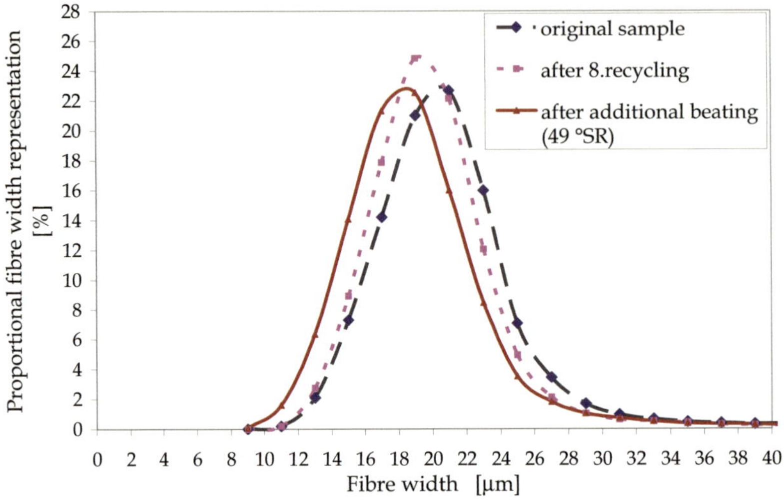

Fig. 25. Relative distribution of pulp fibres width (49 °SR, 120 °C).

Additional beating of 8-times recycled fibres at 49 °SR evoked that the average value of breaking length increased 2.6-times and tearing index 1.5-times.

Intensive additional beating influenced also porosity. Increased inter-fibres bonding decreased the area between fibres thereby the air permeability of sheet structure decreased. The porosity of experimental sheets prepared after beating at 49 °SR demonstrated significant decrease from 5 935 ml.min^{-1} to 1 425 ml.min^{-1} (fig. 26).

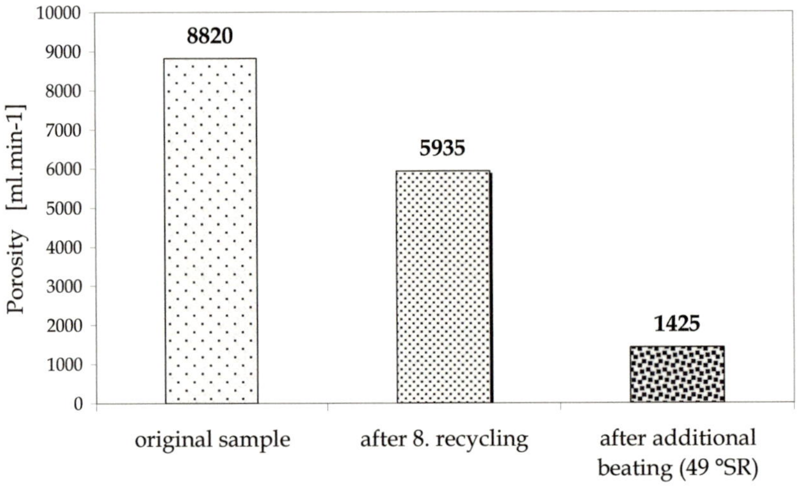

Fig. 26. Sheets porosity.

Negative effect of intensive additional beating at 49 °SR displayed on sheets brightness too (see fig. 27).

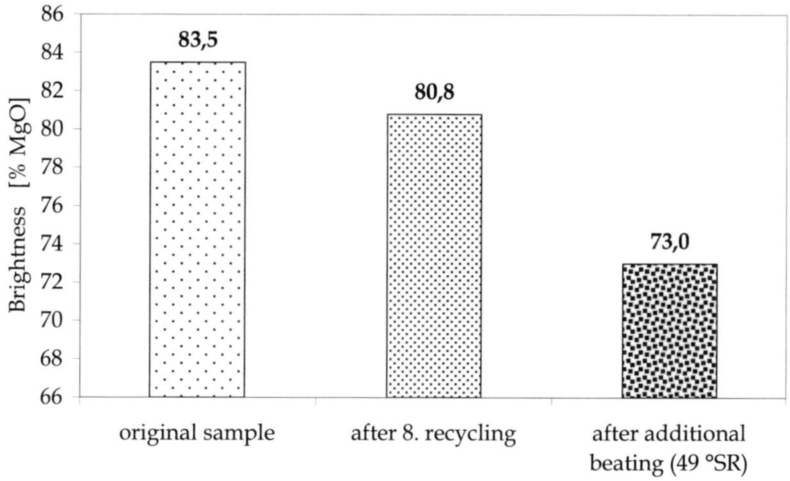

Fig. 27. Sheets brightness.

Beating at 29 °SR and drying temperature 120 °C induced the brightness decrease of kraft hardwood pulp about 2,7 % MgO. Additional beating at 49 °SR and following drying decreased the brightness value about 7,8 % MgO. This effect could be explained by enlargement of fibres surface due to the beating which influenced diffusion reflection of fallen luminous flow.

4. Conclusion

Achieved results of kraft hardwood pulp recycling demonstrate that all monitored characteristics deteriorate with increasing recycling level.

Although the pulp fibres were refined only at 29 °SR during 8-times recycling, it was observed the decrease of average fibre length and the increase of relative distribution of shorter and thinner fibres.

Negative influence of recycling combined with increased drying temperature manifested on mechanic characteristics. Increased fibre fragility due to the hornification caused the decrease of breaking length and tearing index during recycling.

Repeated drying during recycling increased the pulp sheets porosity due to the fibre hornification and shrinkage. Decreased swelling ability was caused by cell wall pores closing and their limited re-opening ability in water contact.

Repeated refining and drying influenced negatively average polymerization degree of cellulose. This fact displayed by the decrease of relative longer fibres distribution.

The negative effect of recycling on optical characteristics of bleached pulp (brightness, colourity) was amplified by increased drying temperature. Pulp colour moved into yellow shape during recycling.

Additional beating at 49 °SR of 8-times recycled pulp fibres showed that these fibres retained sufficient potential of strength properties with parallel decrease of porosity and brightness.

In practice the paper contains beside of pulp fibres, also many auxiliary paper agents that are combined with other factors (way of treatment, warehousing) and could negatively influenced some recycled fibres properties.

The growth of paper production in the future will not be possible without utilization of recycled kraft pulp fibres. The majority of them are originated from hardwood. The treatment of these recycled fibres should take into account the dominant influence of beating (additional beating). Achieving of optimal beating degree 28 - 29 °SR assures the extension of fibres life cycle (increase of recycling levels).

Drying temperature is other important recycling factor and is closely linked on the changes of fibres swelling ability and paper porosity. This effect influences significantly technological parameters of paper production and should be take into account during the designing of paper machines using recycled fibres.

5. Acknowledgment

This research was supported by the Scientific Grant Agency of Ministry of Education of the Slovak Republic under contract No. 1/0272/11.

6. References

Ackermann, Ch., Gottshing, L., Pakarinen, H. (2000). Papermaking potential of recycled fiber. In *Recycled fiber and deinking*, Papermaking Sci. Technol.Ser., Fapet Oy, Helsinki, Ch. 10, 2000. ISBN 952-5216-07-1, p. 358-438

Albersheim (1965). In Kačík, F., Kačíková, D. (2007). Charakteristika a analýza celulózy a jej derivátov. Monografia.1. vyd. Zvolen: TU vo Zvolene, 2007. 94 s. ISBN 978-80-228-1819-3

Alince, B. (2002). Porosity of swallen pulp fibers revisited. In *Nordic Pulp and Paper*. ISSN 0283 2631, 2002, vol. 17, (1) p. 71-73

Alince, B., van de Ven, T.G.M. (1997). Porosity of swollen pulp fibers evaluated by polymer adsorption. In *The Fundamentals of Papermaking Materials*, Trans. 11th Fundamental Researh symp., Cambridge, 1997, baker, C.F.(ed.) Pira International, Leatherhead, Surrey, UK

Attwood, D. (1983). Fiber recycling. In *Recycling in the production of pulp and paper*. Eucepa symposium, Praha 1983, p.91

Blažej, A. et al. (1975). Chémia dreva, 1.vyd. Bratislava:Alfa, 1975. 221 s

Blažej, A., Krkoška, P. (1989). Technológia výroby papiera. 1.vyd. Bratislava : Alfa, 1989. 584 s. ISBN 80-05-00119-3

Blažej, A., Šutý, Š. (1973). Rastlinné fenolové zlúčeniny. 1.vyd. Bratislava: Alfa, 1973. 236 s

Blechsmidt, O. (1979). Altpapier-Faserstoff für die Papiererzeugung. VEB Fachbuchverlag, Leipzig 1979

Bos, A. (1972). Reflectance spectra of cellulose. Appl. polym. Sci. 16, 2567-2576. In *Wood Chemistry, Ultrastructure, Reactions*, Gruyter. Berlin, New York, 613 p. ISBN 0-89925-593-0

Čabalová, I., Geffert, A. (2009). Zmeny priemerného polymerizačného stupňa listnáčových a ihličnáčových buničín v procese recyklácie. In *ACTA FACULTATIS XYLOLOGIAE*, ISSN 1336-3824, 2009, 51 (2): 79-85

Čabalová, I., Kačík, F., Sivák, J. (2009). Zmeny distribúcie mólových hmotností celulózy pri recyklácii buničinových vlákien. In *ACTA FACULTATIS XYLOLOGIAE*, ISSN 1336-3824, 2009, 51 (1): 11-17

Daňhelka, J., Kössler, I., Boháčková, V. (1976). Determanitionof molecular weight distribution of cellulose by conversion into tricarbanilate and fractionation. In *Journal of polymer Science: Polymer Chemistry Edition*, 14, s. 287-298

De Ruvo, A., Htun, M. (1983). Fundamental and practical aspects of paper-making with recycled fibers. In: *The Role of Fundamental Research in Paper Making*, Brander(ed.), Mechanical Engineering Pub., Ltd., London, Vol.1, s. 195-225

Dupont, A. L. (1996). Degradation of Cellulose at the Wet/Dry Interface. In *Restaurator.* International Journal for the Preservation of Library and Archival Material. ISSN 0034-5806, 1996, 17, s. 145-164

Dupont, A. L., Mortha, G. (2004). Comparative evaluationof size-exclusion chromatography and viscometry for the characterisation of cellulose. In *Journal Chromatogr. A.*, ISSN 0021-9673, 2004, 1026 (1-2), p. 129-141

Falkehag, S.I., Marton, J. (1961). Lignin structureand reactions. Advances in Chemistry series, R .F. Gould (Ed.), American Chem. Soc. Publications, Washington, p.75

Göttsching, L. (1976). Einfluss des Altpapier-Recycling auf Suspensions und Papiereigenschaften. In *Sekundárne vlákna a ich využitie v papierenskom priemysle.* Eucepa symposium, Bratislava 1976

Göttsching, L., Stürmer, L. (1975). Wochnbl. Für Papierfabrik 103, 1975,č. 23/24, s. 909

Hnětkovský, V. (1983a). Speciální otázky zpracování sběrového papíru. PPC, Praha 1983

Hnětkovský, V. (1982). Zužitkování sběrového papíru jako základní vlákniny v papírenském průmyslu. PPC, Praha 1982

Hnětkovský, V. et al. (1983b). Papírenská příručka. 1. vyd. Praha : SNTL, 1983. 864 s

Hon (1994). In *Kačík, F., Kačíková, D. (2007). Charakteristika a analýza celulózy a jej derivátov.* Monografia.1. vyd. Zvolen: TU vo Zvolene, 2007. 94 s. ISBN 978-80-228-1819-3

Howard, R.C. (1990). The effects of recycling on paper quality. In *J. Pulp paper Sci.* 16(5), p 143-149

Howard, R.C. (1994). Fundamental problem in recycling. In *Prog. Paper Recycling* 3 (3), p. 66-70

Howard, R.C. (1995). The effects of recycling on paper quality. In *Technology of paper recycling*, R.W.J. Mc Kinney (ed.), blackie Academic, London, Ch. 6, p. 180-203

http://swivel.com/charts/2381-Paper-Consuption-per-capita-by-Country

http://www.cepi.org/content/default.asp - Key Statistics 2009. European Pulp and Paper Industry

Hubbe, M. A. (2006). Bonding between Cellulosic Fibers in the Absence and Presence of Dry-Strenght Agents-A review, In *BioResources*. ISSN 1930-2126, 2006, 1 (2), s. 281-318

Hubbe, M. A. et al. (2003). Changes to unbleahced kraft fibers due to drying and recycling. In *Prog. Paper Recycling*. 2003, vol 12 (3) s. 11-20

Hubbe, M. A. et al. (2007). What happens to cellulosic fibers during papermaking and recycling? A Review. In *BioResources*. ISSN 1930-2126, 2 (4), p.739-788

Hubbe, M. A., Zhang, M. (2005). Recovered kraft fibers and wet-and dry-strenght polymers. In *TAPPI 2005 Practical Papermarkes Conf..*, TAPPI Press, Atlanta

Hubbe, M.A., Heitmann, J.A. (2007). Rewiew of factors affecting the release of water from cellulosic fibers during paper manufacture, In *BioResources*. ISSN 1930-2126, 2 (3), s. 500 -533

Chen, Y., Wang, Y., Wan, J., Ma Y. (2009). Crystale and pore structure of wheat straw cellulose fiber during recycling. In Cellulose. Vol.17 (2) s.329-338 www.SpringerLink.com

Chovanec, D. (1975). Výskum rozdielov v morfológii vlákien listnatých drevín. In IV celulózo-papierenské dni. Vranov 1975, s. 113

ISO 5351/2-1981 (E) (1981): Cellulose in dilute solutions - Determination of limiting viscosity number. Part 2: Method in iron (III) sodium tartrate complex ($EWNN_{mod\ NaCl}$) solution

Jayme, G., Büttel, H. (1968). The determination and the meaning of water retention value (WRV) of various bleached and unbleached cellulosic pulps, Wochenbl. Papierfabr. 96 (6) s. 180-187

Kačík, F., Geffertová, J., Kačíková, D. (2009). Charakterizácia celulózy a buničín metódou gélovej permeačnej chromatografie a viskozimetrie. In *Acta Facultatis Xylologiae*. ISSN 1336-3824,2009, 51 (2) s. 93-103

Kačík, F., Kačíková, D. (2007). Charakteristika a analýza celulózy a jej derivátov. Monografia.1. vyd. Zvolen: TU vo Zvolene, 2007. 94 s. ISBN 978-80-228-1819-3

Karlsson, H. Fibre Guide. (2006). Fibre analysis and process applications in the pulp and paper industry. AB Lorentzen & Wetre, Box 4, SE-164 93, KISTA, Sweden, p.120, ISBN 91-631-7899-0

Kato, K.L., Cameron, R.E. (1999). A review of the relationship between thermally-accelerated ageing of paper and hornification. In *Cellulose*, ISSN 0969-0239, 6 (1), p. 23-40

Khantayanuwong, S. (2002). Effects of Beating and Recycling on Strength of Pulp Fibers and Paper, *Kasetsart J.* (Nat. Sci.) 36, s. 193-199

Khantayanuwong, S. (2003). Determination of the Effect of Recycling Treatment on Pulp Fiber Properties by Principal Component Analysis. In *Kasetsart Journal*. Bangkok (Nat. Sci.) 37, 2003, s. 219-223

Khantayanuwong, S. et al. (2002). Changes in Crystallinity and Re-swelling Capability of Pulp Fibers by Recycling Treatment. In *JapanTappi Journal* 56 (6), s. 103-106

Klofta, J.L., Miller, M. L. (1993). Effects of deinking on the recycle potential of papermarking fibers. In *Res. Forum Recycling*, DAPPI, 1993, s. 207-220

Kučerová, V., Halajová, L. (2009). Sledovanie zmien recyklovaných buničín metódou gélovej permeačnej chromatografie. In *Acta Fakultatis Xylologiae*, ISSN 1336-3824, 2009, 51, (2), 87-92

Laivins, G. V., Scallan, A. M. (1993). The mechanism of hornification of wood pulps. In *Products of Papermaking*, Oxford, 1993, C.F.Baker (ed.) Pira International, Leatherhead, Surrey, UK, Vol. 2 p. 1235-1259

Lešikár, M. (2006). Perspektívy balení potravin do obalů z papírů a lepenek. In *Papír a celulóza*. ISSN 0031-1421, 2006, *61, (2)*, s. 59

Margutti, S., Conio, G., Calvini, P., Pedemonte, E. (2001). Hydrolytic and oxidative degradation of paper. In Restaurator. ISSN 0034-5806, 2001, (22) 2, s. 67-131

Marx et al. (1966). Naturwissenshaften 53, 1966, s. 466. In *Blažej et al. Chémia dreva*. 1.vyd. Bratislava:Alfa, 1975. 221 s

Matsuda et al. (1994). Effects of Thermal and Hydrothermal Treatments on the Reswelling Capabilities of pulps and Papersheets. In *Journal of Pulp and paper Science*, ISSN 0826-6220, 1994, 20 (11), p. J323-J327

MD (2010). Stav a vyhlídky průmyslu papíru a celulózy v některých vybraných zemích světa. In *Papír a celulóza*. ISSN 0031-1421, 2010, *65 (12)*, s. 352-354

Milichovský, M. (1994). Recirkulace a recyklace - výzva současnosti. In: *Papír a celulóza*. ISSN 0031-1421,1994, 49 (2), s. 29-34

Nazhad, M. M. (2005). Recycled fiber quality-A review, In *J. Ind. Eng. Chem.* 11 (3), s. 314-329

Nazhad, M. M., Paszner L. (1994). Fundamentals of strength loss in recycled paper. In *Tappi Journal*, ISSN 0039-8241, 1994, 77 (9), p. 171-179

Nordman, L. (1976). The applicability of recycling fiber to different papwe qualities. In Sekundárne vlákna a ich využitie v papierenskom priemysle. Eucepa symposium, Bratislava 1976

Okayama, T. (2001). Characteristicsof pulp fiber with high recycling performance. *Journal Shigen Junkan* s. 39-40

Okayama, T. (2002). The Effects of Recycling on Pulp and Paper Properties. *Japan TAPPI Journal*. ISSN 0022-815X, 2002, 56 (7), p. 986-992

Paavilainen, L. (1993). Conformability-flexibility-of sulfate pulp fibers. In *Paperi Puu*.ISSN 0031-1057, 1993, 75 (9-10), p. 896-702

Pauler, N. (2002). Paper optics. AB Lorentzen&Wettre, Elanders Tofters, Östervála, Sweden 2002, p. 93, ISBN 91-971765-6-7

Peng,Y.X., Valade, J.L.,Law, K.N. (1994). Effekcts of recycling and blending of virgin fibers on paper properties, In *China Pulp and Paper* 13 (2), p. 3-9

Phipps, J. (1994). Effects of recycling on the strength properties of paper. In *Paper Technol.* 35 (6), p. 34-40

Požgaj, A. et al. (1997). Štruktúra a vlastnosti dreva. 2. vyd. Bratislava : Svornosť, 1997. 485s. ISBN 80-07-00960-4

Putz H.-J. (2000). Recovered paper grades, quality control, and recyclability. In: Göttsching, L & Pakarinen, H (eds.) Papermaking Science and Technology, Book 7, Recycled Fiber and Deinking. ISBN 952-5216-07-1, 2000, p. 61-87

Rowell, R.M., (2005). Handbook of wood chemistry and wood composites. 2005, Taylor & Francis Group. ISBN 0-8493-1588-3

Shao, S.Y., Hu, K.T. (2002). Hornification of recycled fiber, In China Pulp Paper (2), p. 57-60

Solár, R. (2001). Chémia dreva. 1. vyd. Zvolen : TU vo Zvolene, 2001. 101 s. ISBN 80-228-1007-X

Solár, R., Mamoň, M., Kurjatko, S., Lang, R., Vacek, V. (2006). A simple method for determination of kinetics of radial, tangential and surface swelling of wood. In *Drvna industrija.* ISSN 0012-6772, 2006, 57 (2), p. 75-82

Solár, R. et al. (2009). Alkaline and alkaline/oxidation pretreatments of spruce wood. Part 1: Chemical alterations of wood and its digestibility under conditions of kraft cook. In *WOOD RESEARCH.* ISSN 1336-4561, 2009, *Vol.* 54, (4) 2009. p. 142

Souček, M. (1977). Zkoušení papíru. 1. vyd. Praha:SNTL, 1977. 339 s

Souček, M. (2009). Grafické papíry na bázi recyklovaných vláken, In *Papír a celulóza.* ISSN 0031-1421, 2009, 64 (6), s.164-165

Stone, J. E., Scallan, A. M. (1966). Influence of drying on the pore structures of the cell wall. In: *Consolidation of the Paper Web,* Trans. Symp. Cambridge, Sept. 1965, f.Bolam (ed.), Tech. Sec. British Paper and Board Makers´Assoc. Inc. London, Vol. 1, s. 145-174

Stone, J. E., Scallan, A. M. (1968). A structural model for the cell wal of water-swollen wood pulp fibers based on their accessibility to macromolecules. In *Cellulosa Chem. Technol.* ISSN 0576-9787, 1968, *2 (3), 343-358*

Stürmer, L., Göttsching, L. (1979). Physical properties of secondaryfiber pulps under the influence of their previous history. Part 3: Influence of the paper manufacturing process, In *Wochenbl. Papierfabrik.* 107 (3), p. 69-76

ŠK (2010). Rast cien buničiny o 40 % a zberového papiera o 80 % tlačia na ceny výrobcov papiera na Slovensku. In *Papír a celulóza.* ISSN 0031-1421, 2010, 65 (6), s. 162

Takayuki O. (2002). The Effects of Recycling on Pulp and Paper Properties. In *Japan TAPPI Journal,* ISSN 0022-815X, 2002, vol. 56 (7), p. 986-992

Woodward, T., W. (1996). Recycled fiber types, processing history affect pulp behavior during papermarking. www.paperrecovery.eu, 2007

Zugenmaier, P. (2008). Crystaline Cellulose and Derivatives: Characterization and Structures. Springer Series. In Wood Science, Springer-Verlag Berlin Heidleberg 2008, 286 s. ISBN 978-3-540-73933-3. http://www.springer.com/978-3-54-73933-3

Part 3

Potential Uses of Recycled Wastes

12

Reuse of Waste Shells as a SO_2/NO_x Removal Sorbent

Jong-Hyeon Jung[1], Jae-Jeong Lee[2], Gang-Woo Lee[2], Kyung-Seun Yoo[3] and Byung-Hyun Shon[4]
[1]*Daegu Haany University*
[2]*Yoosung Co. Ltd., R&D Center*
[3]*Kwangwoon University*
[4]*Hanseo University*
Korea

1. Introduction

Lately, many environmental pollution problems are taking place due to advancement of science and technology, enhancement of industry, rapid economic development, change of life style, and population increase. Not only in Korea but also worldwide consciousness of crisis over environmental pollution problems is spreading and especially countermeasures to solve the environmental pollution come into the limelight as concerns of the entire globe beyond any regions and nations. In the past, environmental pollutions including abnormal weather changes, global warming, El Nino, Ra Nina, ozone layer destruction, and marine pollution were issues of a limited area, but now they appear in fact as issues of the entire globe [Jung, 1999, 2008].

In the southern coast of Korea, a lot of oyster shells are dumped as a by-product of marine aquaculture industry. A large amount of oyster shells is a general waste fishermen should take care of but it seems difficult to handle it effectively due to the problems of securing of landfill sites and collection/transportation of oyster shells [Jung, 2005, 2007]. This waste piles up at coastal areas and causes many environmental problems including pollution of coastal fisheries, management problem of public water surface, damage of natural landscape, and health/sanitation problem.

In Japan, 200,000 tons of oysters were produced in 2007 [Asaoka et al., 2009]. And Table 1 shows oyster production of Korea from 1997 to 2006. On the basis of these data, generation of oyster shells is estimated on the average at 270,000 tons/year [Kim, 2007], and more than 50~70% of which was dumped into public waters and reclaimed lands, which cause an unpleasant fishy smell as a consequence of the decomposition of fresh remnant attached to oysters (Kim, 2007; Yoon et al., 2003; Shin et al., 1998). Approximately 30~50% of shells from harvested oysters was utilized and the remainder was disposed (Yoon et al., 2003; Kwon et al., 2004). Thus, recycling of waste oyster-shells has arisen as an imminent issue in the mariculture industry. As a recycling process, a lot of studies on the application of waste oyster-shells to construction materials (Yoon et al., 2003), laver farming, fertilizer (Nippon

Steel Corp., 1993), sludge conditioners (Lee et al., 2001), eutrophication control (Kwon et al., 2004), filtering medium (Park and Polprasert, 2008), catalyst (Nakatani et al., 2009), soil conditioner (Lee et al., 2008a,b), and desulphurization sorbents (Jung et al., 2000) have been reported. The desulphurization system can be divided into three groups such as dry adsorption (Garea et al., 2001), wet scrubbing (Chu et al., 1997), and wet/dry system (O'Dowd et al., 1994).

Year	Oyster Production(ton)	Estimated Generation of Oyster Shells(ton)
1997	17,210	258,150
1998	9,905	148,575
1999	11,690	175,350
2000	15,939	239,085
2001	10,056	150,840
2002	7,950	119,250
2003	20,201	303,015
2004	25,690	385,350
2005	27,320	409,800
2006	31,016	465,240

Table 1. Oyster production and estimated generation of oyster shells in Korea.

The research on application of oyster shells to a sorbent for incineration and desulfurization is judged to be very helpful in preserving marine eco-system, preventing the damage of natural landscape and solving health/sanitation problem. Therefore, new applications utilizing these wasted oyster shells are expected to contribute towards recycling consciousness within the society [Asaoka et al., 2009]. The application of many kinds of waste shells, which have been dried, crushed and calcined, to sorption of acidic gases and nitrogen oxides is not only economically valuable but also very significant in the aspect of waste recycling. In addition, considering that chlorine which has been found to be a precursor of such toxic organic substances as dioxin and furan lately creating social concerns is an acidic gas, the use of waste shells as mixed sorbents is judged to be feasible.

In this research, we are going to provide a basic data to a process for removing both sulfur oxides and nitrogen oxides at the same time from exhaust gases. For this purpose, first of all, the basic physical and chemical properties of waste oyster shells were investigated. In addition, the calcination and hydration reaction of waste oyster shells were experimented and the preparation method of sorbents was investigated. In order to investigate the feasibility of using oyster shells as a sorbent for removal of sulfur and nitrogen oxides, the performance of prepared sorbents was compared for understanding reaction characteristics using a fixed bed reactor.

2. Environmental problems of oyster shells and utilization plans for each field

2.1 Environmental problems caused by oyster shells

Environmental problems caused by waste oyster shells were as follows [Kim, 2007]; i) increase of waste, ii) pollution of marine eco-system due to illegal landfill, iii) increase of bad smell due to negligence, iv) a huge amount of treatment expense, and v) weak demand on recycled materials (fertilizer, etc.) from oyster shells. Fig. 1 shows the generation process of waste oyster shells. Oyster meat is consumed and a significant amount of waste oyster shell is discarded.

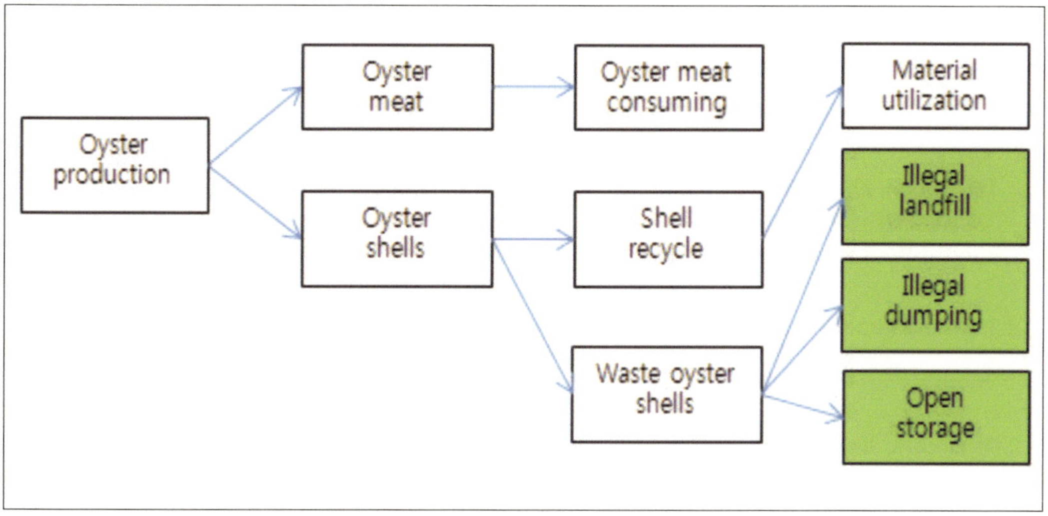

Fig. 1. Generation steps of the waste oyster shells.

2.2 Recycling plan of oyster shells on the aspect of environmental engineering

Environmental engineering research on recycling of oyster shells has been somewhat conducted in China and Japan, but most of the research has focused on the purpose of water purification in a limited scope [Kim, 2007].

2.2.1 Removal of air pollutants (SO_2, H_2S, CO_2, HCl, NOx, etc.) from the exhaust gases

- As a desulfurization sorbents of power plants, the oyster shell is more efficient and less expensive than limestone and so its commercialization is judged to highly feasible.
- The oyster shell used in the desulfurization process exhibits a high efficiency without calcination processing.
- The treatment problem of calcium sulfate formed after the reaction should be resolved, and the cost involved in calcination should be estimated.

2.2.2 Soil improvement (acidity reduction, organics adsorption) and adsorption of heavy metals

- In case oyster shell is used as a soil conditioner, it has higher adsorption and desorption of heavy metals than general soil, and so it may bring in stable and economical effects.
- Injection of calcined and processed oyster shells would be very effective in growing plants due to increased exchange capacity of cations used by plants.
- The potential risk cannot be evaluated without long-term research and the effect on plant growth should be reviewed.

2.2.3 Application to waste water treatment (nitrogen, phosphorous, heavy metals, pH neutralization, etc.)

- The oyster shell can neutralize acidic waste water from mines very fast and remove 99% of heavy metals, and so it is judged to be a useful material to replace limestone.
- A plan for the secondary treatment after adsorption of heavy metals should be prepared.

2.2.4 Improvement of sludge dehydration capacity (digested sludge etc.)

- Dehydration capacity of the digested sludge is about the same as that of the existing limestone, which suggests that the oyster shell may be an inexpensive substitute.
- In case of dehydration of waste water sludge, injection of oyster shell/slaked lime mixture reduces the required dehydration time.

2.2.5 Others (calcination, red tide, coagulation, fertilizer, etc.)

- Oyster shell powder can coagulate and remove red tide organism.

2.3 Recycling limitation

Recycling of waste oyster shells has much limitation in environmental engineering. First, the quantity of waste oyster shells is too much to be recycled in environmental engineering. Second, a plan to retreat the waste oyster shells contaminated after adsorption of heavy metals should be prepared above all. Third, recycling of waste oyster shells would require removal of impurities through proper processing and treatment through a calcination process, which decreases the economic feasibility. Fourth, the repeated use lowers the efficiency below commercial products.

3. The kind and characteristics of sorbents

It is well known that alkali sorbents such as CaO, $CaCO_3$, $Ca(OH)_2$, NaOH, Na_2CO_3, $NaHCO_3$, KOH, MgO, $Mg(OH)_2$, dolomites, and dolomite limestones($CaCO_3/MgCO_3$) are used to remove SO_2 and acid gases in the flue gas cleaning processes (Kunio et al., 1994; Jonas et al., 1984; Jung et al., 2000). Under these sorbents, calcium-based alkali sorbents are the most widely used to remove SO_2. Because calcium-based sorbents have good reactivity toward acidic gases and lower cost compared to other sodium-based alkali adsorbent.

The sorbent should be selected in consideration of the most important factors, hygroscopicity and price. Sodium-based sorbents have better removal efficiency of pollutant gases but is more expensive than calcium-based sorbents. While calcium-based sorbents exhibit hygroscopicity, but sodium-based sorbents deliquescency. In Table 2, various sorbents frequently used in flue-gas desulfurization processes are compared and the physical properties of the sorbents are also described as a part of basic research to understand whether they are utilized in flue-gas desulfurization processes. In addition, in Table 3 the molecular weight, specific gravity, density, decomposition temperature, melting point, particle size, solubility, and maleficence of the sorbents frequently used in flue-gas desulfurization processes are presented.

Sorbent	Deliquescent	Hygroscopic	Application
NaOH	Yes	Yes	Spray Drying, Wet Scrubbing, Dry Scrubbing
Na_2CO_3	Yes	Yes	Spray Drying
$NaHCO_3$	Yes	Yes	Dry Scrubbing
$CaCO_3$	No	Yes	Spray Drying, Wet Scrubbing, Dry Scrubbing
$Ca(OH)_2$	No	Yes	Wet Scrubbing, Dry Scrubbing
CaO	No	Yes	Spray Drying, Wet Scrubbing, Dry Scrubbing
Waste oyster shells (WOS)	No	Yes	FGD absorbent

Table 2. Characteristics of selected sorbents.

Sorbent Description	$CaCO_3$	CaO	$Ca(OH)_2$	$NaHCO_3$	Na_2CO_3	NaOH
Molecular Weight	100	56.08	74.09	84	106	40
Specific Gravity	2,93	3.37	2.24	-	2.53	2.13
Bulk Density(ton/m^3)	1.11	-	1.7 ~ 2.3	-	2.08	-
Decomposition Temp.(℃)	898	-	580	-	-	-
Melting Point(°C)	-	2,572	-	-	851	318.4
Size of Particle(μm)	4 ~ 44	-	4 ~ 100	-	>100	-
Solubility(g/100cc , at 20℃)	0.0014	-	0.185	7.1	6.9	-
Impact of Human	low	low	low	low	low	large

Table 3. Characteristics of sorbents in FGD process.

4. Materials and methods

4.1 Physicochemical analysis of oyster shells

The seashells of oyster, hard-shelled mussel, clam, and seashell from Tong Young province around South Sea in Korea were used as a main material. Salts and other organic substances were removed by washing and drying the waste seashells. Limestone from Danyang and Jungsun province in Korea was adapted for comparison of physicochemical properties of oyster shells. All the materials were crushed 2 times by Jaw crusher and Ball mill after drying enough. The physical and chemical characteristics of the waste oyster shells were analyzed by ICP (ICPS-7500 Shimadzu, Japan), SEM (JEOL superprobe JSM-5400, USA), XRD (SIMENS, Deutsche), and BET surface area (Micromeritics Co., USA). ICP was applied to analyze the atomic properties of the materials. SEM was used to observe the microtissue of the surface of wasted shells. Surface area of the sorbents was measured by BET technique after pretreating to remove vapor in vacuum 1x10^{-3} and 180°C for 2 hours. The crystal state of shell sorbent and the products before and after reaction was assured by X-ray diffractometer under the condition of 30 kV and 20 mA in the ranges of 10-70 degrees.

4.2 Hydration of oyster shells

Hydration apparatus was manufactured in order to investigate the effects of properties of calcination sorbent and hydration condition on the reactivity of the hydrated lime [EPRI, 1988]. The reactivity of calcined waste shell was compared with that of imported lime samples. Also, hydration state according to initial temperature was conformed. These were conducted by measuring the reactivity of quicklime. The samples were placed in a stirred vacuum flask containing deionized water and the temperature was measured at time intervals. Rate of heat release was a measure of reactivity. The faster temperature increases the faster the slaking rate. This method can be used in flasker design, to improve slaker performance, to evaluate and compare different limes, or test incoming lime shipments for quality control purposes.

It was calculated for the hydration speed to measure the hydrated reaction using the device of Fig.2 with calcinated shell sorbents suggested by EPRI(1988), respectively. This device was designed to react under same temperature according to the change of hydration condition and was consisted of Dewar flask wrapped dual rubber plate and a mixer made with teflon. K-type thermocouple (1/8", 30 cm) and a thermoscope were installed on the upper side of flask and conformed the variation of inner temperature.

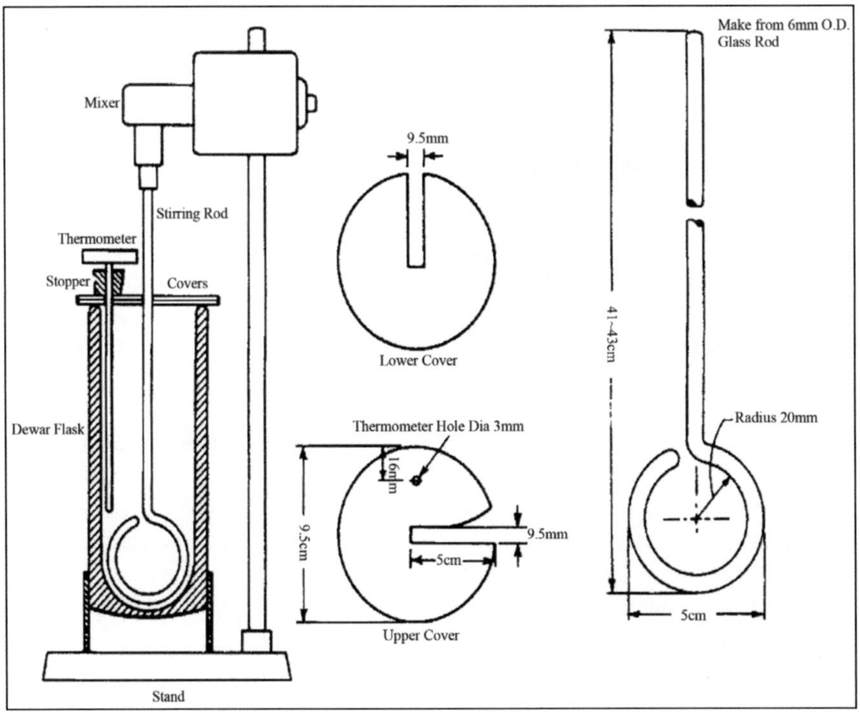

Fig. 2. Experimental apparatus for measuring the hydration rate of absorbents [EPRI, 1988].

In order to check hydration rate, the crushed seashell sieved with the degree of 6 mesh was adapted to calcination reaction. It was added 900 ml water of 25°C to Dewar flask. And a mixer was set to 400 rpm [EPRI, 1988]. After feeding of CaO (100 g), the data system was continuously operated to check the temperature variation in every 5 seconds.

As shown in Fig. 3, the reactor used to study the calcination reaction of oyster shells was made of 15 mm ID x 450 mm L quartz glass tube which is corrosion-resistant at high temperatures. The reactor was installed inside a tubular furnace insulated with ceramic wool to reduce heat loss from the inside to the outside and to maintain a constant and uniform temperature inside the electric furnace. The samples were injected where the temperature gradient of the measured reactor temperature was constant and the desired operating temperature was reached, and the flow rates were maintained constant at 2 ℓ/min, 3 ℓ/min, and 4 ℓ/min (1 atm). In addition, the temperature (700 ~ 1000°C) and flow rate were varied to study the effect of calcination temperature on the BET specific surface area, and the experimental conditions are given in Table 4.

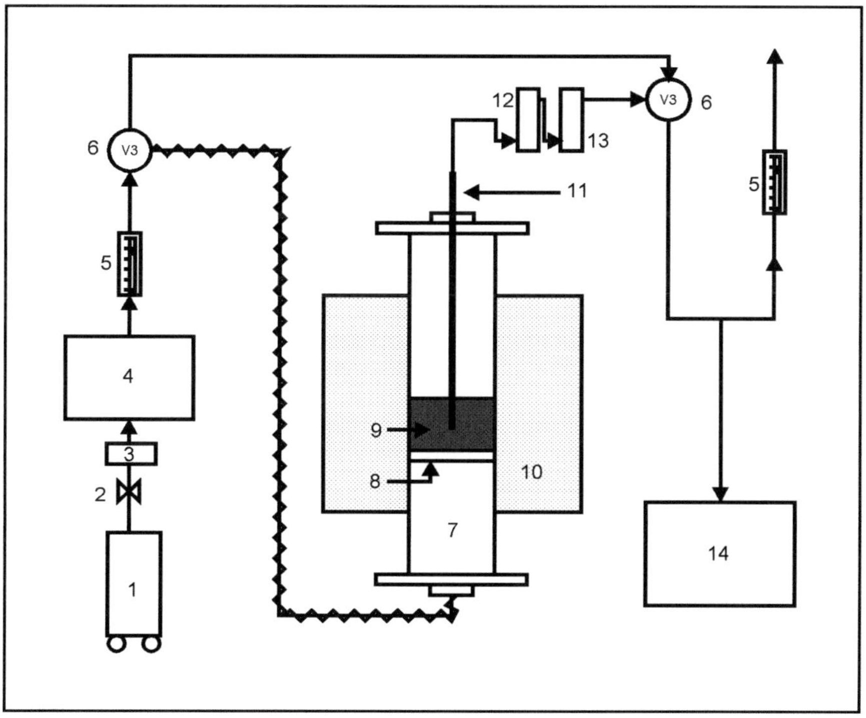

1. Air compressor	8. Porous quartz disk
2. Two way valve	9. Waste shell/Alkali absorbent
3. Mass flow controller	10. Furnace
4. Mixing chamber	11. Thermocouple
5. Flowmeter	12. Mist eliminator
6. Three way valve	13. Condensor
7. Quartz reactor	14. Gas sampling system

Fig. 3. Schematic diagram of the calcination apparatus.

Experimental variables		Conditions
Pretreating	Calcination temperature (℃)	700 ~ 1000
	Hydration time (hr)	24
	Hydration temperature (℃)	90
	Slurrying velocity (rpm)	200
	Absorbent drying time (hr)	24

Table 4. Pretreating and experimental conditions.

A thermo gravimetric analysis (TGA) was used to analyze the activation energy and calcination rates, and the experimental equipment was composed of a gas supply system, reactor system, and data treatment system. The specimen measurement dish of the analyzer was installed vertically inside the quartz tube reactor in the cylindrical electric furnace whose temperature can be raised up to 1,200°C. About 10 to 20 mg of 40/60 mesh crushed shells was placed on the dish and nitrogen gas was passed at a flow rate of 30 ml/min for 10 minutes through the analyzer to replace air with nitrogen in the reactor. A corrosion-resistant stainless pipe was used from the gas mixer to the entrance of the reactor, and the inside pressure of the reactor was maintained a little over the ambient pressure in the entire experiments. When the reaction conditions were stabilized, the reactor was heated to 900°C while nitrogen passed at a flow rate 30 ml/min. The reaction was judged to be completed in case of no further weight change during the heating period. The heating rate of the calcination reaction was 50°C/min over the temperature range of 100 to 600°C and 10°C/min over the range of 600 to 900°C.

4.3 Analysis of waste oyster shells as a SO_2/NO_x removal absorbent

To enhance the physicochemical properties of the waste oyster-shells, pretreating techniques were applied to the samples before SO_2/NOx removal test. The goal of pretreating process was to convert the relatively low reactive calcium component (in the form of calcium carbonate) into a form of calcium oxide and calcium hydroxide that readily reacts with acid gases. These processing referred to calcinations and hydration, respectively.

To calculate the sorbents capacity, the SO_2/NOx removal experiments were carried out using a fixed-bed reactor system (Jung et al., 2005) under atmospheric pressure at 150.0°C. The fixed-bed quartz reactor (0.025 m in diameter, 0.25 m in height) was placed in a hot air bath and the temperature was controlled by PID type controller with the precision of ± 1.0°C. After the sample put on the reactor (1 g of samples) and the temperature was stabilized in N_2 flow, reacting gas containing SO_2, O_2, and NO_x was injected into the reactor using mass flow controllers (MFC, BROOKS instrument inc., Model 5850E, England). At the same time water was also injected to the reactor by syringe pump to keep a steady vapor concentration in the simulated gas. Hygrometer and SO_2/NOx analyzer were used to measure the gas concentration and signals from the measuring instrument were recorded at a personal computer with RS-232C interface. Table 5 shows experimental variables and experimental conditions [Jung, 1999; Jung et al, 2005, 2009].

Experimental variables		Conditions
Gas reactivity	SO_2 concentration (ppm)	1800
	NOx concentration (ppm)	250
	O_2 concentration (%)	6
	Reaction temperature (℃)	150
	Water content (%)	10

Table 5. Experimental variables and conditions.

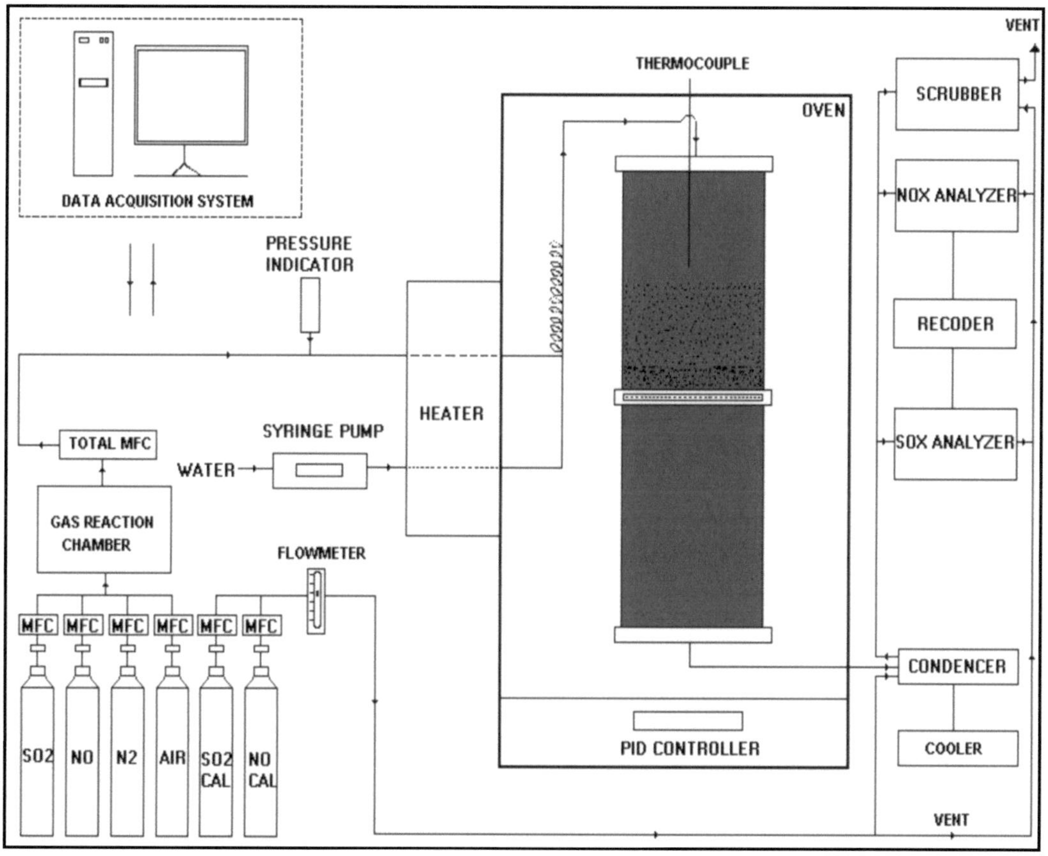

Fig. 4. Schematic diagram of packed-bed experimental apparatus.

5. Results and discussion

5.1 Physicochemical characteristics of oyster shells

The physicochemical compositions of the waste oyster shells, mussel, clam, seashell, and commercial Jungsun limestone were summarized in Table 6 [Jung et al., 2007, 2005, 2000; Kwon et al., 2003]. To perform XRF (X-ray fluorescence spectroscopic) analysis, samples were powdered after dehydrated in a drying oven at 105.0°C for 24 hours. From the composition analysis, it was found that oyster-shells consist of mostly CaO, some of SiO_2,

MgO, Al_2O_3, and Fe_2O_3. The composition of CaO in the oyster-shells was around 53.81 ~ 52.94 wt.% which is comparable to that of commercial Jungsun limestone and was in good agreement with the results of Yoon et al. (2003) reporting the CaO content of oyster-shells was about 53.7 wt.% [Jung et al., 2007, 2005, 2000].

X-ray diffraction (XRD) patterns of waste oyster shells, limestone, and calcined waste oyster shells (C-WOS) are shown in Fig. 5. Fig. 5 shows the results of XRD analyses of the waste oyster shells and limestone with and without calcination. The patterns for waste oyster-shells and limestone were nearly similar and the diffraction peaks of $CaCO_3$ as major phases are identified (Fig. 5a, 5b). The patterns for calcined waste oyster shells were exhibited peaks characteristics of CaO (Fig. 5c) [Jung et al, 2007]. And, the intensity of these peak increases with increasing the calcination temperature indicating CaO phase has been formed enough after calcination followed by hydration reaction. Comparing the results of Fig. 5 with Table 6, the waste oyster shells can be utilized as FGD absorbent [Jung et al, 2000].

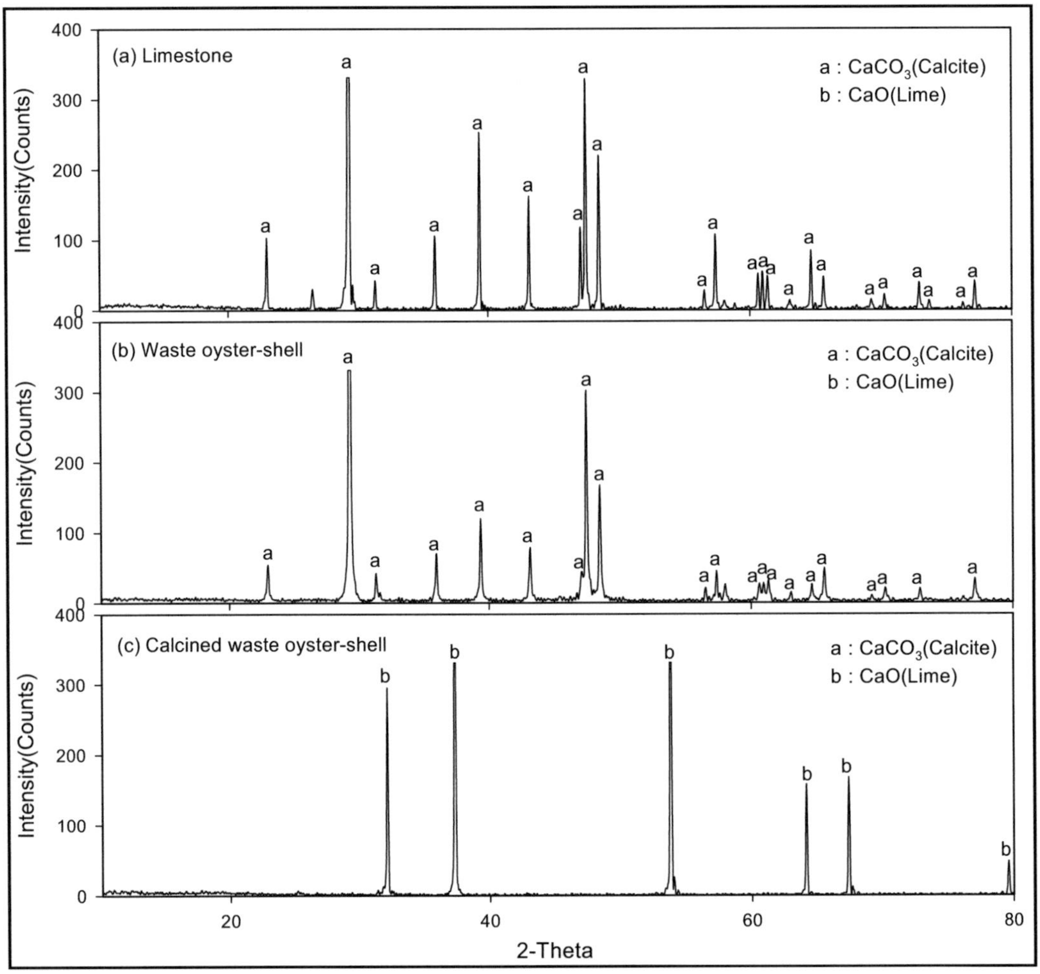

Fig. 5. XRD profiles of oyster-shells and limestone; (a) limestone, (b) oyster shells, and (c) calcined oyster shells.

Sorbents	Chemical composition [wt.%]						Pore volume [cc/g]
	SiO_2	Al_2O_3	Fe_2O_3	CaO	MgO	Ignition Loss	
Waste oyster shells (WOS)	0.40 0.62	0.22 -	0.04 0.32	53.81 52.94	1.70 0.78	44.87 44.02	0.0869 -
Mussel	0.20	0.13	0.03	53.70	0.33	45.61	0.0129
Clam	0.46	0.20	0.04	53.92	0.22	45.16	0.1025
Seashell	0.66	0.40	0.04	53.58	0.20	45.12	0.0888
Jungsun limestone	2.43	0.25	0.14	53.80	0.85	42.50	0.0697

Table 6. Physicochemical properties of tested absorbents and seashells [Jung et al, 2005].

5.2 Characteristics of calcination and hydration reaction with oyster shells

Fig. 6 shows the change of specific surface area of calcined hydration limestone. As shown in Fig. 6A, specific surface area of calcined hydration limestone particle increases rapidly with increasing hydration time up to 6 hours and then slowly increases to 12 m^2/g. Effects of hydration temperature on the specific surface area of the calcined limestone has been shown in Fig. 6B. When the initial water temperature was in the range of 30~90°C with every 10°C, up to the 70°C BET surface area was not changed largely. However, rapid increase of BET surface area of absorbent appears above the temperature of 80°C, indicating the rapid hydration reaction occurs. We can conclude the optimum range of temperature and time for hydration reaction of absorbent is about 80~90°C and 24 hours, respectively [Jung et al, 2000].

Fig. 7A~B shows the adsorbed volume with relative pressure on the waste oyster shells and limestone. This figure shows the difference of amount of nitrogen gas according to the change of relative pressure absorbed to waste shell and limestone. We can conclude each of the sorbent shows a similar BET value in the condition of raw material.

Fig. 8 shows pore size distribution of waste oyster shells, clam, and seashell according to pore diameter. The results lead to the finding that shells have larger average pore size with the value of pore volume of 0.013~0.024 cm^3/g, which is higher than that of limestone. And, other result is pore size distribution according to the change of pore diameter of oyster manufactured as an absorbent with calcination and hydration. It was appeared that the average pore diameter of oyster was much bigger than that of raw materials. This result also indicates that calcination and hydration processes can enhance the removal capacity of acid gases [Jung et al, 2005].

To clarify the effect of temperature, TGA analysis of the waste oyster shells and limestone were examined under N_2 atmosphere by Automatic Derivative Differential Thermo-balance (Fig. 9). Heating rate in the calcination, which was conversion from calcium carbonate to calcium oxide, was 10.0 °C/min in the temperature range of 600.0~ 900.0°C. As can be seen in Fig. 9, calcinations of the oyster-shells started at 645.0°C and completed at 780.0°C, whereas Jungsun limestone started over the temperature 680.0°C and completed at 845.0°C. The mass of waste oyster shells and limestone in the calcinations under N_2 atmosphere decreased 44.8% and 44.0%, respectively. From this analysis, it could be confirmed that mass decreasing resulted from the reaction $CaCO_3 \rightarrow CaO + CO_2$. Also, in the case of Jungsun limestone, the temperature of the calcinations should be over about 850.0°C [Jung et al, 2007].

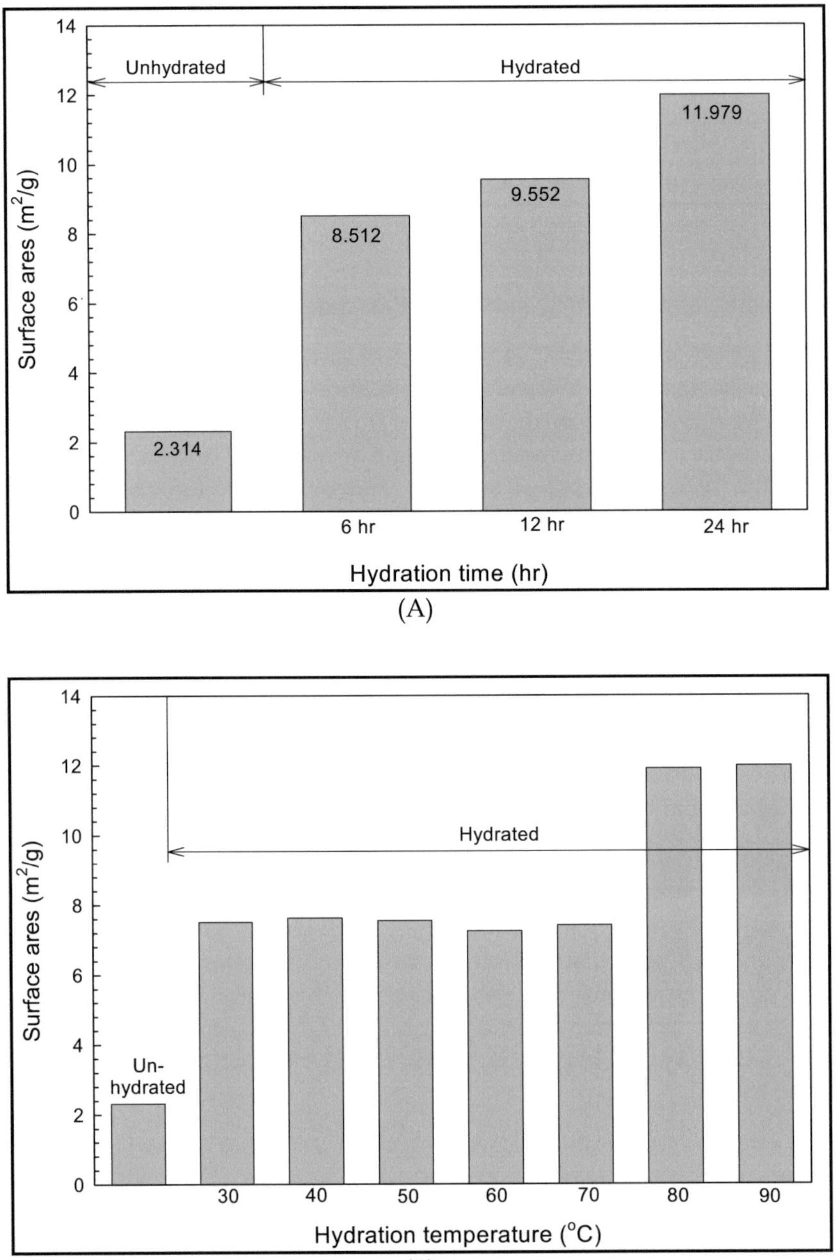

Fig. 6. (A) Variation of surface area of calcined limestone as a function of hydration time. (B) Effect of hydration temperature on the surface areas of calcined limestone.

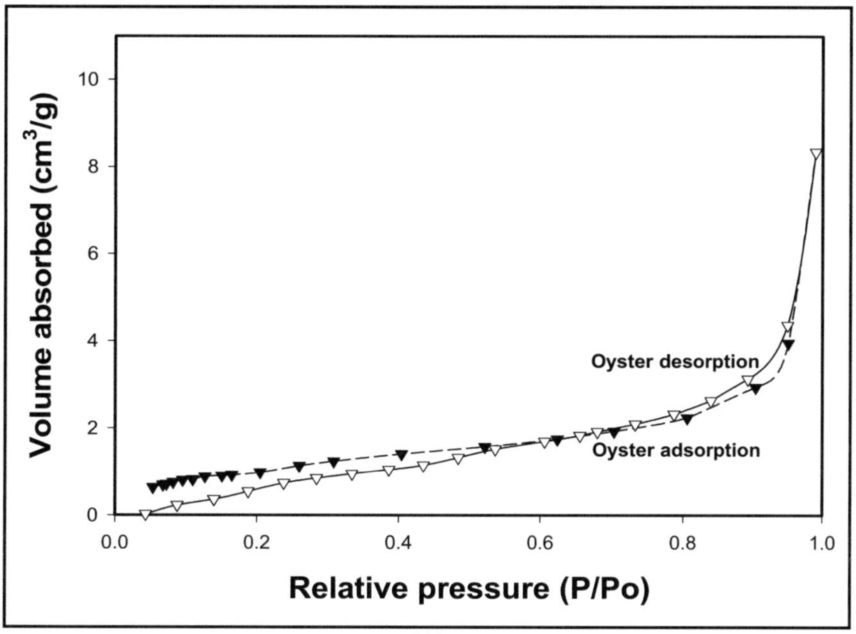

(A)

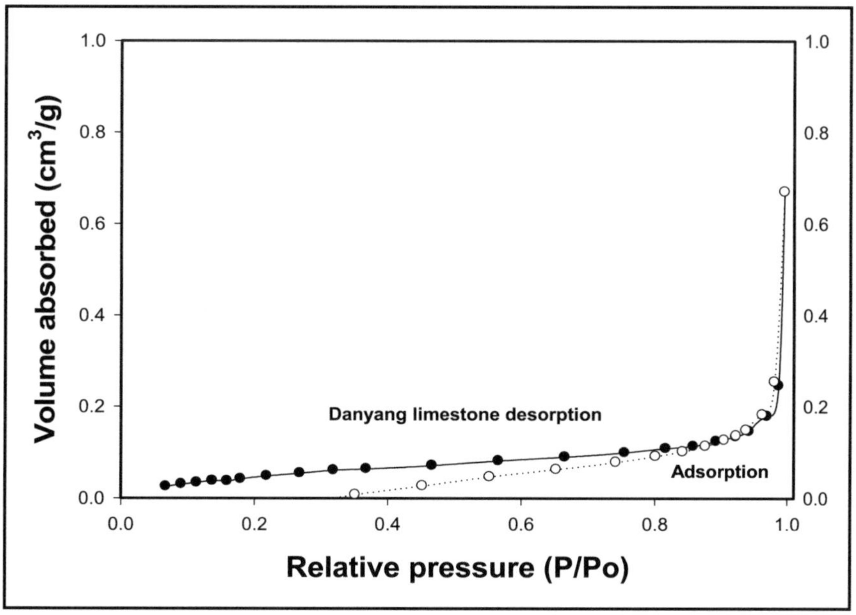

(B)

Fig. 7. N_2 BET results of various waste oyster shells, seashell and limestone. (A) waste oyster shells, (B) limestone.

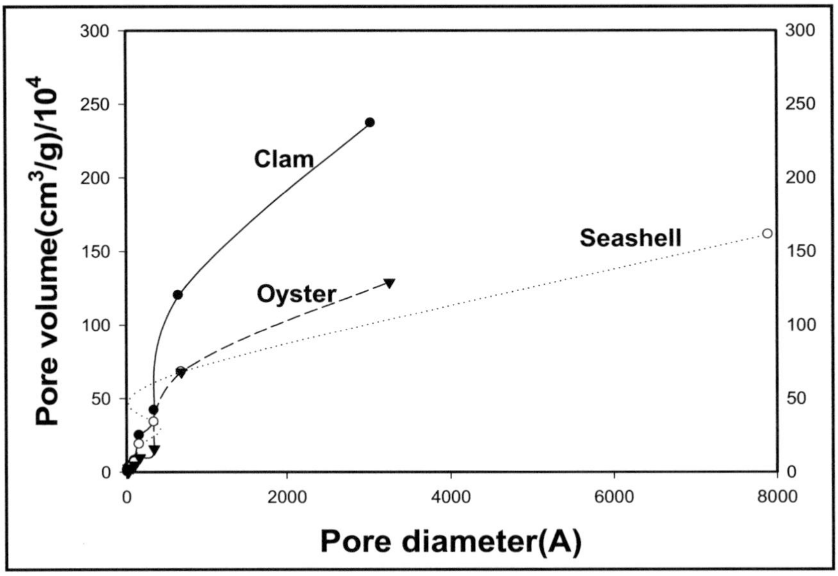

Fig. 8. Pore volume of waste oyster shells, seashell and limestone.

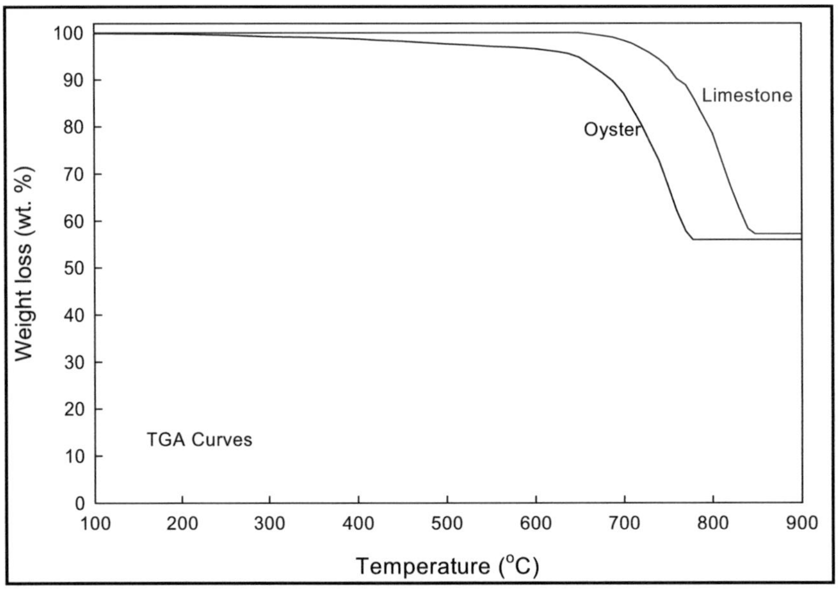

Fig. 9. TGA profiles of the waste oyster shells and limestone in the temperature range of 600 - 900°C.

The calcination reaction of waste oyster shells and limestone, main components is $CaCO_3(s)$, is gas-solid reaction, in which we assumed that the calcinations reaction is n-order reaction. The activated energy and reaction order of waste oyster shells and limestone are calculated using the following equations and TGA experimental results. Where x=mass of waste oyster shells and limestone, k=reaction rate, X=conversion, n=reaction order, A=preexponential factor, E_a=activation energy, R=gas constant, T=absolute temperature.

$$CaCO_3(s) \xrightarrow{k} CaO(s) + CO_2(g) \tag{1}$$

$$dx/dt=k(1-X)^n \tag{2}$$

$$k=A \cdot \exp(-E_a/RT) \tag{3}$$

$$dx/dt= A \cdot \exp(-E_a/RT)\,(1-X)^n \tag{4}$$

Table 7 shows kinetic data of the waste oyster shells and limestone. It was calculated from the results of calcinations experiment (Fig. 9). Activation energies, reaction constants, and reaction orders were 176±8.90 kJ/mole, 15.07±1.14 sec^{-1}, and 0.42±0.10 for oyster shells and 201.72±5.17 kJ/mole, 16.47±0.63 sec^{-1}, and 0.37±0.08 for Jungsun limestone, respectively. These values were similar to those of other calcium-based sorbents (Jung et al., 2000). The activation energy of the oyster shells was smaller than that of Jungsun limestone. As can be seen in Fig. 10, this is because structural elements of natural limestone is inorganic materials, whereas oyster shells was comprised of thin $CaCO_3$ layer and created by living things thus its surface was irregular and porous.

Absorbents	E_a [kJ/mole]	Order[n]	Sample wt.[mg]	Flow gas	Conditions
Waste oyster shells (WOS)	176.1±8.90	0.42±0.10	15 – 16	N_2	Isothermal
Jungsun limestone	201.72±5.17	0.37±0.08	15 – 16	N_2	Isothermal

Table 7. Kinetic data of tested absorbents.

Fig. 10 shows scanning electron micrographs (SEM) of the various solid Jungsun limestone and waste oyster shells. Parts (a), (a'), (b), and (b') of Fig. 10 show the scanning electron micrographs of the fresh and calcined samples, respectively. SEM pictures for the hydrated and calcined/hydrated samples are shown in Parts (c), (c'), (d), and (d') of Fig. 10, respectively. Observation of the waste oyster shells morphology indicates some agglomeration of the particle/grains in the hydrated waste oyster shells (H-WOS) as compared to the fresh oyster. However, as shown in Fig. 10(b'), in the SEM photograph of the calcined sample, enormous amount of agglomerations by sintering were observed.

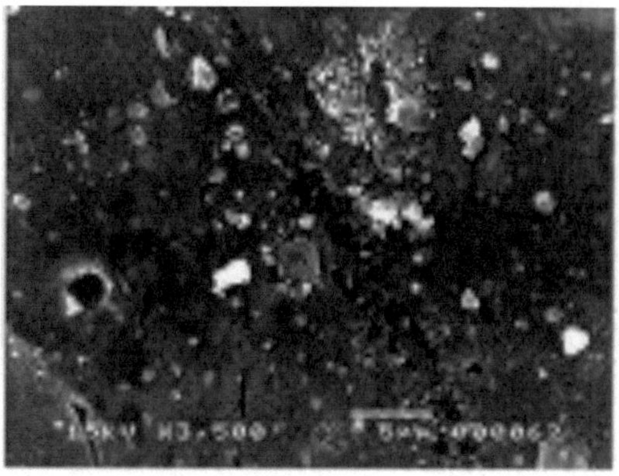

(a)

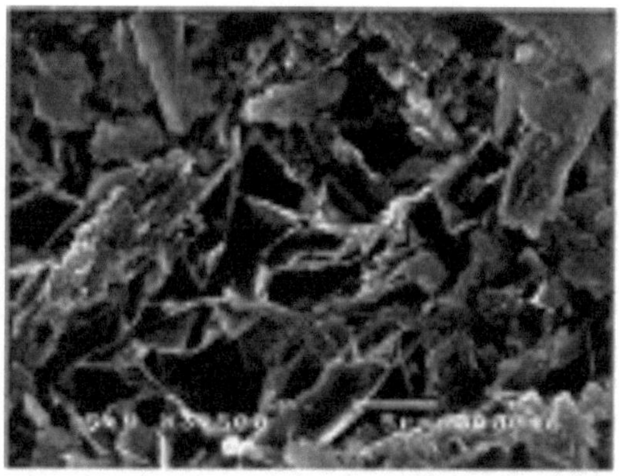

(a')

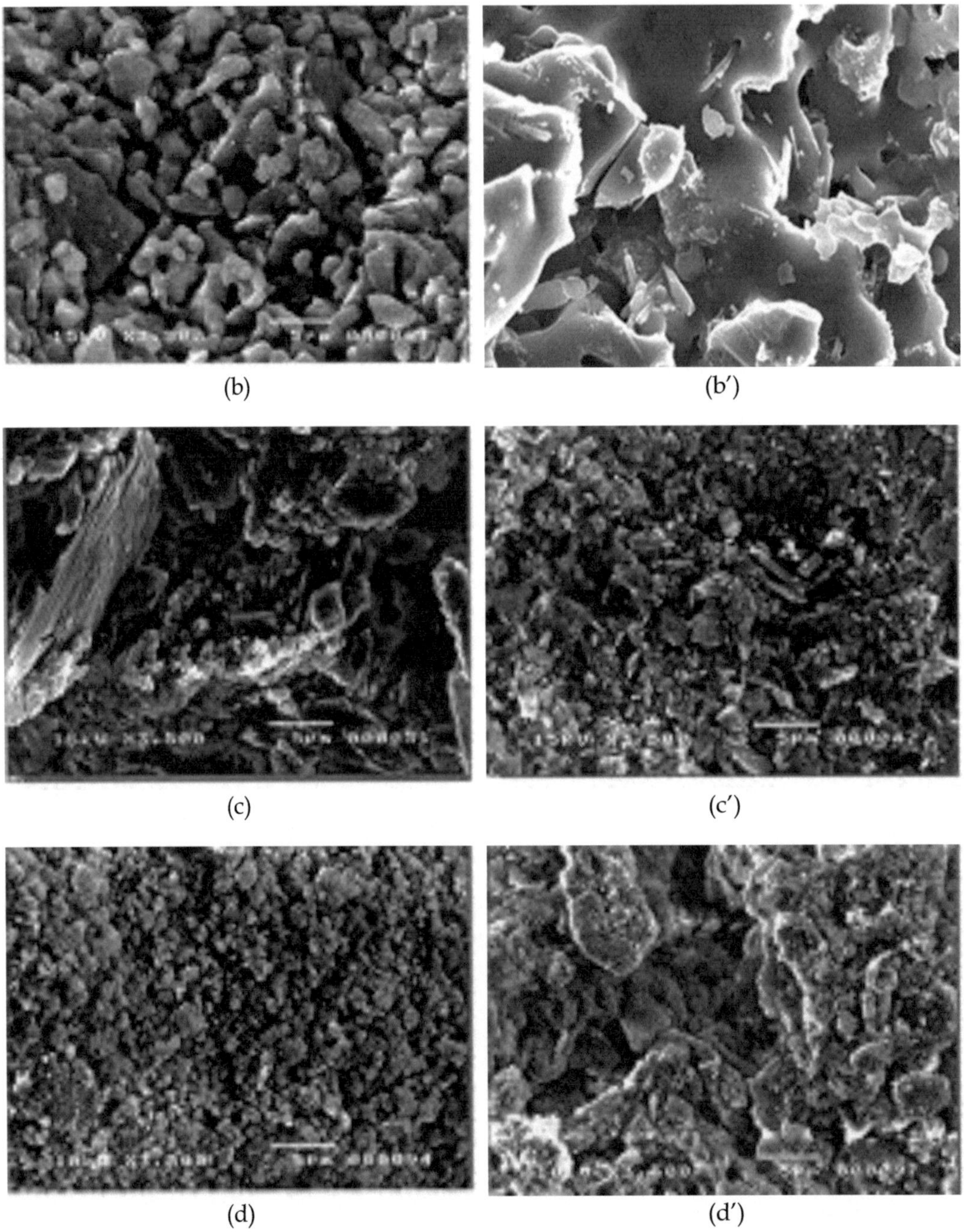

Fig. 10. SEM micrographs of limestone and oyster shells ; (a) fresh limestone, (a′) fresh oyster shells, (b) calcined limestone at 850°C, (b′) calcined oyster shells at 850°C, (c) hydrated limestone, (c′) hydrated oyster shells, (d) hydrated limestone after calcinations, and (d′) hydrated oyster shells after calcinations.

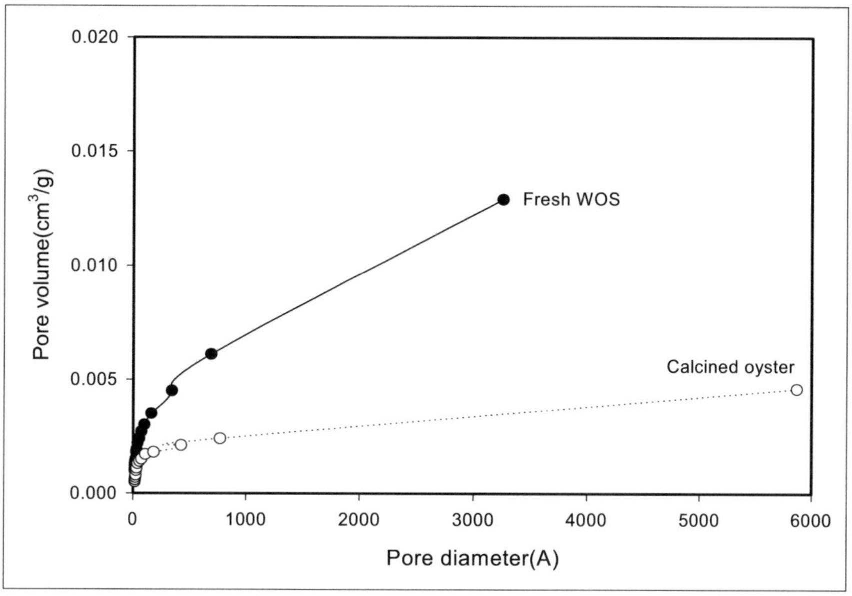

Fig. 11. Cumulative pore volume curves for two types of waste oyster shells as a function of pore diameter.

Fig. 11 shows pore volume with respect to pore diameter. The calcined oyster had a decreased pore volume relative to that of the fresh oyster. The specific surface area of oyster shells and limestone as a function of different calcination temperatures is shown in Fig. 12. The specific surface area of the oyster shell changed from 2.4465 m^2/g before calcination to 2.3950 m^2/g at 700.0°C, 2.2810 m^2/g at 750.0°C, 2.2120 m^2/g at 800.0°C, 2.1209 m^2/g at 850.0°C, 1.9510 m^2/g at 900.0°C, 1.8000 m^2/g at 950.0°C, and 1.7000 m^2/g at 1,000.0°C. The specific surface area of waste oyster shells was larger than that of calcined oyster shells and it was decreased with increasing calcination temperature. It is indicated that some agglomerations by sintering were blocking the specific surface area of calcined oyster shells. On the contrary, specific surface areas of limestone increased from 1.2368 m^2/g before calcinations to 1.3100 m^2/g at 700.0°C, 1.3200 m^2/g at 750.0°C, 1.5100 m^2/g at 800.0°C, 2.1544 m^2/g at 850.0°C, 2.3205 m^2/g at 900.0°C, 2.0210 m^2/g at 950.0°C, and 1.2124 m^2/g at 1,000.0°C. The specific surface area of limestone increased with increasing calcination temperature up to maximal 900.0°C and then a little decreased.

The specific surface area at 850.0°C was similar to that at 900.0°C. Therefore, we conclude that the calcination temperature had a positive effect on the development of the specific surface area for limestone, unlike waste oyster-shells. The change of surface area in oyster shells and limestone by processing (calcination/hydration) is shown in Fig. 13. As Fig. 13 shows, specific surface areas of oyster shells and limestone were changed from 2.4465 m^2/g and 1.2368 m^2/g before calcination/hydration to 12.9780 m^2/g and 11.3380 m^2/g by pretreating process, respectively. From this result, we could expect that the sulfating reactivity of oyster-shells sample increases to about 5 times by calcination/hydration reaction due to the increase of specific surface area and pore volume. Because of acid gas removal capacity of absorbents was proportional to the specific surface area (Jung et al., 2005).

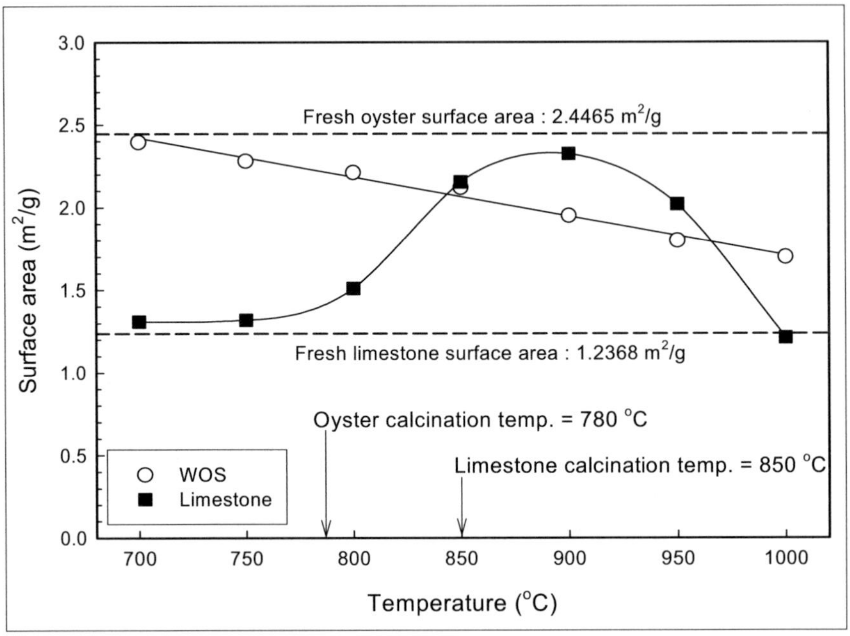

Fig. 12. Effect of calcinations temperature on the surface area.

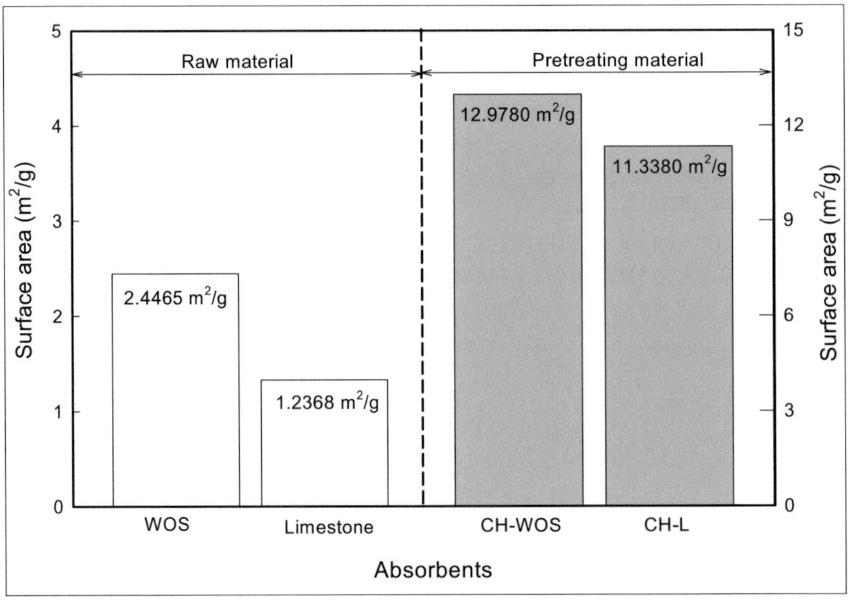

Fig. 13. Variation of surface area of oyster shells and limestone with and without pretreatment.

5.3 Characteristics of waste oyster shells as a SO_2/NO_x removal reaction

The desulfurization efficiency of the raw material was shown in Fig. 14. From a comparison of SO_2 removal quantities between waste oyster shells and limestone, the desulfurization capability of waste oyster shells was higher about 50% than that of Jungsun limestone. This means that the SO_2 removal capacity of oyster shells was superior to the limestone due to the specific surface area as can be seen in Fig. 13.

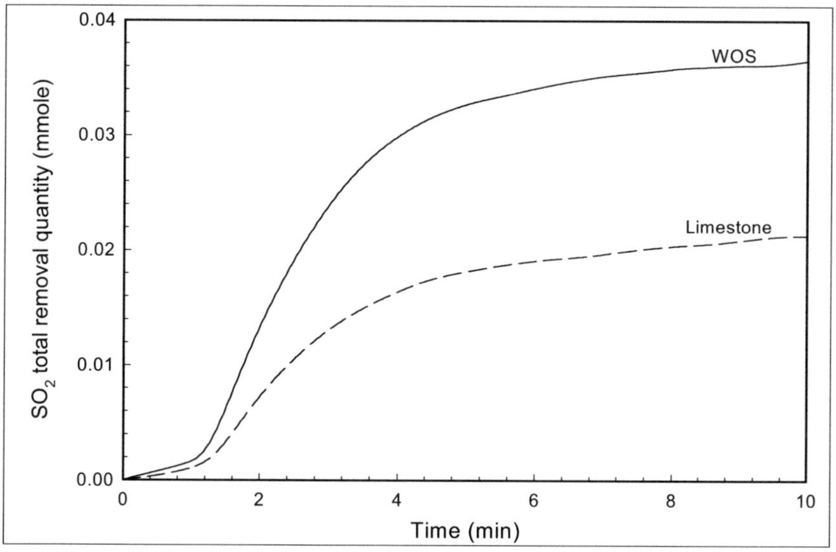

Fig. 14. Comparison of SO_2 removal quantities between waste oyster shells and limestone.

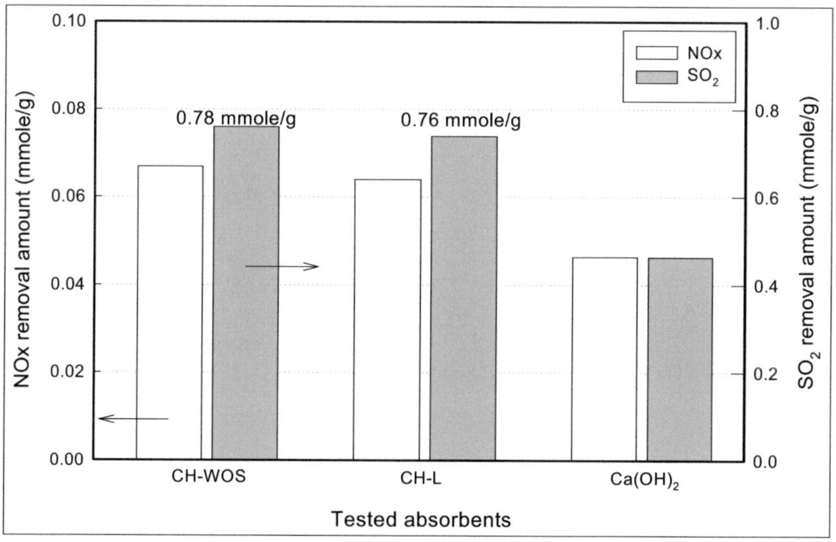

Fig. 15. SO_2/NOx removal amounts of tested absorbents.

In general, a power plant discharges SO_2 of 1800 ~ 1900 ppm, O_2 of 6%, CO_2 of 13%, N_2 of 74%, water content of 10%, and NO of 600 ppm to the air during the combustion. The

SO_2/NOx removal capacity of the calcined/hydrated limestone (CH-L) and calcined/hydrated waste oyster shells (CH-WOS) was carried out in a fixed bed reactor. As can be seen in Fig. 15, desulphurization capacity of the adsorbent was one order of magnitude higher than the denitrification capacity regardless of absorbent species. This is because the Henry's Law constant and diffusion coefficient in the gas phase for SO_2 is much higher than those for NO (Yuan, 1990). And, SO_2 and NOx removal quantities of CH-WOS were higher just a little than that of CH-L. It can be inferred that waste oyster shells is a good sorbent for the removal of SO_2 and NOx in the flue gas cleaning processes. The SO_2 and NOx absorption mechanism on absorbent can be explained by combining equations listed below (Nakamura, 1995).

$$CaCO_3(s) + SO_2(g) + 2H_2O(l) \leftrightarrow CaSO_3 \cdot 2H_2O(s) + CO_2(g) \quad (5)$$

$$CaO(s) + H_2O(l) \leftrightarrow Ca(OH)_2(s) \quad (6)$$

$$Ca(OH)_2(s) + SO_2(g) + H_2O(l) \leftrightarrow CaSO_3 \cdot 2H_2O(s) \quad (7)$$

$$CaCO_3(s) + 2NO(g) + 1/2O_2(g) \leftrightarrow Ca(NO_2)_2(s) + CO_2(g) \quad (8)$$

$$Ca(OH)_2(s) + 2NO(g) + + 1/2O_2(g) \leftrightarrow Ca(NO_2)_2(s) + H_2O(l) \quad (9)$$

Removal of SO_2 ($Ca(OH)_2$ conversion ratio) in simulated flue gas by the CH-L and CH-WOS absorbents was examined under different reaction conditions to study effects of the coexistence of NOx in a flue gas (Fig. 16). As depicted in Fig. 16, the reaction rate of desulphurization by CH-WOS increased about 30% higher than that of CH-L. In addition, SO_2 removal activity was enhanced in the presence of NOx, which might be due to its oxidizer role of SO_2 (Tsuchiai et al., 1996).

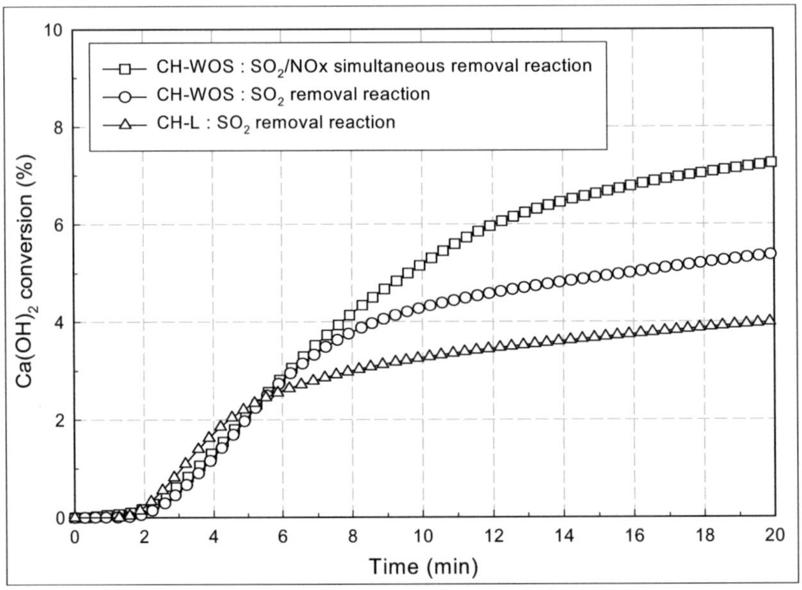

Fig. 16. $Ca(OH)_2$ conversions by SO_2 and SO_2/NOx reaction for limestone and waste oyster shells.

6. Conclusions

Enormous amount of waste oyster-shells were dumped into public waters and landfills, which cause a bad smell as a consequence of the decomposition of organics attached to the shells. Also, marine pollution by waste oyster shells has become one of the serious problems in mariculture industry in Korea. The present study has conducted to develop a means of converting waste oyster shells into useful absorbent for removal SO_2/NOx from industry. In this study, feasibility of waste seashell as absorbents for the control of air pollution has been investigated in a fixed bed reactor. To seek for a feasibility to recycle the waste oyster shells as desulfurization/denitrification sorbent, pretreating experiments and SO_2/NOx removal activity were investigated. Physicochemical properties of waste oyster shells have been characterized using the XRD, SEM, and BET.

By pretreating process, such as calcinations and hydration, the specific surface area and pore volume of waste oyster-shell were increased than that of the fresh particles, which makes it possible to enhance the removal capacity in acid gases. XRD analysis of calcined waste oyster shells were exhibited peaks characteristic of calcium oxide, whereas raw waste oyster-shells showed that the main peaks were characteristic of calcium carbonate.

And it was concluded that the optimal temperatures for calcination and hydration were 800.0 ~ 850.0°C and 90.0°C respectively. Pores of absorbent are formed by the emission of CO_2 during the high temperature calcination but it was agglomerated by sintering. Therefore, the specific surface area decreased and it was completely different from limestone. SO_2/NOx removal experiments have been carried out to test the reactivity of absorbents in a fixed bed reactor. SO_2 removal activity and reaction rate of calcined/hydrated waste oyster-shells were higher than that of calcined/hydrated limestone. It is clearly indicated that absorbent prepared by waste oyster-shells are substituted for commercial limestone and can be directly applied to industries which try to reduce their emissions of SO_2 and NOx.

7. References

Asaoka, S. et al. (2009). Removal of hydrogen sulfide using crushed oyster shell from pore water to remediate organically enriched coastal marine sediments. *Bioresource Technology*. 100, 4127–4132.

Chu, K-J. et al. (1997). Characteristics of gypsum crystal growth over calcium-based slurry in desulfurization reaction. *Materials Research Bulletin*, 32(2), 197.

EPRI, (1988). FGD chemistry and analytical methods handbook, CS-3612.

Garea, A. et al. (2001). Kinetics of dry flue gas desulfurization at low temperatures using $Ca(OH)_2$: competitive reactions of sulfation and carbonation. *Chem. Eng. Sci.*, 56(4), 1387.

Jung, J-H. (1999.2). A study on reaction characteristic of SO_2/NOx simultaneous removal for alkali absorbent/additive in FGD and waste incinerator Process, Ph.D thesis, *Pusan National University*, Korea.

Jung J.-H. et al. (2000). Physicochemical Characteristics of Waste Sea Shell for Acid Gas Cleaning absorbent. *Korean J. Chem. Eng.*, 17(5), 585-592.

Jung, J-H. et al. (2005).Reactivity of Bio-sorbent Prepared by Waste Shells of Shellfish in Acid Gas Cleaning Reaction. *The Korean Journal of Chemical Engineering*, 22(4), 566-568.

Jung, J-H. et al. (2007). Reuse of waste oyster-shells as a SO_2/NO_x removal absorbent. *Journal of Industrial and Engineering Chemistry*, 13(4), 512-517.

Jung, J-H. (2008.2). Effects of air pollutants on the health/environmental risk assessment and weathering of stone cultural properties in Gyeongju and its Vicinities, Ph.D thesis, *Daegu Haany University*, Korea.

Jonas K. et al. (1984). Similities between lime and limestone in wet-dry scrubbing. *Chem. Eng, Process*, 18, 239-247.

Kwon, Y-S. et al. (2003). Sorption characteristics of heavy metals for oyster shell and fly ash. *J. Korea Society of Waste Management*. 20(4), 337-345.

Kim, H-S. (2007). The study of application of discarded oyster shell powder as an architectural material, Dong-A University, Master thesis, Korea.

Kwon, H-B et al. (2004). Recycling waste oyster shells for eutrophication control. *Resources Conservation and Recycling*, 41, 75-82.

Kunio K. and Hidenori S. (1994). Effective dry desulfurization by a powder-particle fluidized bed. *Journal of Chemical Engineering of Japan*, 27(3), 276-278.

Lee, D-Y. et al. (2001). Using seafood waste as sludge conditions. *Wat. Sci. Tech.*, 44(10), 301-307.

Lee, C-H. et al. (2008a). Effects of oyster shell on soil chemical and biological properties and cabbage productivity as a liming materials. *Waste Manage*. 28, 2702–2708.

Lee, S-W. et al. (2008b). Nano-structured biogenic calcite: a thermal and chemical approach to folia in oyster shell. *Micron* 39, 380–386.

Nippon Steel Corp. (1993). Production of metal or alloy refining flux from oyster shell and other shells by pulverizing shells, Japan pat., J 05025524.

Nakamura, H et al. (1995). Pilot-scale test results of simultaneous SO_2 and NOx removal using powdery form of LILAC absorbent, *1995 SO_2 Control Symposium,* March, Florida, U.S.A.

Nakatani, N. et al. (2009). Transesterification of soybean oil using combusted oyster shell waste as a catalyst. *Bioresource Technology* 100(3), 1510–1513.

O'Dowd, W.J. et al. (1994) Characterization of NO_2 and SO_2 removals in a spray dryer/baghouse system. *Ind. Eng. Chem. Res.*, 33, 2749.

Park, W-H., Polprasert, C., (2008). Roles of oyster shells in an integrated constructed wetland system designed for P removal. *Ecol. Eng*. 34, 50-56.

Shin, C-H et al. (1998). Sterilization effect of silver ion-exchanged oyster shell powder on underwater microorganism. *Environ. Eng. Res.*, 3(3), 123-129.

Tsuchiai, H. et al. (1996). Removal of sulfur dioxide from flue gas by the absorbent prepared from coal ash: Effects of nitrogen oxide and water vapor. *Ind. Eng. Chem. Res.*, 35(3), 851.

Yoon G-L. et al. (2003). Chemical-mechanical characteristics of crushed oyster-shell. *Waste Management*, 23, 825-834.

Yuan C.S. (1990). Simultaneous collection of SO_2 and NOx via spray drying : Using sodium-based and calcium-based sorbents with select additives, Ph.D. thesis, Illinois University, U.S.A.

13

Carbon Steel Slag as Cementitious Material for Self-Consolidating Concrete

Yu-Chu Peng
Graduate Institute of Construction Engineering, National Taiwan University of Science and Technology,
Depart of Leisure Management, Taiwan Hospitality & Tourism College, China,Taiwan

1. Introduction

In Taiwan, self-consolidating concrete (SCC) exhibiting high-flow behaviour is a widely used concrete material to prevent conventional concrete problems such as honeycomb structures that occur as a result of poor practice. SCC is also used as the material of choice for heavily reinforced concrete structures located in seismic zones [Paczkowski Piotr,Kaszynska Maria,2007.]. Pozzolanic materials are important ingredients for making SCC [Mihashi H, Yan X,1995.]. For many years, pozzolanic admixtures, such as blast furnace slag (BFS), pulverized coal ash (fly ash), silica fumes, and copper slag have been recycled to partially replace Portland cement in concrete mixtures. The main advantages of using pozzolanic materials are improvements in performance and significant reduction in the life-cycle costs of concrete structures; the latter, in particular, continues as a significant problem for engineers [Khalifa AJ, Ramzi T,2002. Li G, Zhao X.,2003. Zhang MH, Bilodeau A, Malhotra VM, Kim KS, Kim JC,1999.]. Materials such as steel slag, normally considered as waste, have promising applications as partial Portland cement replacements in concrete mixtures. Considerable research and development has been conducted to develop new concrete technologies such as SCC. Further, the construction of durable concrete has also been pursued. Initially, pozzolanic admixtures were solid waste and it was extremely costly to treat and dump them into a final storage area. Today, however, in the concrete industries in Taiwan and elsewhere, these admixtures are important materials for the production of low-cost durable concrete, and an example of environmental protection and resource conservation.

In Taiwan, carbon steel slag (CSS) is a by-product of the reduction during the production of refining carbon steel in an arc furnace, and is seldom recycled. On average, the production of one ton of carbon steel yields 10 kg of CSS waste, and hence, in Taiwan, more than 56,000 tons of CSS is produced each year. Due to the relatively small amounts of CSS relative to blast furnace slag (BFS), environmental protection agency (EPA) regulations had previously permitted the dumping of CSS. Today, the dumping of such waste is not permitted, and the proper disposal of CSS has become a huge problem. Since lime, coke and silicon iron are added to promote the reducing process during high-temperature-refinery scrap steel procedures, the CSS contains large amounts of CaO, SiO_2 and Al_2O_3. This waste composition, however, is similar to BFS or Portland cement [Chiang CC, Chenn YY, Lin TY,

Hwang CL,2004.Yu-Chu Peng,2009.]. Hence, CSS can be considered for use as a pozzolanic admixture to partially replace Portland cement in a concrete mixture.

Rather than use CSS for backfill soil or as material to be retained in the plant, steel slag can be regarded as a low-quality clinker and can be used to partially substitute the clinker of composite Portland cement [Wu X, Zhu H, Hou X, Li H,1999.Sakuraya T,1999.]. In Japan and other industrialized countries, steel slag has already been applied for use in civil engineering applications such as road base construction and soil stabilization [Geiser J.,1999. Roy DM, Idorn GM.,1982.]. In Germany, anbout 17.1% of steel slag is used for highway construction, 5.4% is recycled, and 40.5% is used in agricultural fertilizer production [Luxán MP, Sotolongo R, Dorrego F, Herrero E.2000. Monshi A, Asgarani MK,1999. Mihashi H, Yan X, Arikawa S.,1995. Hogan FJ, Meusel JW.,1981. ACI Committee 211.,1993.]. The mineralogical composition of steel slag is as follows: anhydrous calcium silicates and silicoaluminates; gehlenite, larnite and bredigite; magnetite and magnesioferrite and manganese oxides [Esfahani M. Reza,Kianoush M. Reza,2005. Hwang Soo-Duck,2008. Koehler Eric P. ,Fowler David W.,2008.]. Thus, some researchers have tested the effects of mixed iron slag (36%~45%), steel slag (6%~22%) and limestone (40%~64%) on the setting time of cement paste and the compressive strength at 3, 7 and 28 days [Schindler Anton K.,Barnes Robert W.,Roberts James B.,2007. Whitcomb Brent L., Kiousis Panos D.,2008.]. Nevertheless, other than documenting the chemical composition of CSS, there are few studies on the pozzolanic reactions after the addition of CSS.

2. Research plan

2.1 Material

The aggregate was quarried from I-Lan River, Northern Taiwan, and consisted of large amounts of elongate slate and fragile particles. The cement and superplasticizer (SP) used corresponded to ASTM C150 type I Portland cement and ASTM C494 type F high range water reducing agent (HWRA), respectively. A naphthalene lingo-sulfonate base was used for promoting the flow ability of SCC; the specific gravity was 1.18; ph, 6.93 and chloride ion content, less than 50 ppm. As a by-product of the carbon steel manufacturing process, while the carbon steel settles in the smelter (since its density is high), impurities remain on top. The carbon steel is then transported to a water basin maintained at a low temperature for solidification. The end product (CSS) is a hard solid material that is then sent to a crusher for further processing. The CSS is powdered to pass through sieve No. 4 (4.76 mm). Subsequently, it is re-ground for 3 h at a speed of 60 rpm to pass through sieve No. 200 (75 μm). In this study, type I Portland cement has been used. Class F fly ash and BF slag were obtained from Taiwan Power Company and China Steel Corporation, respectively. The SP was Glenium 51 obtained from Taiwan Durusle Company, Taiwan. In Table 1, the specific gravities of Portland cement and CSS are listed as 3.14 and 2.67, respectively; further, CSS powder and Portland cement (type I) have specific surface areas of 2504 cm^2/g and 3622 cm^2/g, respectively. Hence, CSS has the least fineness, which is characteristic of materials with low surface areas. As shown in Table 2, CSS is highly alkaline, with a pH of 11.50, an absorption capacity (SSD) of 7.60%, fineness modulus (FM) of 1.76 according to ASTM C136, and dry loose density of 1266 kg/m^3 according to ASTM C29. Figure 1 shows the relationships of CSS with BFS and Portland cement; the percentage of the main composition (SiO_2 and CaO) of CSS lies between that of BFS and Portland cement.

	Item	OPC	CSS
Physical properties	Specific gravity	3.14	2.67
	Specific surface area (cm^2/g)	3622	2504
	pH	—	11.50
	Absorption capacity (%)	—	7.60
	Fineness modulus (FM)	—	1.76
	Dry loose density (kg/m^3)	—	1266
Chemical compositions (%)	SiO_2	21.46	26.52
	Al_2O_3	4.84	5.95
	Fe_2O_3	3.12	3.78
	CaO	62.34	46.45
	MgO	2.87	13.27
	SO_3	2.06	0.65
	f-CaO	0.88	2.11
	Na_2O	0.22	0.26
	K_2O	0.70	0.11
	CaO/SiO_2	2.91	1.75

Table 1. Physical properties and chemical composition of OPC and CSS.

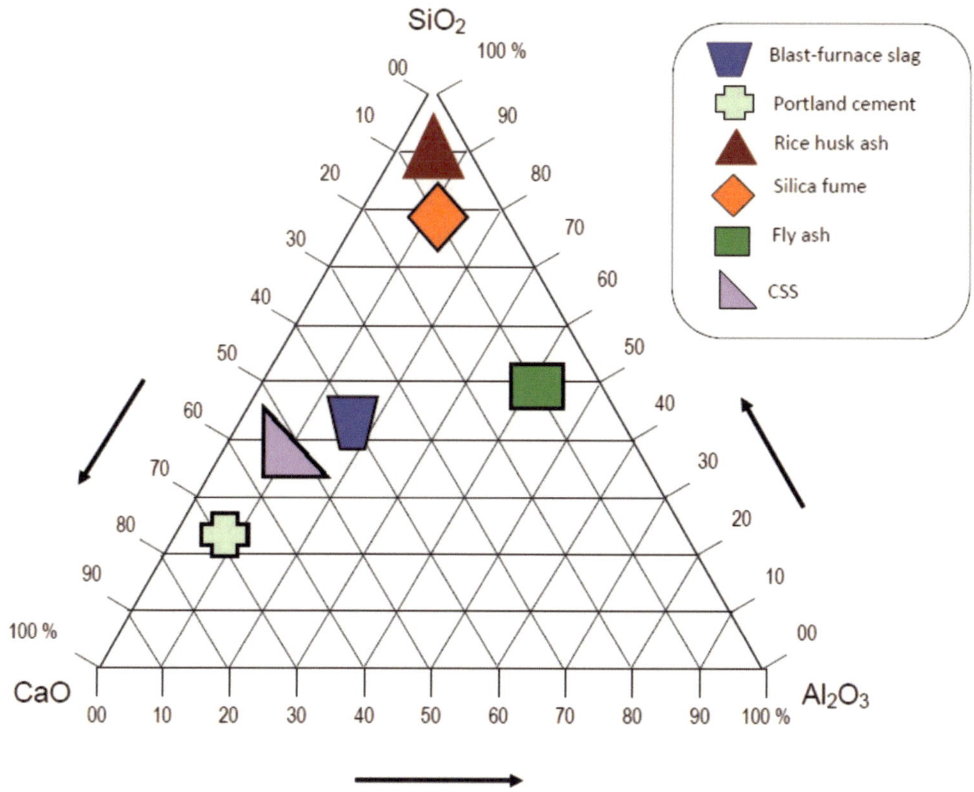

Fig. 1. Comparison of compositions of CSS, BFS and Portland cement.

2.2 Mixture design

In order to obtain high-strength SCC with lower water content (160 kg/m³), in this study, w/cm ratios (water/(cement+CSS)) of 0.28, 0.32 and 0.40 were selected. Further, large amounts of SP were added to achieve better flow behaviour. CSS powder was used to replace the 5.0%, 7.5% and 10% weights of Portland cement. Mixtures with three different w/cm ratios were prepared for ordinary plain concrete (OPC) and carbon steel slag concrete (CSC), as shown in Table 2; designated as OPC28, OPC32 and OPC40 and CSC28, CSC32 and CSC40, respectively (Whitcomb Brent L., Kiousis Panos D.,2009. Kwan Albert K. H.,Ng, Ivan Y. T.,2008.) .

Designation of concrete	w/c ratio[a]	w/cm ratio[b]	Mix Proportion (kg/m³)						
			Cement	CSS/ cement (%)	Fine aggregate	Coarse aggregate	Water	SP[c]	Water + SP
OPC28	0.28	0.28	572	--	757	901	145	15	160
OPC32	0.32	0.32	500	--	783	932	148	12	160
OPC40	0.40	0.40	400	--	820	976	152	8	160
CSC28	0.28	0.28	545	5.0	832	820	146	14	160
CSC32	0.34	0.32	465	7.7	861	849	149	11	160
CSC40	0.44	0.40	364	10.0	901	888	153	7	160

[a] w/c ratio = water/cement
[b] w/cm ratio = water/(cement + CSS)
[c] SP = Superplasticizer

Table 2. Mixture proportion of SCC.

3. Results and discussions

3.1 Workability of SCC

Figure 2 illustrates slump vs. different w/cm ratios for both OPC and CSC; the figure indicates that all slump values are greater than 230 mm, the specification for SCC with high flow. Concrete with a lower w/cm ratio than 0.28—implying significantly high binder content—may lead to higher slump and satisfactory flowability. For an identical dosage of SP mixtures, CSC has higher slump than OPC mixtures. Hence, it is clear that the use of CSS might improve the workability of SCC; and this is similar to the test results of the research papers referenced. Hence, the use of CSS in SCC can lead to high flow properties.

3.2 Setting time of SCC

Figure 3 shows the effects of CSS on the penetration resistance of concrete, indicating that as the w/cm ratio of OPC or CSC increases, the penetration resistance decreases and the setting time increases. Further, the setting time of CSC is longer than that of OPC irrespective of the w/cm ratio. This is due to the low PAI values of CSS. Thus, the setting time increases with the amount of CSS. This result is similar to the results in a previous study, which showed that the addition of BFS decreased the setting time of SCC [Mihashi H, Yan X, Arikawa S.,1995.].

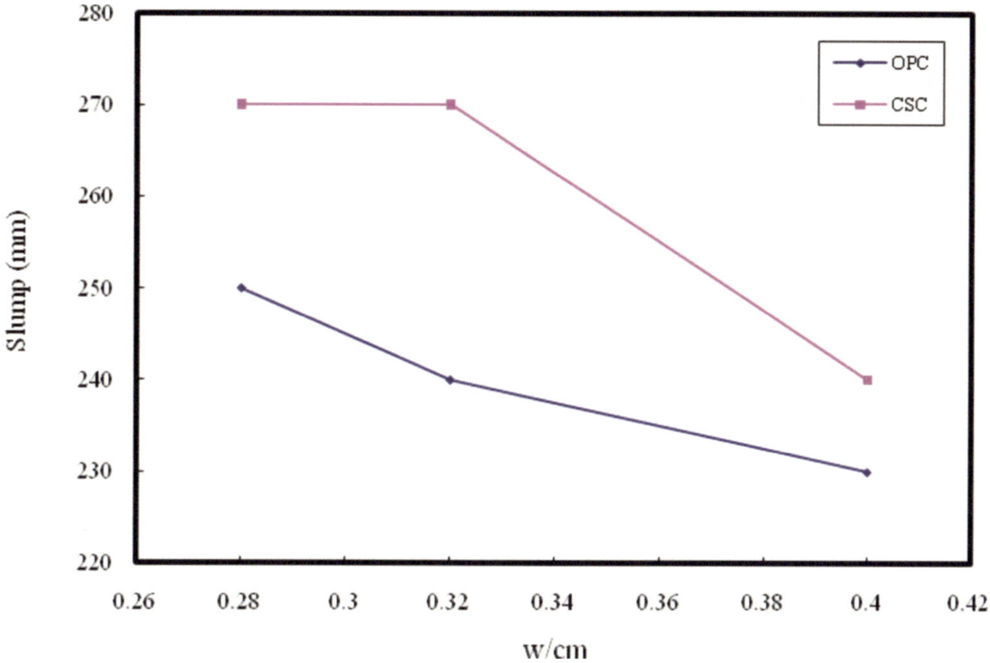

Fig. 2. Slump vs. various w/cm ratios of concrete.

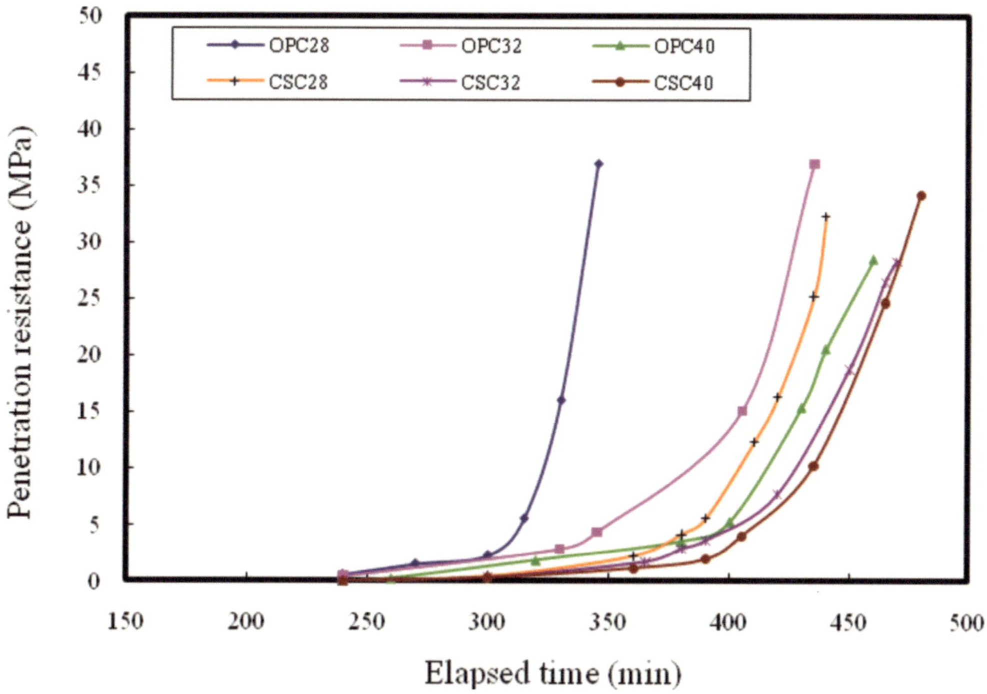

Fig. 3. Comparison between the OPC and CSC with respect to penetration resistance of concrete.

3.3 Compressive strength of SCC

The compressive strength and percentage of concrete mixtures with different w/cm values at the specified age are shown in Table 3. The compressive strength of each mixture is greater than 41 MPa at 56 days. This satisfies the requirement that SCC must have high strength [Hogan FJ, Meusel JW.,1981.]. The compressive strength of OPC and CSC with a w/cm ratio of 0.28 is either equal to or higher than 83 MPa at 90 days; however, that of CSC28 at any age is lower than that of OPC28. In contrast, the percentage of compressive strength is higher than 90% at 28 days, and it is reduced to 76% at 90 days. This corresponds to a 15% reduction in the PAI at 28 days. The compressive strength of CSC with a w/cm ratio of 0.32 or 0.40 at any age, however, is higher than that of OPC, and the percentages of compressive strength are from 106% to 134% and from 108% to 121%, with respect to the w/cm ratio. This clearly indicates that the addition of CSS improves the strength development of cement paste as long as the water-to cement (w/c) ratio is greater than 0.32 or w/cm is higher than 0.28. This means that the reactions of strength development of cement with CSS will be enhanced with sufficient water contents. It is suggested, however, that the total water content of concrete, including the moisture in liquid admixture be maintained as low as possible to avoid large shrinkage and sedimentation. Figure 4 shows the influence of CSS content on compressive strength: higher CSS content mixtures will cause lower compressive strength.

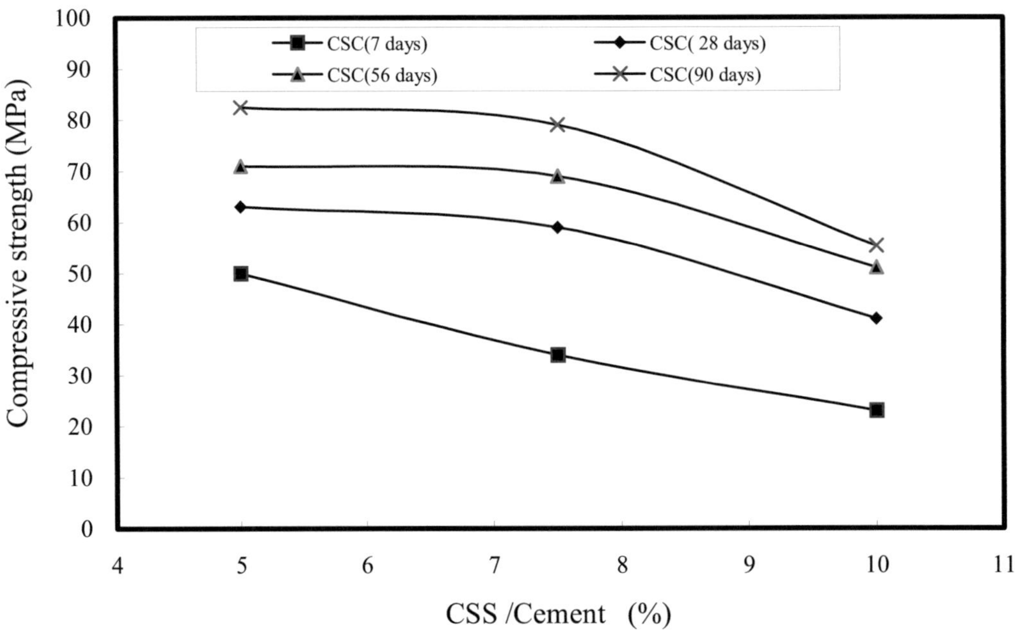

Fig. 4. Effect of CSS content at different ages on the compressive strength of SCC.

Designation of concrete	w/c ratio	w/cm ratio	Compressive strength, MPa (% compressive strength[a])				
			3 days	7 days	28 days	56 days	90 days
OPC28	0.28	0.28	46.5 (100)	53.9 (100)	69.6 (100)	93.7 (100)	95.6 (100)
CSC28	0.29		42.9 (92)	50.0 (93)	63.2 (91)	71.0 (76)	82.5 (86)
OPC32	0.32	0.32	25.0 (100)	32.0 (100)	49.6 (100)	57.3 (100)	59.1 (100)
CSC32	0.34		26.5 (106)	34.1 (107)	58.5 (118)	69.4 (121)	78.9 (134)
OPC40	0.40	0.40	16.3 (100)	22.2 (100)	37.1 (100)	45.1 (100)	45.8 (100)
CSC40	0.44		17.6 (108)	24.1 (109)	40.7 (110)	50.6 (112)	55.3 (121)

Table 3. Compressive strength and percentage compressive strength of concrete.

3.4 Ultrasonic pulse velocity (UPV) of SCC

Theoretically, the ultrasonic pulse velocity (UPV) of a solid object is higher that of air, and a high-density solid will have high UPV. Therefore, the UPV is a good measure of the soundness of hardened concrete. It is generally acknowledged that the UPV increases with concrete density. In our study, the UPV of all mixtures was greater than 4200 m/s. Table 4 shows the UPV and the difference in UPV between CSC and OPC (as a percentage) at each w/cm ratio, from 3 to 90 days. The UPV of CSC28 at all ages is lower by 1% to 2% than that of OPC28; however, the UPV of CSC32 and CSC40 is higher by 3% than that of OPC32 and OPC40, respectively. This result is similar to trend in compressive strength—the addition of CSS enhances the pozzolanic reaction with high w/c or w/cm ratios, i.e., sufficient water. Figure 5 shows a good linear relationship between the compressive strength and UPV of concrete for both OPC and CSC. In other words, UPV is a good method for evaluating the performance and homogeneity of SCC.

Designation of concrete	w/c ratio	w/cm ratio	UPV of concrete, m/s (% UPV[a])				
			3 days	7 days	28 days	56 days	90 days
OPC28	0.28	0.28	4606 (100)	4787 (100)	4859 (100)	4876 (100)	4890 (100)
CSC28	0.29		4525 (98)	4683 (98)	4828 (99)	4830 (99)	4835 (99)
OPC32	0.32	0.32	4381 (100)	4523 (100)	4760 (100)	4777 (100)	4785 (100)
CSC32	0.34		4466 (102)	4669 (103)	4814 (101)	4821 (101)	4825 (101)
OPC40	0.40	0.40	4211 (100)	4274 (100)	4595 (100)	4635 (100)	4651 (100)
CSC40	0.44		4295 (102)	4361 (102)	4659 (101)	4709 (102)	4728 (102)

[a] Percentage UPV = (CSC/OPC) × 100 at fixed w/cm ratios

Table 4. UPV and percentage UPV of SCC.

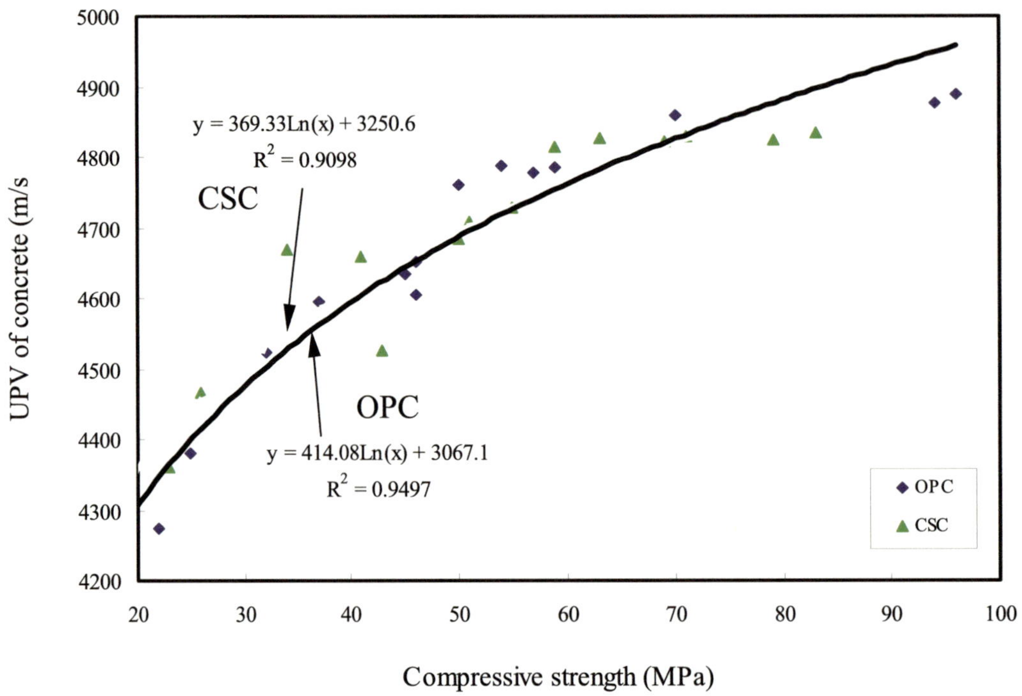

Fig. 5. Relationship between UPV and compressive strength of OPC and CSC.

3.5 Microstructure observation

Scanning electron microscopy (SEM) observations are conducted with specimens at the ages of 3 and 28 days. The image characteristics of concrete at 3 days are shown in Figs. 6(a)–8(a). As shown in Fig. 6(a), at the early age of 3 days, considerable amounts of hexagonal-shaped calcium hydroxide ($Ca(OH)_2$), spherical-shaped C-S-H gel in CSC28 (w/cm = 0.28) and certain amounts of fine pores (dark zone) exist. Figure 7(a) shows the presence of numerous rosette-shaped mono-sulfoaluminate (AFm) and small amounts of needle-shaped ettringite (AFt) in CSC32 (w/cm = 0.32). Figure 8(a) also shows that there are rosette-shaped AFm in CSC40 (w/cm = 0.40). Here, the w/cm ratio is greater than or equal to 0.32 as a result of increase in CSS amounts and the existence of high Al_2O_3 and Fe_2O_3 contents. The primary hydration products are hexagonal-shaped $Ca(OH)_2$, spherical-shaped C-S-H gel and a certain amount of rosette-shaped AFm. This observation confirms the conclusions made for both strength and UPV that the reaction of CSS with cement paste requires sufficient water to aid hydration. At a later age, as shown in Fig. 6(b), the microstructure of CSC28 is extremely dense. Figure 7(b) also shows the presence of numerous rosette-shaped AFm, but no needle-shaped Aft, while Fig. 8(b) shows large pores in CSC40 with a large amount of hexagonal-shaped $Ca(OH)_2$ in the reaction process. While this is advantageous for the hydration reaction of CSS, it also indicates that more pores are observed with higher w/cm ratios.

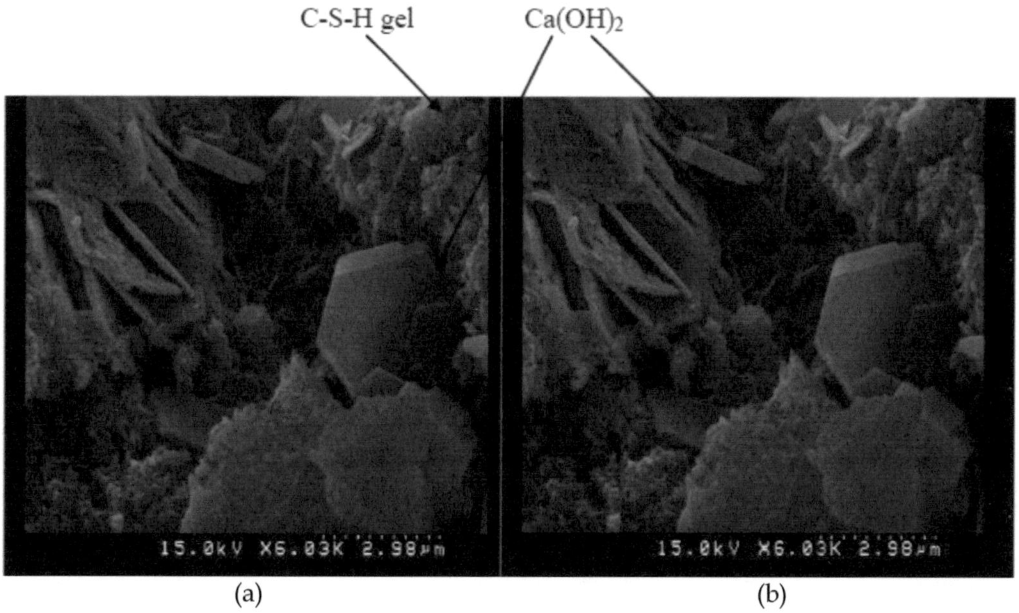

Fig. 6. SEM micrograph of CSC28: (a) at 3 days; (b) at 28 days.

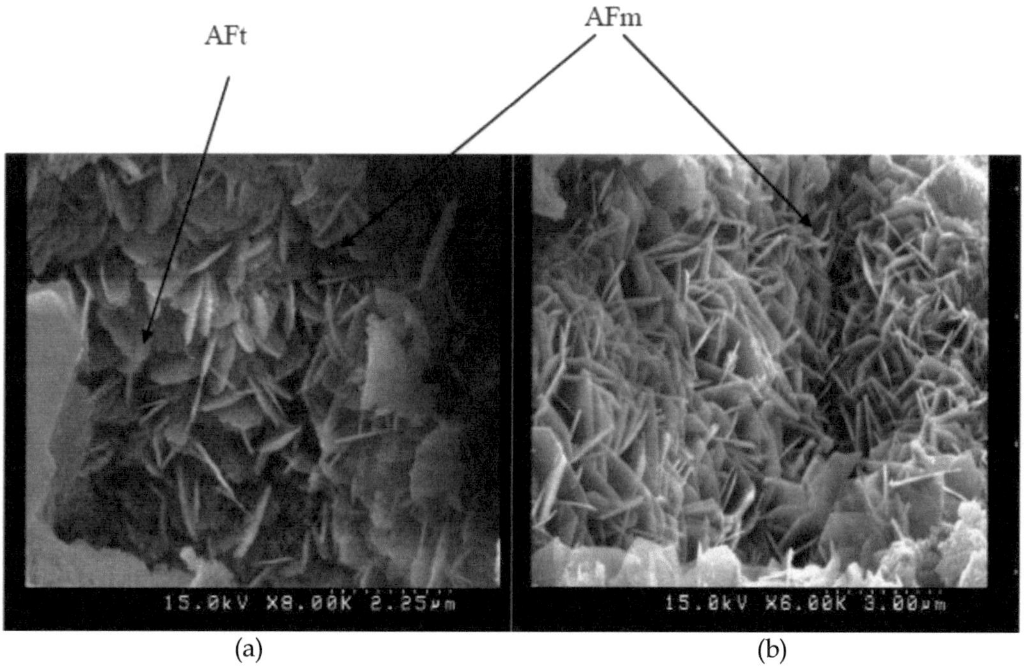

Fig. 7. SEM micrograph of CSC32: (a) at 3 days; (b) at 28 days.

(a) (b)

Fig. 8. SEM micrograph of CSC32: (a) at 3 days; (b) at 28 days.

4. Conclusions

In this study, we have conducted investigations on the recycling of carbon steel slag CSS to produce SCC. Our conclusions are as follows:

- The major chemical compositions of CSS are CaO, Al_2O_3 and SiO_2, a composition similar to that of Portland cement and BFS. The PAI of CSS is 85% at 28 days. Hence, CSS can be expected to have good cementitious properties and effects.
- CSS can be designed as easily as SCC. In comparison with OPC, increasing the CSS content will increase the setting time.
- Concrete using CSS has a higher compressive strength than that using OPC. If w/cm ratios of 0.32 or 0.40 are used, the percentage of compressive strength increases by more than 21% at 90 days.
- As the amount of CSS in concrete increases, the compressive strength decreases. The strength is similar to other concrete, however, when CSS of 5.0–7.5% is used, except at 7 days.
- The SEM images show that the hydration rate of CSS is lower than that of OPC. Further, large amounts of $Ca(OH)_2$ and AFm are present in CSS as a cementitious material.

In this manner, we have shown that CSS can potentially be used as a cementitious material in self-consolidating concrete.

5. References

[1] ACI Committee 211. Guide for selecting proportions for high-strength concrete. ACI Mater J 1993;90:272–283.

[2] Chiang CC, Chenn YY, Lin TY, Hwang CL. The application of electric arc furnace reduction slags to high performance concrete. Journal of the Chinese Institute of Civil and Hydraulic Engineering 2004;16:167–178.

[3] Esfahani M. Reza, Kianoush M. Reza, Bond strength of glass fibre reinforced polymer reinforcing bars in normal and self-consolidating concrete, Canadian Journal of Civil Engineering, v 32, n 3, p 553-560, June 2005.

[4] Gregori Amedeo, Ferron Raissa, Sun Zhihui, Shah Surendra, Experimental simulation of self-consolidating concrete formwork pressure, ACI Materials Journal, v 105, n 1, p 97-104, January/February 2008.

[5] Geiser J. Metallurgical slag-create a future construction material. The International Associate of Metallurgical Slag Recycle and Utilization. Beijing, 1999. p. 1–8.

[6] Hogan FJ, Meusel JW. Evaluation for durability and strength development of a ground granulated blast furnace slag. Cem Concr Aggregates 1981;3:40–52.

[7] Hwang Soo-Duck, Khayat Kamal H., Effect of mixture composition on restrained shrinkage cracking of self-consolidating concrete used in repair, ACI Materials Journal, v 105, n 5, p 499-509, September/October 2008.

[8] Koehler Eric P., Fowler David W., Dust-of-fracture aggregate microfines in self-consolidating concrete, ACI Materials Journal, v 105, n 2, p 165-173, March/April 2008.

[9] Kwan Albert K. H., Ng, Ivan Y. T., Performance criteria for self-consolidating concrete, Transactions Hong Kong Institution of Engineers, v 15, n 2, p 35-41, June 2008.

[10] Khalifa AJ, Ramzi T, Mohammed AG. Use of Copper slag and cement by-pass dust as cementitious materials. Cem Concr Aggregates 2002;24:7–12.

[11] Luxán MP, Sotolongo R, Dorrego F, Herrero E. Characteristics of the slags produced in the fusion of scrap steel by electric arc furnace. Cem Concr Res 2000;30:517–519.

[12] Li G, Zhao X. Properties of concrete incorporating fly ash and ground granulated blast-furnace slag. Cem Concr Compos 2003;25:293–299.

[13] Mihashi H, Yan X, Arikawa S. Strength properties and frost damage resistance of high performance concrete using blast furnace slag and silica fume. In: Schwesinger P, editor. Proceedings of the Fourth Weimar Workshop on High Performance Concrete. Germany, 1995. p. 195–204.

[14] Monshi A, Asgarani MK. Producing Portland cement from iron and steel slag and limestone. Cem Concr Res 1999;29:1373–1377.

[15] Mihashi H, Yan X, Arikawa S. Strength properties and frost damage resistance of high performance concrete using blast furnace slag and silica fume. High Performance Concrete: Material Properties and Design. Germany, 1995. p. 195–204.

[16] Paczkowski Piotr, Kaszynska Maria, Self-consolidating concrete for on-site bridge applications, International Conference organised by the Institution of Civil Engineers, ICE, p 312-320, 2007.

[17] PENG Yu-chu, HUANG Chau-long. Engineering properties of sintered waste sludge as lightweight aggregate in a densified concrete mixture [J]. J Chongqing Univ: Eng Ed [ISSN 1671-8224], 2009, 8(4): 231-238.
[18] Roy DM, Idorn GM. Hydration, structures, and properties of blast furnace slag cements, mortars, and concrete. ACI J 1982;82:444–457.
[19] Schindler Anton K., Barnes Robert W., Roberts James B., Rodriguez Sergio, Properties of self-consolidating concrete for prestressed members, ACI Materials Journal, v 104, n 1, p 53-61, January/February 2007.
[20] Sakuraya T. The utilizing condition of metallurgical slag and steel slag for Japanese Refinery Steel Industry. The International Associate of Metallurgical Slag Recycle and Utilization, Beijing, 1999. p. 15–20.
[21] Wu X, Zhu H, Hou X, Li H. Study on steel slag and fly ash composite Portland cement. Cem Concr Res 1999;29:1103–1106.
[22] Whitcomb Brent L., Kiousis Panos D., Development of self-consolidating concrete for thin wall applications including validation, Journal of Materials in Civil Engineering, v 21, n 10, p 587-593, 2009.
[23] Zhang MH, Bilodeau A, Malhotra VM, Kim KS, Kim JC. Concrete incorporating supplementary cementing materials: effect on compressive strength and resistance to chloride-ion penetration. ACI Mater J 1999;96:181–189.

14

Possible Uses of Steelmaking Slag in Agriculture: An Overview

Teresa Annunziata Branca and Valentina Colla
Scuola Superiore Sant'Anna - PERCRO - TeCIP Institute
Italy

1. Introduction

Slags are the main by-products generated during iron and crude steel production and the steel industry is committed to increasing and improving their recycling.

Over the past decades, the steel production has increased and, consequently, the higher volumes of by-products and residues generated have driven to the reuse of these materials in an increasingly efficient way. In recent years new technologies have been expanded, and some of them are still under developing, in order to improve the recovery rates of slags. On this subject material separation technologies and carbon sequestration could dramatically reduce CO_2 emissions from steelmaking processes. On the other hand, the increase of slags recovery and use in different fields of application, such as in agriculture, allowed to reduce landfill slags and to preserve natural resources. In addition to the environmental achievements, these practices produced economic benefits, by providing sustainable solutions that can allow the steel industry to achieve its ambitious target of "zero-waste" in the incoming years (worldsteel, 2008).

Steel is produced by mean two main ways:

1. the iron ore based steelmaking (Fig. 1), which represents about 60-70% of the world steel production. The main raw materials are: iron ore, coal, limestone and recycled steel scrap. The main production routes are: the ironmaking iron ore based on Blast Furnace (BF) followed by steelmaking in the Basic Oxygen Furnace (BOF), and the ironmaking based on Direct Reduction of Iron ore (DRI), followed by steelmaking in the Electric Arc Furnace (EAF). In the BF coke is the reducing agent of iron ore. Limestone or dolomite (fluxes) are added into the blast furnace where they react with iron ore impurities, such as silica. Steel is produced from pig iron, scrap and lime in the BOF, where oxygen is blown to burn off the carbon.
2. the scrap-based steelmaking (Fig. 2), which represents about 30% of the world steel production. This way is based on the scrap recycling in the EAF, where the main input are steel scrap and electrical energy that is needed to melt the scrap into steel.

In both BOF and EAF the reactions between oxygen, carbon (carbon as gaseous carbon monoxide), silicon, manganese, phosphorus and some iron as liquid oxides produce oxidized compounds that react with lime or dolomitic lime to form slag. At the end of the refining operation, after steel pouring into a ladle, the slag is poured into a vessel and is subsequently tapped into a slag pot.

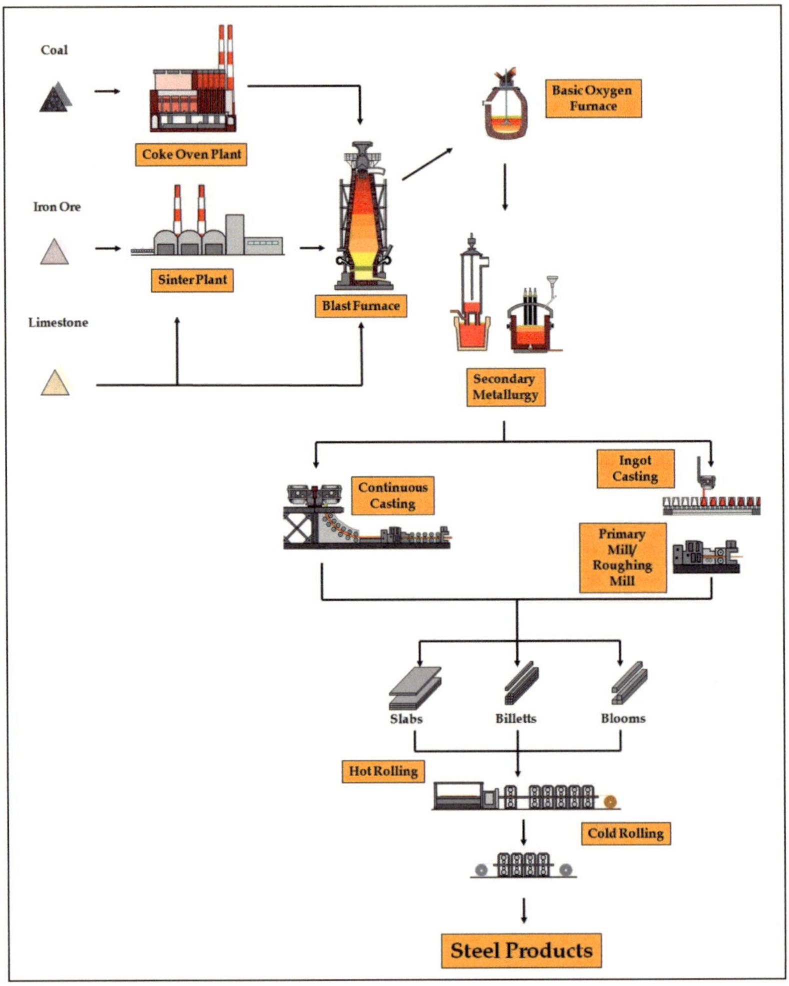

Fig. 1. The iron ore based steelmaking.

The main by-products resulting by ironmaking and steelmaking are slags (that represent 90% of the total by-products), dusts and sludges. On the average about 200 kg of by-products per ton of steel result from the steel production through electric arc furnace, while about 400 kg of by-products per ton of steel production through BF/BOF (World Steel Association, n.d.).

The use of by-products from steel industry goes back to many centuries ago. In 350 BC Aristotle already stated "When iron is purified by fire, there forms a stone known as iron slag. It is wonderfully effective in drying out wounds and results in other benefits". In later centuries slag has been used as construction material. The discovery of the hydraulic properties of granulated BF slag gave birth to a new era in slag exploitation: slag has been used as binding agent and/or addition for concrete. The use of steelmaking slag, from Basic-

Bessemer or Thomas processes, as phosphating and/or liming agent started in 1880 (Geiseler, 1996).

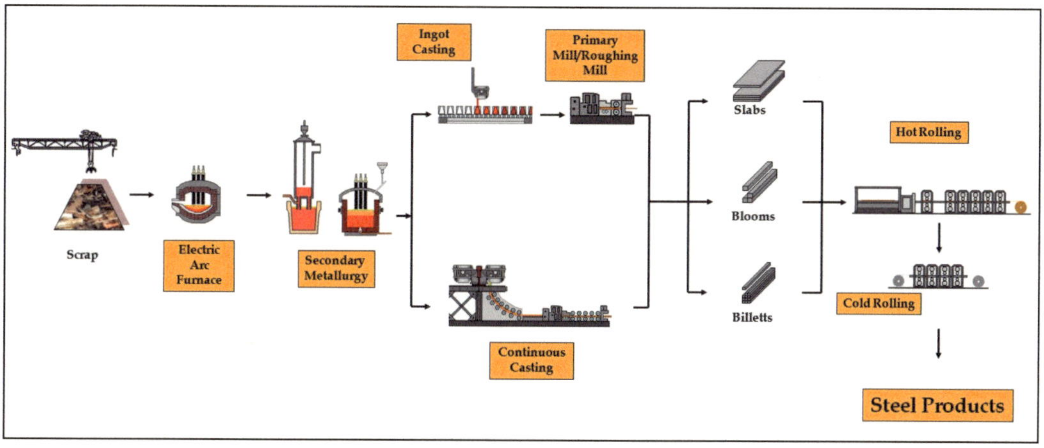

Fig. 2. The scrap-based steelmaking.

The traditional use of slag as landfill material, after the increase of steel production since the mid-1970's, has reached its limit and the pressure for natural resources and energy saving have driven steel industry to increase the recycling of this material, by facing other important challenges (such as technologies development, production facilities maintenance and ferrous slag products certification) in order to improve their application in different sectors (The Japan Iron and Steel Federation - Nippon Slag Association, 2006).

Slags coming from both BF and steelmaking processes are consumed at steelworks, used in cement, roadbed material, ground improvement material, civil engineering material and fertiliser. Once slags were dumped; nowadays they are considered marketable products and only a small percentage is processed as industrial waste in landfills.

In the past steelmaking processes were exclusively designed for the production of specific quality of iron and steel. One of the current steelmakers' goal is to design processes to produce high quality slags, both prior to and during the slags production, according to the market requirements, in order to satisfy environmental and technical requirements of international and national standards (Euroslag, 2006). On one hand, selling by-products produces revenues for steelmakers, that in turn generate economic development for the worldwide industry. On the other hand, the sustainable use of slags contributes to natural resources saving, to CO_2 emissions reduction, to energy consumption reduction, to the formation of a society founded on the recycling practice (as landfilling is avoided) and to the promotion of the steel industry sustainability. Therefore potential economic and environmental benefits make slags by-products that can be further recovered and used. For all these reasons, the effective utilisation of slag turns it into high value added product and allows to improve competitiveness of the steel industry.

This chapter intends to review the state of the art related to the use in agriculture of steel slag, mainly coming from BOF process using basically Linz-Donawitz (LD) converter, in different worldwide contexts. The review covers different aspects by summarizing its use as fertiliser and as liming agent, its potential use as amending material for soils, and by paying attention also to different technologies and methodologies aiming to improve the quality of the slag, in order to increase and make progress in its use in agriculture. On one hand, studies based on the use of slag in agriculture will be considered, which treat the use of steel slags for amending acid soils and as source of important factors and growing agents (not only calcium and magnesium compounds, but also other elements such as silicon, providing important beneficial effects for some crops and increasing the plant yields) and its use as Fe source for reducing Fe chlorosis in different crops. Moreover, investigations will be described concerning the heavy metals contained into the slags and their behaviour on the soil, in order to evaluate possible harmful effects after slag application for agricultural purposes and to avoid their possible negative environmental impacts, as well as the use of steel slags for metal stabilisation in contaminated soils. On the other hand, investigations focused on the obtaining a slag with high phosphorus content to be used as fertiliser (together with other slags with a low content in phosphorus to be recycled inside the steelmaking process) will be discussed.

2. Ironmaking and steelmaking slags and their use

Every year more than 40 million tonnes of iron and steel slags are produced in Europe. The production of iron and steel slag in Europe in 2008 was of 45.6 million tonnes[1] and decreased in 2009 and 2010 only due to the steel production slow down caused by the economic crisis. The main compounds contained into the slags are calcium, silicon, magnesium, aluminium and iron oxides.

Slags are generated in all stages of steel production and the following four different kinds of slags from different routes can be identified: BF slag, BOF slag, EAF slag and Secondary Steelmaking Slag. The slagging agents with fluxes, such as limestone, dolomite and silica sand, are added into BF or steelmaking furnaces in order to remove impurities from ore, scrap and other ferrous charges during smelting. The slag formation is the result of a complex series of physical and chemical reactions between the non-metallic charge (lime, dolomite, fluxes), the energy sources (coke, oxygen, etc.) and refractory bricks. Because of the high temperatures (about 1500°C) during their generation, slags do not contain any organic substances. The slags protect the metal bath from oxygen and maintain temperature through a kind of lid formation. Due to the fact that slags are lighter than the liquid metal, they float and may be easily removed. Slag is generated in a parallel route of the main processes of hot metal production in ironmaking and steelmaking and therefore the slag generation process is considered as a part of the whole steel production process (EC, 2001).

The uses of different types of slags in Europe are shown in Figure 3. Ironmaking and steelmaking slags are used in different ways with high added value. While steelmaking slags are mainly used for road construction, hydraulic engineering, and fertilizer, BF slag is mainly used for cement production[2].

[1] Source: European Slag Association, EUROSLAG
[2] Source: European Slag Association, EUROSLAG

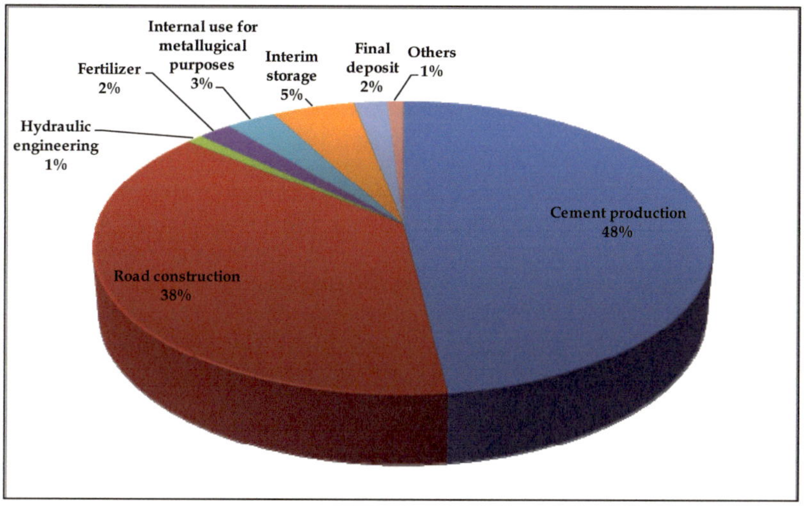

Fig. 3. The uses of different types of slags in Europe.

The marketed BF slag can be subdivided into three main types, depending how they are cooled:

- The Granulated BF Slag (GBFS) is produced by injecting high pressure water, followed by quenching and granulating it. It is used as material in cement, as fine aggregate for concrete and in civil engineering works, exhibiting better long-term strength, better resistance to chemical attack, and it brings down the cost of cement.
- The Air-cooled BF Slag (ABFS) is discharged to a cooling yard and naturally cooled with moderate sprinkling. The crystallized slag, after crushing, sieving and removing magnetic matter, can be used as construction aggregate, concrete products, road bases and surface and clinker raw material.
- The third type of cooled BF slag is represented by Pelletised Slag, with a vesicular texture, used as a lightweight aggregate and, when it is fine grounded, as cementitious material.

Steelmaking slags include slags from BOF and EAF. Since at this stage the steel production processes vary, depending on the steel being made, the slags chemical properties change as well. This results in a more difficult use of steel slags compared to the BF slag. They are discharged to a cooling yard or to a slag ladle and they are naturally cooled with moderate sprinkling. After crushing, sieving and removal of magnetic matter, they achieve granularity appropriate to different applications. Because of their lime contents they expand in reaction with water. After this expansion they are stabilised by "natural ageing" for long periods outdoors in natural rainfall and other weather or "steam ageing", through high-temperature vapour.

Steelmaking slag deriving from BOF process (using the Linz-Donawitz (LD) converter) comes from the pig iron refining process, which converts molten pig iron and steel scraps into high quality steel. Most slags from steel plant derive from this process, with an average of 150-200 kg of slag generated per tonne of steel produced. X-ray diffraction studies have shown that the major phases present in LD slag are dicalcium ferrite, calcium alluminate and wüstite, but it contains also some reactive mineral phases, such as $2CaO{\cdot}SiO_2$, $3CaO{\cdot}SiO_2$ and free CaO e MgO (Das et al., 2007).

The EAF slag utilisation is quite similar to the one of BOF slag, but, due to its lower lime content, EAF slag is very stable and can be used in asphalt without any problems.

Secondary Steelmaking Slag disintegrates into a powder due to instability of the dicalcium silicate, causing an increase in dust emissions to the environment. Some studies have successfully been carried out in order to reduce this phenomenon and to make this slag suitable for use and valuable (Rozman et al., 2007), (Branca et al., 2009).

Recently, due to a better understanding of the slag formation mechanisms and of the overall BF process, a large number of different slags has been designed, and it is also currently possible to control, optimise and minimise slags production. This progress applies also to BOF slag. The silicon, phosphorous and sulphur removal, before the refining into the basic oxygen furnace, has led to the reduction of tap-to-tap time, to the costs decrease, to the reduction of the amount of slag produced and to the production of higher quality steels. Moreover, through the vacuum degassing process, after decarburisation in the BOF, the hydrogen and nitrogen contents have been reduced. The ladle treatments produced significant reduction in impurities in steel and the use of selected slags minimised the formation and modification of *inclusions* (usually they are non-metallic particles contained in steel, that, depending on their number, their size and their distribution, can have a detrimental effect to mechanical properties of steel), with the result of improving mechanical properties of steel (Dippenaar, 2004).

Some slags are internally used in steelmaking furnaces or in sinter plants, while about 50% of this kind of slag is used outside the plant in the construction sector, mainly for road construction, as an aggregate in bituminous pavements, as a binding agent in base courses, for strengthening the subsoil and for soil conditioning. Free lime, after separation, can be used as fertiliser, in cement and concrete production, for waste water treatment and in coastal marine blocks. In soil conditioning slags are efficient in soil neutralisation. In addition, the siliceous liming materials improve soil structure and reduce fungal infections. Blast furnace slag can be used also in agriculture because of its high sorption capacity of phosphorus, which remains into the available form for the plants. Negative effects, resulting from steel slags use, could derive from their heavy metal concentrations, but such metals tend to bound to the slag matrix and thus they are not available for plants. All these factors contribute to underline positive effects of using slag as liming materials, that lead to better yield of the crops, soil protection and reduction of natural resources consumption (R. Hiltunen & A. Hiltunen, 2004).

2.1 The main legislation about the use of slags

Over the past decades the by-products recovery and reuse have significantly increased also because of the stringent legislation for environmental protection, that differs all over the world. In Europe, slag is mainly bound to the Waste Framework Directive (WFD), but the main legislation concerning the use of slag includes other laws, as follows: the Kyoto protocol, the Reference Document of Best Available Techniques, Harmonisation Committees TC 351 Dangerous Substances and TC 154 Aggregates, the REACH directive.

The main issue concerning the use of ferrous slags is the question whether it is a waste or a by-product. In order to market them the better way is to consider them by-products because the term "waste" indicates a material to be deposited instead to be used (Kobesen, 2009).

The Waste Framework Directive (WFD) (2006/12/EC) is the most important document governing the use of slag. Until recently, slag was considered as waste, but, after many years of discussion, the WFD has been amended by adding the term by-product (2008/98/EC). On one hand, the WFD provides the main concepts and definitions concerning the waste management; on the other hand, it sets the principles of waste management. Moreover it clarifies the definition of waste by introducing definitions of by-product and *end-of-waste* status, thanks also to the work of steel industry in supporting the EU Commission. The most important articles for the slag use are, as follows:

- Article 3, which provides the definition of waste and hazardous waste;
- Article 4, which describes in what order wastes are to be discarded;
- Article 5, which concerns by-products and provides the following conditions to meet in order a by-product is not considered a waste: direct use of a by-product, without any further processing other than normal industrial practice; by-product production as an integral part of a production process; any further use has to be lawful and certain; the by-product use has to be consistent to the principles of the EU waste policy, such as environment and human health protection.
- Article 6, End-of-waste status, regulates when a substance is classified as waste, ceases to be waste. The article 6 of the WFD defines the end-of-waste (EOW) criteria and what requirements have to be met. These criteria are developed in accordance with the following four conditions:
  1. The substance or object is commonly used for specific purposes
  2. A market or demand exists
  3. The substance or object fulfils the technical requirements and meets the existing legislation and standards
  4. The use of the substance or object will not lead to overall adverse environmental or human health impact.

The EOW criteria shall embrace limit values for pollutants where necessary and shall take into account the possible adverse environmental effects of the substance or object (Eloneva et al., 2010).

The Kyoto Protocol (Conference of the Parties, 1997), drawn in 1997 and come into force since 2005, regulates and decreases CO_2 emissions down to the 1990 CO_2 emission levels, because of its key role in the global warming.

The use of slag in different fields of application, for example the granulated BF slag (GBFS) as substitute for cement and for clinker in cement and the steelmaking slag for soil conditioning, can allow the reduction of the tonnages of CO_2 emission. Furthermore, in order to reduce the CO_2 emissions from the ironmaking and steelmaking processes, some studies have been carried out to sequestrate CO_2 in slag, through the free lime (CaO) and the di- and tricalciumsilicates carbonation (Abassapour et al., 2004).

The BREF document (EC, 2001) provides the best available techniques concerning environment, health and safety for iron and steel industry. It has been recently revised (EC,

2009) and, among other things, it gives the guidelines for EU Member States about the best available techniques for producing, treating, processing and using slag.

The new European Regulation No 1907/2006 for Registration, Evaluation, Authorisation and Restriction of Chemicals (REACH), adopted by the European Parliament and the European Council in December 2006, came into force on 1st June, 2007 (EC, 2006). This is not a directive, but a regulation, which replaces some national regulations and directives with a single system. It aims at: providing a high level of human health and environmental protection; ensuring that people are responsible to understand and manage the risks linked with the use of chemical substances that they put on the market; consolidating innovation and competitiveness of the EU chemical industry; encouraging the implementation of alternative methods for evaluating of the hazards of substances; promoting a free circulation of substances on the internal market while enhancing competitiveness and innovation (Kobesen, 2010). The registration concerns only products or by-products, while wastes are excluded from registration obligation. The regulation comprises new substances (substances which are put on the market after 18th September 1981) and phase-in-substances (substances which have been put on the market before 18th September 1981). Phase-in-substances are already registered in the so-called EINECS register (European Inventory of Existing Commercial Chemical Substances), whereas new substances will be registered in the ELINCS register (European List of Notified Chemical Substances).

Since the steel industry has been committed to clarify that iron and steel slag is produced and sold as by-product but not as waste, it was clear that ferrous slag had to be registered under REACH as a substance before 1 December 2010. In this respect the FEhS-Institute initiated the formation of a Consortium "Ferrous Slag", open to all European producers of iron and steel slag, in order to make a joint registration (Bialucha et al., 2011).

With regard to the Harmonisation Committees, the Technical Group (TG) 13, within the Technical Committee (TC) 154 concerning aggregates, deals with Dangerous Substances, by producing standards about release of some dangerous substances from aggregates. These include natural aggregates, ironmaking and steelmaking slags, defined of manufactured lightweight aggregates and recycled aggregate from material previously used in construction. The standards have to be compared with geologically similar deposits and Regulated Dangerous Substances (RDS) released are compared and identified. For slags the relevant RDS are, as follows: mineral oil, metals like V, Cr, Zn, Pb, Mo, As, Hg, Cd, other inorganic substances such as chlorides and sulphates (Kobesen, 2009).

Member states used different methods of investigation, which have to be harmonised. For example, the objectives of the Technical Committe 351 ("Construction Products – Assessment of release of dangerous substances") WG 1 are to:

- enable product TC to select the appropriate test;
- determine the release performance;
- ensure methods are scientifically sound;
- be relevant to CE marking of the product.

The following three technical specifications have been proposed (Bialucha et al., 2011):

1. TS-1, describing principles for selecting appropriate leaching tests for a specific product;

2. TS-2, describing a dynamic surface leaching test (DSLT);
3. TS-3, containing information on horizontal up-flow percolation test.

3. The use of converter slag

LD slag can be used in different fields of application such as fertilizer, soil conditioners, recovery of metal values. Because of its hard characteristics it is also used as aggregates for road construction, for the base and sub-base layer in road construction and for hydraulic engineering structures. On this subject tests have been carried out in order to assess technical properties. In particular the volume stability, which is the key aspect for using steel slags as a construction material, has been evaluated, by comparing the behaviour of slag under practical conditions, such as in road constructions; on the other hand, the assessment of environmental compatibility of aggregates as building material has been tested through leaching tests in order to continuously control quality (Motz & Geiseler, 2001).

Since 1880 steelmaking slag from Basic-Bessemer or Thomas process has been used as a phosphatic fertiliser, but also the current LD slag composition (mainly containing CaO, MgO, SiO_2, Mn and other valuable micronutrients, such as copper, zinc, boron and cobalt) makes it suitable as liming materials. On one hand, calcium and magnesium compounds, because of their basicity, improve soil pH; on the other hand, they are also plant nutrients and stabilisers for soil aggregates. Physical treatments of slag as well as its mineral composition influence the solubility and plant availability of the nutrients.

Silicate has a special bond in the slag minerals and it is useful for plant nutrition and soil quality. In fact silicate provides beneficial effects on plant health and soil structure, increase the phosphate mobility in the soil and the efficiency of phosphate fertilisation (Rex, 2002).

3.1 The use of steelmaking slags as fertilisers and as liming agents

Although the by-products recycling has always been a commitment of the steel industry, the growth of steel production in recent years has pressed the sector for increasing their use in a more effective way, in order to achieve a sustainable steel production. Even though steelmaking slags are continuously studied in order to improve their recycling, there are some limiting factors for their use. In particular a small amount of slags is used as fertiliser in agriculture and this use depends on the market situations. Due to the low market value of fertilisers, the long distances transportation is a limiting factor. In addition natural lime stone fertilisers are in competition to the slag use. Therefore the development of new markets for the slag, in order to ensure its utilisation in the future, is required. In this respect the steel industry is committed to minimize the amount of slag which has to be deposited, by improving its use through the increase of its properties (Drissen et al., 200).

Until the eighties steel was produced via the Thomas-Bessemer process, through the open hearth furnaces. The resulting slag containing phosphate has been used as fertiliser for about 70 years. The current steelmaking process is based on the Basic Oxygen Steelmaking process, where a basic slag is produced in the Linz-Donawitz converter. The LD slag contains about 1-3 wt% of P_2O_5, which is too low to be used as phosphate fertilizer, but, at the same time, it is too high to be used in the BF or recycled in the sinter plants.

Nevertheless the LD slag contains high levels of lime (CaO) and MgO that make it a potential liming agent, may improve soil pH and can be used as plant nutrients. Particularly free lime, which is one the main slag constituents, can partially dissolve by reacting with water to produce calcium hydroxide, $Ca(OH)_2$, as shown in Eq. (1):

$$CaO + H_2O \rightarrow Ca(OH)_2 \quad (1)$$

The calcium hydroxide dissolves into Ca^{2+} and OH^-, resulting in a pH increase.

All these factors characterizing this material, can allow to recycle an industrial residue and to improve the fertility of acid soils.

On this subject some studies have been performed in several countries. Among these a three-years research carried out in the Basque Country of northern Spain, by using LD converter slag on pasture land, has produced significant results (Besga et al., 1996). The comparative study analyzed soil modifications produced by LD slag and those produced by traditional liming agents. In particular, the influence on soil pH, soil Ca and Mg content, the percentage of Al saturation and the yields have been taken into consideration in this study. The achieved results concern different aspects that have been considered in it. Firstly, increasing rates of LD slag application have increased the soil pH linearly, that, as a consequence, has led to Al solubility decrease; this has allowed the P absorption, due the changing of insoluble forms to soluble ones. In addition the Ca and Mg soil contents increase, resulting in an increased yield, while the consequence of the pH has reduced the toxic effects of Mn (e.g. on the white clover). On the other hand, the Cd, Cr and Ni monitoring has shown that, after LD slag application, there was not heavy metals accumulation in the soil.

A research project that was funded by the Research Fund for Coal and Steel (RFCS) programme (Kühn et al., 2006), led to relevant results about the liming and fertilising effects of fine grained iron and steel slags, such as BF slag, LD slag or ladle slag, compared to other liming materials, such as burnt lime or carbonate limestone, in field trials investigations as well as in greenhouse pot experiments. The research aimed to investigate the fertilising effects on both the soil and the plants. These investigations have been carried out in arable land (Germany, Austria and Spain), in green land meadows (Germany and Spain), and in forestry (Spain). Furthermore the behavior of trace elements, such as Ca, Mg, P, Cr, V, Zn and Pb, in soils and plants has been investigated, in particular, the behavior of heavy metals, especially of Cr and V, their mobility and their bonds in the soil.

Investigations carried out on soil and yield have proven that the yields of experimental crops, in long term experiments by using iron and steel slags, were higher than those achieved after using different liming materials; on the other hand, pH has increased in the same way as a result of the use of both kind of materials. The use of basic slag in soil tests has produced the same fertilising effects, while the results of P content and soil pH has been higher, compared with super phosphate or rock phosphate use.

The comparison between the converter slag and the converter sludge applications to soils and tea plants, conducted in northern Iran, produced effects on soil properties and on the tea plant nutrient concentrations (Jamali, S. F. K. et al., 2006). Results have shown that converter slag application increased soil pH more than converter sludge treatment, probably because of the higher original pH of the slag. In addition they have supplied Ca and Mg to soil and, consequently, their concentrations have increased in tea leaves.

The use of alkaline slag for amending acid soil and improving plant growth has been analysed in a recent study carried out in Iran (Ali & Shahram, 2007). After the application of increasing amount of slag, the soil pH proportionally increases. Moreover, at pH values between 7.4 and 8.5, the Fe availability decreases, while at higher pH values it increases; on the other hand, the P and Mn availability proportionally increases. The greenhouse studies have shown that the slag application (1% and 2% (w/w)) in tea garden soil and (0.5, 1 and 2%) in rice field soil leads to the increase of plant yield and the P and Mn uptake; an increase of Fe and K uptake has been detected in rice field, a decrease of K uptake in tea garden has been observed, while Fe uptake has not been changed.

The same results have been achieved after the application of basic slag in acid sulphate soils in an incubation study. The investigation aimed at assessing the ability of basic slag to neutralize acid and its effectiveness on the solubility of basic cations in the soils, in order to achieve a sustainable use of acid sulphate soils in coastal areas of Bangladesh (Shamim et al., 2008). A wide range of processes can lead to the addition of acid cations and the removal of basic ones in the soils. The acid water, penetrating into the ground, through the leaching process, tends to increase the acidification of the soil, except if bases are compensated by different sources, such as atmospheric deposition. When acid cations are in soil solution, they tend to replace basic cations. This can also affect the metals and metalloids mobility in the groundwater, with the result that this can be a threat to groundwater and the health of aquatic and terrestrial ecosystems. Incubation experiments showed that basic slag increases soil pH, mainly due to its neutralizing effect that releases basic elements in the acid sulphate solution. This process reaches the highest value after 180 days of incubation under saturated moisture condition, probably due to slow releasing of basic ions from basic slag. In addition the application of basic slag increases K, Ca and Mg in soils, although, in some cases, over the course of time, a small decrease of this trend has been observed, probably because of the formation of insoluble compounds of Ca and Mg.

3.2 Steelmaking slags as a silicon source for plant

After oxygen, silicon (Si) is the most abundant element in the earth's crust. Along with some other elements that are not considered essential, under particular agro-climatic conditions, it can increase the crop yields by promoting some physiological processes. Silicon sources for agricultural purposes must display some important features, such as high soluble Si content, low cost, availability for plants, balanced ratios and amounts of Ca and Mg, increase of phosphate mobility, suitable physical properties, easy application, and absence of heavy metals.

In order to have an effective use of Si fertilization in agronomic practices, an adequate knowledge of physical and chemical characteristics of Si sources and of the rates and methodologies for applying them are needed. A number of field and greenhouse studies have demonstrated that the use Si soil amendments increases crop production and quality. The application of Si fertiliser has beneficial effects on both rice and sugarcane. Although the mechanism of response of sugarcane to Si fertilization is not yet well understood, some studies have shown that the yield increase of the sugarcane may be associated with different factors, such as Al, Mn and Fe toxicity alleviation, increased P availability, reduced lodging, improved leaf and stalk erectness, freeze resistance and improvement in plant water economy. Furthermore the Si accumulation protects plants from certain diseases, such as a resistance to biotic and abiotic stresses (Savant et al., 1999). The use of Si in the plants can help the plants against pathogen attacks (Motz & Geiseler, 2001).

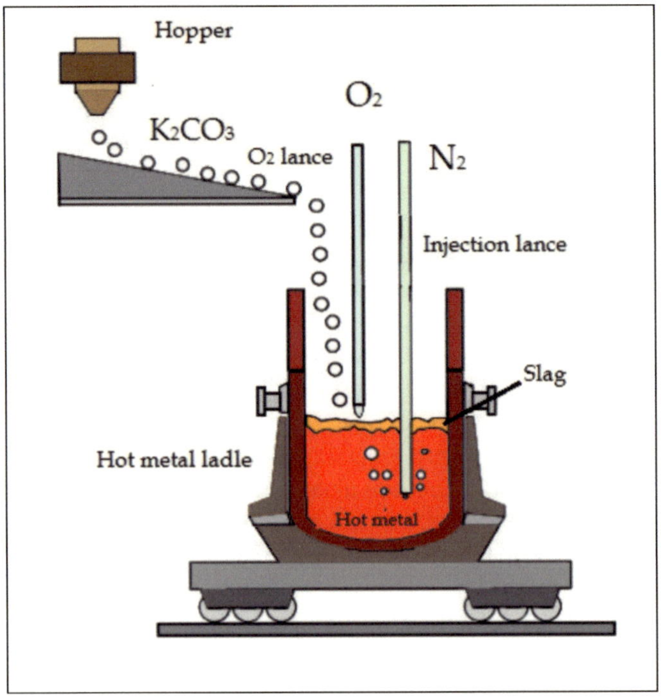

Fig. 4. Production of potassium silicate fertilizer from steelmaking slag (Takahashi, 2002).

The use of silica (SiO_2) as a fertiliser increases the rice resistance to diseases and vermin. For this purpose the slag produced in the hot metal desiliconization process, which contains mainly silica, has been taken into consideration in order to develop a potassium silicate fertiliser. This is an example of a new steelmaking process developed by NKK (Nippon Kokan Kabushiki Kaisha) and referred as ZSP (Zero Slag Process), focused on the reduction of the amount of generated slag and also to the stabilization of the composition of slag generated through hot metal pre-treatment (Takahashi, 2002). This fertiliser, developed by adding potassium to the desiliconization slag, dissolves with difficulty in the water and slowly dissolves in the weak citric acid released by plant roots. The potassium contained into this fertiliser is slowly released and is effectively adsorbed by plant. The process consists in the desiliconisation process of the hot metal and subsequently the potassium carbonate (K_2CO_3) is continuously added into the ladle containing the hot metal (Fig. 4). Then the uniformly melted slag, that is recovered from the hot metal ladle, is solidified by cooling and crushed into a granular form. This fertilizer has been demonstrated to be as effective as other commercial potassium silicate fertilisers and combined potassium chloride-calcium silicate fertilisers when it was applied to some vegetables, in particular to rice. Its marketing started in December 2001 in Japan.

As described above, since the silicon fertilisation has been turned out to be beneficial for plant growth, such as rice and sugar cane, the identification the most promising and potential available Si sources to plant has been studied. In particular, in a greenhouse experiment several Si sources have been evaluated in order to test their ability to supply Si to rice crops. Among different Si-rich materials, metallurgical slags have been evaluated, because the high temperatures used in ironmaking and steelmaking processes release Si from crystalline form to reactive and as consequence more soluble forms, with the result to supply it to plants (Pereira et al., 2004). In the comparative study differences between silicon sources in relation to Si uptake have been observed. Furthermore steel slags (LD, AOD, electric, and stainless steel furnaces) have shown higher Si availability than BF slag, and differences depending on the type of steel produced and on the type of furnace used to produce steel. Phosphate slags provided the highest Si uptake.

On the other hand, recent studies have shown that Si concentration is negatively related to As content in straw and polished rice, that is Si in the soil available for plant reduces the uptake of As (Bodgan & Schenk, 2009).

4. Use of steel slags for metal stabilization in contaminated soils

Some investigations about the addition of steel slags in contaminated soils have been carried out. The stabilization technique is based on the incorporation of amendments, in order to minimize metals and metalloids, such as As, Cr, Cu, Pb, Cd and Zn that can be found in contaminated soils at wood treatment plants. In particular, when the copper sulphate and chromate copper arsenate are used to protect wood from insects and fungi, they can cause the soil phytotoxicity. While the As can be stabilized by sorption on Fe oxides and also by the formation of amorphous Fe (III) arsenates, the Cr immobilization takes place through Cr reduction from Cr (VI), which is mobile and toxic, to Cr (III), which is stable. The Cu stability in soil is pH dependent, because its mobility increases with decreasing pH. Carbonates, phosphates and clays can reduce the mobility and availability of Cu in soil. The proposed mechanism consists in precipitation of Cu carbonates and oxy-hydroxides, ion exchange and formation of ternary cation–anion complexes on the surface of Fe and Al oxy-hydroxides. While Pb can be stabilized by using phosphorus-containing amendments, that reduce the Pb mobility, Zn can be immobilized in soil by using phosphorus amendments and clays (Kumpiene et al., 2008, as cited in Negim et al., 2009). To this aim some chemical and mineralogical agents, such as industrial by-products have been applied. For instance, the use of alkaline materials, organic matters, phosphates, alumina-silicates and basic slag has been shown to limit the accumulation of Cu in plants cultivated in Cu-contaminated soils.

Among recent studies, the use of slag with basic properties into a Cu-contaminated soil has led to relevant results in soil composition (Negim et al., 2010). Because of its Ca and P content, the basic slag, on one hand, is a fertiliser, as it improves the physico-chemical properties of the soil and by increasing plant growth; on the other hand, it is a liming material, as it increases the precipitation and sorption of metals such as Cu. For this reason the investigation concerned the effects on soil pH, soil conductivity, plant growth and chemical composition of bean plants (*Phaseolus vulgaris*) in pot experiments, by mixing soil with increasing basic slag addition rates, from 0% to 4%, in controlled conditions. This material affects the soil solution composition through acid-base, precipitation and sorption reactions. On the other hand, the foliar concentration can be influenced by soil solution changes, competitions for root uptake and root-to-shoot transfer. In particular the soil pH has increased from 5.6 to 9.8, and the soil conductivity has proportionally risen from 0.14 mS cm^{-1} to 0.82 mS cm^{-1} by applying increasing rates of basic slag, probably due to the basic slag composition, particularly to the Ca content. Furthermore the foliar Cu concentration has probably caused a phytotoxic effect in plants grown in Cu-contaminated soil. After the basic slag addition at 1% rate the bean growth, along with the decrease of foliar Cu concentration, has been observed. Moreover while the Ca foliar concentration has increased after applying increasing rates of basic slag, the foliar P concentration has not been improved. These results suggested that the use of basic slag at 1% addition rate is effective as a liming material but, is not effective as P fertiliser. Furthermore, the basic slag addition in contaminated soil does not increase the foliar concentrations and accumulations for Cd, Cr, and Zn.

5. Use of slag as an iron fertilizer

The problem of iron (Fe) chlorosis can affect many crops on calcareous soils, resulting in substantial yield losses. Generally it has been corrected through the addition of Fe synthetic chelates, but these have resulted very expensive. Various studies have been focused on applying different Fe sources, in order to reduce the economic burden and to recycle some industrial by-products, such as converter slag (Wallace et al., 1982) (Sikka & Kansal, 1994). They are used not only as soil amendment but also as source of important plant nutrients, such as P, K, Mg and Fe oxides.

An investigation pursued on 1984 showed that the application of a steel by-product (dust containing 430 g Fe kg^{-1}) as fertilizer to alkaline soils, with o without sulphuric acid, increased dry matter yield of sorghum (Anderson & Parkpian, 1984). A similar treatment, through a mixture of sulphuric acid and iron sulphates, has allowed to correct Fe chlorosis in corn and alfalfa (Stroehlin & Berger, 1963).

While converter sludge has been used as Fe fertiliser in calcareous soils with positive results, recently the use of converter slag as source of Fe fertiliser in some calcareous soils incubation studies has led to relevant achievements. On this subject, pot experiments in a greenhouse have been carried out in China (Wang & Cai, 2006). Relevant results of this study have shown that the use of moderate steel slag or acidified slag as Fe fertilizer leads to the increase in Fe uptake and corn dry matter yield. This phenomenon is proportional to the application rate and is enhanced by the acidification of slag, although increasing application rates do not produce further improvements in yield and in Fe uptake. This suggests a possible optimized rate of these applied substances. On the other hand, in experiments conducted with the sandy loam dry matter yield significantly decreased. This can be explained because the Fe availability decreases with salt levels increase, resulting in a yield decrease and an increase of chlorosis in plants.

Although further studies still have to be conducted in order to investigate the correct rates of converter slag for different crops and its possible residual environmental impacts to the soil, important results have been achieved by using this by-product as a source of available Fe (Torkashvand, 2011). In an incubation study, by adding to the soil converter slag (from Isfahan steel factory, Isfahan, Iran), containing about 24% of Fe oxides, along with elemental sulphur and organic matter, the soil pH has increased, due to the alkaline pH of slag. But during the incubation time the pH decreased. This can be due, according to some previous studies, either to the precipitation of the free carbonates as calcium carbonate (Abassapour et al., 2004) or to the hydrolysis of Fe^{3+} in the soil (Rodriguez et al., 1994). The decrease of soil pH probably results from the decomposition of organic matter applied and subsequent organic acids and CO_2 release as well as the buffering ability of the calcareous soils. The observed yield increase in these soils may be due to the some nutrients availability as a consequence of pH increase.

The converter slag application has proportionally increased ammonium bicarbonate-diethylenetriamine pentaacetic acid (AB-DTPA) extractable Fe, although in some incubation soils Fe extractable decreased, maybe due to the temporary fixation of iron by organic matter. Further in the pot experiment converter slag has been shown to be very effective in correction of Fe chlorosis in calcareous soils.

6. Recovery of phosphorus from steelmaking slags

Phosphorus is an essential element not only for plants and animals but also, along with nitrogen (N) and potassium (K), for fertilisers production. Nevertheless P has a detrimental effect on steel, by reducing the low-temperature toughness of iron and steel products. For this reason during the hot metal pre-treatment and in the steelmaking processes the dephosphorisation process takes place. The result is that most of the phosphorus in hot metal, originating from raw materials (e.g. iron ore and coal), is removed through the steelmaking slag.

The requirement of reducing environmental impact and disposal costs of slags has led to study and develop different strategies in order to internally recycle them. The slag recycling in steel production processes allows to realize a waste-free steelmaking process as well as the consumption reduction of iron and lime.

For its high content of lime (CaO) the LD slag can be used as fluxing material, replacing limestone, with the result that iron and steelmaking costs are reduced. Nevertheless free CaO reduces the effective use of LD slag, because this compound makes low hydraulic properties of slag. However the slag recycling in internal processes is limited, due to its high amount of S and P (about 1-3% of P_2O_5 content in the converter slag) that negatively affect them. These values are too high for using it into sinter process (where fine grain raw material is processed into coarse grained iron ore sinter for charging the BF), BF and in pre-dephosphorisation process; on the other hand they are too low for slag use as phosphatic fertiliser. Therefore some studies have been carried out in order to remove phosphorus from steelmaking slag. One of them concerns a waste-free steelmaking process in which the slag is recycled within the steelmaking process and a slag containing high phosphorous is produced and it is suitable as fertiliser (Li et al., 1995). This study has been carried out as a process modelling, by computer simulations from the thermodynamic perspective and mass balance, and results indicate the possibility to develop this process inside steelworks. The proposed steelmaking route mainly consists in a de-siliconisation process in the De-Si furnace of hot metal produced in the blast furnace and in a de-phosphorisation process in the De-P furnace before the refining into the conventional converter. The process is outlined as follows and shown in Figure 5:

- The whole slag produced into the converter (at 1923K), is totally returned to the Pre-De-P furnace, where the hot metal is dephosphorised;
- The whole slag from the Pre-De-P furnace (at 1623K) is transferred to the Regenerator;
- The Regenerator (at 1873K) contains the carbon-saturated hot metal and the most of the phosphorous is transferred from the slag to the hot metal;
- The hot metal containing phosphorous is transferred from the regenerator to the De-P-II unit (at 1623 K), which contains hot metal and synthetic slag. The phosphorous is transferred from the hot metal to this slag, by obtaining the final slag, which contains more than 10% of P and thus it can be used as fertilizer;
- A part of dephosphorised slag from the regenerator is transferred to the Pre-De-P furnace and the remainder to the De-Si furnace;
- The slag from the De-Si furnace (at 1623K) can be used in the sinter plant or in the blast furnace.

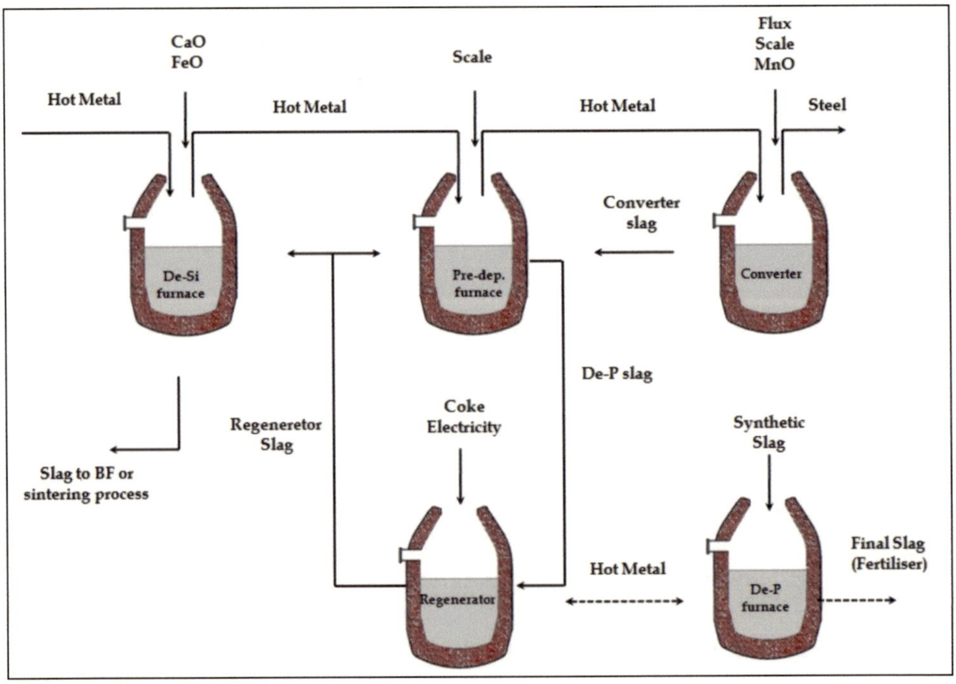

Fig. 5. Proposed method of converter slag utilization (Li et al., 1995).

The simulation results of the proposed process show that it is possible to develop a waste-free steelmaking process, through the internal slag recycling. The slag obtained in the de-siliconisation furnace can be used into the BF and/or into the sinter plant, due to its low content in phosphorus ((mass% P) = 0.48). The slag resulting from the de-phosphorisation furnace, with high P content (about 10%), can be used as fertiliser. The simulation results show that steel with low P content could be produced and that the slag generated in each unit could be completely recycled. The energy required for regenerator reactions is supplied by electric power and the total energy required is 130 kWh/t; the lime consumption is about one half of the current process. The proposed method of converter slag utilisation could allow the whole slag recycling (by thus reducing its environmental impact and contributing to save natural resources) and the production of slag with a high P content, that will be available for fertilising use.

Among different studies concerning phosphorus recovery from steelmaking slag, a research conducted in Japan, based on a method using a strong magnetic field, has been implemented for the separation and recovery of crystalline phases containing P from steelmaking slag (Yokoyama et al., 2007).

The phases of hot metal pretreatment slag, resulting by the composition analysis, can be subdivided, as follows:

- *phase A*, where P is concentrated at 10% or more and hardly any Fe content;
- *phase B*, that is composed mainly of Ca and Si, with a few percentage of P and hardly any Fe;
- *phase C*, which is composed closely to pure FeO;
- *phase D*, that is composed of CaO-SiO_2- FeO.

Thus the pretreatment slag can be subdivided in general into two groups: the crystalline phases containing P but not Fe (*phase A* and *phase B*) and the phases containing Fe but not P. Through measurements of magnetic properties of phases components, different behaviours in a magnetic field have been found. This has allowed to magnetically separate them, by applying a strong magnetic field of several teslas to the crushed slag.

The P recovery has resulted in obtaining a raw material for the production of fertiliser in the chemical industry. On the other hand the slag obtained after P recovery is used as building material, for roads construction and can be recycled in sintering process. Therefore the expected results of the magnetic separation method will have positive effects on environment, by reducing the generation of slag and CO_2 emissions and by reducing natural resources exploitation and landfill consumption. Moreover, the implementation of this technology can allow the use of iron ore with a higher P content, which is cheaper, and thus it will produce substantial economic benefits.

7. Environmental concerns about the slags use in agriculture

According to its environmental policy, the European targets concern the environment preservation, protection and its quality improvement, the human health protection and the efficient use of natural resources, by adopting measures for handling environmental issues on global and local level. The use of slags has a crucial significance as regard the environmental aspects. The main problem concerning the utilization of steel slags in agriculture consists of the possible leaching of heavy metals.

Heavy metals are broadly distributed in the Earth's crust and some of their chemical forms can be a potential risk to biosphere, in particular to the water life, because of their solubility. Their bioavailability depends on the plants ability to uptake them from soil and water, due to the secretion by plants roots of chelators compounds; furthermore many heavy metals are transported by sulphur ligands, such as glutathione, and organic acids. Moreover some heavy metals are insoluble and they often interact with soil particles, and therefore they are not available to plants (Babula, P., et al., 2008).

Chromium (Cr) is used in different industrial field of applications such as steel industry, wood preservatives, electroplating, metal finishing, leather tanning, textiles and chemical manufacture and it is a frequent contaminant of both surface ground waters. In oxidizing conditions is highly soluble and forms Cr(VI) anions, such as chromates CrO_4^{2-} or dichromates $Cr_2O_7^{2-}$. Under reducing conditions, through a process involving a chemical reduction and a precipitation, Cr(VI) converts to Cr(III) that is insoluble. Both forms are stable in the environment. The roots plants can absorb both forms Cr^{3+} and CrO_4^{2-}, but, according to some date, the Cr(III) forms stable compounds (e.g. hydroxides, oxides and sulphates). Therefore it is less soluble and, consequently, less bioavailable (Srivastava et al., 1994, as cited in Babula, P. et al., 2008). However Huffman et al. has shown as there are not uptake differences between Cr(III) and Cr(VI) by bean (*Phaseolus vulgaris, Fabaceae*) and wheat (*Triticum aestivum, Poaceae*).

Although Cr is an essential element for animal and human health, hexavalent Cr salts have toxic and carcinogenic effects. The plant mechanism of toxic effect of Cr is due to the reaction between Cr-complexes and hydrogen peroxide that produces hydroxyl radicals. They can trigger off DNA alteration (Shi & Dalal, 1990a, b, as cited in Babula, P. et al., 2008), by affecting, for example, its replication and transcription.

Among heavy metals, steelmaking slags contain Vanadium (V). The V content in the processed ore is about less of 2%. During the blowing process into the LD converter the V is transferred to the converter slag as V_2O_5 (about 5%), which represents the main source for some procedures aiming to extract V from LD converter slag. Due to its heavy metals content and the environmental problems resulting to their release to earth, LD slag is often subjected to treatments, aiming to extract these harmful but also precious elements from it. Because of its physical properties, such as high tensile strength, hardness, and fatigue resistance, V is used in ferrous and non-ferrous alloys. For all these reasons it is desirable to recover this valuable element.

Among some studies about this topic, a recent research aims to investigate on the extraction procedure of V by using salt roasting and sulphuric acid leaching and how some leaching parameters, such as particle size, acid concentration, reaction temperature and solid:liquid ratio (S/L), may influence the kinetics process (Aarabi-Karasgani et al., 2010). The found optimum condition of leaching allows to achieve a maximum V recovery of 95%. Furthermore the size fraction of below of 0.850 mm has shown to be mostly effective in order to attain the maximum extraction. Two leaching stages have been proposed: the first one (the first 15 minutes), when the V leaching is faster, and a second stage (more than 30 min), when the leaching becomes slower. In addition, the leaching rate is controlled by chemical reaction at low temperature while at high temperature it is controlled by the solid product diffusion.

The increasing interest concerning the slags use for soil conditioning has focused the attention on the heavy metal concentrations in these materials. Several investigations carried out in Finland have shown that the concentration of some elements, such as Cr and Zn, are low because of the high temperatures of the processes. On the other hand, long-term experiments in Germany have shown that the application of steel slag as liming material does not increase the content of mobile chromium into the soil and, after using steelmaking slags as fertiliser, significant increases in Cr content have not been found in plants (R. Hiltunen & A. Hiltunen, 2004).

Nevertheless it is important to carry out further investigations focused on the heavy metals behaviour on the soil in order to better understand the effects of long-term use of steelmaking slags in agriculture. In the above-mentioned research project (Kühn et al., 2006), the preliminary investigation, based on the origin of heavy metals in the BF-BOF route, has led to the result that Cr is originated from ore, but it is also influenced by the scrap used, whereas V content is only affected by the iron ore input in the BF. In the second part of the investigation, the analysis of long term effects of Cr and V has showed significant accumulation of their content in the cultivated layer of the soil, after application of converter slag.

Results on soil analysis have shown that the highest values for Cr and V have been detected during the 50 years test conducted in St. Peter site (Black Forest), after using of basic slag, where Cr has increased of 40-50 mg Cr/kg of aqua regia soluble and total Cr, whereas V has increased of 60-70 mg V/kg of aqua regia soluble and 63-80 mg V/kg of total V.

It has been shown that both Cr and V, even though they have increased in the top soil, they are stable and immobile in the soil. In addition, after more than 50 years tests, they did not move into the deeper soil and therefore they cannot adversely affect the groundwater and consequently the human and animal health.

Plant analysis conducted in pot experiments have shown no significant differences for Cr and V uptake, but in different crops they have shown different results for Cr and V. In particular, the Cr concentrations in rye, rape and winter wheat was about lower of the detection limit (< 0.035 mg kg^{-1}). As far as V is concerned, it could not be measured in rye, rape, spring barley and winter wheat, while it has shown the highest concentration in potatoes after basic slag fertilisation of the Austrian field trials.

Furthermore it has been pointed out that the metals uptake by the plants is affected by the soil properties. For example the Cr and V contents into potatoes are reduced in soils with higher content of organic matter and with an heavy texture. In addition, the uptake of Cr, V and Cd by potatoes is favoured by low pH.

8. Conclusion

The steel industry is committed to increasing the way for recycling slags generated during the steel production. Since their use as landfill material has almost reached its limit, the pressure for saving natural resources and energy has led steel industry, along with other important technological challenges, to improve and increase the recycling of this by-product. While in the past steelmaking processes were exclusively design for the production of specific qualities of iron and steel, one of the today's goals for steelmakers is to design processes to produce high quality slags, according to the market requirements. New technologies and/or the improvement of existing technologies have been investigated and developed in order to achieve the ambitious target of "zero-waste" in the incoming years. To this aim, the effective utilisation of slags turns it into high value added product and allows to improve the steel industry competitiveness. On the other hand, the sustainable use of slags contributes to natural resources saving, to CO_2 emission reductions and to consolidate a society founded on the recycling practice.

The use of steel slags in agriculture produces not only economic but also ecological advantages. A more effective exploitation of natural resources can be achieved in both the steelmaking processes and in the agriculture. Obviously soil fertilisers have to supply nutrients, but should not have negative effects on the environment and on the human, animal and plant health. Therefore many studies have been particularly focused on the behavior and immobilization in the soil of the main heavy metals (e.g. Cr and V, contained in converter slag at higher concentrations), in order to achieve a more effective and sustainable use of steel slags in agriculture and thus improve its recycling.

9. References

Aarabi-Karasgani, M., Rashchi, F., Mostoufi, N. & Vahidi, E. (2010). Leaching of vanadium from LD converter slag using sulfuric acid. Hydrometallurgy, Vol. 102, No. 1-4, pp. (14-21), ISSN 0304-386X

Abbaspour, A., Kalbasi, M., & Shariatmadari, H.. (2005). Effect of Steel Converter Sludge as Iron Fertilizer and Soil Amendment in Some Calcareous Soils. *Journal of Plant Nutrition*, Vol. 27, No. 2, pp. (377 - 394), ISSN 0190-4167

Ali, M.T. & Shahram, S.H. (2007). Converter slag as a liming agent in the amelioration of acidic soils. International Journal of Agriculture and Biology, Vol. 9, No. 5, (2007), pp. (715-720), ISSN 1560-8530

Anderson, W.B. & Parkpian, P.(1984) Plant availability of an iron waste product utilized as an agricultural fertilizer on calcareous soil. *Journal of Plant Nutrition*, Vol.7, No.1, pp.(223–233)

Babula, P., Adam, V., Opatrilova, R., Zehnalek, J., Havel, L. & Kizek, R. V. (2008). Uncommon heavy metals, metalloids and their plant toxicity: a review. Environmental Chemistry Letters, Vol. 6, No. 4, pp. (189-213), ISSN 1610-3661

Besga, G., Pinto, M., Rodríguez, M., López, F. & Balcázar, N. (1996). Agronomic and nutritional effects of Linz-Donawitz slag application to two pastures in Northern Spain. Nutrient Cycling in Agroecosystems, Vol. 46, No. 3, (1996), pp. (157-167), ISSN 1385-1314

Bialucha, R., Merkel, T. & Motz, H. (2011) European environmental policy and its influence on the use of slag products, Proceeding of the Second International Slag Valorisation Symposium, Leuven, Belgium, April 18-20, 2011

Bogdan, K., Schenk, M. K. (2009). Evaluation of soil characteristics potentially affecting arsenic concentration in paddy rice (Oryza sativa L.). Environmental Pollution, Vol. 157, No. 10, (2009), pp. (2617-2621), ISSN 2617-2621

Branca, T.A., Colla, V. & Valentini, R. (2009). A way to reduce environmental impact of ladle furnace slag. *Ironmaking & Steelmaking*, Vol.36, No.8, (November 2009) , pp. (597-602(6)), ISSN 0301-9233

Conference of the Parties, 1997. Kyoto Protocol to the United Nations framework convention on climate change. Report of the Conference of the Parties, Third Session, Kyoto, 1–10 December, FCCC/CP/1997/L.7/Add.1, http:/www.unfccc.de.

Das, B., Prakash, S., Reddy, P.S.R. & Misra, V.N. (2007). An overview of utilization of slag and sludge from steel industries. *Resources, conservation and Recycling,* Vol.50, No. 1, (2007), pp. (40-57), ISSN 0921-3449

Dippenaar, R. Industrial uses of slag - The use and re-use of iron and steelmaking slags, *Proceedings of VII Intenational Conference on Molten Slags, Fluxes and Salts,* ISBN 1–919783–58–X, The South African Institute of Mining and Metallurgy, 2004

Directive 2006/12/EC of the EuropeanParliament and of the Council of 5April2006 on waste

Directive 2008/98/EC of the European Parliament and of the Council of 19 November 2008 on waste and repealing certain Directives

Drissen, P., Ehrenberg, A., Kühn, M. & Mudersbach, D. (2009) Recent Development in Slag Treatment and Dust Recycling. steel research international, Vol. 80, No. 10, (October 2009), pp. (737-745), ISSN 1869-344X

EC (2001), Integrated Pollution Prevention and Control (IPPC) - Best Available Techniques Reference Document on the Production of Iron and Steel - December 2001, European Commission, Brussels

EC (2006), Commission of the European communities, regulation (EC) no 1907/2006 of the European Parliament and of the Council of 18 December 2006 concerning the Registration, Evaluation,Authorisation and Restriction of Chemicals(REACH), establishing a European Chemicals Agency, amending Directive 1999/45/EC and repealing Council Regulation (EEC) No 793/93 and Commission Regulation (EC) No 1488/94 as well as Council Directive 76/769/EEC and Commission Directives 91/155/EEC,93/67/EEC, 93/105/EC and 2000/21/EC,Available from: http://eurlex.europa.eu/LexUriServ/LexUriServ.do?uri=CELEX:32006R1907:en:NOT

EC (2009), Integrated Pollution Prevention and Control—Draft Reference Document on Best Available Techniques for the Production of Iron and Steel - July 2009, European Commission, Brussels

Eloneva, S., Puheloinen, E.-M., Kanerva, J., Ekroos, A., Zevenhoven, R. & Fogelholm, C.-J. (2010). Co-utilisation of CO_2 and steelmaking slags for production of pure $CaCO_3$ - legislative issues. Journal of Cleaner Production, Vol.18, No. 18, (2010), pp. (1833-1839), ISSN 0959-6526

Euroslag (2006) Legal Status of Slags. Position Paper. January 2006. The European Slag Association - *EUROSLAG*. Duisburg, Germany

Geiseler, J. (1996). Use of steelworks slag in Europe. *Waste management*, Vol. 16, No. 1-3,(1996), pp. (59-63), ISSN 0956-053X

Hiltunen, R. & Hiltunen, A. (2004). Environmental aspects of the utilization of steel industry slags, *Proceedings of VII Intenational Conference on Molten Slags, Fluxes and Salts*, ISBN 1-919783-58-X, The South African Institute of Mining and Metallurgy, 2004

Jamali, S. F. K., Forghani, A. & Shirinfekr, A.(2006).Effect of Steelmaking Slag and Converter Sludge on Some Properties of Acid Soil under Tea Planting, Proceeding of 18th World Congress of Soil Science, Philadelphia, Pennsylvania, USA, July 9-15, 2006

Kobesen, H. (2009). Legal Status of Slag Valorisation, *Proceeding of the First International Slag Valorisation Symposium*, Leuven, Belgium, April 6-7, 2009

Kobesen, H. (2010). The registration of Ferrous Slags within REACH, *Proceeding of the 6th Euroslag Conference*, Madrid, Spain, October 19-22, 2010

Kühn, M., Spiegel, H., Lopez, A. F, Rex, M. & Erdmann, R. (2006). Sustainable agriculture using blast furnace and steel slags as liming agents, European Commission, ISBN 92-79-01702-0, Luxembourg, INTERNATIONAL

Li, H.-J., Suito,H. & Tokuda, M.(1995).Thermodynamic Analysis of Slag Recycling Using a Slag Regenerator. *ISIJ Intenaltional*, Vol. 35, No. 9 (1995), pp.(1079-1088), ISSN 0915-1559

Motz , H. & Geiseler, J. (2001). Products of steel slags an opportunity to save natural resources. Waste Management. Vol. 21, No. 3, (2001), pp. (285-293), ISSN 0956-053X

Negim, O., Eloifi, B., Mench, M., Bes, C., Gaste, H., Montelica-Heino, M. & Le Coustumer, P. (2010). Effect of basic slag addition on soil properties, growth and leaf mineral composition of beans in a Cu-contaminated soil. Journal Soil and Sediment Contamination, Vol. 19, No. 2, (2010), pp. (174-187), ISSN 1532-0383

Pereira, H. S., Korndörfer, G. H., Vidal, A. A., and Camargo, M. S. (2004). Silicon sources for rice crop, Scientia Agricola (Piracicaba, Brazil). Vol. 61, No. 5, (Sept./Oct. 2004), pp. (522-528), ISSN 0103-9016.

Rex, M. (2002). Environmental aspects of the use of iron and steel slags as agricultural lime, *Proceeding of The Third European Slag Conference*, Keyworth, Nottingham, UK, October 2002

Rodriguez, M., F. A. Lopez, Pinto, M., Balcazar, N., & Besga, G. (1994). Basic Linz-Donawitz slag as a liming agent for pastureland. *Agronomy Journal*, Vol. 86, No. 5, pp. (904-909), ISSN 0002-1962

Rozman, A., Lamut, J., Debelek, M., & Knap, M. (2007). Stabilisation of ladle refining slags with borax, *Proceedings of 5th European Slag Conference*, Luxembourg, Edited by Euroslag, Duisburg, Germany, 19th-21st September 2007, 137-145

Savant, N. K., Korndörfer, G. H., Datnoff, L. E., and Snyder, G. H. (1999). Silicon nutrition and sugarcane production: a review. Journal of Plant Nutrition, Vol. 22, No. 12 (1999), pp. (1853-1903), ISSN 0190-4167

Shamim, A. H. Khan, H. R. & Akae, T. (2008). Neutralizing capacity of basic slag in acid sulfate soils and its impacts on the solubility of basic cations under various moisture regimes. Journal of the Faculty of Environmental Science and Technology. Okayama University, Vol. 13, No. 1, (2008), pp. (67-74), ISSN 1341-9099

Sikka, R & Kansal, D. B. (1994). Effect of fly-ash application on yield and nutrient composition of rice, wheat and on pH and available nutrient status of soils. *Bioresource Technology*, Vol. 51, No.2-3, pp. (199-203), ISSN 960-8524

Stroehlein, J. L. , & Berger, K. C. The Use of Ferrosul, a Steel Industry Byproduct, as a Soil Amendment. *Soil Sci. Soc. Am. J.*, Vol. 27, No. 1, (1963), pp. (51-53)

Takahashi,T., Yabuta,K.(2002). New Application of Iron and Steelmaking Slag. NKK TECHNICAL REPORT-JAPANESE EDITION, No. 87, (2002), pp.(43-48), ISSN 0915-0536

The Japan Iron and Steel Federation - Nippon Slag Association (July 2006). The Slag Sector in the Steel Industry, Available from http://www.slg.jp/e/index.htm

Torkashvand, A. M. (2011). Effect of steel converter slag as iron fertilizer in some calcareous soils. *Acta Agriculturae Scandinavica Section B – Soil and Plant Science*, Vol. 61, No. 1, pp. (14-22), ISSN 0906-4710

Wang, X. & Cai, Q.-S. (2006). Steel Slag as an Iron Fertilizer for Corn Growth and Soil Improvement in a Pot Experiment. *Pedosphere*, Vol. 16, No. 4, pp. (519-524)

Wallace, A., Samman, Y. S. & Wallace, C. A. (1982). Correction of lime-induced chlorosis in soybean in a calcareous soil with sulfur and an acidifying iron compound. *Journal of Plant Nutrition*, Vol.5, No.4-5, (1982), pp. (94-97)

World Steel, 2008. Sustainability report of the world steel industry. 34 pp.(Available from: http//www.worldsteel.org/pictures/publicationfiles/Sustainability%20Report%202008_English.pdf

World Steel Association, (n.d.) Available from http://www.worldsteel.org

Yokoyama, K., Kubo, H., Mori, K., Okada, H., Takeuchi, S. & Nagasaka, T. (2007)Separation and recovery of phosphorus from steelmaking slags with the aid of a strong magnetic field. *ISIJ international*,Vol.47, No.10,(2007), pp.(1541-1548), ISSN0915-1559

15

From PET Waste to Novel Polyurethanes

Gity Mir Mohamad Sadeghi and Mahsa Sayaf
Dep. Polymer Engineering & Color Technology,
Amirkabir University of Technology, Tehran,
I.R. Iran

1. Introduction

It is well known that Poly (ethylene terephthalate) (PET) is a semi-crystalline thermoplastic polyester widely used in the manufacture of apparel fibers, disposable soft-drink bottles, photographic films, etc. The world production of PET in 2002 was 26 million tons which is expected to rise to 58 million ton in 2012 (Kloss J et al, 2006 & Shukla SR, 2009). The majority of the world's PET production is for synthetic fibers (in excess of 60%) with bottle production accounting for around 30% of global demand. The polyester industry makes up about 18% of world polymer production and is third after polyethylene (PE) and polypropylene (PP). Large numbers of post-consumer PET products, especially bottles and containers, do not create a direct hazard to the environment, but are being concerned due to their substantial volume fraction in the solid waste streams, their high resistance to the atmosphere, their poor biodegradability and photo degradability. Recently, recycling of PET has received a great deal of attention. Although the nontoxic nature, durability and crystal clear transparency of PET during use are major advantages, its non biodegradability is the serious cause of concern to the environmentalists. Since land filling of such non biodegradable waste has severe limitations, chemical recycling is the best possible alternative. Therefore, chemical recycling of PET leads to various advantages: consuming waste to get new useful materials and changing of a non- biodegradable polymer to a biodegradable one. Chemical recycling of PET includes chemolysis of the polyester with an excess of reactants such as water (hydrolysis) (Pusztaszeri SF, 1982, Mishra S et al, 2003; Schwartz J, 1995; Lamparter RA et al, 1985; Tindall GW et al, 1991 & Doerr ML, 1986) alcohols (alcoholysis), glycols (glycolysis) (Akiharu F et al,1986; Ostrowski HS,1975; . Güçlü G et al, 1998, Andrej K, 1998; Berti C et al, 2004; Manfred K et al,1993), amines (aminolysis) (Shukla SR et al,2006 ; Fabrycy E et al,2000;Zahn H et al,1963; Popoola V,1998) and ammonia (ammonolysis) (Blackmon KP et al,1990). Aminolysis has been little explored as chemical degradation of PET for synthesis of useful products. The use of ethanolamine for aminolytic degradation of PET waste has been investigated. (Shukla SR et al, 2006) The product obtained BHETA has potential for further reactions to synthesize useful products such as polyurethanes. There are few reports on the usage of recycled BHETA from PET to synthesis of polyurethanes. Depolymerization of the PET waste, using ethanolamine to obtain BHETA and BHETA-based polyurethanes, has been investigated in our works (Shamsi R et al, 2009; Mohammadi M et al, 2010; Mir Mohamad Sadeghi G et al, 2011). This chapter aims at the study on synthesis of novel polyurethanes based on PET waste. Firstly, PET and polyurethanes are concisely reviewed, with emphasis on the methods of synthesis, their structures, properties and applications. Then, various chemical decomposition methods of PET are introduced. Using aminolysis in the presence of Ethanolamine, applying of aminolysis product (BHETA) as chain extender or ring opening agent to obtain new polyurethanes are demonstrated. Mechanical, thermal properties, biodegradability, chemical resistance, adhesion of novel synthesized materials are studied. Thirdly, effective parameters such as structural hard segment, chain length, chemical structure, and crystallinity on final properties as well as biodegradability are investigated.

2. Poly (ethylene terephthalate)

Poly (ethylene terephthalate) is a thermoplastic polymer resin of the polyester family and is used in synthetic fibers; beverage, food and other liquid containers; thermoforming applications; and engineering resins often in combination with glass fiber. Depending on its processing and thermal history, polyethylene terephthalate may exist both as an amorphous (transparent) and as a semi-crystalline polymer. The semicrystalline material might appear transparent (particle size < 500 nm) or opaque and white (particle size up to a few microns) depending on its crystal structure and particle size. Its monomer , BHET can be synthesized by the esterification reaction between terephthalic acid and ethylene glycol with water as a byproduct, or by transesterification reaction between ethylene glycol and dimethyl terephthalate with methanol as a byproduct (Fig. 1-a). Polymerization is through a polycondensation reaction of the monomers (done immediately after esterification/ transesterification) with water as the byproduct (Fig.1-b).

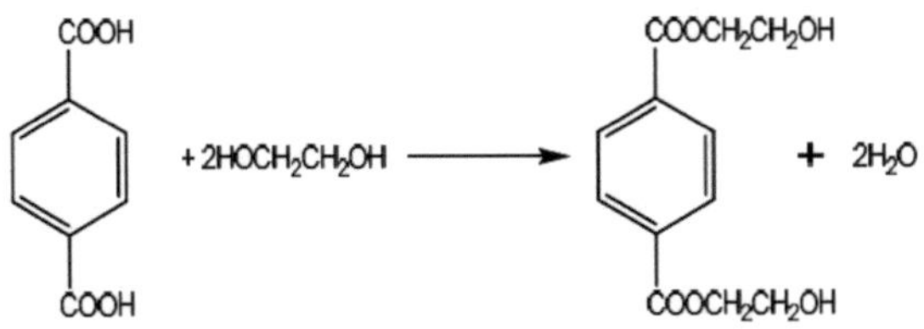

Fig. 1. a. Chemical Reaction between ethylene glycol and terephthalic acid yields BHET.

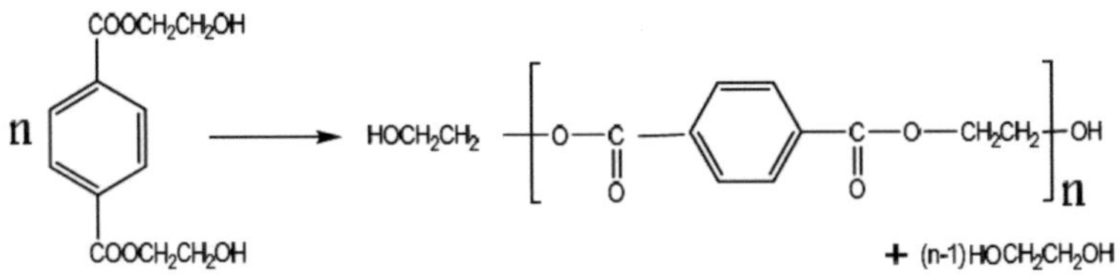

Fig. 1. b. Polycondensation of BHET yields PET.

3. PET waste, as an opportunity instead of a problem

PET is used in the preparation of a variety of products differing widely in their physical characteristics and hence, their end uses. The varieties of prominence are fibers and filaments, sheets and disposable soft-drink, soda drinks, juice, mineral water, soy sauce bottles, photographic films, etc. The world production of PET in 2002 was 26 million tons which is expected to rise to 55 million ton in 2010. PET resin, or bottle grade, is one of the fastest growing plastics markets. Polyester fiber is the second largest segment, but the market is mature. The third use, film, is also a mature market, for example PET market in USA in 2008 could be shown in Fig.2.

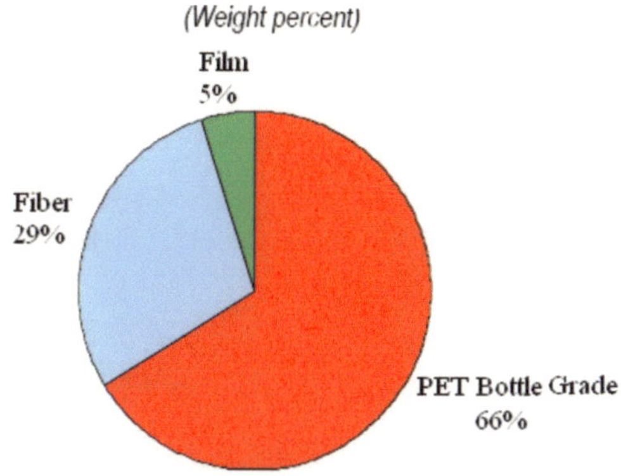

Fig. 2. PET market in USA in 2008(chemsysytem.com,nexant ,PERP 07/08-5:1-8).

Large numbers of Post-consumer PET products especially bottles and containers do not create a direct hazard to the environment, but as a problem due to its substantial volume fraction in the solid waste streams, its high resistance to the atmosphere, its poor biodegradability and photo degradability. PET accounts for more than 8% by weight and 12% by volume of the world's solid waste (Shamsi R et al, 2009).

An estimated billion plastic bottles are disposed of each year, while recycling a single plastic bottle can conserve enough energy to light a 60 W light bulb for up to 6 h.

Recently, recycling of PET has received a great deal of attention and many attempts are currently directed toward recycling of post-consumer PET products because of both environmental protection and economic benefits. Also necessity for recycling of this product is felt more (Fig 3). In Singapore, 684,400 tones of plastic waste were generated in 2008 and the recycling rate was 9%. Although the nontoxic nature, durability and crystal clear transparency of PET during use are major advantages, its non biodegradability is the serious cause of concern to the environmentalists. Because it isn't appropriate to dispose of waste PET by land-filling, alternative methods of recycling of waste PET products include physical and chemical recycling have been developed.

To minimize the fast buildup of PET waste, different mechanical, thermal, and chemical methods to separate, recover and recycle PET from post-consumer waste stream have been used (Mohammadi M et al, 2010). Products made from recycled PET bottles include carpeting, concrete, insulation and automobile parts. Recycled PET bottles are also used in drainage filtration systems, asphalt concrete-mixes and road stabilizations. Recycling rate of such polymer products is however still low comparing to that of paper, glass and metals. Currently only 3.5% of generated polymeric products is recycled whereas these percentages for paper, glass and metals are, respectively, 34%, 22% and 30%.

Physical recycling of PET consists of the collection, separation, digestion, granulation of polymer waste and then recirculation into production. Blending of materials with PET waste has also been studied. The effect of waste PET addition on thermal transmission (or insulation) property of ordinary concrete has been studied and reported that corresponding percentages for PET bottle pieces vary between 10.27% and 18.16%, depending on the geometries of added pieces. Moreover, concrete-PET blends due to their ability in water absorption, a possible application could be in sports courts and pavements which need good water drainage. M.C. Almaza'n and co worker proposed a different method to obtain activated carbon using the actual waste plastic commercial vessels made of PET as raw material (Mohammadi M et al, 2010).

Fig. 3. PET waste as a threat, serious cause of concern to the environmentalists.

In a recyclability analysis determination of a global index which takes into account social, economic and environmental aspects is believed to be an interesting approach for industrial organizations. Thus, in this case following aspects may be analyzed:

- Social: A stronger and wide-spread PET recycling sector (market) would generate employment and contribute to reduce the volume of municipal solid wastes.
- Environmental: PET recycling contributes to reduce mass and energy consumption.
- Economic: Technical/economic analyses could demonstrate the viability of the chemical recycling of PET regarding costs, with adequate technical applications. Since land filling of such non biodegradable waste has severe limitations, chemical recycling is the best possible alternative. Therefore, chemical recycling of PET leads to various advantages: consuming waste to get new useful materials and changing of a non- biodegradable polymer to a biodegradable one.

3.1 Chemical recycling of PET waste

Chemical recycling of PET includes chemolysis of the polyester with an excess of reactants such as water (hydrolysis), alcohols (alcoholysis), glycols (glycolysis), amines (aminolysis) and ammonia (ammonolysis). (Shamsi R et al, 2009; Mohammadi et al, 2010 & Mir M. Sadeghi G et al, 2011).

3.1.1 Hydrolysis

PET is polyester, and the functional ester group can be hydrolyzed by water. Hydrolysis of PET can be carried out in an acid, alkaline or neutral environment to produce monomers terephthalic acid (TPA) and ethylene glycol. During hydrolysis reaction, PET hydrolyzes to a carboxylic acid and an alcohol as follows:

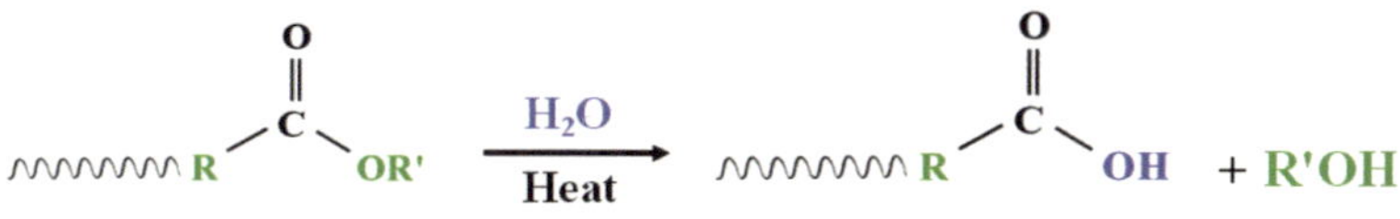

Fig. 4. Hydrolysis Reaction.

Life time alternative in a PET chemical recycling plant is depicted in Fig 5. Kinetics of hydrolysis of PET Pellets in Nitric Acid (Mohammadi M et al, 2010), kinetics and Thermodynamics of Hydrolytic acidic and neutral depolymerization of poly (ethylene terephthalate) at high pressure and temperature has been investigated.

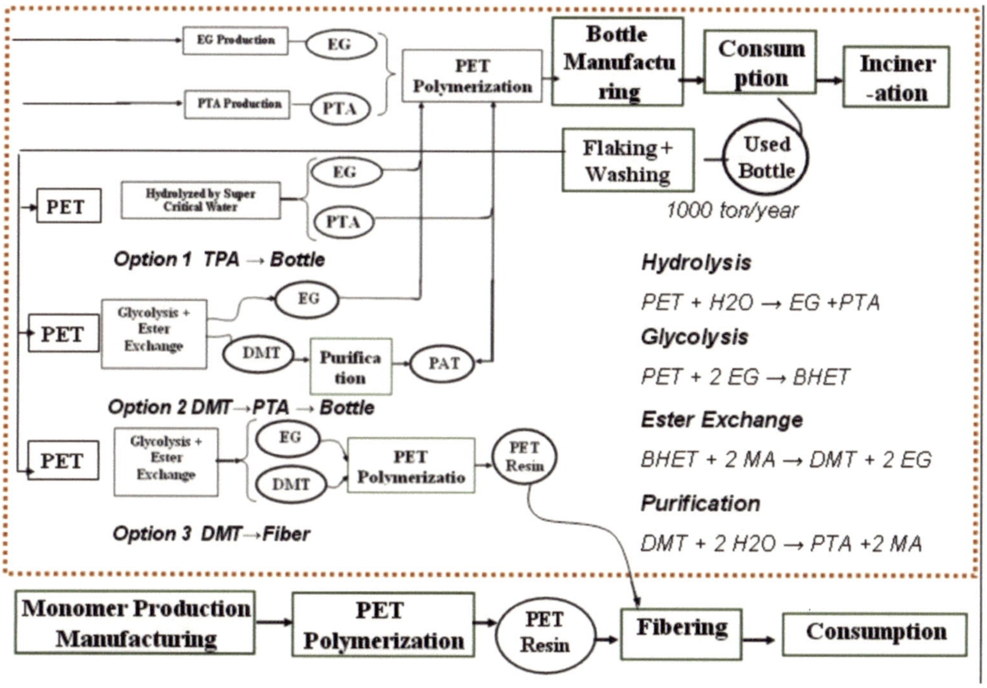

Fig. 5. Life time alternative for PET chemical recycling, DMT (Dimethyl terephthalate), TPA (Pure terephthalic acid) commonly uses PTA, EG (Ethylene Glycol).

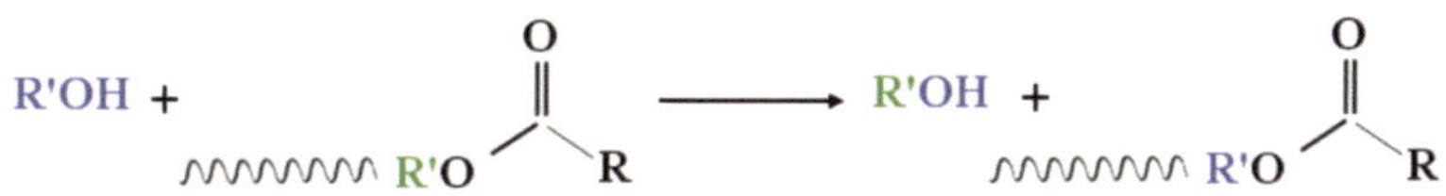

Fig. 6. Glycolysis Reaction.

3.1.2 Glycolysis

Glycolysis is breakdown of ester linkages by a glycol, resulting in oligomers or oligoester diols/polyols with hydroxyl terminal groups. Oligoesters coming from the glycolysis of PET waste have been well known to be utilized as a starting material in the manufacture of unsaturated polyesters, vinyl ester resins, epoxy resins, alkyd resins and polyurethanes. Glycolysis is carried out using different glycols like; ethylene glycol, propylene glycol, 1, 4-butanediol and triethylene glycol, diethylene glycol (DEG), dipropylene glycol (DPG), glycerol (Gly) and etc. During glycolysis reaction, the organic group R″ of an ester with the organic group R′ of an alcohol exchanges. (Shamsi R et al, 2009; Mohamadi et al, 2010 & M. M. Sadeghi G et al, 2011)

3.1.3 Methanolysis

Methanolysis is the degradation of PET using methanol at high temperatures and high pressures with the main products being dimethyl terephthalate (DMT) and ethylene glycol.

Methanolysis is the recycling process which has been practiced and tested on a large scale for many years in the past. In this case, polyester waste is transformed with methanol into DMT (Dimethyl terephthalate), under pressure and in presence of catalysts. Finally the crude DMT is purified by vacuum distillation. Degradation of PET using ethylene glycol at high temperatures and high pressures with the main products being BHET is used to produce some different materials such as polyester, polyurethane resins and esteric plasticizer, as shown in Fig. 7.

3.1.4 Aminolysis

Aminolysis is another method of chemical degradation of PET, which has been relatively little investigated, compared to the other techniques.

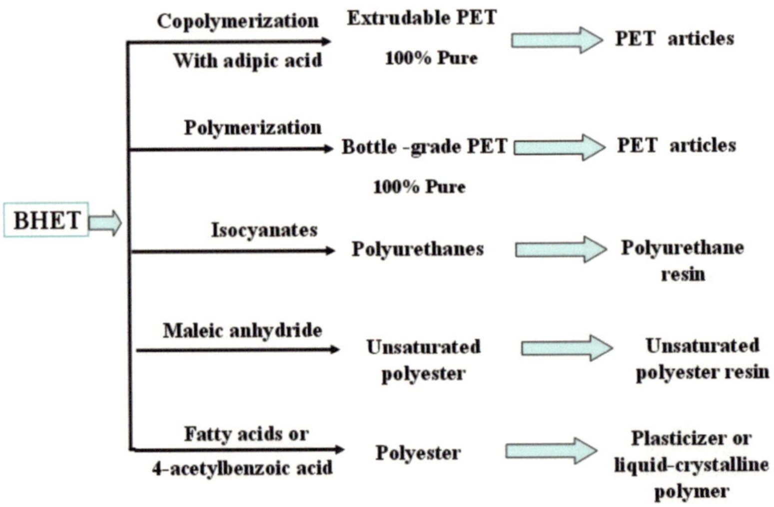

Fig. 7. Using of BHET (glycolysis product of PET waste) to obtain other materials.

Depolymerization of PET waste has been carried out using various amines, such as ethanolamine, benzylamine, hexamethylenediamine, aniline, methylamine hydrazine monohydrate and some polyamines. Catalysts such as lead acetate, glacial acetic acid, sodium acetate and potassium sulfate are usually used to facilitate the reaction. Aminolysis products, such as BHETA, have the potential to undergo further reactions to yield secondary value-added products. In this direction, very recently the synthesis of unsaturated polyesters polyurethanes, epoxy resin hardeners and non-ionic polymeric surfactants has been reported.

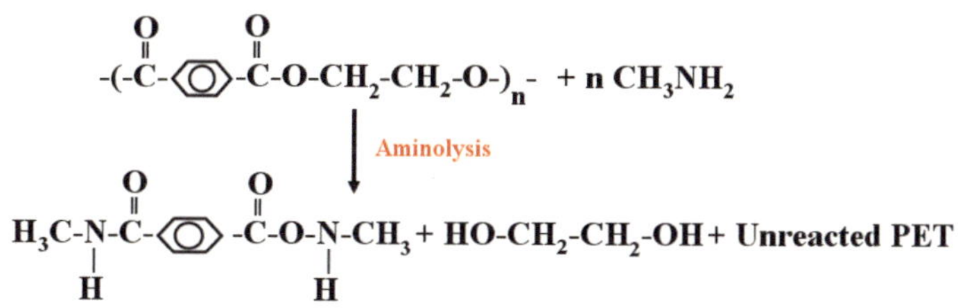

Fig. 8. Aminolysis reaction.

Zahn and Pfeifer carried out aminolysis of PET with solutions of hydrazine, benzyl amine, ethylene diamine, hexamethylene diamine, piperidine and aniline. They obtained different reaction products as the diamides of terephthalic acid, which do not possess any potential for further chemical reactions. According to Popoola the basicity of an amine relative to water as well as its steric hindrance due to size determines the rate of degradation of PET. During aminolysis of PET with methylamine, the methyl terephthalamide is obtained, which isn't enough reactive for its recycling into any useful product through further reactions. Shukla and Harad have been investigated the use of ethanolamine for the aminolytic degradation of PET waste in the presence of different simple chemicals such as glacial acetic acid, sodium acetate and potassium sulphate as catalysts. The product obtained, BHETA has potential for further reactions to obtain useful products.

4. Polyurethane

Polyurethane is any polymer composed of a chain of organic units joined by carbamate (urethane) links. Polyurethane polymers are formed through step-growth polymerization, by reacting a reactant (with at least two isocyanate functional groups) with another reactant (with at least two hydroxyl or alcohol groups) in the presence of a catalyst. Generalized formation reaction of the urethane group is:

$$R-NCO + R'-OH \rightarrow R-NH-COO-R'$$

Thermoplastic polyurethanes (TPUs) are linear polymers formed by the polymerization reaction of three basic components:

1. A diisocyanate ($NCO-R-NCO$)
2. A short-chain diol, so-called chain extender ($OH-R'-OH$)
3. A long-chain diol (OH OH)

As shown in the above reactionurethane linkage is produced by reacting an isocyanate group, $-N=C=O$ with a hydroxyl (alcohol) group, $-OH$. In fact, polyurethanes are produced by the polyaddition reaction of a polyisocyanate with a polyalcohol (polyol) in the presence of a catalyst and other additives. The reaction product is a polymer containing the urethane linkage, -RNHCOOR'-. A broad range of physical properties can be achieved by varying the chemistry and molecular weight of the various components, and through manipulation of the ratios in which they are reacted in polyurethanes. Therefore polyurethanes have received recent attention as regards the development of wide family of polymeric materials (paints, adhesives, elastomers, flexible, and rigid foams, etc.) and thus play an important and increasing role in our daily life. The greatest advantage offered by polyurethane is their versatility, both in finished product properties and ease of production and application. By the proper choice of isocyanate and polyol, products can be made with properties ranging from low viscosity resins used in printing to high modulus solids used in industrial parts. Polyurethanes are applied to the manufacture of flexible, high-resilience foam seating; rigid foam insulation panels; microcellular foam seals and gaskets; durable elastomeric wheels and tires; automotive suspension bushings; electrical potting compounds; high performance adhesives; surface coatings and surface sealants; synthetic fibers (e.g. Spandex); carpet underlay; and hard-plastic parts (i.e. for electronic instruments) and any other industrial parts.

4.1 Components

Polyols are higher molecular weight materials manufactured from an initiator and monomeric building blocks. They are most easily classified as polyether polyols, which are made by the reaction of epoxides (oxiranes) with active hydrogen containing starter compounds, or polyester polyols, which are made by the polycondensation of multifunctional carboxylic acids and hydroxyl compounds.

Isocyanates two or more functional groups are required for the formation of polyurethane polymers. Volume wise, aromatic isocyanates account for the vast majority of global diisocyanate production. Aliphatic and cycloaliphatic isocyanates are also important building blocks for polyurethane materials, but in much smaller volumes. The two most important commercial, aromatic isocyanates are toluene diisocyanate (TDI) and diphenylmethane diisocyanate (MDI). TDI consists of a mixture of the 2,4- and 2,6-diisocyanatotoluene isomers. The most important product is TDI-80 (TD-80), consisting of 80% of the 2,4-isomer and 20% of the 2,6-isomer. This blend is used extensively in the manufacture of polyurethane flexible slabstock and molded foam. TDI, and especially crude TDI and TDI/MDI blends can be used in rigid foam applications, but have been supplanted by polymeric MDI. TDI-polyether and TDI-polyester prepolymers are used in high performance coating and elastomeric applications.

Chain extenders (f =2) and cross linkers (f =3 or greater) are low molecular weight hydroxyl and amine terminated compounds that play an important role in the polymer morphology of polyurethanes. The choice of chain extender also determines flexural, heat, and chemical resistance properties. The most important chain extenders are ethylene glycol, 1,4-butanediol(1,4-BDO or BDO),1,6-hexanediol,cyclohexane,dimethanol and hydroquinone bis (2-hydroxyethyl) ether (HQEE).

4.2 Microstructure of polyurethanes

Segmented polyurethanes that consist of alternating soft and hard segments offer unique possibilities of tailor-made polymers by varying block length and composition. The structure of the lineal polymeric chain of thermoplastic polyurethane is in blocks, alternating two different types of segments linked together by covalent links, forming a block copolymer. These segments are:

4.2.1. Hard Segments which are segments formed by the reaction of the diisocyanate and the short-chain diol. They have a high density of urethane groups of high polarity, and for this reason, they are rigid at room temperature (high hardness).

4.2.2. Soft Segments which are segments formed by the reaction of the diisocyanate and the long-chain diol. They have a low polarity as they have a very low density of urethane groups, and therefore, they are flexible at room temperature (very low hardness). A general structure of thermoplastic polyurethane's chain would be as Fig. 9:

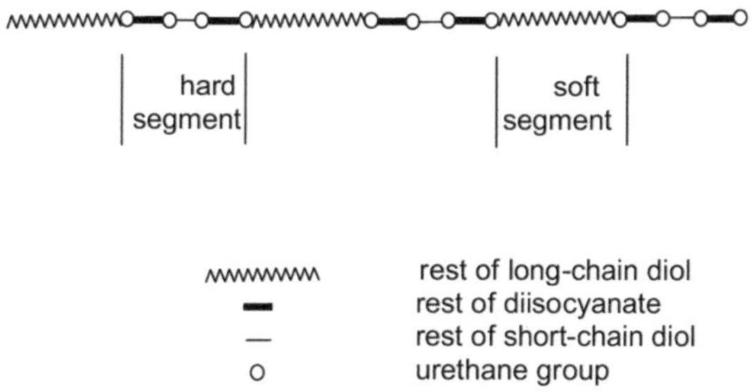

Fig. 9. General structure of polyurethane's chain.

The polarity of hard segments produces a strong attraction between them, which causes a high degree of aggregation and order in this phase, forming crystalline or pseudo-crystalline areas located in a soft and flexible matrix. This so-called phase separation between both blocks will be more or less important, depending on the polarity and molecular weight of the flexible chain, the production conditions, etc. The crystalline or pseudo-crystalline areas act as a physical crosslink, which accounts for the high elasticity level of TPUs, where as the flexible chains will impart the elongation characteristics to the polymer. The schematic representation of the segmented micro structure and two-phase morphology of polyurethane are shown in Figs. 10 and 11. These "pseudo crosslinks" , based on hydrogen bonding between carbonyl groups and -NH groups of various chains , however, disappear under the effect of heat, and thus the classical extrusion, injection molding and calendaring processing methods are applicable to these materials.

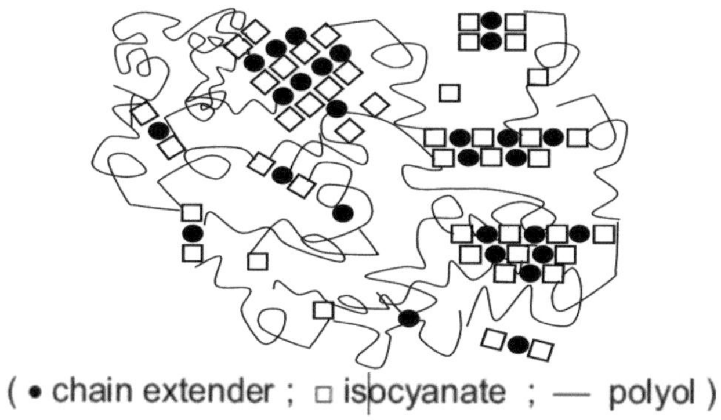

Fig. 10. Schematic represent of microstructure of segmented polyurethane chains.

Consequently -and not less importantly- TPU scraps can be reprocessed. When TPUs are cold, the "pseudocrosslinks" reappear again, providing the elastic properties to the obtained article. When TPUs are dissolved in a proper solvent, the "pseudo crosslinks" are also broken up by the solvent, and therefore, disappear. Due to this it is possible to apply a TPU in solution by classical methods of coating applications; when the solvent evaporates the "pseudo crosslinks" are formed again.

The soft domains provide the thermoplastic polyurethane with a very low Tg, in comparison with other polymers of the same hardness, maintaining the elasticity at very low temperatures. The presence of polar and non polar counter balanced microdomains is the cause of the good chemical resistance of TPUs, particularly oil and grease resistance. Thermoplastic polyurethanes are very versatile items, since a variety of soft and hard segments can be combined, with their respective ranges of molecular weights, and considering also the variety of molecular weights of the final polymer.

So that it is possible to obtain from very soft (60 Shore A) to very hard polyurethanes (80 Shore D), with different degrees of crystallinity, to be used in many applications and market segments which require high performance. This peculiar structure which differentiates thermoplastic polyurethanes from other polymers provides polyurethanes with the following main properties:

- Very high elasticity
- Excellent abrasion resistance
- Very good tear strength
- Good oil and grease resistance
- Excellent mechanical properties, combined with a rubber-like elasticity
- Outstanding low-temperature performance
- High transparency

4.3 Morphology

Chain extenders play an important role in the polymer morphology of polyurethane fibers, elastomers, adhesives, and certain integral skin and microcellular foams. The elastomeric properties of these materials are derived from the phase separation of the hard and soft copolymer segments of the polymer, such that the urethane hard segment domains serve as cross-links between the amorphous polyether (or polyester) soft segment domains.

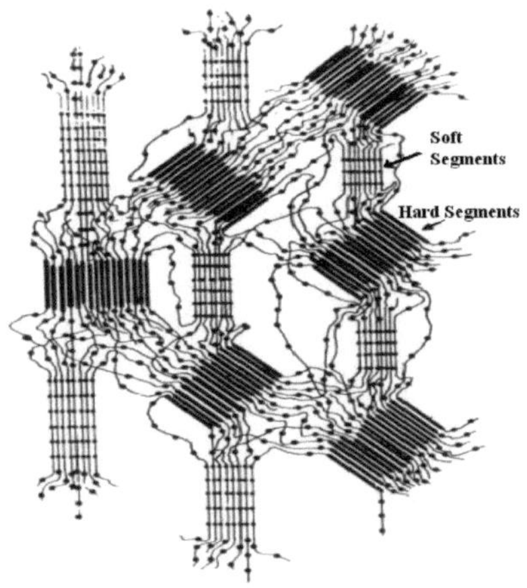

Fig. 11. Schematic representation of two-phase morphology in polyurethanes.

Schematic representation of the superstructure (domain and chain formation) and dimensions in polyurethanes is shown in Fig.12. This phase separation occurs because the mainly non-polar, low melting soft segments are incompatible with the polar, high melting hard segments. The soft segments, which are formed from high molecular weight polyols, are mobile and are normally present in coiled formation, while the hard segments, which are formed from the isocyanate and chain extenders, are stiff and immobile. Because the hard segments are covalently coupled to the soft segments, they inhibit plastic flow of the polymer chains, thus creating elastomeric resiliency.

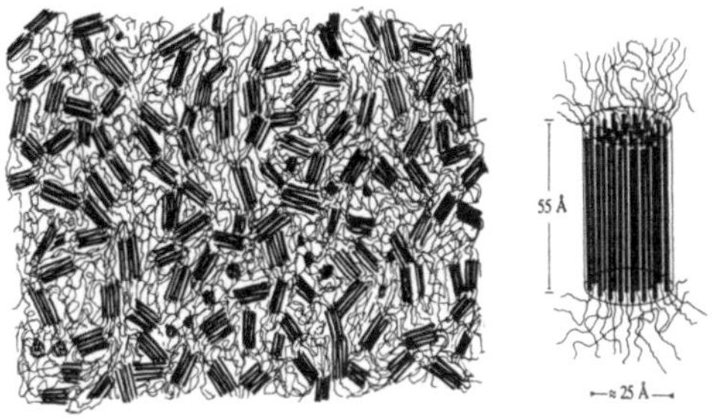

Fig. 12. Schematic representation of superstructure (domain and chain formation) in polyurethanes.

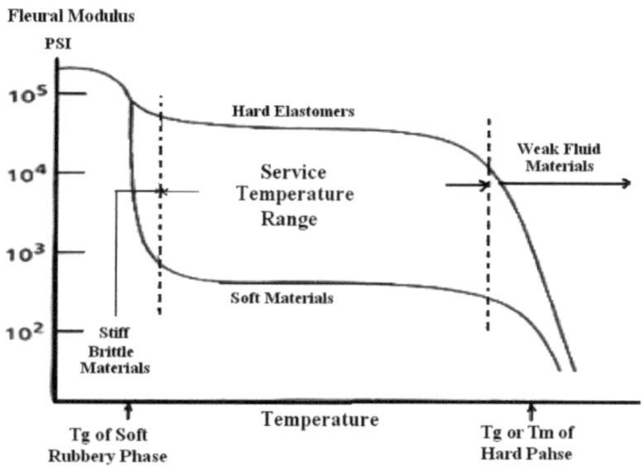

Fig. 13. Schematic representation of soft and hard segments thermal transitions.

Upon mechanical deformation, a portion of the soft segments are stressed by uncoiling, and the hard segments become aligned in the stress direction. This reorientation of the hard segments and consequent powerful hydrogen bonding contributes to high tensile strength, elongation, and tear resistance values. Phase separation in polyurethanes can be studied by dynamic thermal analysis. Thermal transitions of hard and soft segments of a typical polyurethane consisted of a soft rubbery phase and hard phase are presented in Fig. 13. According to nature and physical chemistry properties of polyurethanes, increasing of hard segment content affects on mechanical as well as their thermal properties definitely.

4.4 Wide variety of soft & hard segments in synthesis of polyurethanes

A broad range of physical properties can be achieved by varying the chemical structure and molecular weight of the various components and also through manipulation of the ratios in which they are reacted in polyurethanes. Therefore, polyurethanes have received recent attention with regard to the development of degradable polymers because of their great potential in tailoring polymer structures to achieve mechanical properties and biodegradability to suit variety of applications such as biodegradable polymers, soft tissue adhesives, clinical demand and meniscus scaffold. Multiblock copolymers based on caprolactone and lactic acid, polyglycoles, polyesters and multifunctional aliphatic carboxylic acids as soft segments have been investigated used in various applications such as medical or industrial fields. Commercial polycaprolactones with different molecular weights as a soft segment, Polycaprolactone-based polyurethanes using diols such as Ethyleneglycole BHET, 1,4-butanediol or other diols as chain extender for ring opening polymerization of caprolactone have been studied. Using of nature-based polyols to prepare polyurethane foams is common but there are few reports on elastomeric polyurethane. Presence of aromatic ring in PET structure caused to improve mechanical and thermal properties of polyurethane structure; while presence of esteric bonds leads to biodegradation. (Yeganeh et al, 2007; Heijka R et al, 2005)

5. Novel polyurethanes based on aminolysis product of PET waste

The product obtained, BHETA has potential for further reactions to synthesize useful products. Using ethanolamine to obtain BHETA and BHETA-based polyurethanes has been investigated in author's works. Three types of polyurethanes have been synthesized using BHETA. In the first case, BHETA is used as ring opening agent in caprolactone polymerization, and then novel biodegradable polyurethane has been synthesized. In the second and third cases, BHTEA is used as chain extender to synthesis of high modulus and special polyurethanes which is discussed in the next topics.

5.1 First step: Aminolysis of PET waste to obtain BHETA

In aminolysis with ethanolamine, the obtained product, BHETA is in its pure. Mechanism of synthesis of BHETA which proposed by Shukla and Harad is shown in Fig. 14. (Shukla SR, et al (2006). Ethanolamine was used for the aminolysis of PET waste materials in the molar ratio 1:6 (PET: ethanolamine) under reflux in the presence of catalyst for time period up to 5 h. The catalyst, sodium acetate, was used in concentration 1% by weight of polymer. At the end of the reaction, distilled water was added in excess to the reaction mixture with vigorous agitation to precipitate out the product, BHETA.

The filtrate contained mainly unreacted ethanolamine and little quantities of a few water soluble PET degradation products. The precipitate obtained was filtered and dissolved in distilled water by boiling for about 20 min. White crystalline powder of BHETA was obtained by first concentrating the filtrate by boiling and then chilling it. It was further purified by recrystallisation in water. It was then dried in an oven at 80°C. Different techniques of analysis were used for characterization of BHETA.

After synthesis and high purification, BHETA has been characterized using ^{1}HNMR and Fourier transform infrared (FTIR) spectroscopy depicted in Figures 15-16. BHETA has been shown in Fig. 15, it may be clearly seen that the spectrograph contains peaks at 1056 and 3288 Cm^{-1} indicating the presence of primary alcohol. The peaks for secondary amide stretching are observed at 1311, 1554 and 3369 Cm^{-1}. Fig. 16 shows the 1H NMR spectra of chain extender. The purified BHETA was characterized by FTIR, ^{1}HNMR, and melting point. Synthesized BHETA has been melted at 227°C.

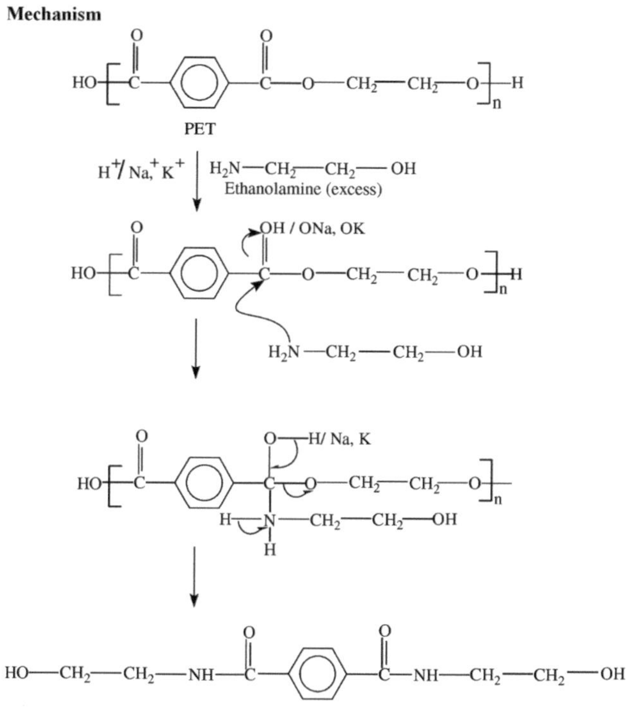

Fig. 14. Depolymerization mechanism of (2-hydroxy ethylene) Terephthalamide.

The FTIR spectrograph of the purified As can be seen, the shift and splitting pattern of ^{1}H NMR at 8.52 ppm, 7.91ppm , 4.73 ppm, 3.52 ppm and 3.34 ppm, corresponding to H of the amine group, aromatic ring, hydroxyl Group, CH_2 bonded to hydroxyl group, and CH_2 bonded to amine group respectively. These obtained results confirmed that the PET fibers depolymerized and expected product (BHETA) was synthesized successfully.

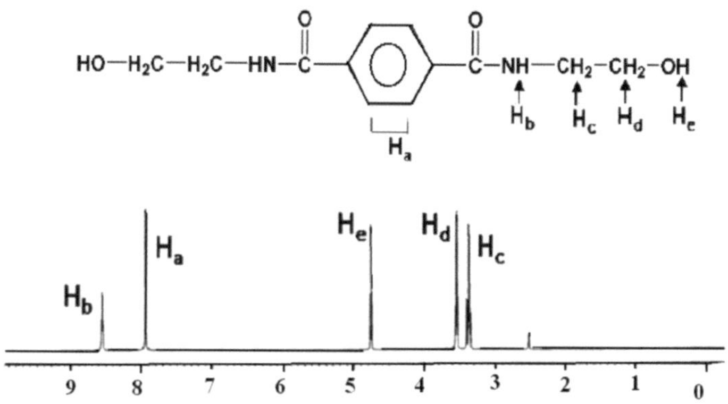

Fig. 15. ^{1}HNMR spectrum of BHETA.

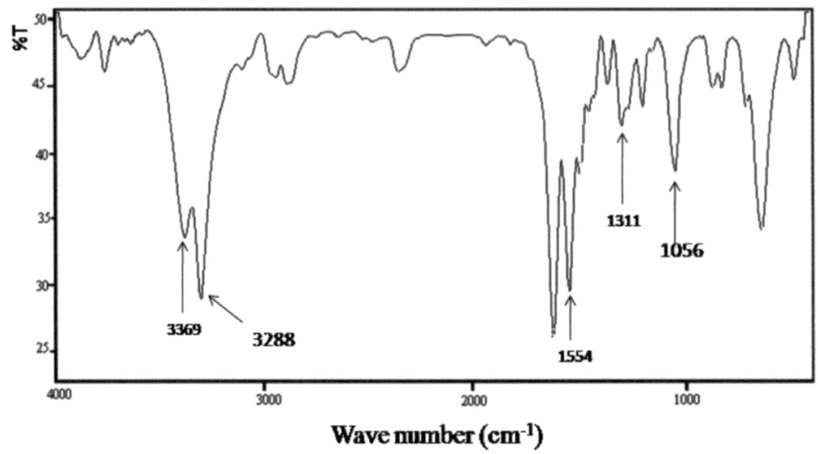

Fig. 16. FTIR spectrum of BHETA.

5.2 Synthesis of polyurethanes based on BHETA

Three types of polyurethanes have been synthesized based on BHETA. In the first case, BHETA is used as ring opening agent in caprolactone polymerization, and then novel biodegradable polyurethane has been synthesized. In the second and third cases, BHTEA is used as chain extender to synthesis of special and high modulus polyurethanes as follows:

6. Biodegradable polyurethanes based on ε-Caprolactone

Polyols with different molecular weights have been synthesized through ring opening polymerization of caprolactone by BHETA. Polyurethanes with different soft segment chain-lengths have been synthesized using above-mentioned polyols.

6.1 Ring opening polymerization: Using of BHETA

Ring opening polymerization of caprolactone by BHETA is a unique method which is used to synthesis of biodegradable polyurethanes. BHETA synthesized locally was reacted with ε-caprolactone through ring opening polymerization at 130°C using 1 wt% DBTDL as the catalyst for 3.5 h in a round-bottom flask equipped with a condenser, stirrer, thermometer and nitrogen gas-inlet tube. The reaction and HNMR spectrum of the synthesized polyol is shown in Fig. 17. Polyols with different molecular weights through ring opening polymerization of caprolactone by BHETA have been synthesized. (Mir M. Sadeghi et al, 2011).

The various molar ratios of ε-Caprolactone to BHETA used in the synthesis of polyols, named Polyol-8 to Polyol-142, are shown in Table 1. Then urethane linkages were formed using Di-isocyante without chain extender. For synthesis of polyurethanes (PU-8 to PU-142) at first, polyols were extended with HDI. Calculated amount of HDI/DMF solution was added drop-wise at 110°C. The homogeneous mixture was then poured slowly into a Teflon mold and maintained at 60ºC for 12 h. The films were then removed and placed in a desiccator for testing.

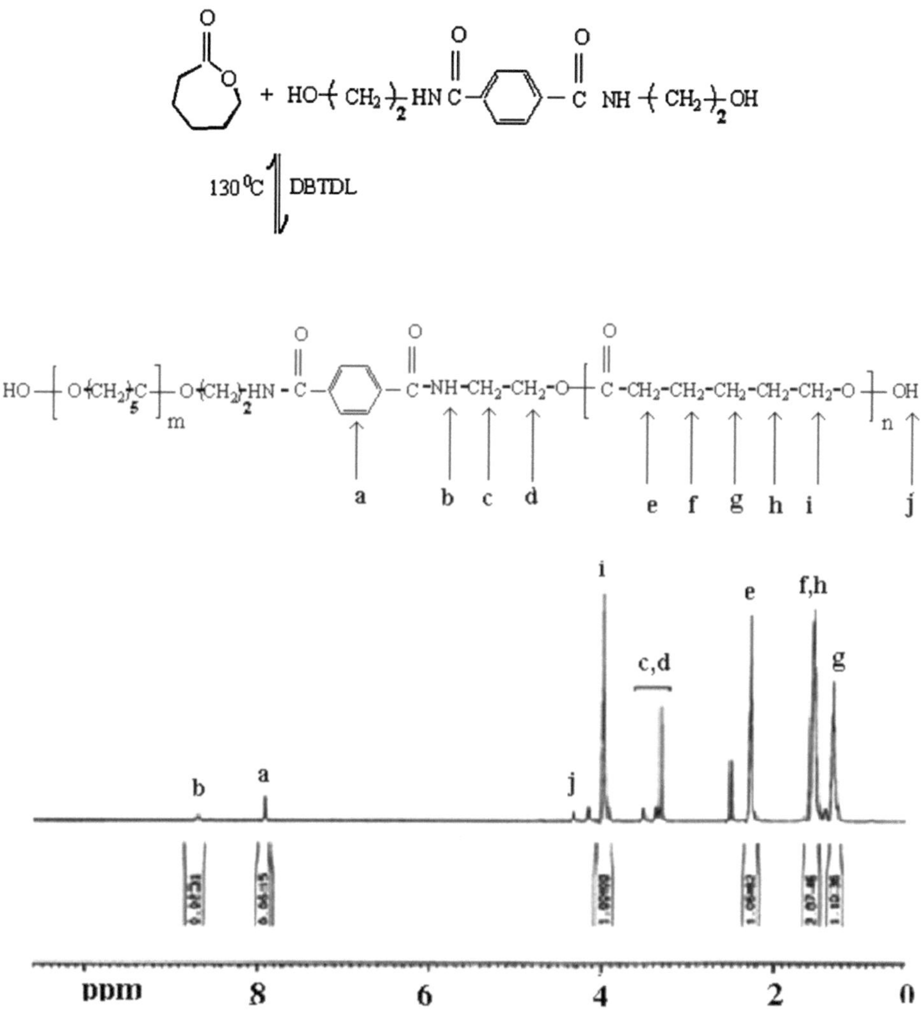

Fig. 17. Ring opening polymerization of caprolactone by BHETA to obtain polyol.

Sample	Molar ratio of ε-caprolactone to BHETA	n (theory)	M_n (theory)	n (experimental)[a]	M_n (experimental)[a]
Polyol-8	8	4	1196	3.97	1189
Polyol-22	22	11	2792	10.3	2632
Polyol-38	38	19	4616	17	4160
Polyol-76	76	38	8948	37	8720
Polyol-142	142	71	16472	71	16472

[a] Obtained from HNMR spectrum

Table 1. Molar ratios, theoretical and experimental "M_n" and "n" of used polyols.

6.2 Polyurethane characterization

6.2.1 FTIR spectroscopy

FTIR spectra of polyurethanes with different polyol chain lengths are shown in Fig. 18. There are distinctive absorbance bands at *ca* 3435 Cm^{-1}(nonbonded -NH groups), *ca* 3330 Cm^{-1}(bonded -NH groups), *ca* 2940 and *ca* 2865 Cm^{-1} (asymmetric and symmetric CH_2 groups), *ca* 1730 Cm^{-1}(overlapped carbonyl groups of polycaprolactone and urethane linkages), *ca* 1620 Cm^{-1} (C=C stretching vibration in aromatic ring), *ca* 1540 Cm^{-1} (-CO-N amide II), *ca*(1468,1418,1390,1368) Cm^{-1}: various modes of CH_2 vibrations, *ca* (1240,1169,1040) Cm^{-1}: stretching vibration of the ester groups and stretching of the ether group (~1100 Cm^{-1}).

Mechanical properties of polyurethanes and Mn of used polyol are shown in Table 2.

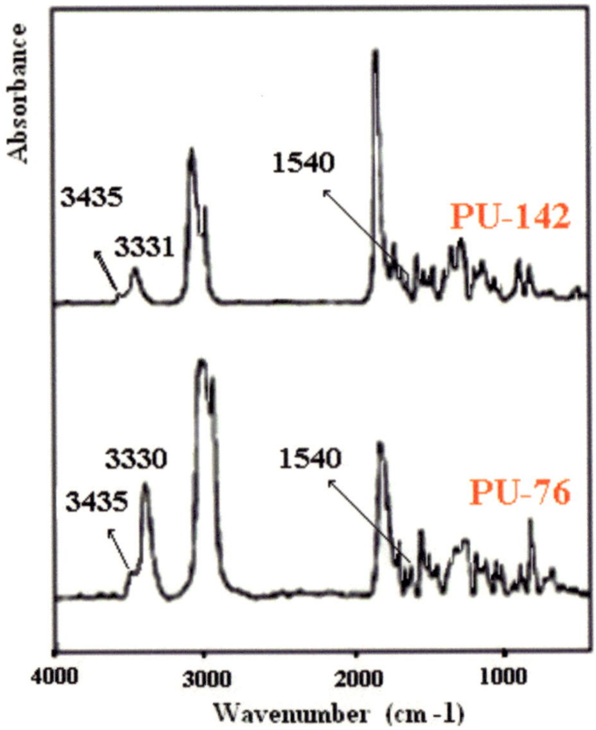

Fig. 18. FTIR spectra of PU-22 to PU-142.

Sample	Mn	Tg °C	Tm °C	ΔHf J/g	χ_c* %	Tensile strength (MPa)	Elongation @ break %
PU-22	2,632	−46.1	36	−26.55	18	16	440
PU-38	4,160	−48.4	55.3	−41.4	29	14	520
PU-76	8,720	−50.5	60.8	−64.4	45	10	6
PU-142	16,472	−56	63.6	−74.35	52	16	7

Table 2. Thermal and mechanical properties of polyurethanes and Mn of used polyol.

6.2.2 DSC results

DSC thermograms and the related data such as Tg, T_m, ΔH_f and degree of crystallinity (α_C) are shown in Fig. 20 and Table 2. The melting curves in Fig. 21 clearly show an endotherm for the melting, indicating the presence of one distinct crystalline zone in all polymers, which could be ascribed to ordering or size in the crystallites. As shown obviously in Table 2, increasing of molecular weight of synthesized polycaprolactone leads to a regular increase in the observed melting point from 36°C to 63 °C and also in heat of fusion of samples.

As shown in Table 2, the crystallinity, α_c , of polycaprolactone (PCL) phase increases with increasing molar ratio of polycaprolactone to BHETA. α_c is calculated using a value of ΔH_f for completely crystalline PCL, equal to 142 J g^{-1}. T_m of PU containing PCL with lower molecular weight (Mn= 2632) shows lower T_m than T_m of PU with higher Mn. The results indicate that the crystals obtained in PU-142 during crystallization are larger than those developed in PUs with lower content of soft segments (Table 2). The presence of hydrogen bonding between the hard and soft segments restricts the phase separation and ordering (crystallization) of PUs. When the soft segment length increases from 2632 to 16472, the reduction in the degree of connectivity between the hard and soft segments should make the phase separation process and crystallization easier. These peaks could be related to HDI/BD hard segments which have a regular repeat structure capable of high degree of hydrogen bonding, exhibit very sharp endothermic peaks at higher temperatures about 290°C. DSC thermograms for all samples indicate that an increase in molar ratio of ε-caprolactone to BHETA changes the size and peak position temperature of the endotherms above 280°C.

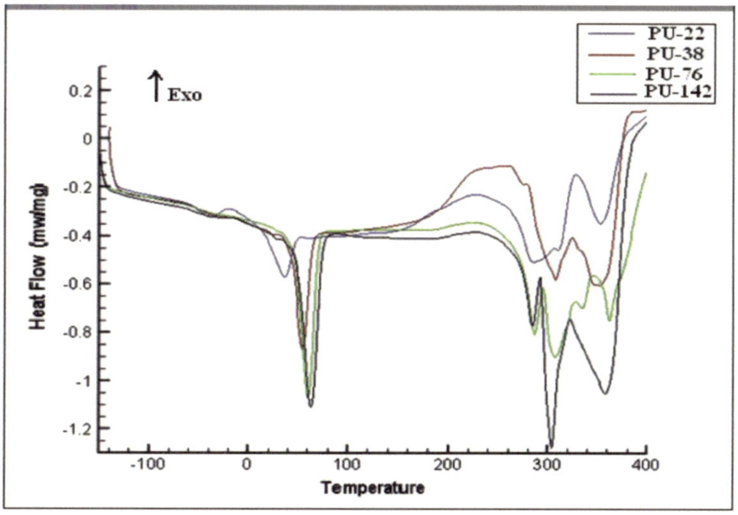

Fig. 19. DSC thermograms of PU-22 to PU-142.

Longer soft segments produce better phase-separated systems which caused to form more readily sharp peaks as indicated by the trend in phase separation with increasing soft segment length at 290°C. However, these endothermic peaks may be thought to arise from the disruption of ordered, non-crystalline hard-segment aggregates. As shown in Fig. 19, as the lengths of soft segments increases, Tg of polyurethane decreases due to more flexibility of soft segments chains. Caprolactone-based polyurethane shows Tg at -54°C by DSC method for polycaprolactone with M_n about 2000 g/mol. Presence of BHETA in the chemical structure results in increasing of Tg in our work.

6.2.3 DMTA results

DMTA Results are shown in Figs. 20 and 21 for PU-22 and PU-38 respectively. Two main transitions are present in the DMTA spectra: a first peak (T_β) located in the low-temperature region (at about -75°C and - 85°C) and a second peak (T_α) seen in a higher temperature region (at about -25°C). Increasing in soft segments length caused to decrease T_β of polyurethanes. The observed T_β at –75°C and –85°C which was assigned to methylene sequence local relaxations, in analogy to results reported previously, is due to relaxation of caprolactone-based soft segments. It is found clearly that T_β decreases with increasing of -CH_2- units in caprolactone chains in sample PU-38 rather than PU-22.

For the assignment of T_α, we have considered a mixing-transition temperature (T_{mix}), which would be the result of the various degrees of mixing between the ester and urethane blocks.

According to this interpretation, the matrix would be formed by a PCL-rich continuous phase, in which PCL crystallites would be embedded, and amorphous PCL segments emerging from entangled domains with urethane segments would be connected to these crystallites. Some damping or fluctuations at higher temperatures is seen which corresponds to the hard-segments glass transition. As Shown in Figs. 20 & 21 E' in glassy region decreases with increasing of crystalline PCL soft segments length, whereas increases in rubbery plateau. The polymer with a higher content of PCL crystalline soft segments gives a higher E' in rubbery region for PU-38 in comparison with PU-22. TGA results shows that $T_{90\%}$ and char residue decreases with increasing of molecular weight of soft segments for samples PU-22 to PU-142, which confirms decreasing in concentration of urethanes and aromatic groups in the samples. Also, presence of aromatic ring due to BHETA led to increasing of thermal satiability.

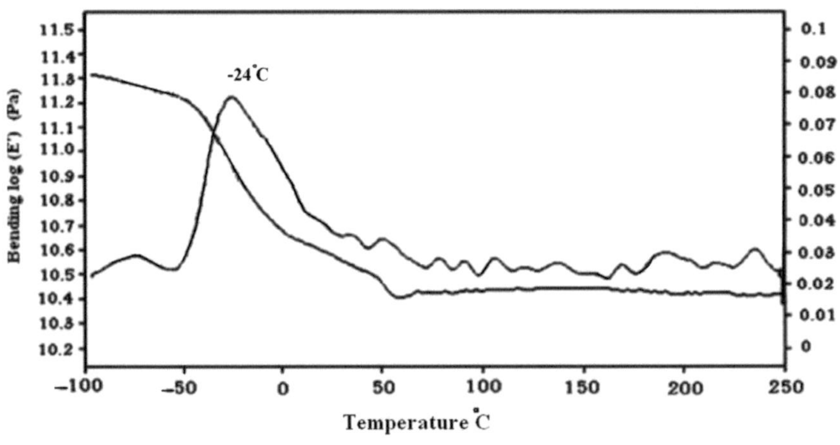

Fig. 20. DMTA thermogram of PU-22.

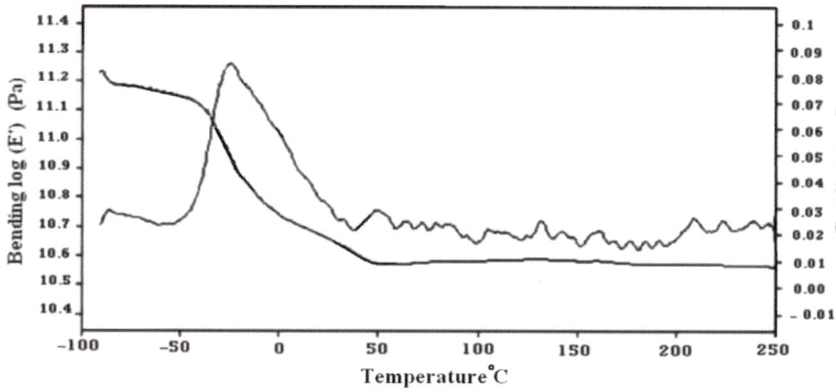

Fig. 21. DMTA thermogram of PU-38.

6.2.4 Optical Microscopy (OM) and Scanning Electron Microscopy (SEM)

Optical microscopy (OM) images shown in Fig. 22 illustrate the morphology of the synthesized PUs (scale bar shows 30 microns). OM results show rather smooth and rather rough structures for samples PU-22 and PU-38, while are fibrous-looking for PU-76 and PU-142.

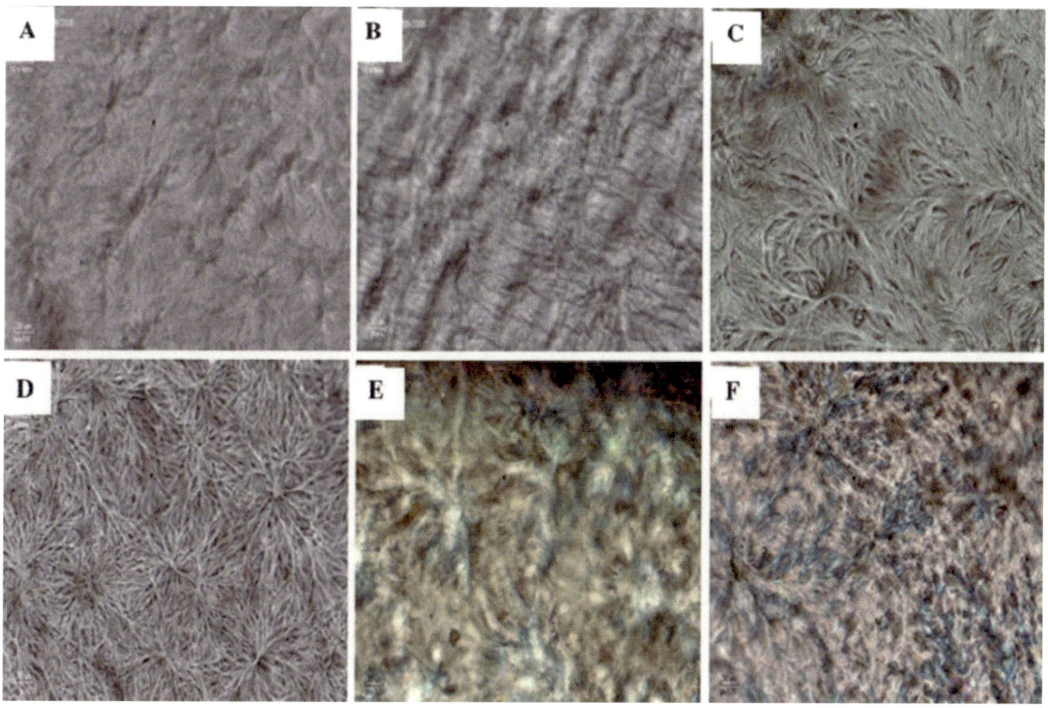

Fig. 22. Optical microscopy images: (a) PU-22; (b) PU-38; (c) PU-76; (d) PU-142; (e) Disordered PU-76; (f) disordered PU-142.

Comparison shows as the soft segment length increases, the tendency to crystallization increases and formation of different structures is observed clearly. Circular regions with 200 μm diameters are seen for PU-142. DSC results confirm the presence of crystallites with sharp melting points from 36°C-63.6°C, which could be correlated to these structures. Rough structures are seen for samples PU-142 and PU-76, whereas PU-22 and PU-38 are less high. SEM results confirm the observed phenomena in OM images. Structures formed in the order of magnitude of *ca*10-100 nm and as M_n of soft segments increases, domains are larger.

6.2.5 Tensile strength and biodegradability study

Tensile strength and elongation at break of samples, before and after biodegradability tests, is shown in Figs. 23.a and 23.b which shows that elongation at break of PU-22 and strength of PU-38 has the highest values respectively.

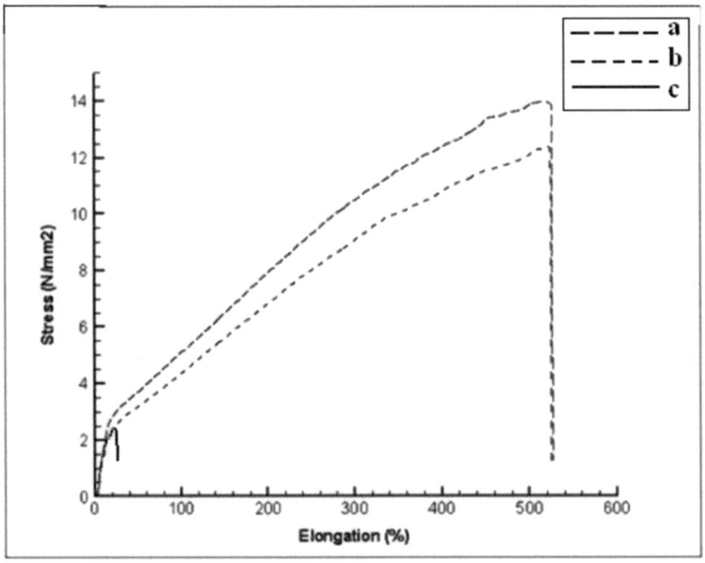

Fig. 23. a-Stress-Strain curves of PU-22: (a) Before biodegradability test; (b) after biodegradability test (3 days in compost); (c) after biodegradability test (20 days in compost).

In contrast, elongation at break of PU-142 and PU-76 are lower than others dramatically. As shown in Fig. 24, PU-22 and PU-38 show elastomeric behavior. Of course, PU-38 shows necking; this isn't seen for PU-22, PU-76 and PU-142 show brittle behavior. Fig. 25 shows schematic chemical structures of PU-22 to PU-142. The observed necking could be correlated to extension of chain folding in chemical microstructure of PU-38 under constant force (Fig.25b). Very low strengths in samples PU-76 and PU-142 compared to PU-22 and PU-38 (Fig.25) are due to low concentration of urethane linkages and hydrogen bonding. More strength for PU-142 than PU-76, which is related to the soft segment molecular weight of this sample (M_n=16472 g/mol) that is more than entanglement molecular weight (M_e) of polycaprolactone diol as *ca* 8000 g/mol.

The presence of BHETA leads to lower flexibility in polyol chain and finally higher M_e in comparison to virgin polycaprolactone equal to *ca* 8000g/mol. The presence of BHETA leads to lower flexibility in polyol chain and finally higher M_e in comparison to virgin polycaprolactone. Therefore, M_n of Polyol-76 (8720 g/mol) is lower than M_e of BHETA containing polyols. The presence of entanglement in polymeric chains provided higher strength for this sample (Fig.25.c and Fig.25.d). It is interesting to obtain good tensile properties for synthesized polyurethanes in this study (especially for PU-22 and PU-38 which show higher hydrogen bonding confirmed by FTIR. This is explained by the fact that ordering degree of PCL blocks increases, and they are capable of forming crystalline structures, which promotes growth of their tensile strength and Young's modulus. It is noticeable that PU-22 and PU-38 were formulated without chain extender. Mechanical properties of novel polyurethanes in this work are comparable with similar commercial polyurethanes that already available. In this research we choose tensile test because of rapid biodegradation of samples and impossibility to weigh during and after decomposition. In order to compare

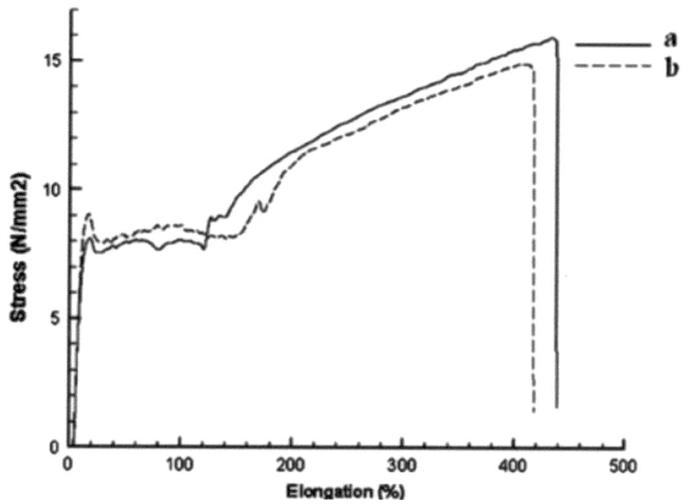

Fig. 24. b-Stress-Strain curves of PU-142: (a) before; (b) after biodegradability.

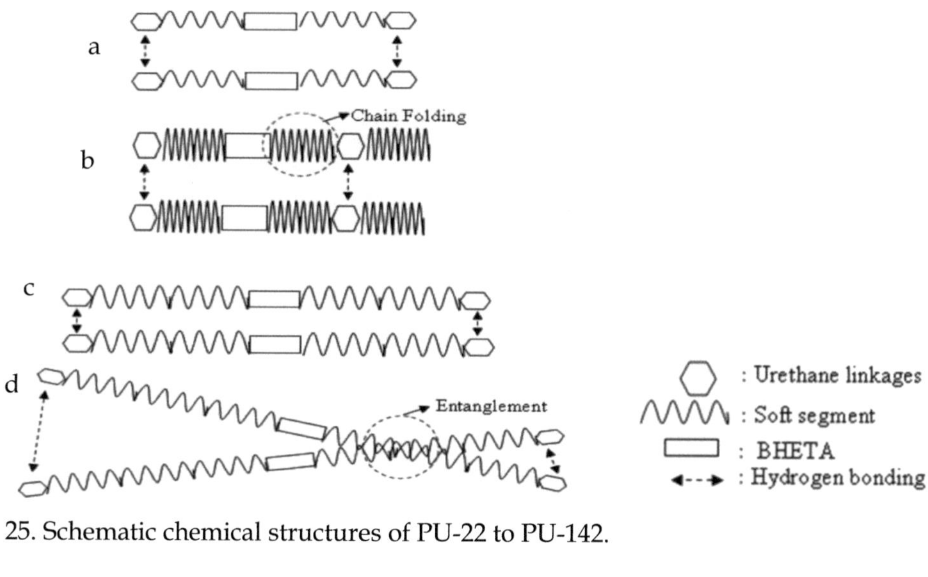

Fig. 25. Schematic chemical structures of PU-22 to PU-142.

biodegradability of polyurethanes, dumbbells of various samples were maintained in compost (soil, straw, leaves, etc.) at 45°C and 95% moisture and tensile strength tests have been achieved after 3 and 20 days. Stress-elongation curves of PU-22 to PU-142, before and after biodegradability test, are shown in Figs. 23 to 24 respectively. As shown in the figures strength and elongation at break have been reduced after 3 days in all samples. After 20 days all of polyurethanes decomposed except PU-22. However, mechanical properties of PU-22 have been reduced dramatically. More resistance of PU-22 to degradation is due to more concentration of urethane groups and less concentration of carbonyl groups in the polymer chains in this sample. Biodegradability of samples is comparable to polymers based on caprolactone; however presence of BHETA affects on biodegradability obviously.

6.3 Synthesis of special polyurethanes

Novel polyurethanes were synthesized based on prepared BHETA, 1,4-Butanediol (BD),Ether type Polyol and various molar ratio of Hexamethylene Diisocyanate (HDI). To evaluate the effect of BHETA, properties of polyurethanes without and with BHETA have been compared. FTIR, thermal transitions (DSC), degradation (TGA) of synthesized PUs have been investigated. (Shamsi et al, 2009).

6.3.1 Materials and synthesis method

PET staple waste fiber consists of short fibers with density of 1.45 g cm-3. PET staple waste fibers were boiled with methanol for 3 h to remove any surface finishing and dirt present in the fiber mass. Ethanolamine (EA), Ether type polyol (Polyol): Mn=2000 (Bayer), Sodium acetate,1,4-Butanediol (BD), Dibutyl Tin dilaurate (DBTDL), Hexamethylene Diisocyanate (HDI) were used as received. Polyurethanes were synthesized using a one-shot polymerization method. BHETA (0.0277 mol), Polyol (0.0119 mol), BD (0.198 mol) and DBTDL (7.78×10^{-4} mol) were dissolved in 200 mL of DMSO in a three-necked flask equipped with a condenser and stirrer. The temperature was raised to 90°C. Then desired amounts of HDI were added and the reaction mixed vigorously. In order to study the effect of BHETA on the polyurethane properties, two samples were synthesized without BHETA.

Sample	NCO/OH ratio (mol)	BD (mol)	Polyol (mol)	HDI (mol)	Reaction Time(min)	Gel Time (min)
PU-3[b]-W[c]	3	1	0.2	3.6	-	70
PU-10[b]-W[c]	10	1	0.2	12	6	-

[a] $\frac{BD(mol)}{BHETA(mol) + Polyol(mol)} = 5$ $\frac{BHETA(mol)}{Polyol(mol)} = 2.33$ [b] NCO/OH ratio in the polyurethane

Table 3. Molar ratios[a] of reactants to synthesis of PUs (with BHETA).

Polyol, BD and DBTDL were dissolved in 100 mL of DMF in a three-necked flask equipped with a condenser and stirrer. Then the desired amounts of HDI were added. After removing the mixtures from the reactor, they were post-cured and dried at 100°C for 8 h.

Tables 3 and 4 give the various molar ratios used in the synthesis of PU-1 to PU-10 (with BHETA) and PU-3-W and PU-10-W (without BHETA).

Sample	NCO/OH Molar ratio	BD (mol)	BHETA (mol)	Polyol (mol)	HDI (mol)	Reaction Time (min)	Gel Time (min)
PU-1[b]	1	1	0.14	0.06	1.2	6	-
PU-5[b]	1.5	1	0.14	0.06	1.8	-	35
PU-3[b]	3	1	0.14	0.06	3.6	-	50
PU-4[b]	4	1	0.14	0.06	4.8	6	-
PU-7[b]	7	1	0.14	0.06	8.4	6	-
PU-10[b]	10	1	0.14	0.06	12	6	-

a $\frac{\text{BD(mol)}}{\text{Polyol(mol)}} = 5$ b: NCO/OH ratio in the polyurethane C: polyurethane without BHETA

Table 4. Molar ratios of reactants to synthesis of PUs (without BHETA).

6.3.2 FTIR analysis of polyurethanes

As seen in the spectra of PU-3, PU-10 and PU-3-W (Fig. 26), strong inter-urethane hydrogen bonding is developed for all samples. Participating N–H group the hydrogen bond and non-bonded N–H group absorption displays a characteristic absorption band between 3300 and 3446 cm^{-1} and 3446 cm^{-1} respectively. FTIR spectrum of polyurethanes would display two carbonyl bands: one at 1707 cm^{-1} assigned to bonded C-O groups, and a second at 1731 cm^{-1} assigned to free C-O groups.

6.3.3 Thermal analysis

DSC and TGA results are shown, respectively, in Figs. 28a and 28b for samples PU-3 and PU-3-W. Comparison of DSC thermograms of PU-3 and PU-3-W shows the first endothermic peak at 150°C for PU-3-W, whereas it is at 190°C for PU-3 due to presence of BHETA in the chemical structure. DSC Thermograms in Figures 28a and 28b show exothermic peaks for PU-3-W and endothermic peaks for PU-3.

Exothermic crosslinking reactions are due to the thermodynamically favorable conformation that such interchain covalent bonds would promote. Conversely, the destruction of interchain hydrogen bonding, chain scission and pyrolysis reactions cause a DSC endotherm. The bond dissociation energy for a carbon–carbon single bond is relatively high (*ca* 375 kJ mol^{-1}) and bond scission is endothermic.

Conversely, for PU-3-W, crosslinking reactions predominate over destruction of urethane hydrogen bonding and chain scission. Dissociation or chain scission reactions in thermal degradation mechanisms of polyurethanes were summarized by Saunders *et al.* in four types of reactions that may take place during thermal degradation: (i) Dissociation to isocyanate and alcohol; (ii) formation of primary amino and olefin; (iii) formation of secondary amine; and (iv) transesterification-type bimolecular displacement. Therefore chain scission reactions in samples PU-3-W and PU-3 are indeed as shown in Fig 27. It has been shown

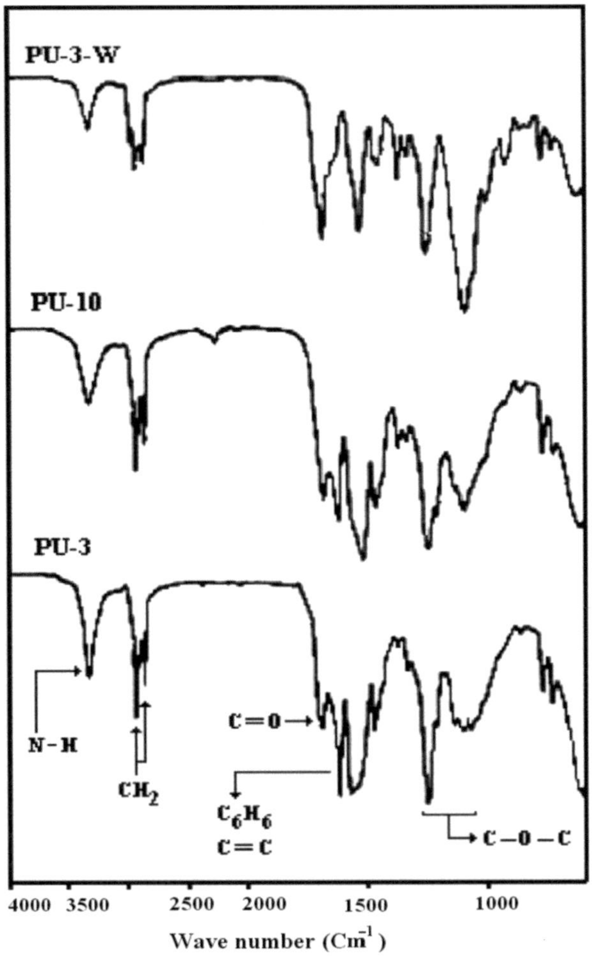

Fig. 26. FTIR spectra of PU-3, PU-10 and PU-3-W.

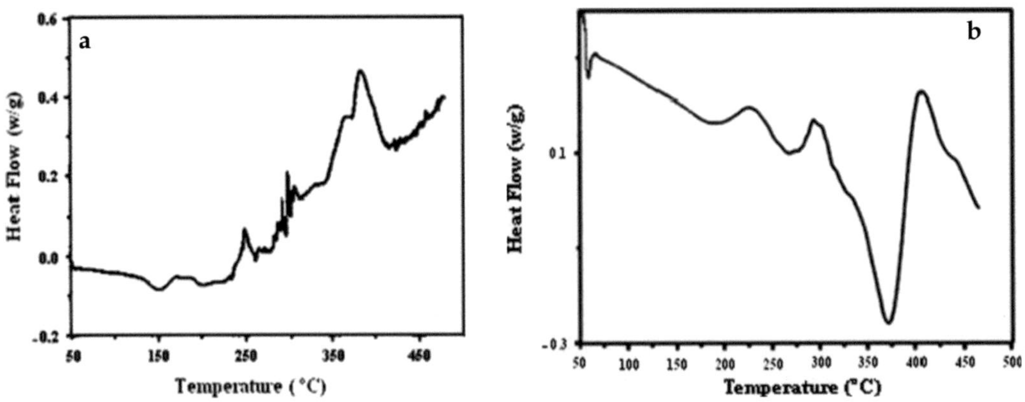

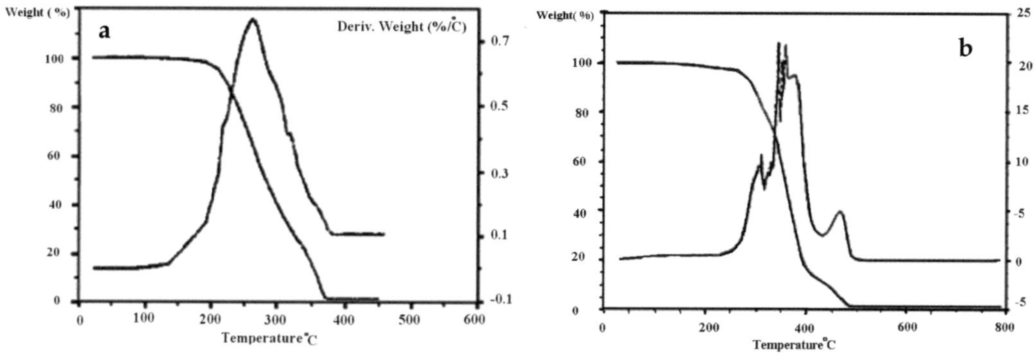

Fig. 27. DSC and TGA thermograms of (a) PU-3-W and (b) PU-3.

that the endothermic behavior (T2 peak) is dissociation and degradation, resulting from long-range ordering and disordering of the hard segment domains in segmented polyurethanes. Some endothermic thermal transitions in DSC analysis of polyurethanes are due to structural decomposition of the polymer. As seen from the TGA and differential TGA (DTGA) thermograms (Figs 27(a) and (b)), weight reduction in PU-3-W begins at 140°C while it starts at about 235°C for PU-3. The first, second and third peaks in the DTGA trace of PU-3 are related to chain scission of urethane linkages, Polyol and BHETA, and isocyanurate and carbodiimide, respectively. 38 Also, the composition of various products was calculated as 20, 68 and 10.3 wt%, respectively. TGA of sample PU-3-W shows continuous degradation, while in PU-3 three-step degradation is observed.

Sample	$T_{initial}$	$T_{25\%}$	$T_{50\%}$	$T_{75\%}$	$T_{90\%}$	Char residue at 500 C (Wt. %)
PU-3[a]	232	330	360	385	445	1.7
PU-3[a]-W[b]	138	249	284	333	360	0.8

Table 5. Initial, 25%, 50%, 75% and 90% decomposition temperatures, and char residues.

The presence of BHETA in the chemical structure of polyurethane (due to longer chain extender and more hydrogen bonding in polyurethane chains) results in a shift of the beginning of degradation from 140 to 235°C and also three-step degradation. As shown in Table 5, the presence of the BHETA aromatic ring causes retardation of degradation. Also, a char residue of about 1.7 wt% is seen for PU-3 and 0.8 wt% for PU-3-W.

6.3.4 Tensile shear strength

Strength measured as maximum load (*F*m) of samples PU-4, PU-7, PU-10 and PU-10-W (applied on aluminum and iron substrates) is given in Table 6. As can be seen, addition of BHETA to polyurethane caused an increase of *F*m and elongation for both Fe-Fe and Al-Al substrates specifically. The surface preparation method (hand abrasion) for both Fe-Fe and Al-Al substrates was identical; therefore it can be concluded that the presence of BHETA causes stronger bonding as opposed to samples without BHETA. Strong bonds between the surface of the metal and polyurethane films, such as hydrogen bonds, are likely to be due to -NH groups in BHETA.

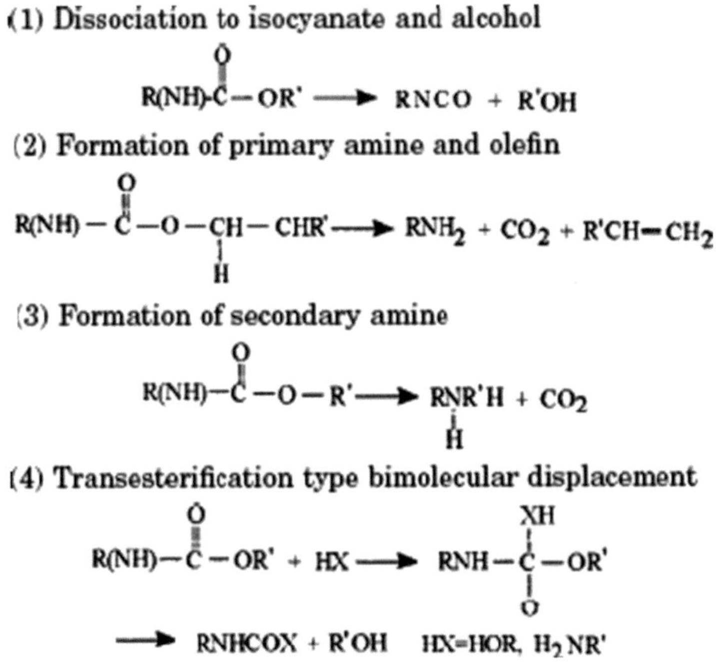

Fig. 28. Probable reactions in the degradation of polyurethanes.

Sample	NCO/OH Ratio (mol)	Maximum load on Fe-Fe (N)	Maximum load on Al-Al (N)
PU-1[a]	1	-	-
PU-5[a]	1.5	-	-
PU-3[a]	3	-	-
PU-4[a]	4	2822.63	2168.13
PU-7[a]	7	2378	1959
PU-10[b]	10	3054.63	1968
PU-3[a] -W[b]	3	-	-
PU-10[a] -W[b]	10	1092.18	806.42

Table 6. shows that the maximum loads for substrates Al-Al.

Figs 29-30 show load–deformation curves for, respectively, PU-4, PU-7 for both substrates. The load–deformation curves of the polyurethanes show brittle failure for Al-Al substrate as well as for Fe-Fe and Fe-Fe are different. The highest value of *F*m (3054 N) is observed for sample PU-10 on Fe-Fe substrate. Comparison of the maximum load for sample PU-10 with commercial epoxy and polyester-type adhesives shows a 2.03- and 2.34-fold increase, respectively. As can be seen in Table 6, the lowest *Fm* value is for PU-3-W on Al-Al substrate. Comparison of Fm for Fe-Fe and Al-Al samples shows higher values for the former.

This relates to higher mechanical interlocking due to the higher porosity of iron. The data obtained in the adhesion evaluation tests show a variation in *Fm* for both substrates. The polyurethanes used have various NCO/OH molar ratios. Therefore *Fm* can be related to free isocyanate groups. However, other parameters such as surface preparation method, moisture, evaporation of solvent, post-curing conditions, also affect the results.

6.3.5 Swelling behavior

The swelling behavior of samples PU-3 to PU-10 with different solvents (DMSO, DMF, EA and Tol) was investigated at room temperature. The measured values of the polymer densities were in the range 1.13–1.2 g cm^{-3}.Fig. 31 shows that the swelling ratio decreases with increasing NCO/OH ratio. Increasing the NCO/OH ratio also increases the crosslinking density, consequently causing a decreasing of the swelling ratio. As regards the effect of the solvent on the swelling ratio, it is seen that an increase of solubility parameter increases the swelling ratio for all NCO/OH ratios (solubility parameters (in (cal.$cm^{-3)}$0.5) are: DMSO, 12.87; DMF, 12.1; EA, 9.1; Tol, 8.9).

6.4 Synthesis of high modulus polyurethanes

Polyurethanes have been synthesized based on BHETA, HDI and polyethylene glycol via prepolymer method. Since catalyst and raw materials have low price, synthesis of BHETA is economical and could be used as diol to synthesis of polyurethanes. In this search, polyurethanes have been synthesized based on BHETA, HDI and polyethylene glycol via prepolymer method. TGA and DSC were carried out to study thermal stability, thermal transitions, Tm and Tg of synthesized polyurethanes. Effect of BHETA content in the main chain on thermal stability of polyurethanes, strength and stiffness has been evaluated. (Mohammadi et al,2010)

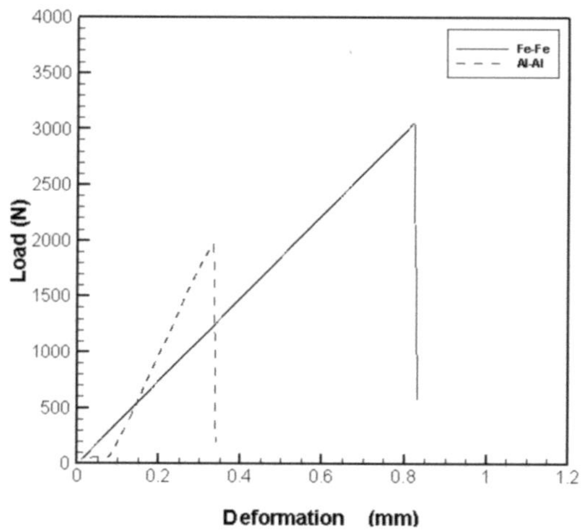

Fig. 29. Load–deformation curves for PU-4 on both substrates.

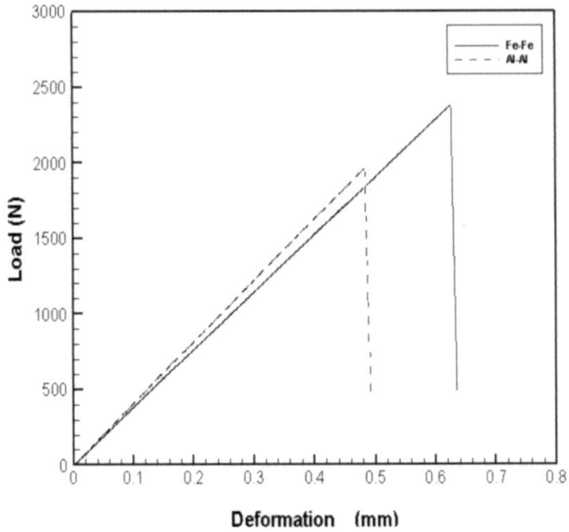

Fig. 30. Load-deformation curves for PU-7 on both substrates.

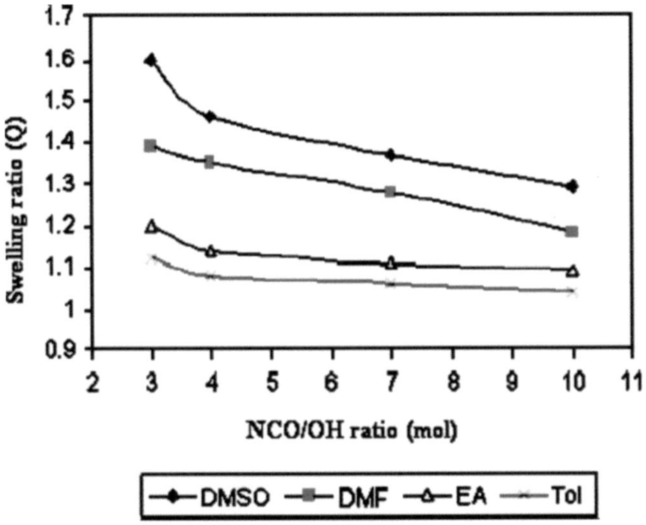

Fig. 31. Swelling ratio as a function of NCO/OH ratio for various solvents.

6.4.1 Synthesis and characterization

A 250 mL round-bottom flask equipped with a temperature controller, magnetic stirrer, reflux condenser, an N_2 inlet, charged with Hexamethylene diisocyanate isocyanate(HDI), polyethylene glycol 1000 (PEG), DMF and Di-butyl Tin Dilaurate (DBTDL) (catalyst 1 wt%). HDI and PEG were reacted for 2 h at 75°C. The obtained prepolymer, then subjected to further reaction with BHETA. The reaction time was 3 h at 70°C. Molar ratio was fixed at 1.1 .The mixtures then were immediately cast on Teflon plates and then were kept in oven for 72 h at 70°C. Details of synthesis and method of synthesis are given in Table 7 and Fig. 32.

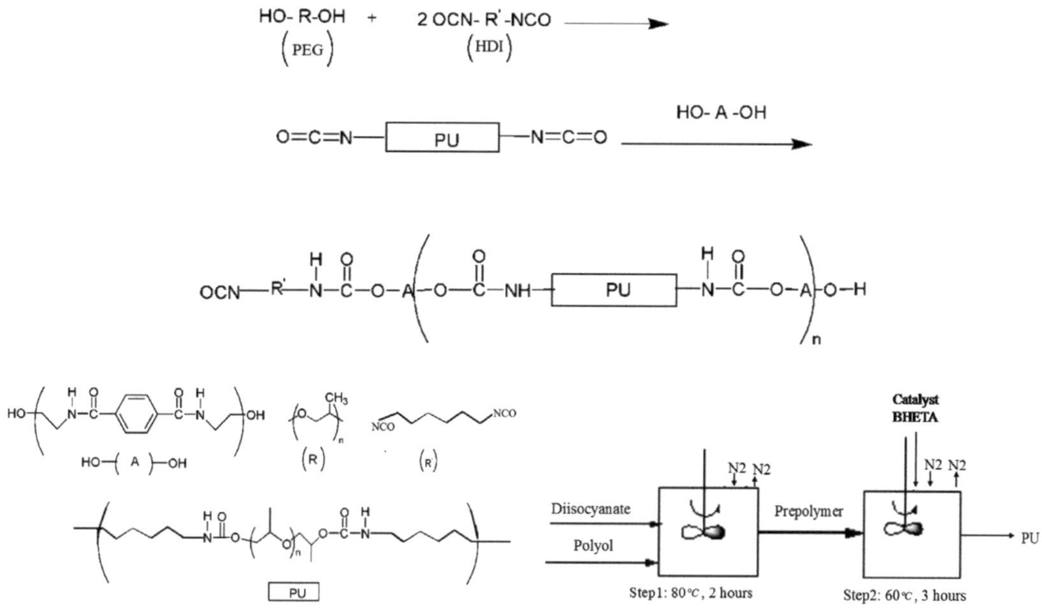

Fig. 32. Synthesis method for PUs.

6.4.2 Chemical resistance & solubility tests

The test results of chemical resistant are shown in Table 8. Polyurethanes were soluble in DMF and DMSO and they were resistant in basic media (50wt %), of course with increasing BHETA, chemical resistance of polyurethane decreases. Aromatic ring leads to increasing distance between chains, therefore chemicals can penetrate in polymer matrix easily and the polyurethane decomposes rapidly.

Sample	PEG (mol)	HDI (mol)	BHETA (mol)	Hard Segment Content
PU53[a]	1	2.75	1.5	53.35
PU46[a]	1	2.2	1	46.54
PU42[a]	1	1.91	0.739	42
PU32[a]	1	1.42	o.29	32.49

Table 7. Description of Samples and a indicates hard segment content of polyurethanes.

Sample	H_2SO_4 98 wt%	Dense HNO_3	50 wt% NaOH	23 wt% NaCl
PU53	---	---	++	++
PU46	---	---	++	++
PU42	--	--	++	++
PU32	-	-	++	++

++: resistant −: un resistant, which number of −shows rate of decomposition

Table 8. Results of chemical resistance tests.

6.4.3 Mechanical properties

Mechanical properties of synthesized polyurethanes are shown in Table 9. The results indicate that increasing of chain extender leads to increasing of strength and stiffness of polymer and decreasing in elongation at break. For example modulus increases from 106.37 MPa to 296.16 MPa in the samples PU46 and PU53 respectively. In fact, the BHETA has an important role in strengthening of polyurethane by increasing of hydrogen bonding (since BHETA has many sites for formation of hydrogen bond) between polyurethane chains effectively. As seen in the Fig. 33 modulus and strength of polyurethanes increases with increasing of BHETA content. As shown in Fig. 34, content of BHETA affect on behavior of stress-strain curve for synthesized polyurethane under tension. It seems elastic region of the curves increases as BHETA content increases.

Sample	Young's modulus (MPa)	Max Stress (MPa)	Elongation at Break (%)
PU 53	296.16	9.18	45.82
PU 46	106.37	6.94	31.13
PU 42	62.68	5.72	51.99
PU 32	28.569	2.98	128.56

Table 9. Mechanical properties of samples contain different hard segment content.

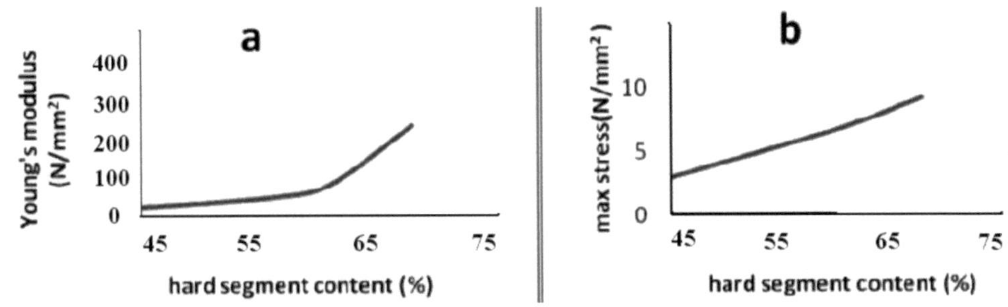

Fig. 33. Young's modulus (a) and max stress (b), as a function of hard segment content.

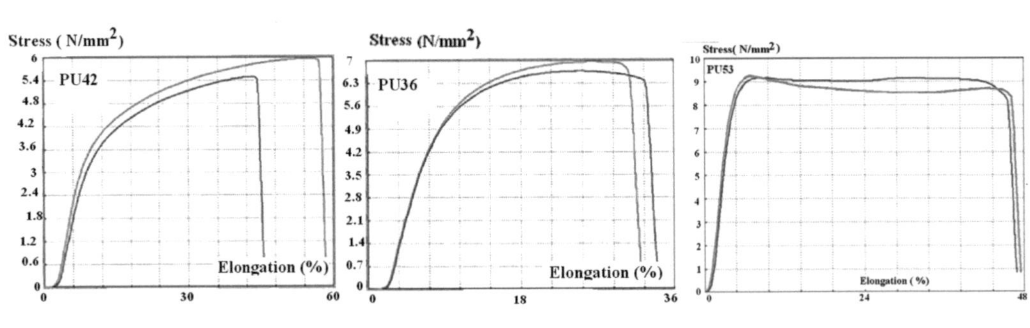

Fig. 34. Stress-Strain curves for PU42, PU46, and PU53: Two samples for each test).

6.4.4 DSC analysis

Thermoplastic segmented polyurethanes display several thermal transitions. The soft phase, responsible of the properties at low temperatures, shows glass and melting transition (if semicrystalline), while the hard phase is responsible of the properties at high temperatures showing multiple melting transitions depending on the hard segment content in the matrix. As can be seen DSC graphs of the PU42 and PU46 shown in Fig. 35, one peak at -21.6 8C and -21.18 8C are appearance for PU42 and PU46, respectively. The peak appearing at the lower temperature might be associated with a soft segment glass transition temperature (T_{gss}). The destruction of interchain hydrogen bonding caused to a DSC endotherm. It seems that the peak at about 142°C in both PU4 and PU46 is related to breaking of hydrogen bonding which has been occurred in PU42 more than PU46. This phenomenon can be attributed to more aromatic rings exist in PU42 per unit length of chains whereas is less in PU46, thus higher flexibility in PU42 caused to higher possibility for hydrogen bonding formation. The third peak in 179°C and 173°C shows formed restructuring in PU42 and PU46, respectively. Apparent endothermic peaks could result from crystal structures. As hard segment content increases, position of the endothermic peaks is shifted to higher temperatures, which is indicative of better ordered hard domain .The peak at around 237°C and 240°C is related to PU42 and PU46 respectively.

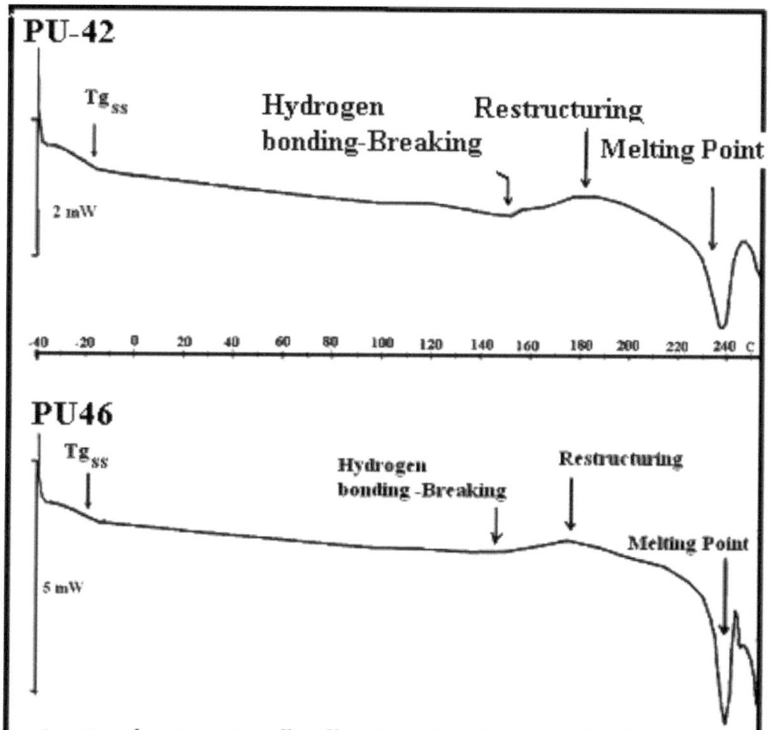

Fig. 35. DSC thermograms of PU42 and PU46.

6.4.5 TGA analysis

Polyurethanes are comparatively thermally unstable polymers; decomposition temperature of the polyurethane depends on the polyurethane structure. Polyurethane degradation usually starts with the dissociation of the urethane bond, CO_2 and isocyanate evaporation. Normally, three mechanisms of decomposition of urethane bonds have been proposed and reactions may proceed simultaneously: dissociation to isocyanate and alcohol, formation of primary amine and olefin and formation of secondary amine and carbon dioxide. Fig. 36 shows TGA curves of PU42 and PU46. The shapes of the weight loss curves of both polyurethanes are almost identical and degradation profiles of polyurethanes depend on the content of BHETA. It could be described with different values which are bringing in Table 10. Initial degradation temperature of PU42 is much higher than that of PU46. Generally, reduction of T_{id} for PU46 may be attributed to these facts: at first flexibility can affected on hydrogen bonding formation; PU42 is more flexible which leads to increase probability of hydrogen bonding formation as, mentioned in DSC tests.

Sample	Tid °C	T10% °C	T20% °C	T30% °C	T50% °C	Char Residue (%)	DTG (Max Temp.) °C
PU 42	201	232	245	256	368	2.534	240
PU 46	172	203	220	233	351	5.100	250

Table 10. Results of the thermo gravimetric analysis of samples.

Second reason; as seen in DSC thermograms restructuring phenomenon occurred at about 170°C. It seems, in the case of PU42 restructuring is predominant rather to decomposition reactions, in restructuring phenomenon new bonds formed, that leads to higher thermal stability in PU42.

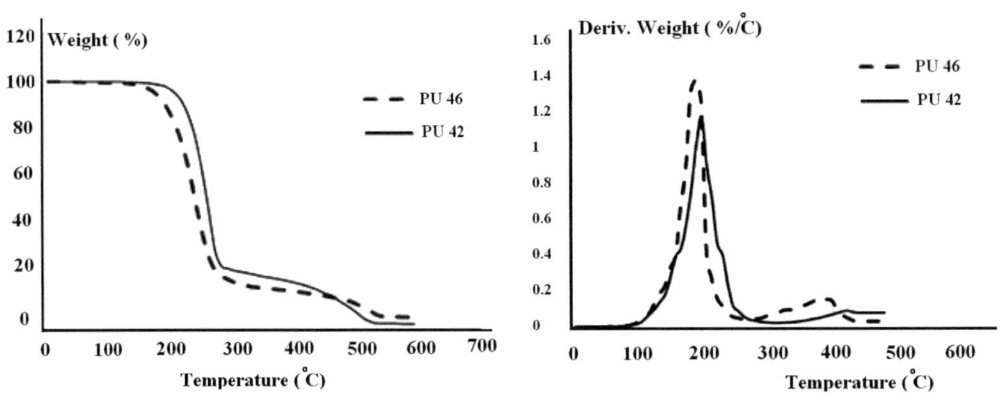

Fig. 36. TGA and DTG thermo grams for PU46 and PU42.

7. Conclusions

Recycling of polyethylene terephthalate (PET) by aminolysis breeds environmental benefits. There are few reports on the usage of recycled BHETA from PET to synthesis of polyurethanes. At first use of ethanolamine for aminolytic degradation of PET waste has been investigated. Obtained product, BHETA has potential for further reactions to synthesize useful products such as polyurethanes which have important industrial applications. In our studies, BHETA has been used as an intermediate to produce useful materials based on PET waste. At first study, ring opening polymerization of caprolactone by BHETA was carried out and polyols with different Mn have been synthesized and then polyurethanes have been synthesized using above mentioned polyols. Increasing of Mn polycaprolactone diol leads to regular increasing in melting point, crystallinity and fusion heat of samples, tendency to crystallization and formation of ordered structures is observed clearly which confirm by SEM and OM. Thermal degradation is serious for the sample containing lowest aromatic concentration. Elongation at break of 4.7 to 520% and strength of 9.3 to 16 MPa for synthesized polyurethanes without chain extender has been obtained. Biodegradability tests show high rated biodegradation for all polyurethanes.

In second study, BHETA has been used as an additional chain extender to synthesize novel segmented polyurethanes used in adhesives and coatings. Strong hydrogen bonding was evident from the FTIR spectra for all synthesized polyurethanes (with and without BHETA). Different thermal behavior for polyurethanes with and without BHETA have been observed using TGA due exothermic or endothermic reactions during their degradation. Addition of BHETA to the polyurethanes caused an increase in maximum load (*Fm*) and elongation for both Fe-Fe and Al-Al substrates. Comparison of the *Fm* for the synthesized adhesive with those of commercial epoxy and polyester-type adhesives shows a 2.03 and 2.34-fold increase, respectively. Chemical resistance tests show a high resistance of the polyurethanes to alkaline, NaCl and water media, but a lower resistance in high-concentration acids.

In third study BHETA uses instead of common chain extenders to synthesize of novel segmented polyurethanes. BHETA has an important role in strengthening of polyurethane, increasing of BHETA content caused to obtain modulus as 300 Mpa, Maximum stress as 9.18 Mpa. Using of BHETA in production of polyurethane leads to obtain polyurethanes with suitable phase separation and mechanical properties, decrease in raw material costs as well as green environment based on a PET waste recycled material.

8. References

[1] Kloss J, Fernanda SM, Souza D, Edilsa R, Silva D, Jair Alves D, 656 Leni A, Sonia Faria Z (2006) Macromol Symp 245-246: 651-656

[2] Shukla SR, Harad AM, Jawale LS (2009) Polym Deg Stab 94: 604-609.

[3] Sivaram S, (1997) National seminar on recycling and plastics waste management, Sep:283-8.

[4] Barbozaa ES, Lopez DR, Amico SC, Ferreira CA (2009) Resour Concerv Recy 53:122.

[5] Pusztaszeri SF (1982) US Patent 4355175.

[6] Mishra S, Goje AS, Zope VS (2003) Poly Plastics Technol Eng 42(4): 581-603.

[7] Mishra S, Goje AS, Zope VS (2003) Polym React Eng 11(1): 79-99.

[8] Schwartz J (1995) US Patent 5395858.

[9] Lamparter RA, Barna BA, Johnsrud DR (1985) US Patent 4542239.

[10] Tindall GW, Perry RL (1991) US Patent 5045122.

[11] Mishra S, Goje AS (2003) Polym React Eng 11(4): 963–987.
[12] Doerr ML (1986) US Patent 4578510.
[13] Yang Y, Lu Y, Xiang H, Xu Y, Li Y (2002) Polym Degrad Stabil 75: 185–191.
[14] Motonobu G, Hiroshi K, Akio K, Tsutomu H, Shoji N(2002)J Phys Con. Mat, 14(44):11427–30
[15] Motonobu G, Hiroshi K, Akio K, Tsutomu H, Shoji N, McCoy BJ(2002)Alche J 48(1):136-44
[16] Akiharu F, Minako S, Masashige M (1986) US Patent 4609680.
[17] Ostrowski HS (1975) US Patent 3884850.
[18] Güçlü G, Kasgöz A, Özbudak S, Özgümüs S, Orbay M (1998)J Appl Polym Sci 69(12) :2319.
[19] Andrej K (1998) J Appl Polym Sci 69(6–8): 1115–1118.
[20] Berti C, Colonna M, Fiorini M, Lorenzetti C, MarchesP(2004) Macromol Mater Eng 289:49-5
[21] Manfred K, Wolfgang S, Uwe S (1993) US Patent 5266601.
[22] Shukla SR, Harad AM (2006) Polym Degrad Stab 91: 1850-1854.
[23] Fabryc y E, Leistner A, Spychaj T (2000) Adhesion. 44(4): 35.
[24] Zahn H, Pfeifer H (1963) Polymer 4: 429-32.
[25] Popoola V (1998) J Appl Polym Sci 36: 1677-83.
[26] Blackmon KP, Fox DW, Shafer SJ (1990) US Patent 4973746.
[27] Shamsi R, Abdouss M, Mir Mohamad Sadeghi G, Afshar Taromi (2009) Polym Int 58:22-30
[28] Mohamadi, M;Mir M Sadeghi, G.& AbdoussM.(2010) *Mat-WissU Werkstofftech,* 41, 8:682-88
[29] Mir M. Sadeghi G, Shamsi R, Sayaf M, J Polym.Environ, DOI 10.1007/s10924-011-0283-7
[30] www.chemsysytem.com, nexant (2009), PERP 07/08-5:1-8
[31] Yeganeh H, Jamshidi H, Jamshidi S (2007) Polym Int 56: 41–49
[32] Heijka R, Calck R, Tien T, Buma P, Penni A, Vet RPH(2005) J Schouten: Biomat 26: 4219–28.

16

Valorization of Organic Wastes by Composting Process and Soil Amendment

Hafedh Rigane and Khaled Medhioub
Unité de Recherche: Etude et gestion des Environnements côtier et urbain, University of Sfax Tunisia

1. Introduction

Agricultural wastes disposal is becoming a serious environmental problem. Indeed, these residues may be highly polluting and phytotoxic. The removal of the produced solid and liquid residues is causing serious environmental problems. The direct application of organic wastes, such as olive husks or olive mill wastewaters, to soil has been considered as an inexpensive method of disposal in addition to the recovery of their mineral and organic components. Nevertheless, due to their ligno-cellulosic contents, olive husks are potentially environmentally harmful biomass. Fresh olive husks (not biologically stabilized) may have phytotoxicity due to their monogenic chemical composition (only lignin and cellulose), to phenols coming from olive oil processing, to its high Carbon/Nitrogen (C/N) ratio and to the presence of hormon inhibitors (De Bertoldi et al., 1986). According to several authors (De Jager et al., 2001; Palm et al., 2001), the improvement of soil fertility mainly under low input agricultural systems requires the input of stabilized or mature organic wastes. Many studies (Gallardo-Lara and Nogales, 1987; He et al., 1992; Ouédraogo et al., 2001; Stamatiadis et al., 1999) have shown that application of mature composts at reasonable rates improves plant growth, soil physical properties and increases available soil nutrient levels.

Composting in a controlled biooxidative process that involves a heterogeneous organic solid substrate may resolve this problem. It evolves through a thermophilic stage and the temporary release of phytotoxins, leading to the production of carbon dioxide, water, mineral salts and stabilized organic matter containing humic like substances. These kinds of fertilizers are used to improve soil fertility and plant production either in organic and conventional agriculture.

The main problems that can occur from excessive application of compost are plant toxicity due to salt content (Stamatiadis et al., 1999) and accumulation in plants of trace metals which may pose a health risk when humans or farm animals consume the plant (Petruzzelli, 1996; Cabrera et al., 1989).

Many wastes produced in important quantities are used for composting process such as solid wastes (olive husk, poultry manure) and also liquid wastes mainly the olive mill wastewaters which may be transformed into an organic fertilizer by composting. This is an inexpensive method of disposal leading to important advantages. The composted olive husks are organic fertilizers that improved soil fertility (Fiestas Ros de Ursinos, 1986; Tomati and Galli, 1992). The olive husk composts bring essentially potaasium (K), in addition to Nitrogen (N), Phosphorus (P) and Magnesium (Mg) and obviously organic matter (OM). Soil amendment with compost enhances its microbial activity and improves its physical and chemical properties. Thus, farmer's interest for recycling organic wastes in agriculture as fertilizers or amendments is increasing. The composting studies were performed on olive husks mixed (80 %) with 20% poultry manure. The initial mixture was realized in order to reach a C/N ratio of 30. The obtained compost was characterized physically and chemically in order to define its fertilizing capacities and use safety.

The objectives of the present work were to assess the suitability of a waste compost to supply some essential plant nutrients such as N, P, K, iron (Fe), manganese (Mn), zinc (Zn) and copper (Cu); evaluate and compare the effects of manure and compost on soil chemical properties of studied soils in the same zone and to prove the effect of compost on crop productivity.

2. Composting process

The composting in a controlled biooxidative process that involves a heterogenous organic substrate in the solid state that evolves through a thermophile stage and the temporary release of phytotoxins, leading to the production of carbon dioxide, water, mineral salts and stabilized organic matter containing humic like substances. By using this method, it is possible to transform olive residues mixed with poultry manure into organic fertilizers with no toxicity to improve soil fertility and plant production.

In this work, we presented examples of solid and liquid wastes chosen in composing essays.

2.1 Solid wastes

The raw materials (olive husks) characterized by high organic carbon, very low nitrogen and important ash contents, were co-composted with poultry manure containing high nitrogen and ash rates (Table 1, Hachicha et *al*., 2003). Two combinations of sifted or non sifted olive husks mixed with poultry manure at 8:2 ratio were selected (Hachicha et *al*., 2003). The sifting operation was performed using a densimetric method that allows its separation into 40% pulp and 60% stone fragments. The latter was used as a substitute for raw olive husks in the role of energy supplier. The sifting was retained to compare the qualities of composts (produced from materials with or without sifting) and to test the possibility of utilization of olive husks without sifting to reduce treatment price. In fact, two windrows were constructed one for each formula.

2.1.1 Preparation of windrows

Windrow 1, composed by a mixture of non sifted olive husks and poultry manure and windrow 2 composed by a mixture of sifted olive husks and poultry manure. These will produce respectively compost C1 and compost C2. The windrows had triangular shape with 3 m wide of the base and 2-3 m high. Each windrow was constituted of 10 tons of the mixture. Olive husks are characterized by a very high C/N ratio due to high carbon and low nitrogen contents. The olive husks are mainly composed of lignin and cellulose fibres with other chemical elements at low concentrations (Vlyssides et *al.*, 1996). The direct application of these solid wastes in soil causes problems because of their phytotoxicity (De Bertoldi and *al.*, 1983). The sifting operation uses a densymetric method that allows their separation into 40% pulp and 60% some fragments. The sifted olive husk compost gives more nitrogen and more synthesised and polymerised humic acids than unsifted olive husk compost (Hachicha, 2002; Hachicha and *al.*, 2003). Poultry manure contain relatively high concentrations of total nitrogen (more than 2.5% of dry matter), of ash (more than 40% of dry matter) and relatively low C/N ratio (Table 1).

During the composting process several parameters were followed such as temperature, pH, electrical conductivity (EC) and ash ratio. Maturity was evaluated by the determination of the C/N ratio, cation exchange capacity (CEC) and humic acid concentrations (Mustin, 1987). Moreover, during the composting process, moisture was kept at 45-50% by adding water and aeration was assured by windrow turning. Temperatures were measured daily basis at different positions in the core of windrow using mercury thermometers and the average of all measurements was recorded. Samples for analysis were collected weekly at different points along the windrows.

Component	Non sifted olive husks	Sifted olive husks	Poultry manure	Windrow 1	Windrow 2
pH	4.92	4.73	8.62	6.43	7.27
E.C. (mmhos/cm)	-	-	-	0.5	0.68
C/N	48.74	37.5	10.44	32	28.5
CEC (meq/100g OM)	-	-	-	73.56	92.93
Ash content (% dry solid (*d.s.*))	5.22	10.77	41.65	24.5	30.2
Total Kjeldhal N (%*d.s.*)	1.06	1.31	2.88	1.5	1.56
Total Organic C (%*d.s.*)	51.67	49.13	30.08	48	44.5
Cellulose (%*d.s.*)"	39.75	28.09	23.17	18.6	13.9
Lignin (%*d.s.*) "	24.72	18.73	17.15	14.3	12.2
polyphenols (%*d.s.*)	0	0	0	0.435	0.18
Fatty substances (%*d.s.*)	4.08	4.29	0.8	3.02	3.67
Total P (%*d.s.*)	0.1	0.13	1.51	0.4	0.42
Total K (%*d.s.*)	0.59	0.71	2.98	0.97	1.06
Mg (%*d.s.*)	0.08	0.10	0.50	0.15	0.23
Ca (%*d.s.*)	0.14	0.17	6.52	1.66	2.13
Na (%*d.s.*)	0.21	0.29	3.58	1.32	1.25

Table 1. Properties of the initial solid wastes and mixtures (Hachicha et al., 2003).

2.1.2 Soil amendment

For amendment tests, we applied two types of composts: compost 1 and compost 2. The essays were realized in plots and in field.

Cultures have been achieved in plots containing 15 Kg of soil situated in a zone characterized by arid climate. The prepared plots are distributed in sets of 9 units:

- a control set with only a contribution of manure (100% manure);
- ten sets have been reserved to combinations of different percentages in manure and in composts either 1 or 2.

Every test set includes 9 plots, four of which have been condemned during the analysis cycle. The various plots controlled during the production cycle permitted to follow the beginning of the tuberization and the development of different plant tubers.

We applied compost and poultry manure (PM) in different percentages: 100% C.1 (Compost 1), 50% C.1 or C.2 + 50% PM (Compost 1 or Compost 2 mixed with manure in same quantities), 75% C.1 or C.2 + 25%PM (Compost 1 or Compost 2 mixed with manure at 75%/25%), 25% C.1 or C.2 + 75% PM (Compost 1 or Compost 2 mixed with manure at 25%/75%), 200% C.1 or C.2 (the quantity of applied compost is double compared to 100%) and 100% PM (poultry manure).

In field experiment, the compost was applied soil harvested with tomato cultures (Rio Grande variety) realized in three experiment soils of the same zone. The region is characterized by calcimagnesic soils characterized by calcareous parent rock containing in most cases active lime. The compost retained for tests is the compost 1 which was applied on soil and where studies (Hachicha et *al.*, 2003) showed its agronomic interest. In fact, these authors recorded the same potato yield with both composts, the agronomic field test showed that the sifted olive husk compost (compost 2) acted more positively on the tuber size of potato whereas the non sifted olive husk compost (compost 1) was more beneficial to the plant height and to the leaf weight. In addition, the compost 1 was prepared without sifting the starting materials. In order to improve the soil fertility and to valorise olive husks, tests of amendments by the composts have been exercised on some soil types in this region.

In order to evaluate the quality of soils amended with cow manure and those amended by the compost of olive husks, samples of soils have been appropriated for determination of textural classification, pH, electric conductivity, total lime, active lime, organic matter, organic carbon, total nitrogen, C/N ratio, major elements (Ca, Mg, P, K) and trace elements. Analyses were also realized on leaves of tomato to test the similarity between compost and manure and to testify their values referring to World fertilizer use manual (Halliday and Trenkel, 1992).

2.1.3 Crop production

The assessment of the agronomic quality of produced composts has been achieved through potato cultures (Spunta variety) in plots and tomato cultures in field experiment. The compost is applied on soils by spreading of 40 tons/ha (the same dose is applied for manure in this zone).

2.1.4 Evolution of composting process

The principal parameters controlling composting process are pH, Electric conductivity, temperature, cation exchange capacity (CEC), C/N ratio.

- **Temperature**

The data related to the effect of temperature on the composting process indicate that optimum decomposition takes place between 55 and 60°C (Morel et al., 1984). For both windrows, high core temperatures (60-70°C) remained for three months in windrow 2 and for four months in windrow 1. The presence of more easily biodegradable compounds in the sifted olive husks (material 2) and an important fraction of cellulose and lignin in raw material 1, could explain the difference showed in the thermophilic phase length. The characteristics of composts were shown in table 2.

- **pH evolution**

The pH variation ranged from 6.5 to 8, which consists to aerobic composting (Gardena and Wang, 1981). Nevertheless, for both analysed windrows, pH slightly decreased over the first two weeks and gradually increased during the active phase for each substrats with a trend towards stabilization. The initial drop in pH has been reported by other authors (De Bertoldi and *al.*, 1983; Hardy and Sivasithamparam, 1989; Sikora and Sowers, 1983) and is considered to be a consequence of the acid-forming bacteria activity which breaks down complex carbonaceous material to organic acid intermediates. However, the products formed are readily consumed and simultaneous protein degradation starts releasing basic compounds which increase the pH (Godden and Penninck, 1990). In both windrows, the pH showed a progressive increase (slightly more rapid for 1) with similar behaviour. For both produced composts, pH remained at alkaline levels.

- **Electric conductivity (EC)**

For both windrows, EC rose progressively. This increase, which took place especially during higher temperatures, may be attributed to the important mineralization that liberates ions. No significant difference was observed between the two windrows. After 150 composting days, E.C. for each final product was 1.5 ms cm^{-1} against 0.8 for the unsifted olive oil processing solid residue.

- **Maturity assessment**

We have chosen C/N ratio and cation exchange capacity (CEC) as indicators of maturation progress. However, a large number of parameters were tested by different authors (Morel and *al.*, 1984; Stentiford and Pereira, 1985). The aerobic fermentation produces an important quantity of CO_2 noted here by a decrease in organic carbon content. Nitrogen constitutes the second most important element after carbon in the composting process. Over the composting cycle, the Kjeldahl nitrogen concentration (Table 2) increased. For final products, nitrogen content was about 2.5% dry solid. Increases in total nitrogen have also been reported during the composting of sludge and animal manure (Bernal et *al.*, 1996; Tikia et *al.*, 1998). In addition, to the weight losses due to the strong degradation of carbon compounds, a small nitrogen increase during composting could be attributed to fixing bacteria (Jodice and Nappi, 1986) or to inorganic nitrogen immobilizing phenomenon (Tam and Tikia, 1999). Consequently, the decrease of the C/N ratio (Table 2) during composting was mainly caused by the loss of carbon concentrations and the accumulation of nitrogen. Stability was reached within three months for the unsifted husks and two months for the sifted material. In both cases, the C/N ratio remained around 13 after 150 days of composting. During the composting process, microbial activity permitted the increase of cation exchange capacity and humic acid contents (Bernal et *al.*, 1996; Tikia et *al.*, 1998).

Elements	Compost 1	Compost 2	Manure
pH	8.51	8.44	8
organic matter (% dry matter)	40.1	40.25	43.5
C/N	13.65	13.32	16.9
Nitrogen (%dry matter)	1.87	2.10	2.00
P (ppm)	4891	6830	4600
K (ppm)	10660	12190	10250
Ca (g/l)	21.43	22.57	16.4
Mg (ppm)	5472	7747	3800

Table 2. Comparison of fertilizing values of composts 1 and 2.

An increase in CEC (cation exchange capacity) values was observed from the beginning of the composting process. Stabilization was reached five months later. In final products, the CEC recorded was about 140 meq/100g organic matter for compost 1 and 150 meq/100 g organic matter for compost 2, against about 176 meq /100g for manure. The mass of potato obtained in different plots were observed in figure 3 which showed a significant differences between the effects of composts 1 and 2. In fact, all plots amended with compost 1 showed potato masses less than 84 g, whereas in plots amended with compost 2, the potato masses exceed 260 g (Figure 1).

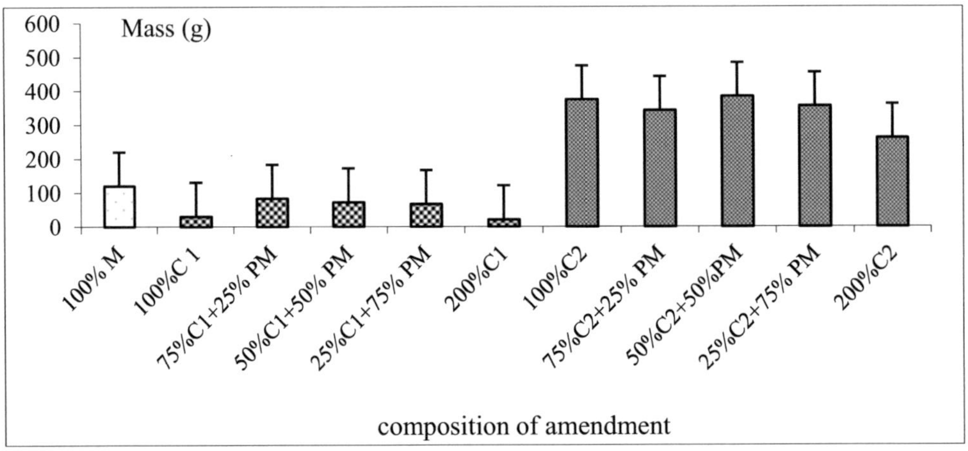

Fig. 1. Mass of potato obtained in different plots.

In field experiment, the productivity of cultivated soils amended with compost 1 which was chosen for economic reasons by removal of sifting operation is presented in Figure 2. The high productions were recorded in soils 2 and 3 amended with compost. This fact is explained by the positive roles of clay and active lime which permit to organic matter retention in soil, and also the role of compost which permits to produce more fertilizing elements related to starting materials.

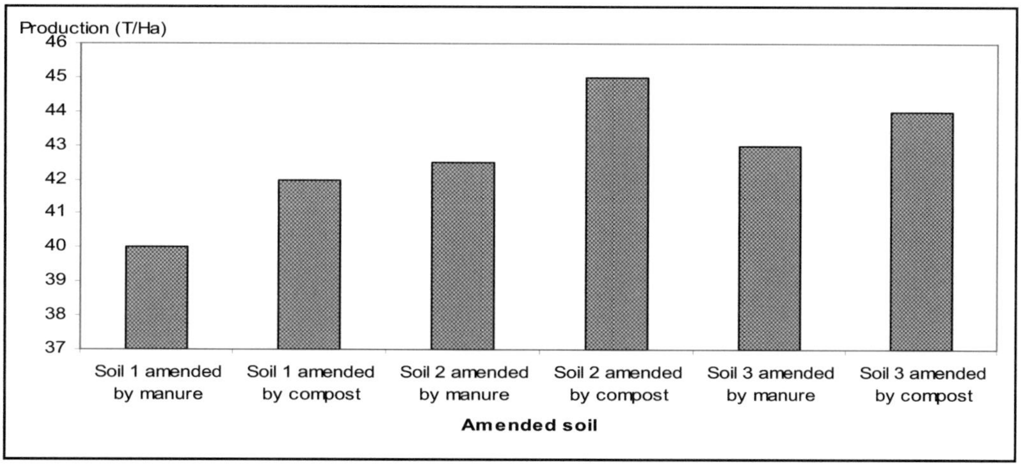

Fig. 2. Soil productivity in different soils.

Clay tenor and organic matter were more important in soil 2 amended by compost, whereas active lime was recorded only in soil 3 (Table 3) (Rigane and Medhioub, 2010). This parameter is important in soil which causes quickly the immobility of the soluble precursors and by the transformation of unsoluble components (such as lignin) caused by an active biological activity, developed with aerated structure in constructed clods of clay-humus-$CaCO_3$ (Duchaufour, 1997).

Parameters	Soil 1 amended with compost	Soil 1 amended with compost	Soil 2 amended with compost	Soil 2 amended with compost	Soil 3 amended with compost	Soil 3 amended with compost
Clay (%)	11	15	10	18	12	14
Sand (%)	69	66	59	44	63	55
Silt (%)	20	19	31	38	25	31
pH	7.77	7.85	7.83	7.91	8.19	8.14
EC (mmhos/cm)	11.43	8.42	10.2	9.82	1.87	1.85
Total lime (%)	4.6	5.3	3.4	2.7	21.6	25.7
Active lime (%)	-	-	-	-	2.7	4.2
Organic carbon (%)	0.74	0.62	0.7	0.87	0.58	0.78
Organic matter (%)	1.2	1.07	1.2	1.5	1	1.34
C/N	12.33	10.68	11.66	11.6	11.6	11.6
Total Nitrogen (%)	0.06	0.058	0.06	0.075	0.05	0.067
Available P (ppm)	169	166	174	108	58	101
Available K (ppm)	225	180	294	235	140	150
Fe (%)	4.7	3.5	2.7	3	2.2	2.5
Zn (%)	0.97	0.82	0.65	0.56	0.4	0.4
Cu (%)	0.37	0.35	0.28	0.31	0.26	0.25
Mn (%)	1.2	1	0.98	0.85	0.72	0.65

Table 3. Soil analyses (Rigane and Medhioub, 2010).

Compared to normal ranges in leaves (Halliday and Trenkel, 1992), the leaves in vegetables in all experiment soils grown under temperate conditions were considered as normal (Table 4) (Rigane and Medhioub, 2010). In fact, the application of composts in experimental soils showed no negative effect on tomato crop. Nevertheless, Basing on deficiency levels presented by Halliday and Trenkel (1992), the leaves vegetables in amended soils showed deficiency in N, P and K elements.

Elements	Soil 1 amended with compost	Soil 1 amended with manure	Soil 2 amended with compost	Soil 2 amended with manure	Soil 3 amended with compost	Soil 3 amended with manure	Normal Ranges in leaves*
N (%)	3.1	3.15	3.1	3.17	3.18	3.14	2.9- 4.9 %
P (%)	0.19	0.15	0.12	0.13	0.2	0.17	0.4-0.7 %
K (%)	1.2	1.3	1	1.2	1.2	1	2.7- 5.9 %
Ca (%)	3.2	3	3	3	3.5	3.3	2.4-7.2 %
Fe (ppm)	63	75	62	60	57	48	101-291 ppm
Zn (ppm)	38	23	39	27	37	31	20-85 ppm
Mn (ppm)	29	30	27	32	0.28	35	55-220 ppm
Cu (ppm)	6.5	6.1	5.8	5	7.3	6.1	10-16 ppm

* World Fertilizer Use Manual (1992)

Table 4. Leaf analysis (Rigane et Medhioub, 2010).

2.2 Liquid wastes

The amounts of liquid wastewaters such as olive mill wastewaters (OMW) which is produced in Mediterranean countries are important in the world. Olive oil production has normally been concentrated in the Mediterranean basin countries: Spain, Portugal, Italy, Greece, Turkey, Tunisia and Morocco. These seven countries alone account for 90% of world production. The high content of mineral salts and the presence of organic compounds, such as fatty acids and polyphenols in the OMW generate difficulties of their disposing and utilization of large amounts of this liquid. The disposal and treatment of this liquid waste are the main problems of the olive oil industry because of its high organic load and content of phytotoxic and antibacterial phenolic substances, which resist to biological degradation (Aktas et al. 2001). The beneficial effects are linked to its high nutrients concentration, especially K, and its potential for mobilizing soil ions, while, negative effects are associated with its high mineral salt content, acidity with low pH and the presence of phytotoxic compounds, mainly polyphenols (Paredes et al. 1999). Besides, other authors have observed negative effects on plants and soil properties when OMW is used directly as an organic fertiliser (Sierra et al. 2001; Casa et al. 2003; Cereti et al. 2004).

The treatment of OMW and their disposal are becoming a serious environmental problem. Different methods were used based on thermal concentration, physical and chemical and biological treatments of the OMW as well as their direct application to agricultural soils as a fertilizer have been widely tested.

The Composting technology is a biological process used for treatment of organic wastes to obtain organic soil fertilisers (Mustin, 1987). Composting experiments with OMW showed that OMW needed lignin–cellulosic wastes as bulking agents and other materials as a nitrogen source for its suitable composting, so that the phytotoxicity could be eliminated and a final product with stabilised and humified organic matter obtained by several authors (Tomati et al. 1995; Vlyssides et al. 1996; Paredes et al. 2000; Paredes et al. 2005). The objective of thist work was to study the effects of OMW on the composting of organic wastes and the behaviour of the compost obtained on soil properties.

2.2.1 Composting operation

To study the possibility of treatment of OMW by composting, two piles were prepared by mixing olive husks (OH) with poultry manure (PM) and both humidified with confectionery wastewaters (CWW). Olive mill wastewater (OMW) was added to one pile (pile 1).

The OH are characterized by the high values of dry matters, C/N ratio, calcium and magnesium contents. The PM is characterized by relatively high pH, mineral matter, phosphorus and potassium. As liquid effluent, the confectionery wastewaters showed the relatively high humidity, electric conductivity, organic matter, nitrogen and sugars. The OMW was acidic (pH 5.3) with conductivity (20 mSm^{-1}) and important concentrations of N, P and K and organic matter (OMW is characterized by black color) and high content of phenolic compounds (8900 mg l^{-1}). The OMW used in the present study were obtained from a OMW disposal site in the city of Agareb in Sfax region (Southern Tunisia), which derived from olive oil production plants.

The piles presented the following compositions:

Pile 1: Olive husks (OH)(75%) + Poultry Manure (PM)(25%) + Confectionery wastewaters (CWW) + Olive Mill Wastewaters (OMW). The final result after composting constitutes compost C1. The volumes of CWW and OMW used were similar (2.8 m^3 for each wastewater).

Pile 2: Olive husks (OH)(75%) + Poultry Manure (PM) (25%) + Confectionery wastewaters (CWW). The final result after composting constitutes compost C2. The volume of CWW used was 5.6 m^3.

- **Temperature**

The air blowing was stopped during the compost maturity period (6 months). In fact, OM degradation was greater in pile 1 in the mixture with OMW which may be explained by the longer thermophilic phase for this pile. The temperature increased quickly at the beginning of the process to thermophilic values, reaching the maximum level (68°C) (Figure 3).

In both piles, the temperature was maintained between 60 and 70 °C for about 100 days, which will contribute to the transformation of highly polymerized substrate (lignin and cellulose) by thermophilic microorganisms and also to the hygienisation of the end-product (compost) due to pathogen, weed and seed reduction. When the temperature started to decrease, the piles were turned in order to improve both the homogeneity of the material and the fermentation process. The thermophilic phase (T > 40°C) lasted approximately 105 and 120 days for mixtures 2 and 1, respectively. The biooxidative phase of composting was considered finished when the temperature of the piles was stable and close to that of the atmosphere. This occurred after 180 and 160 days in piles 1 and 2, respectively.

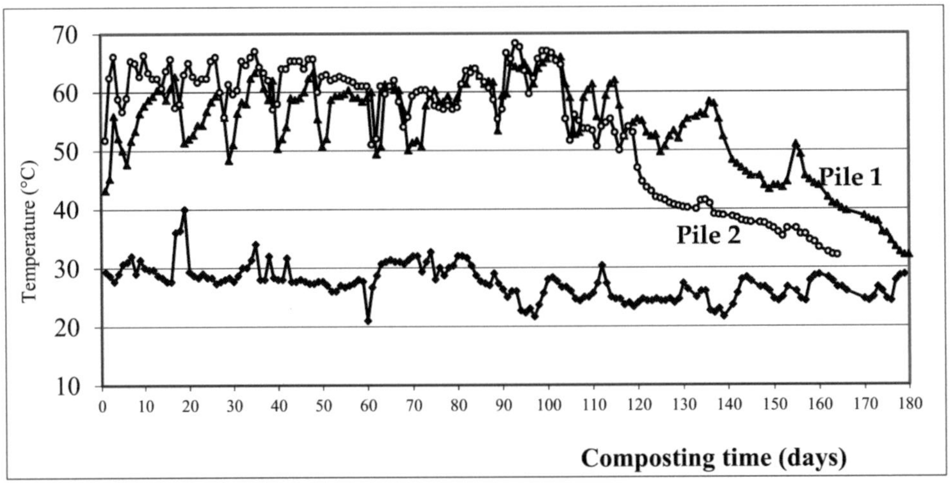

Fig. 3. Temperature evolution during composting process of studied piles (Rigane and Medhioub, 2011).

- **Organic matter evolution**

The initial OM concentrations were approximately equal in both piles (Table 2). However, OM degradation was more important in the mixture without OMW (pile 2), which may be due to the longer thermophilic phase for this pile (Figure 4). According to Paredes et al. (2005), this fact can be explained by the higher content of easily degradable organic compounds provided by OMW.

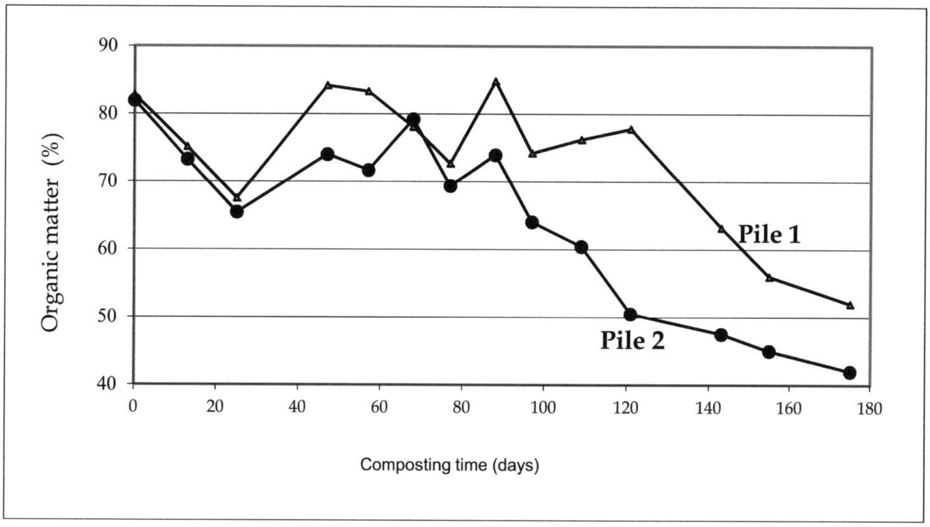

Fig. 4. Organic matter evolution during composting process (Rigane and Medhioub, 2010).

The difference between two composts can be attributed to the higher content of easily degradable organic compounds provided by OMW. OM decomposition bring about an increase in pH and EC in the piles, as recorded by Paredes et al. (2005), which was explained as a consequence of the degradation of acid-type compounds, such as carboxylic and phenolic groups, the mineralisation of compounds, such as proteins, amino acids and peptides, to ammonia and the relative increased concentration of ions, due to the loss of pile weight. This fact was observed in the higher pH and electrical conductivity of both composts in comparison to manure (Table 5). The EC in the compost prepared with OMW (C1) was more important than manure (M) and C2 without OMW which can be explained by the high soluble salt content provided from OMW.

This observation can explain the significant difference in the humification ratio (C_{HA}/C_{FA}) between composts and manure (Table 5). Respect to organic matter humification, a higher humification ratio was recorded in C2 without OMW. Therefore, a positive correlation was observed between CEC and C_{HA}/C_{FA} (Rigane et Medhioub, 2011). In fact, the CEC was more important in compost C2 prepared without OMW which it reached 255 meq/100g, while in compost C1 with OMW, it had 237 meq/100g. Both composts, particularly C1 with OMW, had high levels of macronutrients, especially K and Ca (Table 5), compared with those found in manure. Moreover, a higher EC was recorded in composts than manure and mainly in the OMW compost. This was probably linked to the soluble salts contained in OMW.

Parameters	Manure	Compost C1: OH+PM+ CW+OMW	Compost C2: OH+PM+ CW
pH	8	8.34	8.61
Electric conductivity (mS/cm)	4.55	5.46	5.18
Total organic carbon (%)	33.82	27.47	22.37
Organic matter (%)	63.25	52.24	42.40
C/N	16.93	11.75	13.56
Total Nitrogen (%)	2.15	1.56	1.32
CEC (meq/100 g)	215	237	255
C_{HA}/C_{FA}	0.64	0.88	2.35
Ca (ppm)	16400	68604	40869
Mg (ppm)	3800	5289	4347
K (ppm)	10250	25200	23314
P (ppm)	4600	8539	7184

Table 5. Physical and chemical characteristics of manure and two types of composts.

- **Soil amendment essay**

In soils amended with two composts were recorded a higher fertilizing values than those amended with manure (Table 6). Tomati et al. (1996), in an experiment with potato grown on soil amended by compost or manure mixtures, obtained higher crop yields with OMW compost.

Amendment	Manure	Compost C1	Compost C2
Potato yield production (T/Ha)	42.5	46	47

Table 6. Comparative effects of composts 1 (C1) and 2 (C2) on potato yield production.

The soil analysis showed that the addition of compost produced great changes in soil pH. This observation was also deduced by Gallardo-Lara and Nogales (1987) in a calcareous soil amended with solid waste compost.

In addition, the studied composts increased significantly the soil salinity, according to the EC values. In the most cases, the increases of Corg, Norg, P and K nutrient contents and the CEC values in soil due to organic amendment were also observed in soils with composts.

The levels of chlorides and sodium were for all time higher in the soils with composts and manure. The Corg, Norg, and available K concentrations and the CEC values of the amended soils decreased after cropping compared to value before cropping for each parameter. This fact is considered as a consequence of the OM mineralisation, plant uptake and fixation of phosphorus as calcium phosphates hardly available to plants in calcareous soils (Bernal et al. 1993). However, these parameters were increased significantly by the organic amendment type, particularly with composts. The increases of Corg, Norg and plant-available nutrient contents and CEC values in soil amended with organic fertilisation were also observed by Cabrera et al. (1997), in study of effects of applications of OMW sludge compost. After cropping the compost C1 prepared with OMW showed a residual organic content more important than manure and compost C2. Besides, Mg and P contents were more important in soil amended with C1 with OMW.

3. Conclusions

With the vast amounts of olive and poultry residues production in Mediterranean countries, their treatment and disposal are becoming a serious environmental problems. An attention has been paid to wastes and technologies are available nowadays for reducing their pollutant effects and for their transformations in a final product which can be used without any pollution risk. By using composting technologies, it is possible to transform these residues mixed with appropriate percentages into organic fertilizers (composts) with no phytotoxicity to improve soil fertility and plant production.

The composting of liquid or solid organic agro-industrial wastes offers an important advantages: valorization of wastes, reducing the decrease of farm manure used for soil amendment and absence of negative effects on crop qualities.

The composting is a controlled biological process which involves a heterogeneous organic solid substrate may resolve the problem of wastes which can be liquid or solid.

This technology could have an important repercussion in many countries of the world, since they suffer water restrictions and the increase of organic waste volumes. Thus, it requires new environmental and economically viable management options. In this work, we were interested in agricultural valorisation of composts obtained by mixtures of olive husks, poultry manure, Olive mill wastewaters,...

It can be deduced that composting is a suitable alternative for the recycling wastes. The obtained compost had a stabilised and humified organic matter and an important tenors of macronutrients which were recorded due to OM mineralization. The positive effects on soil fertility and productivity increased with the application of compost. However, the composting of organic wastes depends on raw materials. In fact, the composting of wastes containing OMW for example requires a longer time and this could be the major concern regarding the use of compost in soil with sufficient maturity.

4. References

Aktas, E.S., Imre, S. Ersoym, L. (2001). Characterization and lime treatment of olive mill wastewater. *Water Research* Vol. 35, pp. 2336–2340.

Bernal, M.P., Navarro, A.F., Roig, A, Cegarra, J. and Garcia, D. (1996). Carbon and nitrogen transformation during composting of sweet sorghum bagasse. Biol. Fert. Soils, 22, 141-148.

Cabrera, F, López, R., Martín, P. & Murillo, J.M. (1997). Aprovechamiento agronómico de composts de alpechín. *Frutic Prof.* Vol. 88, pp. 94- 105.

Casa, R., D'Annibale, A., Pieruccetti, F., Stazi, S.R., Giovannozzi Sermanni, G. & Lo Cascio, B. (2003). Reduction of the phenolic components in olive-mill wastewater by an enzymatic treatment and its impact on durum wheat (Triticum durum Desf) germinability. *Chemosphere;* Vol. 50, pp. 959- 66.

Cereti, CF, Rossini, F, Federici, F, Quaratino, D, Vassilev, N & Fenice, M. (2004). Reuse of microbially treated olive mill wastewater as fertiliser for wheat (Triticum durum Desf). *Bioresour. Technol.,* Vol.91:135- 40.

De Bertoldi, M., Vallini, G. and Pera, A. (1983). The biology of composting. Waste management and research. 1, 157-176.

De Bertoldi, P.F., Filippi, C. & Picci G. (1986). Olive residue composting. *Waste management and research.* 1, 157-176.

De Jager et al., 2001. A. De Jager, D. Onduru, M.S. Van Wijk, J. Vlaming & G.N. Gachini , Assessing sustainability of low external input farm management systems with the nutrient monitoring approach: a case study in Kenya. *Agric. Syst.* 69 (2001), pp. 99–118.

Duchaufour, P.1997. *Abrégé de Pédologie : sol, végétation, environnement.* Ed. Masson, 291p.

Fiestas Ros de Ursinos, J. A. (1986). Vegetation water used as a fertilizer. In *International Symposium on Olive By products Valorization.* ed. FAO. Madrid, Spain, pp. 321-330.

Gallardo-Lara, F. & Nogales, R. (1987). Effect of the application of town refuse compost on the soil plant system. *Biol. Wastes* 19, pp. 35–62.

Gardena, P. R. and Wang, L. K. (1981). Handbook for environmental engineering. Publ. By Humana press, Clifon, N.J.

Hachicha, R., Rigane, H., Ben Khodher, M., Nasri, M. & Medhioub, K. (2003). Effects of partial stone removal on the co-composting of olive-oil processing solid residues with poultry manure and the quality of compost. *Environmental technology,* vol.24. pp. 59-67.

Halliday D.J. & Trenkel M.E. (1992). *World fertilizer use manual.* International fertilizer industry association. 632 p.

Hardy, G.E. and Sivasithamparam, K. (1989). Microbial, chemical and physical changes during composting of a eucalyptus. In Biol. ferti. Soils 8. 260-270.

Jodice, R. and Nappi, P. (1986). Microbial aspects of compost application in relation to mycorrhizae and nitrogen fixing microorganisms. In compost : production quality and use, edited by De Bertoldi M. and Manios, B.I. (1979). Investigation for soil conditioner production from olive oil processing soild residues. Ph D Thesis. Agriculture University of Athens, Department of Agricultural Biology and Biotechnology.

He, X.T., Traina, S.J. & Logan, T.J. 1992. Chemical properties of municipal solid waste composts. *J. Environ. Qual.* 21 (1992), pp. 318–329.

Morel, J.L., Colin, F., Germon, J.C., Godin, P. & Juste, C.(1984). *Methods for the evaluation of the maturity of municipal refuse compost.* In composting of agricultural and other wastes. Edited by Gasser J.K.R., CEC Workshop, pp.56-72.

Mustin, M. (1987). *Le compost: gestion de la matière organique.* Ed. François Dubusc. Paris. 973 p.

Ouédraogo, E., Mando, A. & Zombré, N.P. (2001). Use of compost to improve soil properties and crop productivity under low input agricultural system in West Africa. *Agric. Ecosyst. Environ.* 84, pp. 259–266.

Palm, A.C., Gachengo, C.N., Delve, R.J., Cadisch, G. & Giller, K.E. (2001). Organic inputs for soil fertility management in tropical agroecosystems: application of an organic resource database. *Agric. Ecosyst. Environ.* 83. pp. 27–42.

Paredes, C., Cegarra, J., Roig, A., Sa´nchez-Monedero, M.A. & Bernal, M.P. (1999). Characterization of olive-mill wastewater (alpechin) and its sludge for agricultural purposes. *Bioresource Technology* Vol.67, pp.111–115.

Paredes, C., Roig, A., Bernal, M.P., Sanchez-Monedero, M.A. & Cegarra, J., (2000). Evolution of organic matter and nitrogen during co-composting of olive mill wastewater with solid organic wastes. *Biology and Fertility of Soils* 32, 222–227.

Paredes, C, Cegarra, J, Bernal MP & Roig A. (2005). Influence of olive mill wastewater in composting and impact of the compost on a swiss chard crop and soil properties. Environment international. Vol.31, pp. 305-312.

Petruzzelli, G. 1996. Heavy metals in compost and their effect on soil quality. In: M. De Bertoldi, P. Sequi, B. Lemmes and T. Papi, Editors, *The composting science part I*, Blackie Academic and Professional Editor, Glasgow, pp. 212–223.

Rigane H. & Medhioub K.(2010). Effects of amendment of calcimagnesic soils with olive husk compost in Tunisia. Compost Science and Utilization. Vol.18, N°4, pp. 249-254.

Rigane H. & Medhioub K.(2011). Co-composting of olive mill wastewaters with manure and agro-industrial wastes. Agricultural valorisation of composts obtained (Tunisia, Northern Africa). In Press in Compost Science and Utilization.

Stamatiadis, S., Werner, M. & Buchanan, M.(1999). Field assessment of soil quality as affected by compost and fertilizer application in a broccoli field (San Benito County, California). *App. Soil Ecol.* 12, pp. 217–225.

Sierra, J., Marti, E., Montserrat, G., Cruanas, R. & Garau, M.A. (2001). Characterisation and evolution of a soil affected by olive oil mill wastewater disposal. *Sci. Total Environ.* Vol. 279, pp. 207–214.

Sikora, L. J. and Sowers, M. A. (1983). Factors affecting the composting process. In Proceedings of the international conference on composting of soild wastes and slurries. Leeds, U.K.

Stentiford, E.I. and Pereira Neto, T. J. (1985). Simplified systems for refuse/sludge compost. Biocycle, vol 26, N°5, pp.46-49.

Tam, N.F.Y. and Tikia, S.M. (1999). Nitrogen transformation during co-composting of spent pig manure, sawdust litter and sludge under forced aeration. Env. Techn. Vol. 20, pp. 259-267.

Tikia, S.M., Tam, N.F.Y. and Hodgkiss,Y. (1998). Changes in chemical properties during composting of spent litter at different moisture content. Agric. Ecosyst. Environ. 67, 79-89.

Tomati, V. & Galli, E. (1992). The fertilizing value of waste waters from the olive processing industry. In *Humus et Plunta Proceedings.* ed. Elsevier Science, B.V. Amsterdam, The Netherlands, pp.107-126.

Tomati, U, Galli, E, Pasetti, L & Volterra, E. (1995) Bioremediation of olive-mill wastewaters by composting. *Waste Management Res.* Vol. 13, pp. 509-518.

Tomati, U., Galli, E., Fiorelli, F.& Pasetti, L. (1996). Fertilizers from composting of olive-mill wastewaters. *International Biodeterioration and Biodegradation* Vol. 44, pp. 155-162.

Tsikalas, E.P. (1985). Enrichment olive oil processing residues with nurients for using as a soil conditioner. Ph D Thesis. Agriculture University of Athens, Department of Agricultural Biology and Biotechnology.

Vlyssides, AG, Bouranis, DL, Loizidou, M. & Karvouni, G. (1996). Study of a demonstration plant for the co-composting of olive-oil-processing wastewater and solid residue. *Bioresour Technol.*, Vol. 56, pp. 187-193.